IRC-SET 2022

Huaqun Guo · Ian McLoughlin ·
Eyasu Getahun Chekole · Umayal Lakshmanan ·
Weizhi Meng · Peng Cheng Wang · Jiqiang Lu
Editors

IRC-SET 2022

Proceedings of the 8th IRC Conference on Science, Engineering and Technology, August 2022, Singapore

Springer

Editors
Huaqun Guo
Singapore Institute of Technology
Singapore, Singapore

Eyasu Getahun Chekole
Department of Information Systems Technology and Design
Singapore University of Technology and Design
Singapore, Singapore

Weizhi Meng
Department of Applied Mathematics and Computer Science
Technical University of Denmark
Kongens Lyngby, Denmark

Jiqiang Lu
School of Cyber Science and Technology
Beihang University
Beijing, China

Ian McLoughlin
Singapore Institute of Technology
Singapore, Singapore

Umayal Lakshmanan
Agency for Science Technology and Research
Experimental Drug Development Centre
Singapore, Singapore

Peng Cheng Wang
Singapore Institute of Technology
Singapore, Singapore

ISBN 978-981-19-7221-8 ISBN 978-981-19-7222-5 (eBook)
https://doi.org/10.1007/978-981-19-7222-5

This Springer imprint is published by the registered company Springer Nature Singapore Pte Ltd.
The registered company address is: 152 Beach Road, #21-01/04 Gateway East, Singapore 189721, Singapore

Organizing Committee

General Chair

Prof. Ian McLoughlin, Singapore Institute of Technology, Singapore

Technical Program Chairs

Dr. Umayal Lakshmanan, A*STAR Research Entities, Singapore
Dr. Eyasu Getahun Chekole, Singapore University of Technology and Design, Singapore

Publicity Chairs

Dr. Weizhi Meng, Technical University of Denmark, Denmark
Dr. Noori Kim, Newcastle University in Singapore, Singapore

Publication Chairs

Dr. Peng Cheng Wang, Singapore Institute of Technology, Singapore
Dr. Jiqiang Lu, Beihang University, China

ePoster Chairs

Dr. Rong Li, Shanghai University of Sport, China
Dr. Elena Lisova, Malardalen University, Sweden and Volvo Construction Equipment, Sweden

Local Arrangement Chairs

Dr. Kasthurirengan Sampath, JOil (s) Pte. Ltd, Singapore

Membership Chairs

Dr. Yongqing Zhu, Singapore University of Social Sciences, Singapore
Dr. Yu Wang, A*STAR Research Entities, Singapore

Steering Committee

Prof. Lawrence Wai Choong Wong, National University of Singapore, Singapore
Dr. Huaqun Guo, Singapore Institute of Technology, Singapore
Dr. Bhojan Anand, National University of Singapore, Singapore

International Advisory Panel

Prof. Chip Hong Chang, Nanyang Technological University, Singapore
Prof. Robert H. Deng, Singapore Management University, Singapore
Prof. Takako Hashimoto, Chiba University of Commerce, Japan
Prof. Maode Ma, Qatar University, Qatar
Prof. Bharadwaj Veeravalli, National University of Singapore, Singapore
Prof. Chau Yuen, Singapore University of Technology and Design, Singapore

TPC Members

Dr. Aishwarya Bandla, National University of Singapore, Singapore
Dr. Eyasu Getahun Chekole, Singapore University of Technology and Design, Singapore
Dr. Boon Seng Chew, Singapore Polytechnic, Singapore
Dr. Yvonne Chow, Singapore Institute of Food and Biotechnology Innovation, Singapore
Dr. Aldy Gunawan, Singapore Management University, Singapore
Dr. Huaqun Guo, Singapore Institute of Technology, Singapore
Dr. Kiam Cheng How, Singapore University of Social Sciences, Singapore
Dr. Noori Kim, Newcastle University in Singapore, Singapore
Dr. Umayal Lakshmanan, A*STAR Research Entities, Singapore
Ms. Jing Lei, Xidian University, China and Singapore University of Technology and Design, Singapore
Dr. Rong Li, Shanghai University of Sport, China
Ms. Xiaoyuan Liu, University of Electronic Science and Technology of China, China and Singapore University of Technology and Design, Singapore
Dr. Elena Lisova, Malardalen University, Sweden and Volvo Construction Equipment, Sweden
Dr. Fang Liu, Singapore University of Social Sciences, Singapore
Dr. Weizhi Meng, Technical University of Denmark, Denmark
Dr. Purnima Murali Muhan, Singapore Institute of Technology, Singapore
Dr. Hongliang Ren, National University of Singapore, Singapore
Dr. Kasthurirengan Sampath, JOil (s) Pte. Ltd, Singapore
Dr. Guoxian Tan, Raffles Institution, Singapore
Dr. Peng Cheng Wang, Singapore Institute of Technology, Singapore
Dr. Yongqing Zhu, Singapore University of Social Sciences, Singapore

Preface

International Researchers Club (IRC) (www.irc.org.sg) was set up in 2001. The vision of IRC is to create a vibrant and innovative research community for Singapore and beyond, through contributions of technical specialties and occupational experiences from its members, and fostering strong networking and social interactions of expatriates and new citizens with the local community.

With the vision of IRC, it is our great pleasure to organize IRC conference on Science, Engineering and Technology (IRC-SET, www.ircset.org) for the younger talents and researchers. IRC-SET 2015 was the inaugural conference of IRC, and IRC-SET 2022 is now in its eighth run. IRC-SET conference aims to provide a platform for young and talented researchers to share their research findings, to obtain comments and feedback from experts in the field, and to exchange innovative ideas of leading-edge research in multi-disciplinary areas. Researchers from universities, junior colleges, polytechnics and secondary schools are warmly welcome to participate in this conference to showcase and present their research projects, results and findings. Unlike other academic conferences, this conference focuses specifically on Education and Youth development and has officially been given technically sponsorship from seven universities, namely National University of Singapore (NUS), Nanyang Technological University (NTU), Singapore Management University (SMU), Singapore University of Technology and Design (SUTD), Singapore Institute of Technology (SIT), Singapore University of Social Sciences (SUSS) and Newcastle University, Singapore. IRC-SET 2022 conference is also supported by IEEE Intelligent Transportation Systems Society (ITSS) Singapore Chapter, IEEE Broadcast Technology Society (BTS) Singapore Chapter and IEEE Singapore Section Women in Engineering (WIE) Affinity Group.

The program of IRC-SET 2022 advocates the importance of innovative technology backed by the strong foundation of science and engineering education. Exposing our young participants to the key technology enablers will encourage more interest into the fields of science, engineering and technology. To select good-quality papers, the IRC-SET 2022 Call for Papers was broadcast to universities, junior colleges, polytechnics and secondary schools. The researchers then submitted their technical papers to the conference online system. To meet the criteria of the conference, the submitted papers (which had to follow the required template) were peer-reviewed by the conference technical program committee, which comprised of a few reviewers coming from the IRC researchers, professors, lecturers and teachers. Comments and recommendations based on novelty of the work, scientific, engineering and technology relevance, technical treatment plausible, and clarity in writing, tables, graphs and illustrations were then provided back to the authors. Through this rigorous review, the technical program committee has selected these papers to be presented in the IRC-SET 2022 conference, which are now published in this proceedings.

IRC-SET 2022 conference was held online and consisted of Opening Speech by General Chair Prof. Ian McLoughlin (Singapore Institute of Technology, Singapore) and Guest Speech by Prof. Tit Meng Lim (Chief Executive of Singapore Science Centre), and nine parallel presentation sessions. The nine presentation sessions covered the key focus areas which included ITSS Session, BTS Session, WIE session, Biomedical Science, Computer Science, Chemical Engineering, Life Science, Computer Engineering and General Engineering. In the closing ceremony, A/Prof. Huaqun Guo (President of IRC) announced the winners of best paper awards and best presenter awards and concluded the conference with a closing speech and an introduction to the International Researchers Club.

This proceedings is dedicated to International Researchers Club and its members.

Singapore
December 2022

Huaqun Guo
Ian McLoughlin
Eyasu Getahun Chekole
Umayal Lakshmanan
Weizhi Meng
Peng Cheng Wang
Jiqiang Lu

Acknowledgments

We would like to extend our utmost gratitude to the people who have, in one way or other, inspired, aided, and contributed to the successful completion of this book.

First of all, we would like to thank Late Prof. Lawrence Wai Choong WONG in Steering Committee for his excellent guidance and valuable contributions to IRC-SET conferences for many years.

We would also like to thank all members of the conference organizing committee who have contributed their time and professional knowledge to make the conference a big success.

Special thanks to all reviewers for their expertise, time, effort, and timely response throughout the peer evaluation process.

Last but not least, our greatest appreciation to all IRC members for their unity.

Contents

Virtual Arm with Multimodal Biased Feedback for Improving EEG Motor Imagery Calibration Training

Jordan Low Jun Yi, Kuai En Kai Ethan, Raghuveer Singh, and Aung Phyo Phyo Wai

Abstract The brain-computer interface (BCI) records electrical signals in the brain. BCI-generated data could be used to develop devices that help people with motor disabilities to move. For this to be done, we must build machine learning (ML) models that are capable of translating brain signals into motor activity. However, it is challenging to build such ML models with current-day BCI data, which are based on motor imagery designs. Herein, we assess the possibility of training ML models based on an alternative form of BCI data that is generated when subjects observe motor movements, as opposed to mere imagery. We demonstrated that this method of data collection improves training accuracy and the ML model's performance. We further improved its performance by providing multimodal biased feedback. Our work provides the foundation for further development of BCI-based ML models.

Keywords BCI · ML · EEG

1 Introduction

1.1 Motor Imagery and Issues

One of the most researched applications of the brain-computer interface (BCI) is its use in motor restoration and motor rehabilitation. By means of assistive technologies such as prosthesis, stimulators or robotic exoskeletons controlled by a BCI, the lost motor function can be regained [1].

As shown in Fig. 1, different motor imagery tasks produce different EEG signals that can be monitored. This information can be interpreted and fed into BCI-controlled prosthetics in order to assist the user in the physical task, encouraging restoration of the lost motor function.

J. L. J. Yi · K. E. K. Ethan (✉) · R. Singh
NUS High School of Math and Science, Clementi Avenue 1, Singapore, Singapore
e-mail: h1710036@nushigh.edu.sg

A. P. P. Wai
Nanyang Technological University, Nanyang Avenue 50, Singapore, Singapore

H. Guo et al. (eds.), *IRC-SET 2022*,
https://doi.org/10.1007/978-981-19-7222-5_1

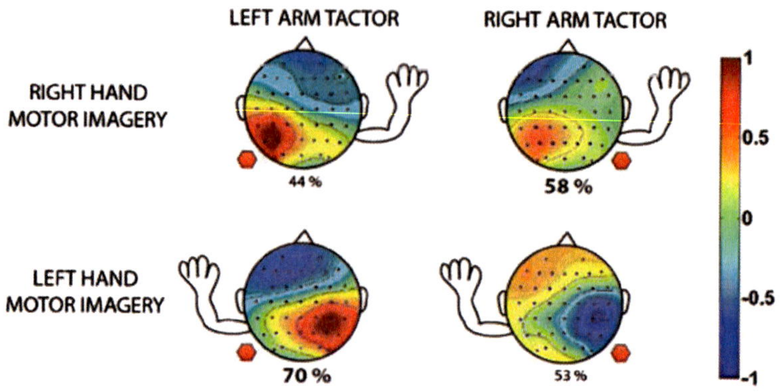

Fig. 1 Different MI tasks produce different brain waves spatially

However, training a person to accurately communicate with a motor imagery BCI is a difficult task that requires several training sessions spanning across weeks. This is known as BCI illiteracy [2]. Training is required as EEGs read event-related-desynchronization (ERD) and event-related-synchronization (ERS) brain waves. These ERD and ERS waves vary across subjects; thus, calibration is needed for each individual.

Motor imagery (MI)-based BCIs demand particularly longer training since the mental rehearsal of a movement without performing it is a counterintuitive task for majority of individuals. Most users cannot visualize a vivid picture of the movement and its kinesthetic experience [3].

Traditional training protocols include execution of a MI task by the user, followed by arrows on a screen as an indicator of which task to carry out, known as visual stimuli. This is shown in Fig. 2. However, an imaginary action can range from the visualization of a self-performed movement from a first-person to a third-person view of the movement, along with other forms of visualization the subject may have [4]. Although such types of MI tasks all involve voluntary actions, they may not involve similar cognitive processes for a myriad of reasons, largely attributed to "vagueness of instruction". This introduces a large amount of inter-subject variability, greatly reducing accuracy of BCIs.

Evidently, to expand the applications BCI has in neurorehabilitation, its training time must be reduced. One such issue contributing to extended training time is that the presentation of feedback is temporally incongruent with the subject's mental image of a bodily movement. This introduces additional variance across subjects, resulting in more data required for BCIs to meet the level of accuracy desired.

In the training paradigm introduced by Pfurtscheller and Neuper [6], subjects imagined either a left-hand or right-hand movement and watched a horizontal feedback bar on a computer screen that was extended to the right or to the left based on the classifier output. This design has been replicated and used in majority of research conducted within this field of study. However, the feedback design currently being

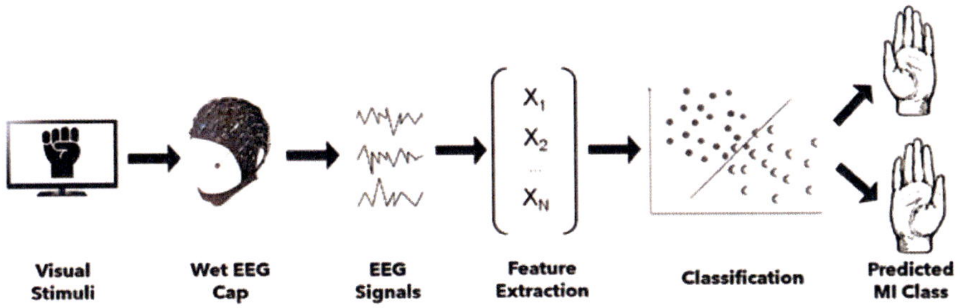

Fig. 2 General open MI calibration system. Modified from [5]

employed has no congruity with the type of image that the subjects held (image of a bodily hand or a foot). Not only does the disparity between the visual feedback and the type of image confuse the subjects during the task, but it also prevents them from imagining the kinesthetic experience and correcting their imagery strategy.

1.2 Ongoing Research

Currently, there has been ongoing recent research of android-like arms being used as a visual stimulus [7]. By observing the real-time movement of the robotic arm, subjects can unambiguously perform the MI task, leading to higher training performance and accuracy. We hypothesize that by combining the use of a virtual robotic arm as visual stimuli together with mirror therapy and biased feedback, we can achieve better MI training performance and accuracy, while making it financially accessible to all.

Mirror therapy exploits the brain's preference to prioritize visual feedback over somatosensory or proprioceptive feedback concerning limb position first devised by Ramachandran and Rogers-Ramachandran [8]. The visual feedback from viewing the reflection of the intact limb in place of the phantom limb made it possible for the patient to perceive movement in the phantom limb. Thus, we use the principle of mirror therapy to aid participants in imagining the kinesthetic experience.

It has been shown that biased self-regulation of motor imagery features can improve in relation to positive bias of feedback [9]. In our experiment, we combine the use of the virtual arm with mirror therapy and biased feedback, to improve subject embodiment and decrease ambiguity to achieve higher MI training performance and accuracy.

2 Hypothesis and Objective

We hypothesize the abstractness of simple visual stimuli, such as arrows, cause heterogeneous imaginations across users, compromising calibration accuracy due to inter-subject variability.

Additionally, we hypothesize using a virtual arm and multimodal biased feedback to demonstrate the task makes training more interesting, intuitive and reduces inter-subject variability and predict a 5–10% accuracy improvement. With similar ERD and ERS responses across individuals, our ML model will be able to train with others' EEG data, resulting in reduced calibration time.

3 Experiment Design and Protocol

Bear in mind methodology is split into 2 sections, experiment design and protocol and data collection and analysis.

The experiment is divided into 2 sessions each lasting ~30 min. In Session 1, participants are shown the standard visual stimuli for their MI tasks. In Session 2, participants are instead shown a video of a virtual arm as visual stimuli. Otherwise, both sessions are identical. Before both sessions, the subject is briefed on what they can expect and shown short snippets of the experiment (Fig. 3).

We grouped the tasks into 2 broad categories: non-MI tasks and MI tasks:

Non-MI tasks	MI tasks
Calibration Deep breathing Flanker attention test	MI task classes Biased feedback Visual stimuli

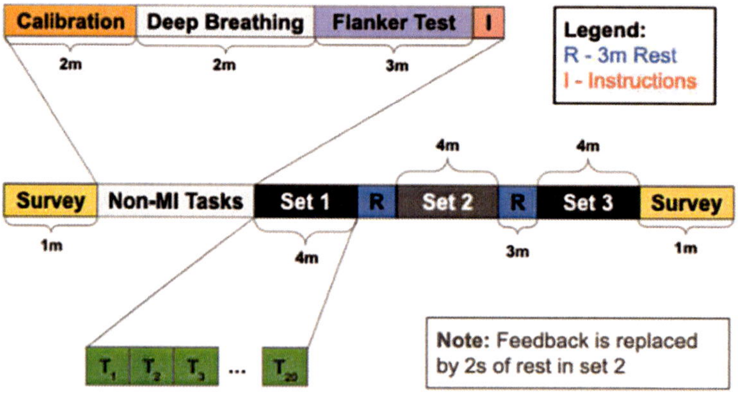

Fig. 3 Block diagram of sessions 1 and 2

3.1 Non-MI Tasks

Calibration The participant closes their eyes and relax for 60 s, thinking of nothing, followed by a short break. They then repeat the same procedure, instead opening their eyes this time. This is required as a baseline reading, to help remove background artifacts from the EEG signals and identify brainwaves related to ERS and ERD.

Deep Breathing A short breathing animation is played. This helps the user relax, clear one's mind of any lingering thoughts and captures the user's current inattention level.

Flanker Attention Test Subjects are told to complete multiple reaction tests, with the aim of accuracy and speed. This activity helps the user focus before MI tasks, get used to visual stimuli on the screen and capture the user's current attention level.

3.2 MI Task Trials

There are 3 sets of tasks, each consisting of 20 trials. Each trial starts with 4 s of fixation cross to maintain focus, 8 s of a random MI task and ends with a 2 s rest. At the end of each set of 20 trials, there is a 3-min break, and participants can continue whenever they want afterward.

We will be using 5 classes of MI tasks:

- **L**: Moving (right hand) leftward
- **R**: Moving (right hand) rightward
- **F**: Reaching (right hand) out/forward
- **C**: Clenching and unclenching (right fist)
- **X**: Resting, no movement.

The 20 trials are split into 5 groups, with each group assigned to a different MI task.

3.3 Biased Feedback

At the end of each trial in Set 1 and Set 3, biased feedback is given to the subject for further motivation. This feedback is pre-computed. Around 30% are positive feedback, 50% are neutral feedback and 20% being negative feedback. The feedback is shown during the 2 s break after every trial. Set 2 has no feedback.

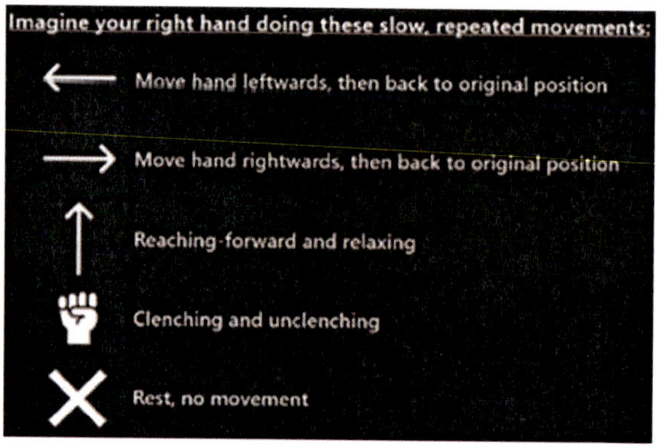

Fig. 4 Visual stimuli

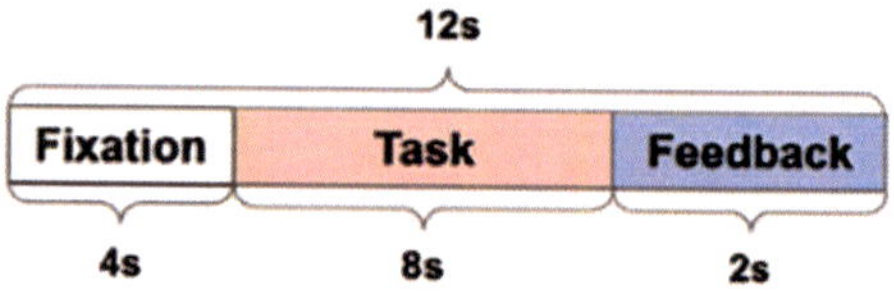

Fig. 5 Block diagram for session 1 MI tasks

3.4 *Session 1 Visual Stimuli: Symbols*

Session 1 serves as our control. We used standard left and right arrows to signal imagining moving their right arm left and right. Due to a lack of symbols to represent reaching out, an up arrow was used, and subjects were reminded to imagine moving their right arm out and not up. A fist emoji/symbol was used to represent clenching, and a X was used to represent rest (imagine holding your arm still) (Fig. 4).

During Set 2, the 2 s of feedback is replaced by 2 s of rest where a blank screen is shown. Otherwise, each trial follows the format of 4 s fixation (fixate on a fixation cross), 8 s of task (the screen just shows a symbol), 2 s of feedback as shown in Fig. 5.

3.5 *Session 2 Visual Stimuli: Virtual Arm*

The virtual arm was constructed using popular and free 3D modeling and animation software Blender. The pre-rigged body was taken from turbosquid.com [10] under the standard 3D Model License [11] (Figs. 6 and 7).

Each trial (imagination session) lasted 12 s, of which the first 4 s were spent fixating on the fixation cross. Close to the last second of fixation, the fixation cross dissolved into the arm performing the motion. This was implemented to minimize cuts which may distress or otherwise jar the subjects.

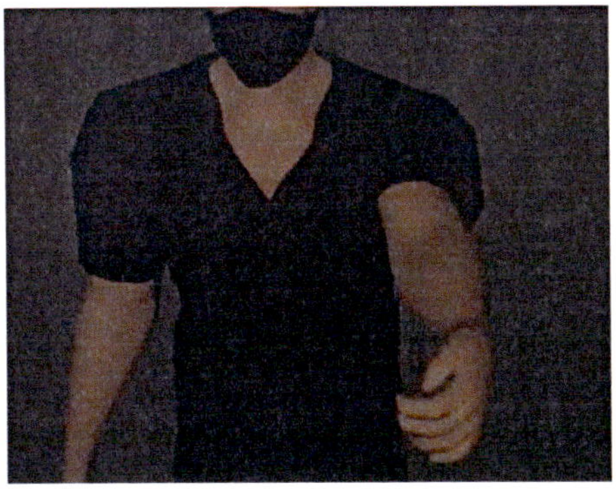

Fig. 6 Screenshot of the "fist" animation

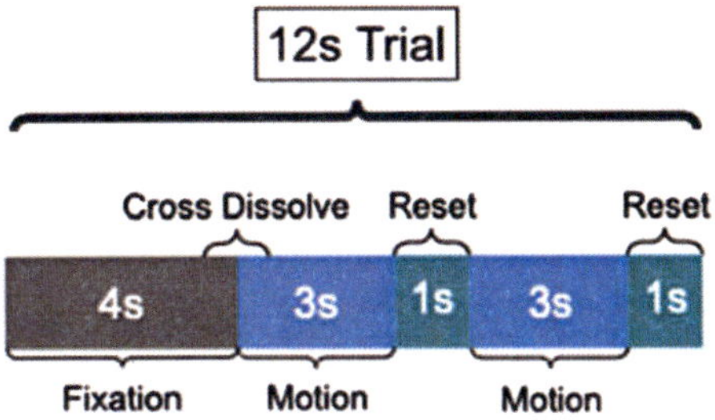

Fig. 7 Block diagram for session 2 MI tasks

The 4 s of fixation was followed by 3 s of the arm performing the motion which was then followed by the arm going back to the original position (reset) in 1 s, which was then followed by another 3 s of motion and the 1 s of the arm resetting. To place emphasis on the desired motion, a 3 s motion to 1 s reset ratio was chosen (Fig. 8).

The animated body was mirrored (subjects are instructed to imagine moving their right arm) to make it easier for the subjects to imagine and to potentially evoke some mirror therapy effects. The body was placed toward the left of the screen to allow a more zoomed in shot while making sure that the moving arm stays in in frame.

Fig. 8 Screenshot of the "right" animation

Inverse kinematics are often used to produce natural looking animations, an attempt was made to implement inverse kinematics, and however, due to a repeated shoulder buckling issues, we opted to use basic key frame animations moving each joint by hand plus some acceleration and deceleration. This technique was sufficient in making natural looking animations for our research as we mainly focused on simple cores movements.

The brightness of the animations was minimized to avoid jarring the subjects and minimize stress on their eyes as lights were turned off during the experiment.

In addition to having a solid gray background, shadows were minimized to improve focus.

Note: Each trial is followed by 2 s of rest/feedback bringing the total trial length to 14 s.

4 Data Collection and Analysis

4.1 Participants and Recordings

We have recruited 13 participants (12 male, 1 female) to take part in both sessions of our experiment, all of which were high school students who gave written informed consent and received monetary compensation for their participation. All subjects were right-handed, and none suffered from any physical disabilities (Fig. 9).

A wet EEG cap was used in this experiment. A total of 14 channels were used—NZ, A1, A2, Fp1, Fp2, FC3, FCz, FC4, C3, Cz, C4, CP3, CPz, CP4. An extra electrode is used to serve as electrical ground. The channels were sampled at 1000 Hz.

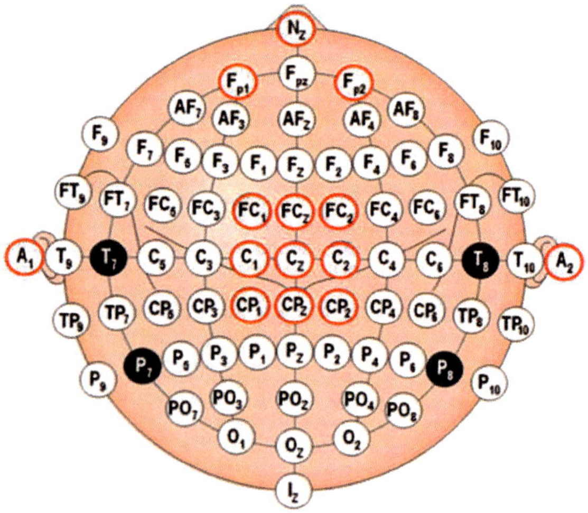

Fig. 9 Diagram of EEG channels

4.2 Experimental Setup

Prior to the experiment, electrically conductive gel was injected into the respective electrodes with a syringe to help improve the resolution of brain signal readings. Then, 14 + 1 electrodes were used, with one taped onto each ear (A1, A2), a third taped onto the forehead of the subject (X1) and the rest positioned on the wet EEG cap. Each electrode is adjusted accordingly to minimize impedance readings and improve signal to noise ratio. Only impedance readings below 50 kΩ were accepted, with most readings < 1 kΩ (Figs. 10 and 11).

Our experimental setup consists of a display, low-latency controls to record timestamps accurate to the millisecond, an EEG cap to record the subject's brain waves and a Tobii eye tracker to ensure the subject is paying attention to the screen. Lights were turned off during the experiment and turned on purely for photographic purposes in the image above.

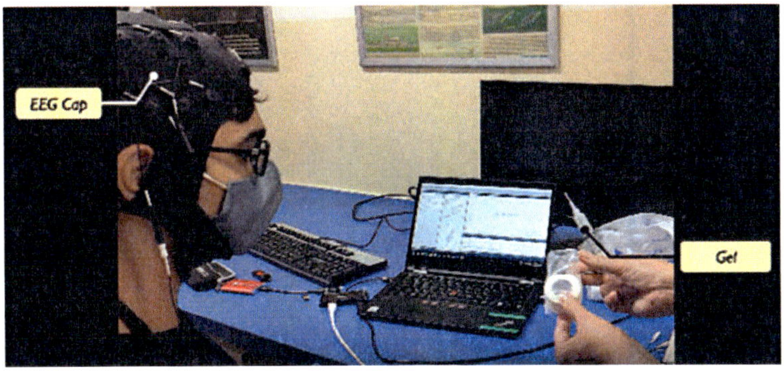

Fig. 10 Setting up wet EEG cap before experiment

Fig. 11 Experimental setup

4.3 Data Collection

The experiment is fully automated and paced by a Python program; we created and made use of a popular open-source library PsychoPy [12]. It displays on-screen instructions and visual stimuli required for MI tasks and communicates with the EEG cap, creating stim code timings to aid data analysis while saving local log files detailing local timestamps correlating with the user's pace. This allows for us to match up the EEG's timestamps with PsychoPy.

Raw data collected:

(1) PsychoPy log files with (computer) timings
(2) EEG time series with stim code timings.

4.4 Data Processing

Only offline processing is carried out after the experiment is completed, and no processing occurs in real time (Fig. 12).

4.4.1 Stim Code Processing

The PsychoPy program sends Stim Code timestamps to the EEG amplifier whenever a new section in the experiment procedure starts and is then used to synchronize the EEG time series data with PsychoPy's timestamps. Afterward, the time series needs to be subdivided into smaller time series, each representing EEG signal data for a MI trial, to be fed into the signal processor.

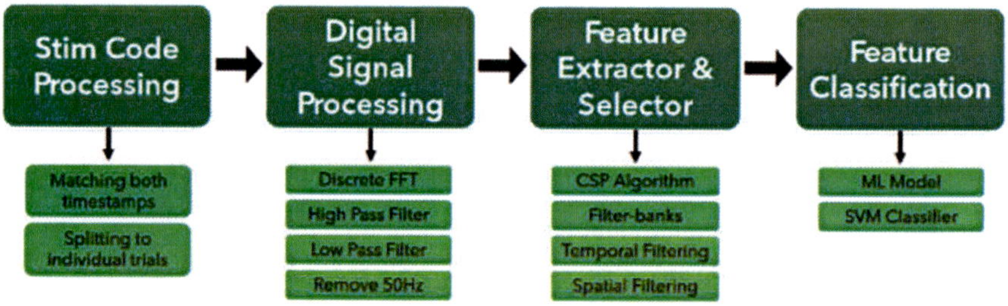

Fig. 12 Post-processing procedure

4.4.2 Digital Signal Processing

Before we can use the EEG data to identify ERD and ERS signals, we must first remove noise. This step is largely handled by the Filter Bank Common Spatial Patterns (FBCSP) Toolbox [13]. A discrete Fast Fourier Transform (FFT) is applied onto the time series to translate it to the frequency domain, and we apply a high and low pass filter, as well as a 50 Hz filter to remove unwanted noise.

Note: 50 Hz is the frequency of Singapore's wall AC power supply, which could generate some EMI, distorting EEG brain signal readings.

4.4.3 Feature Extractor and Selector

The built-in CSP algorithm in the FBCSP Toolbox helps take care of feature extraction and selection from our processed EEG signals, to extract out ERD and ERS brain signals correlated with MI tasks.

4.4.4 Feature Classification

We create a support vector machine (SVM)-based ML model to read the ERD and ERS brain signals and classify them into the 5 classes of MI tasks. This essentially allows our EEG cap to identify which of the 5 classes, the subject is imagining for his/her MI task.

About 75% of the trials were used as a learning dataset to train the ML model, while 25% of the trials were used as a validation dataset to identify the ML model's accuracy.

We compute binary (2-class) classification accuracy between the "relax" class and all 4 other MI classes. This provides a fair comparison of whether feedback improves performances or not for each MI classes (L, R, F, C).

A fivefold CV is used for subject-dependent cross-validation, so each subject's accuracy is computed independently. This means we will have a 2D table of accuracy, with an accuracy term for each subject and each of the 4 MI classes. This allows us to compare accuracies across subjects and MI class and reach various conclusions.

5 Results and Discussion

5.1 Classification Accuracies Across Subjects

From our ML model, we have the table of accuracies for each MI task. Data from both sessions and all 3 trials have been combined for each subject and split between

Table 1 Accuracies of each MI class, averaged across all individuals

Class	C	X	L	R	F
Accuracy (%)	98.8462	97.6923	98.8462	98.2692	99.8077

validation and training dataset for the ML model. After getting the accuracy table for each subject, we average the values to get Table 1.

5.2 *Statistical Significance Testing*

We have run our ML Model 3 more times (starting off without any pre-trained data), this time on each of the 3 sets of 20 trials. We combine the accuracy values from Sets 1 and 3 to get 104 samples of accuracy for "trials with biased feedback" and from Set 2, 52 samples of accuracy for "trials without biased feedback". Since they are sampled from 2 different distributions, we conduct a 2-sample t-test to determine if the distributions have different means.

Variables:
Let μ_2 be the true mean of the distribution of accuracies of trials from Set 2.
Let μ_{1+3} be the true mean of the distribution of accuracies of trials from Sets 1 and 3.
Let the significance level be 0.5%.

Hypothesis:

$$H_0 : \mu_{1+3} = \mu_2$$
$$H_a : \mu_{1+3} > \mu_2$$

SRS: A SRS was conducted to pick the trials used as part of the validation dataset.

Normality: Both populations (Set 2 and Set 1 + Set 3) have a sample size larger than 30. By Central Limit Theorem, both populations are normally distributed.

Independence: It is reasonable to assume all EEG readings are independent of one another.

To test the hypothesis H_0, we compute the 2-sample t-statistic and p-value (Table 2):
As the one-tailed t-value is $0.004 < 0.005$, at the 0.5% significance level, we can conclude that the presence of biased feedback helped improve the accuracy of classification. We theorize this is due to the feedback encouraging the user, causing the subject to focus better and produce more distinct brainwaves.

Table 2 2-sample t-test results

2-sample t-test: assuming unequal variances		
Value	Set 1 + Set 3	Set 2
Mean	0.6039663	0.54367
Variance	0.0217393	0.016123
Sample size	104	52
df	117	
t-stat	2.6464932	
$p(T \leq t)$ one-tail	0.0046262	
t-critical one-tail	1.6579817	

5.3 Limitations

However, as limited MI trials were executed by each user, we cannot conclude with as much certainty as we hoped that our new form of stimuli is significant better.

Additionally, all our subjects are high school students and largely male, unrepresentative of the general population.

We originally planned to use a physical robotic arm; however due to technical issues, we are unable to construct it in time; thus, we resorted to a virtual robotic arm which may not be as interactive and engaging as a physical robotic arm.

Furthermore, our accuracy may have been limited by the use of SVMs, with more time and data we hope to implement deep learning using KCNNs.

6 Conclusion

Our research shows that biased multimodal feedback is extremely effective in boosting MI BCI classification performance in the case of large cores movements such as moving the right arm left and right in large sweeping motions, which may be sufficient to prompt further investigation of its applications in fine motor movements.

7 Future Work

Previously, we have mentioned that all EEG brain signal data processing occurs offline. However, with our already trained parameters for our ML model classifier, with enough computing power, in the future, we could possibly use it to predict which class the subject is executing for the MI task in real time. This could be a big leap forward in technological progress and allow for BCIs to pave the way for motor restoration and motor rehabilitation.

Additionally, with our limited data, we cannot reliably identify the accuracy of each feedback type. This requires more subjects and trials for each section and an additional mapping of feedback type during fivefold CV in our ML model.

References

1. Van Dokkum, L. E. H., Ward, T., & Laffont, I. (2015). Brain computer interfaces for neurorehabilitation—its current status as a rehabilitation strategy post-stroke. *Annals of Physical and Rehabilitation Medicine, 58*(1), 3–8.
2. Guger, C., Edlinger, G., Harkam, W., Niedermayer, I., & Pfurtscheller, G. (2003). How many people are able to operate an EEG-based brain-computer interface (BCI)? *IEEE Transactions on Neural Systems and Rehabilitation Engineering., 11*(2), 145–147.
3. Hwang, H. J., Kwon, K., & Im, C. H. (2009). Neurofeedback-based motor imagery training for brain computer interface (BCI). *Journal of Neuroscience Methods, 179*(1), 150–156.
4. Annett, J. (1995). Motor imagery: Perception or action? *Neuropsychologia, 33*(11), 1395–1417.
5. Lotte, F. (2014). A tutorial on EEG signal processing techniques for mental state recognition in brain-computer interfaces. https://doi.org/10.1007/978-1-4471-6584-2_7
6. Pfurtscheller, G., & Neuper, C. (2001). Motor imagery and direct brain-computer communication. *Proceedings of the IEEE, 89*(7), 1123–1134.
7. Alimardani, M., Nishio, S., & Ishiguro, H. (2018). Brain-computer interface and motor imagery training: The role of visual feedback and embodiment. In D. Larrivee (Ed.), *Evolving BCI therapy: Engaging brain state dynamics*. IntechOpen.
8. Ramachandran, V. S., Rogers-Ramachandran, D. C. (1996). Synaesthesia in phantom limbs induced with mirrors. *Proceedings of the Royal Society of London, 263*(1369), 377–386
9. Alimardani, M., Nishio, S., & Ishiguro, H. (2014). Effect of biased feedback on motor imagery learning in BCI-teleoperation system. *Frontiers in Systems Neuroscience, 8*.
10. 'Basic Bandit By Will Morillas': Pre-rigged 3d blender model https://www.turbosquid.com/3d-models/basic-bandit-3d-1250561
11. Turbosquid 3D Model License. https://blog.turbosquid.com/turbosquid-3d-model-license/
12. Valerdi, L. M. A., & Sepulveda, F. (2015). Implementation of a motor imagery based BCI system using python programming language. In *PhyCS 2015—2nd international conference on physiological computing systems, proceedings*, pp. 35–43.
13. Ang, K. K., Chin, Z. Y., Zhang, H., & Guan, C. (2008). Filter bank common spatial pattern (FBCSP) in brain-computer interface. In *2008 IEEE international joint conference on neural networks (IEEE world congress on computational intelligence), 2008*, pp. 2390–2397.https://doi.org/10.1109/IJCNN.2008.4634130

Extracting Cellulose from Commonly Consumed and Wasted Asian Vegetables

Chen Ye Kai Trevor, Ang Hran Jie Joe, and Wu Dechun

Abstract It is no secret that food waste is a large issue in Singapore, with most of our wasted food not being recycled in Singapore. Among this wasted food, vegetables are the most wasted food among other groups. However, this need not be the case! Cellulose is a vital component of vegetables cells cell walls and makes up 33% of plant matter. This molecule has large varying usage in many industry today. Thus, in this research, we aim to see if extracting cellulose to recycle it can not only be done economically, but also see which vegetable commonly found in Singapore would be the most worth to recycle and to find any relation between their cellulose yield and their characteristics.

Keywords Cellulose · Vegetables · Recycling · Extracting

1 Introduction

Food waste is a major issue in developed countries like Singapore. In the last 10 years, the amount of food waste in Singapore has increased by over 20% [1], and this trend is expected to continue. Among all the food wasted, fruits and vegetables are in the top two spots [2]. Furthermore, only, about 19% of food wasted is recycled in Singapore [3], and this 19% mainly constitutes homogeneous food waste from food manufacturers like yeast/grains and soya bean waste, which are not vegetables.

However, there is large potential in the recycling sector for vegetables. Cellulose is the main content in many plants, with 33% of the plant matter being made of cellulose [4]. Cellulose is mainly found in the walls of plant cells and helps keep plants upright. Cellulose also has a variety of useful applications in many fields. Cellulose is used to produce paper and clothing, and it is also widely used in the pharmaceutical and cosmetic industry [5].

Research has been done on cellulose content of vegetables commonly consumed in western countries, but there is no data on cellulose content for Asian vegetables. Given

C. Y. K. Trevor (✉) · A. H. J. Joe · W. Dechun
Anderson Serangoon Junior College, Singapore, Singapore
e-mail: trevorchen34@gmail.com

H. Guo et al. (eds.), *IRC-SET 2022*,
https://doi.org/10.1007/978-981-19-7222-5_2

the fact that these cellulose are very abundant in vegetables which are commonly wasted in Singapore, this experiment aims to determine what kind of Asian vegetables will yield the highest amount of cellulose. We hope to provide information as to which vegetables have the highest cellulose yield when extracted. This information will support new methods that are in progress which extracts cellulose more efficiently such as the extraction done by Prof. William Chen [6]. Thus, making recycling vegetables by extracting cellulose more viable.

2 Choosing Vegetable

5 vegetables commonly found in Singapore were chosen. The vegetables are Kang Kong, Nai Bai, Cai Xin, Baby Sweet Potato Leaves, and Hong Kong Kai Lan (Refer to Appendix).

These vegetables were chosen as there is a significant difference between their stem to leaf ratio as the stem and leaf have different cellulose and other compound ratios, and also, the stem thickness as cellulose helps contribute to rigidity and density of the stem.

Kang Kong is chosen due to it having hollow stem, while Kai Lan was chosen due to it having a thick and rigid stem.

Nai Bai was chosen due to its low stem to leaf ratio, while sweet potato has a high stem to leaf ratio.

Lastly, Cai Xin was chosen as it has the average stem to leaf ratio and thickness of stem.

3 Method of Extraction

The method [7] used is a modified version of a currently existing plant biomass fractionation method to reduce time spent and prevent acid from fuming so that it is safe to be carried out in school laboratory (Refer to Appendix).

Since the method of extracting cellulose would remove any form of impurities due to production of vegetables, we would safely assume the mass remaining is only the mass of the cellulose.

Step 1:

Vegetables are blended separately with 0.3 dm^3 of distilled water to form a mixture. This is then filtered through a sieve where the residue called pomace is collected and 10 g of pomace is used for each sample of vegetables for extraction.

Step 2:

The pomace is washed in the hot water (100 °C) for 10 min. It is then rinsed and filtered.

This removes sugar, phenolic compounds, and pectins which are soluble in water.

Step 3:

Acid treatment: 30 cm^3 of 1 mol/dm^3 of HCl is added to the pomace. The pomace is then put into a water bath at 60 °C for 20 min. The mixture is then rinsed and filtered through a filter paper. After filtering the solution, Step 3 is repeated 2 more times.

This removes acid-soluble pectins, polysaccharides, and polyphenols.

Step 4:

Alkaline treatment: 30 cm^3 of 1 mol/dm^3 of NaOH is added to the mixture. The mixture is then put into a water bath at 55 °C for 20 min. The mixture is then rinsed and filtered through a filter paper. After filtering the solution, Step 4 is repeated 2 more times.

This removes remaining hemicellulose, lignin, and other alkaline-soluble polysaccharides.

Step 5:

Bleaching process: 6.67 mol/dm^3 NaClO is added to the mixture. The mixture is then rinsed and filtered through a filter paper. This removes residual lignin and phenolics.

Step 6:

The cellulose is washed and dried using filter paper. The mass of the final product obtained is measured to determine the final amount of cellulose obtained. Schultze reagent [8] (saturated solution of $KClO_3$ with nitric acid) is then added to the test. When the indicator turns purple, there is presence of cellulose in the product.

Step 7:

The cellulose is then heated in an oven at 90 °C for 50 min to obtain a dry sample of cellulose. The cellulose residue mass is then weighed and recorded.

4 Control Variables

See Tables 1 and 2.

5 Risk Assessment

Heating of HCl, when HCl with a boiling point of 108.6 °C starts to boil, it will produce HCl fumes which are toxic to humans. HCl fumes can irritate the lungs, causing a cough and shortness of breath. Breathing high levels of the gas or vapor can lead to a build-up of fluid in the lungs, which may cause death [9]. Therefore, we chose the temperature of 60 °C for our hot water bath. This temperature was

Table 1 Control variables and rationale

Control variable	Why should it be controlled?	Method to control it
Mass of pomace used	To ensure that the cellulose collected is extracted from the same mass of vegetable pomace which makes the experiment more accurate	Measuring 10 g of pomace from step 1 for each vegetable and use only those 10 g samples
Duration pomace is left in the respective solutions in each step	To ensure that the amount of cellulose extracted is not increased due to being exposed to the solution longer can cause cellulose to degrade	Place all the pomace into the solutions at the same time and start a stopwatch immediately after to time accurately
Volume and concentrations of all solutions used • 20 cm^3 of 1 mol/dm^3 HCl • 20 cm^3 of 1 mol/dm^3 NaOH • 20 cm^3 of 6.67 mol/dm^3 NaClO	To ensure that the rate of reaction of all the solutions with the pomace is equal, so, no setup reacts more with the acid or alkaline which can degrade the cellulose affecting the yield	Use same concentration of solutions (do not dilute with water) and measure equal volumes to be used using measuring cylinder for each pomace

Table 2 Independent variable and rationale

Independent variables	Why should it be varied?	Method to vary it
Types of vegetables used	To determine different cellulose content between different types of vegetables	Adding 10 g of pomace to different setups

chosen as the higher the temperature the more HCl fumes would be released through evaporation, but heat is needed to overcome the activation energy and increase the rate of the reaction.

Handling of Acid. Acid being corrosive at 1Mol/dm3 is able to irritate the skin, so we wear gloves while handling acid.

Handling of alkaline. It will increase the pH of the skin which would cause the skin to look flaky and red, so we wear gloves while handling acid.

Usage of the oven. The oven has high temperature which can burn the skin, so we wear gloves when using oven.

6 Results

See Fig. 1 and Table 3.

With an ascending order of Nai Bai, Cai Xin, Kang Kong, Baby Sweet Potato Leaves and Hong Kong Kai Lan.

This proves that extracting cellulose from wasted Asian vegetables that are commonly consumed and wasted is the most economical if done to Hong Kong

Fig. 1 End product (cellulose) from extraction

Table 3 Final mass and yield of extracted cellulose

Samples	Mass of cellulose/g	Yield/% [(mass of pomace/mass of cellulose) * 100%]
Kang Kong	0.73	7.3
Nai Bai	0.06	0.6
Cai Xin (control)	0.46	4.6
Baby sweet potato	0.84	8.4
Hong Kong Kai Lan	1.72	17.2

Kai Lan, with it having slightly over double the yield of cellulose as compared to baby sweet potato, which has the second highest yield.

However, there is an anomaly in the results for Nai Bai, which has an unusually low yield of cellulose compared to our other 4 vegetables, suggesting that it is not practical to extract cellulose from wasted Nai Bai. A possible reason for this anomalous value for Nai Bai could be the thicker size of the stem of the Nai Bai. However, we are not able to conclude if this is the case from our current experiment data, and this experiment would have to be redone again to ensure the anomalous data for Nai Bai is not due to experimental errors.

Additionally, by comparing the shape and size of the vegetables together with their cellulose yield, no clear correlation between the 2 can be determined, and the cellulose yield of a vegetable is therefore unrelated to its size.

The Schultze reagent turned purple when added to the five vegetables. Hence, cellulose is present in all 5 vegetables (Fig. 2).

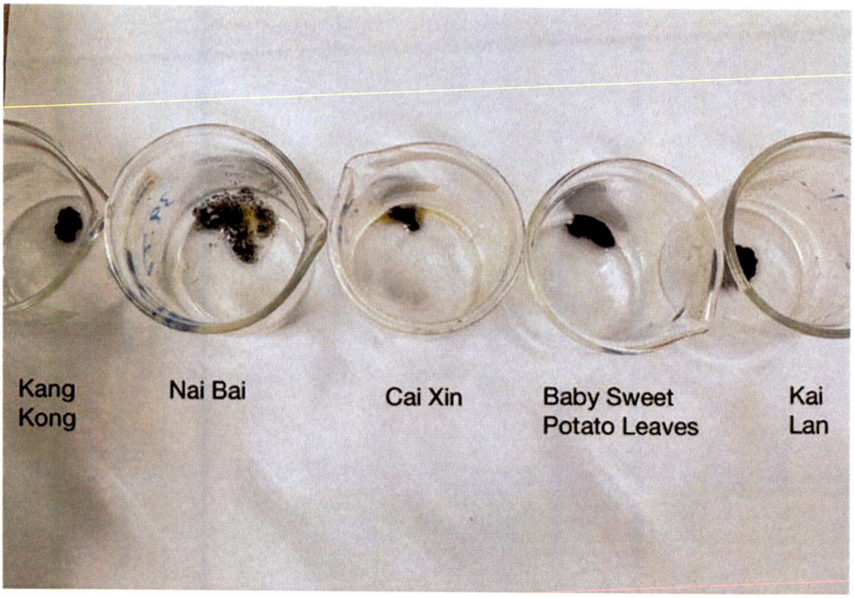

Fig. 2 Testing with Schultz reagent to test for cellulose

7 Limitations

There is a limitation to our research that could be improved on. Filter paper on top of a filter funnel was used to filter the heated pomace solutions to obtain back the residue. However, using this method of filtration would cause a lot of residue of the solution to be lost as it is difficult to remove and collect all the residue on the filter paper. Considering the extraction process requires the pomace solutions to be filtered over 7 times to collect residue, it means that there is significant loss of cellulose throughout the process due to experimental methods, and the actual cellulose yield for all the vegetables should be higher.

8 Conclusion

From our results, the vegetables can be ranked from highest cellulose yield to lowest cellulose yield in the order: Hong Kong Kai Lan (17.2%) Baby Sweet Potato (8.4%) Kang Kong (7.3%) Cai Xin (4.6%) Nai Bai (0.6%).

From this data, we can conclude that for all the vegetables, especially Nai Bai, the cellulose yield is not very high, with none of them having over a 20% cellulose yield. However, if we consider the filtration limitations as mentioned above, it is likely the actual cellulose yield of all the vegetables should be considerably higher. We can also see that the cellulose yield% for Hong Kong Kai Lan is the largest and most significant, having a cellulose yield of 17.2%, which is twice the cellulose yield of the second highest vegetable cellulose yield, which is Baby Sweet Potato. Therefore, if future investors were to look into recycling wasted vegetables, it is most advisable for them to first start recycling Hong Kong Kai Lan due to its significantly larger yield.

Therefore, by considering our project objectives and limitations, there is clearly still more that can be done for a future continuation of our research. Since our goal is to provide data on the highest cellulose yield among various vegetables but our current research only has 5 different Asian samples, future research involving a larger variety of Asian vegetable samples can be done to answer our research objective more holistically. Additionally, future research using a more effective filtration method can be done to increase the accuracy of the average yield of cellulose determined.

Appendix

Equipments

1. Memmert temperature-controlled hot water bath
2. Oven
3. Kettle
4. Blender
5. Filter papers
6. Conical flasks
7. Beakers
8. Filter funnel
9. Glass rod.

Vegetable Descriptions

Kang Kong (Water spinach) • Long stem • Thin stem • Equal stem: leaf ratio	

(continued)

(continued)

Nai Bai (Milk cabbage) • Short stem • Thick stem • Low stem: leaf ratio	
Cai Xin (Choy Sum) • Long stem • Averaged-sized stem • Average stem: leaf ratio	
Baby sweet potato leaves • Long stem • Thin stem • High stem: leaf ratio	
Hong Kong Kai Lan • Average length stem • Average-sized stem • Low stem: leaf ratio	

References

1. Food waste. Ministry of Sustainability and the Environment. https://www.towardszerowaste.gov.sg/foodwaste/. Accessed December 10, 2021.
2. Youjin, L. Every Singaporean household throws away an S$258 worth of food a year: Study. https://www.todayonline.com/singapore/every-singaporean-household-throws-away-average-s258-worth-food-year-study. Accessed December 10, 2021.
3. Food waste management. *National Environment Agency*. https://www.nea.gov.sg/our-services/waste-management/3r-programmes-and-resources/food-waste-management. Accessed December 10, 2021.
4. Cellulose. Definition, uses & facts. *Britannica*. https://www.britannica.com/science/cellulose. Accessed December 10, 2021.
5. Application of cellulose and cellulose derivatives in pharmaceutical industries. *IntechOpen*. https://www.intechopen.com/chapters/45635. Accessed December 10, 2021.
6. NTU scientists transform discarded durian husk into antibacterial gel bandages. https://www.channelnewsasia.com/singapore/durian-husks-into-bandages-ntu-scientists-280736. Accessed December 10, 2021.
7. Monika. Isolation and characterization of cellulose from different fruit and vegetables pomaces. https://www.ncbi.nlm.nih.gov/pmc/articles/PMC6418744/. Accessed December 10, 2021.
8. Srithevi, S., & Sashikesh, G. The effects of acid and cellulose concentrations on the acid hydrolysis efficiency of waste cellulose from Palmyrah fruit husk. https://vingnanam.sljol.info/articles/4197/galley/3451/download/. Accessed 10 December 10, 2021.
9. Hydrogen chloride. In *Medical management guidelines*. Toxic Substance Portal. ATSDR. https://wwwn.cdc.gov/TSP/MMG/MMGDetails.aspx?mmgid=758&toxid=147. Accessed December 10, 2021.

Sustainable Nano-fibrous Scaffold for Dental Tissue Engineering

Sean Lim Shi-An, Harini Manivannan, Karis Yuen Xin-Er, and Nileshkumar Dubey

Abstract In this project, we investigate the possibility of using bacteria cellulose as an eco-friendly nano-biomaterial alternative to culture dental pulp stem cells in 3-dimensional space as compared to the traditional flat-surfaced cell culture flask that is made from non-biodegradable polystyrene for dental tissue engineering. Bacteria nano-cellulose (BNC) is a good candidate in our search for an eco-friendly alternative as it has a unique interconnected network structure and sturdy physical properties, such as sponge-like properties, large specific surface area, and low chemical reactivity are ideal for culturing stem cells. In addition, bacteria nano-cellulose is cheap and easy to produce on a large-scale while being full organic and biodegradable which gives it the potential to replace preexisting non-environmentally friendly cell culture materials. The compatibility of BNC to hold and release drugs was investigated to test its feasibility to be used in the medical field as a way of delivering drugs along with dental pulp stem cells during dental treatments. Cell viability and morphology were investigated using the MTS assay and fluorescence staining. The bone formation from dental pulp stem cells was measured on scaffold loaded with different concentration of dexamethasone using alizarin red-based assay. We choose to investigate dental pulp stem cells as there share with bone marrow stem cells the ability to undergo bone formation and mineralization but are much more accessible and require less invasive treatments to access.

Keywords Stem cells · Environmentally friendly · Sustainable · Dental · Medical

S. L. Shi-An (✉) · H. Manivannan · K. Y. Xin-Er
National University of Singapore High School, Singapore, Singapore
e-mail: seanlimshian@gmail.com

K. Y. Xin-Er
e-mail: H1710060@nushigh.edu.sg

N. Dubey
Faculty of Dentistry, National University of Singapore, Singapore, Singapore

H. Guo et al. (eds.), *IRC-SET 2022*,
https://doi.org/10.1007/978-981-19-7222-5_3

1 Introduction

Dental pulp is found at the center of the tooth and comprises connective tissues, blood vessels and cells. This pulp offers many important functions [1], like protection against infections through immunological surveillance, rapid reparative dentin formation to guard against noxious external stimuli, and resisting breakage under high tension to prevent tooth fractures. When infected, the treatment of choice is usually root canal therapy. Approximately 15 million root canal treatments are performed each year in order to salvage the tooth by removing the infected dental pulp. The use of synthetic materials as substitutes for dental pulp is a practice that dates back almost as far as dentistry itself. Even though this is a well-established strategy, it is based on the substitution of biological tissue for synthetic and, in most cases, inert materials. Furthermore, devitalization of teeth by removing dental pulp increases the frequency of vertical root fracture, apical periodontal lesions, and secondary caries due to the tooth's loss of ability to perceive aggressiveness caused by reinfection, resulting in a higher rate of tooth extraction.

Recent advances in tissue engineering have advanced the way we regenerate the pulp-dentin complex for preservation and possible restoration of the entire structure and function of the tooth. This is done through the improvements of the triad of tissue engineering: mesenchymal stem cells (MSCs), growth factors, cytokines, and migration factors, and the microenvironment (such as the scaffold).

Traditionally, the microenvironment used to culture stem cells are made of polystyrene [2]. Polystyrene being very cheap and easy to produce has allowed it to dominate and become the most extensively used cell culture material. However, polystyrene degrades slowly, and its foam may leach chemicals into the environment, harming water sources, if disposed of improperly [3]. In addition, disposal of non-reusable polystyrene via burning results in a lot of carbon monoxide being produced which harms the environment and accelerates global warming.

Bacterial nano-cellulose (BNC) is a an organic and biodegradable homopolymer of β-1,4 linked glycose. BNC can be collected from Acetobacter using simple inexpensive and environmentally friendly culturing methods that can be expanded for large-scale production [4]. It is a three-dimensional and interconnected nano-structured network of glucan polymers that are semi-transparent, able to withstand strong tension, yet soft and elastic, and possess sponge-like properties making it the ideal environment for culturing cells [5]. The high surface area and biocompatibility allow it to serve as optimal drug delivery system. In this project, we will investigate the effectiveness of bacterial nano-cellulose as a replacement to polystyrene and the efficacy of the scaffold when loaded with different concentrations of dexamethasone, a glucocorticoid medication used to treat rheumatic problems and also enhances DPSCs mineralization and bone formation.

1.1 Aim and Hypothesis

The aim of this project is to test out the effectiveness of using a bacteria nano-cellulose scaffolds to promote the proliferation of DPSCs for the regeneration of dental pulp and how it varies with different concentrations of dexamethasone. We hypothesize that the BNC is a non-toxic alternative to culturing dental pulp stem cells (DPSCs).

2 Materials and Methods

2.1 Production and Purification of the Scaffolds

Twenty liters of coconut juice was boiled with 20 g of di-ammonium hydrogen orthophosphate, 10 g of magnesium sulfate, 30 g of citric acid and 100 g [6]. One and a half liters of the mixture was then mixed with 55 mL of 95% v/v of ethanol in a tray. The mixture stood at 25 °C for 30 min. 100 mL of *Acetobacter xylinum* were then added to each tray with an aseptic technique. The tray was then covered with cloth and paper to prevent external contamination. Coconut gel was obtained from the mixture after ten days. The coconut gel was sliced, boiled in water and dried in a hot air oven for 12 h. It was then grounded using a grinder and screened through a mesh sieve. The nata de coco bacterial cellulose powders were contained in propylene bags until used [7]. Add bacteria nano-cellulose in water and mix with 0.1 M NaOH at 80 °C for 20 min. The powder will then be processed into sheets that can be used as scaffolds. BNC sheets were then cut up into 100 1 cm × 1 cm squares and 36 circles of diameters of 1 cm. Lastly, autoclave the scaffolds.

2.2 Drug Absorbance and Release

First, a standard curve was formed to act as a reference. Dilution of Dexamethasone occurred to form 300 μl of 6 different concentrations each from 0 to 5 μg/ml. The solutions were vortexed for mixing and short spun for it to settle. Afterward, the absorbance of the solutions was measured at 240 nm using a Multiskan GO spectrophotometer and the values were recorded using the SkanIt Software to plot a standard curve. To measure the drug absorbance of the scaffold, scaffold cutouts were transferred to a well plate. Then 1 ml of 1 μg/ml of Dexamethasone was transferred to 30 wells with the scaffold and 1 ml of 5 μg/ml of Dexamethasone were transferred to the other 30 wells with the scaffold. Next, the plates were wrapped with aluminum foil and incubated in a shaker at 25 °C, 100 rpm for 48 h. The scaffolds were then washed with DPBS twice then left to dry in the biosafety cabinet for 24 h. After drying, 1 ml of double distilled water was added to every well and the scaffolds were incubated at 37 °C. 100 μl of water from the wells were transferred to the 96 well UV plate and their absorbance is measured at 240 nm and compared with the standard curve every hour.

2.3 Cell Culture

These scaffolds will be loaded with DPSCs that have been cultured in the Dulbecco's Modified Eagle Medium (DMEM) with 1 g/L D-glucose and 110 mg/L of sodium pyruvate. The medium in which the cells were initially cultured was removed and discarded. Consequently, the cells were washed by Dulbecco's Phosphate Buffer Saline (DPBS) twice to remove any trace of the medium. 4 mL of TryPLE Express, an animal origin-free, recombinant enzyme was then added to dissociate the cells from the walls of the flask under incubation at 37 °C for 2 min. The remaining solution was centrifuged after DMEM was added for 3 min at 200 rpm so that the cells will resuspend and settle at the bottom of the tube. The resulting solution was transferred to the hemocytometer and placed under the microscope to count the number of cells to find out the concentration of cells. One scaffold was placed into each of the 18 wells and immersed in 200 μL of medium before 20,000 cells were added to each of the wells. (Refer to Appendix for detailed procedure).

2.4 MTS Assay

CellTiter 96® AQueous One Solution Reagent was pipetted into each well of the 96 well polypropylene microplates [8] containing the samples in 100 μl of culture medium, Dulbecco's Modified Eagle Medium (DMEM) with 1 g/L D-glucose and 110 mg/L of sodium pyruvate. The plate was incubated at 37 °C for 1–4 h in a humidified, 5% CO_2 atmosphere. The MTS tetrazolium will then be reduced to its formazan product, which is measured by recording the absorbance at 490 nm using a 96 well plate reader. The absorbance is directly proportional to the number of living cells in culture [9]. The absorbance readings were taken on the 1st, 3rd, and 5th day.

2.5 Immunostaining

The DPSCs that were cultured were washed twice with DPBS. The cells were then fixed with 400 μL of 4% paraformaldehyde (PFA) at pH 7.4 for ten min at 37 °C. PFA was consequently removed, and the cells were washed with the buffer solution twice to remove the traces of PFA. To permeabilize the cell, 400 μL of 0.1% Triton X-100 in 1X PBS was added before incubation at 25 °C for fifteen minutes. Triton X-100 was then removed from the cells before they were washed with DPBS solution twice. The cells were then washed with PBS and blocked using 1% BSA for 30 min, followed by staining with TRITC-conjugated phalloidin and DAPI for 1 h according to the manufacturer's instructions; samples were imaged using an upright fluorescence microscope (Leica, USA).

2.6 *Alizarin Staining*

Ten thousand cells were seeded onto every well containing the BNC scaffold with increasing concentrations of dexamethasone (a steroid that enhanced bone formation and mineralization) and 4 wells without the BNC scaffold for positive and negative control, and bone media were added to every well with BNC and 2 of the control wells. The plates were then incubated for fourteen days, then the mineralization of the cells was assessed with Alizarin Red S staining. To do so, the cells were left in 4% formaldehyde for 30 min, then washed twice with distilled water and 40 mM of ARS is added to every well with the seeded cells. Afterward, the plates were incubated at room temperature for 30 min and washed four times with distilled water. 200 μL of 10% acetic acid was then added to each well and incubated at room temperature for 30 min. Lastly, 200 μL of 10% ammonium hydroxide was added to the solution. Absorbance of 100 μL of the supernatant were measured using a microplate reader (Infinite M200, Tecan, Frankfurt, Germany) at a wavelength of 405 nm.

3 Results

See Figs. 1, 2, 3 and Graphs 1, 2 and 3.

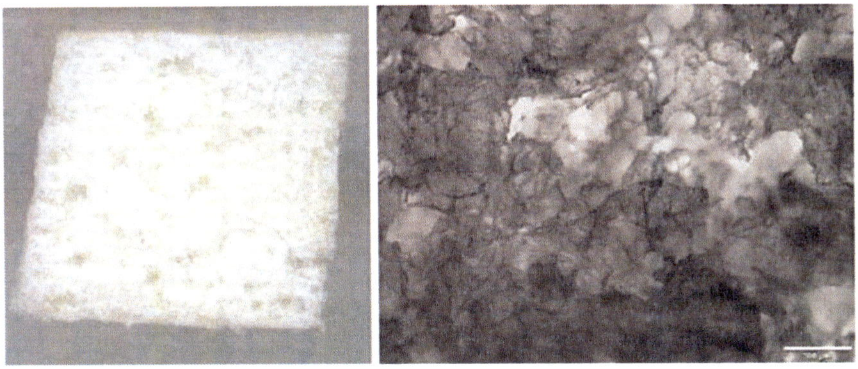

Fig. 1 Stereo and optical microscope images showing structure of BNC scaffold

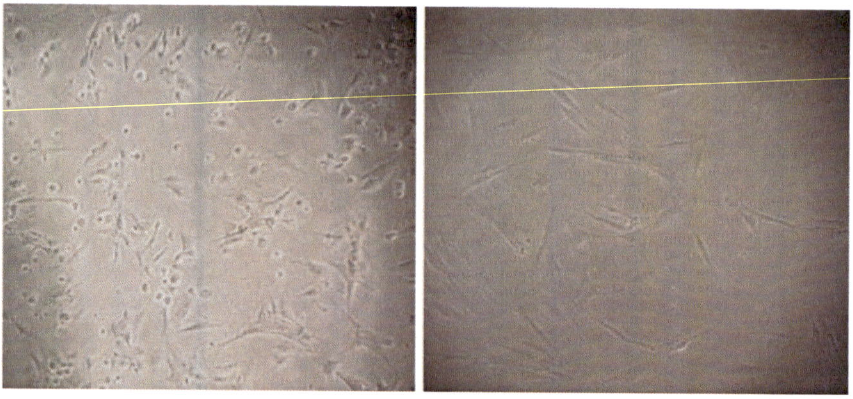

Fig. 2 Representative optical microscopy images of human dental pulp stem cells (hDPSCs)

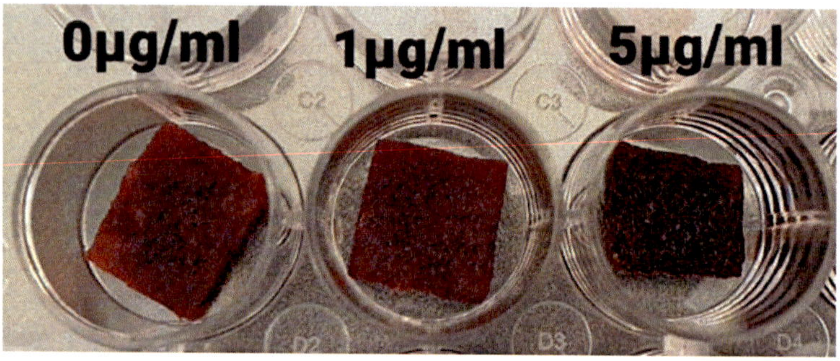

Fig. 3 BNC with different concentrations of dexamethasone after Alizarin staining

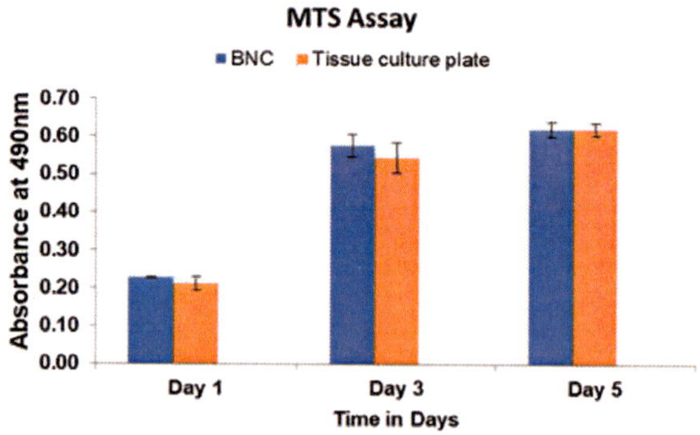

Graph 1 Results of the MTS assay on cells cultured in BNC versus normal cell culture flask

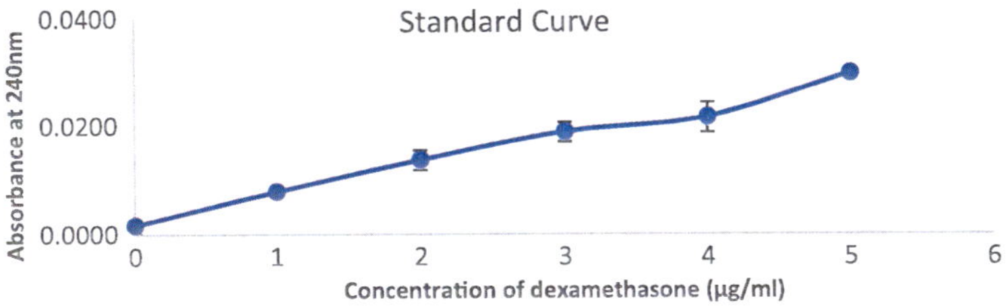

Graph 2 Absorbance values of difference concentrations of dexamethasone

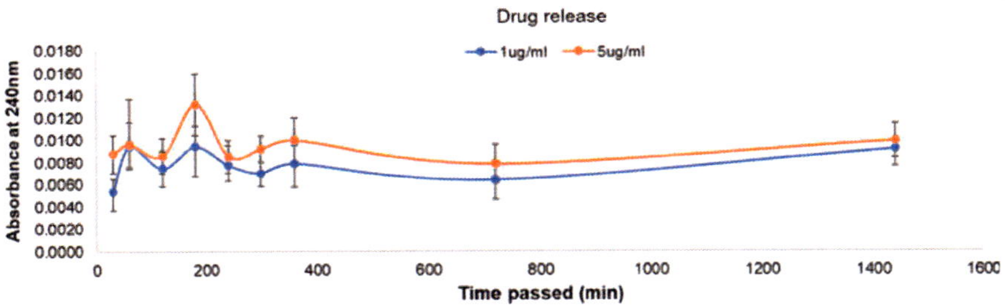

Graph 3 Cumulative release of dexamethasone from the BNC

4 Discussion

4.1 Structure of BNC and Significance Behind DPSCs

The stereomicroscope image (Fig. 1) shows the physical appearance of the BNC scaffold. Under the optical microscope (1B) three-dimensional interconnected network formed by cellulose fibers are visible.

From Fig. 2, we can see the DPSCs cells. DPSCs cells were specifically chosen for this project because of their shared ability with bone marrow stem cells to undergo bone formation while being much more accessible. Unlike bone marrow stem cells which can be obtained via an invasive procedure, DPSCs are much easier to be obtained such as during dental treatments or tooth extraction, hence making DPSCs the main candidate of our study.

4.2 Cell Proliferation in BNC

Quantitative analysis (MTS Assay)

From Graph 1, we can see that the wells with BNC have a similar absorbance value at 490 nm when compared with the wells without the BNC, indicating similar amounts of formazan product was produced hence similar levels of proliferation of cells in well with BNC versus without.

Qualitative analysis (Alizarin Staining)

From Fig. 3, we can see that DPSCs in the BNC can undergo bone formation as shown by the read stains on the BNC after Alizarin staining. Furthermore, we can also see that increasing the concentration of dexamethasone in the BNC results in increased bone formation and mineralization as shown by the increased concentration and darkness of the red color as the concentration of the dexamethasone in the BNC increases.

4.3 Drug Release (Refer to Table 1 in Appendix)

From Graphs 2 and 3, we can see that at the peak, the drug released from BNC soaked in 1 μg/ml is of similar concentration to the actual solution of 1 μg/ml dexamethasone. However, when the BNC is soaked in 5 μg/ml dexamethasone, there is a much larger gap between the concentration of the drug release from BNC and the actual concentration of the 5 μg/ml dexamethasone. This shows us that BNC is much more efficient at releasing drugs at low concentration. Nonetheless, this does not pose as a big issue as drugs like dexamethasone are potent in small volumes. Furthermore, considering that most of the drugs will be used in a small scale following oral surgeries, the ability of BNC to efficiently release drugs at low concentrations is sufficient for it to be used.

4.4 Conclusion, Improvements and Future Work

The MTS Assay and the Immunostaining have shown that the combination of properties of BNC such as its complex interconnected structure and its high-water holding ability does create an ideal environment that is non-toxic to DPSCs and promotes their proliferation. Furthermore, because BNC possess a compact 3-D structure in comparison to the 2-D flat surface of traditionally used cell culture bases, it allows for vertical growth of cells in addition to horizontal growth resulting in a net increase in growth of cells as well. Furthermore, BNC's interconnected structure also allows for increase anchorage of DPSCs.

The Alizarin staining experiment has also shown that DPSCs can and do undergo bone formation in BNC. Furthermore, BNC can also carry drugs such as dexamethasone to influence and enhance the bone formation of DPSCs which is important in dental tissue engineering. This aspect can be further explored in future studies.

BNC has also shown to be able to efficiently release drugs for extended periods of time allowing it to be used in the medical field such as by depositing drugs to reduce inflammation. Future work can expand on this by investigating BNC's potential to deposit other biochemicals such as antibiotics and growth factors to steer the growth of the stem cells.

Appendix

See Figs. 4, 5, 6 and Table 1.

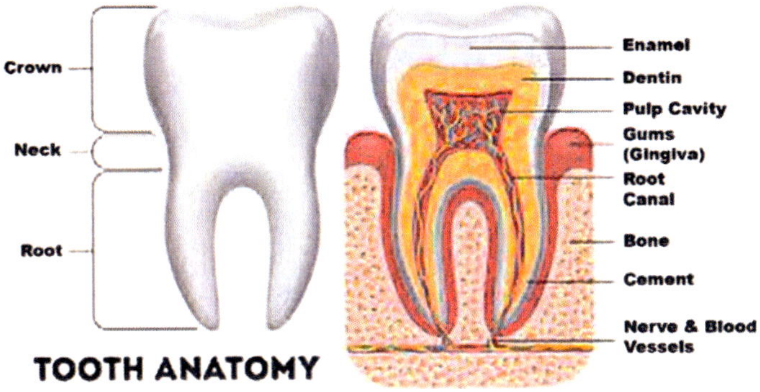

Fig. 4 Anatomy of a tooth

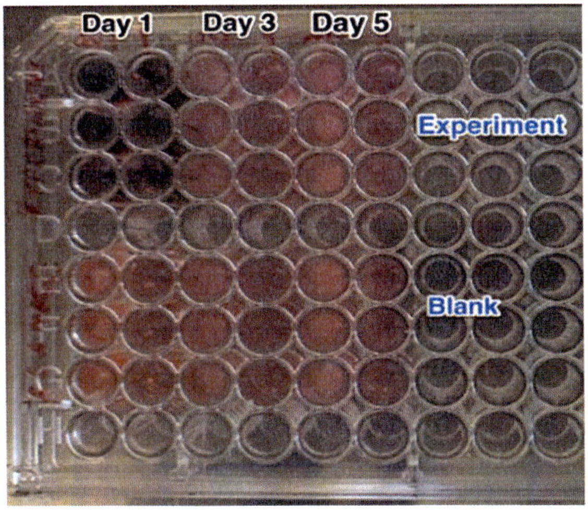

Fig. 5 Cell and BNC in wells during MTS Assay

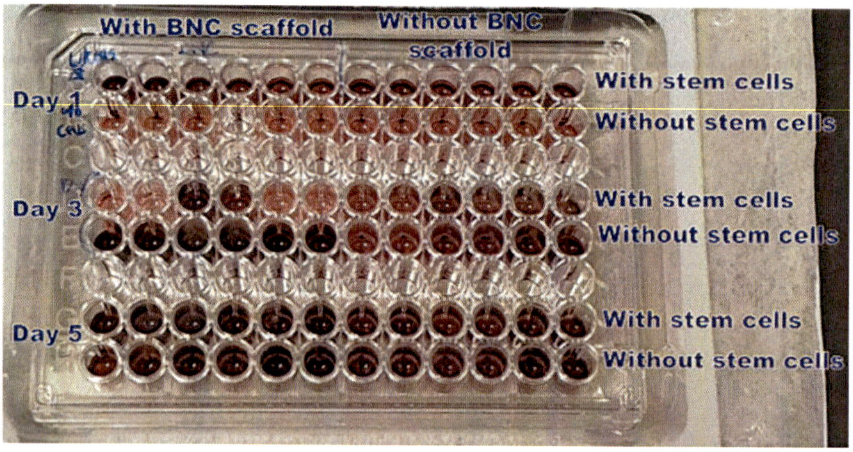

Fig. 6 Well plate used in spectrophotometer to get absorbance values

Table 1 Well plate used in spectrophotometer to get absorbance values

	Concentration of dexamethasone BNC was soaked in (μg/ml)	
	1	5
Time (min)	60	180
Absorbance value	0.0094	0.0132

References

1. Yu, C., & Abbott, P. V. (2007). An overview of the dental pulp: Its functions and responses to injury. *Australian Dental Journal, 52*, S4–S6.
2. Lerman, M. J., Lembong, J., Muramoto, S., Gillen, G., & Fisher, J. P. (2018). The evolution of polystyrene as a cell culture material. *Tissue Engineering Part B: Reviews, 24*(5), 359–372.
3. Farrelly, T. A., & Shaw, I. C. (2017). Polystyrene as hazardous household waste. In *Household hazardous waste management*, 45.
4. Xue, Y., Mou, Z., & Xiao, H. (2017). Nanocellulose as a sustainable biomass material: Structure, properties, present status and future prospects in biomedical applications. *Nanoscale, 9*(39), 14758–14781.
5. Ávila, H. M., Feldmann, E. M., Pleumeekers, M. M., Nimeskern, L., Kuo, W., de Jong, W. C., Schwarz, S., Müller, R., Hendriks, J., Rotter, N., & van Osch, G. J. (2015). Novel bilayer bacterial nanocellulose scaffold supports neocartilage formation in vitro and in vivo. *Biomaterials, 44*, 122–133.
6. Le, B. (2020, July 20). Australian startup develops novel edible scaffold to modernize meat

7. Rachtanapun, P., Jantrawut, P., Klunklin, W., Jantanasakulwong, K., Phimolsiripol, Y., Leksawasdi, N., Seesuriyachan, P., Chaiyaso, T., Insomphun, C., Phongthai, S., Sommano, S. R., Punyodom, W., Reungsang, A., & Ngo, T. (2021). Carboxymethyl bacterial cellulose from Nata de Coco: Effects of NaOH. *Polymers, 13*(3), 348.
8. Greiner Bio-One. (n.d.). 96 well polypropylene microplates. Retrieved January 20, 2022, from https://shop.gbo.com/en/row/products/bioscience/microplates/96-well-microplates/96-well-polypropylene-microplates/
9. Patel, M. I., Tuckerman, R., & Dong, Q. (2005). A Pitfall of the 3-(4,5-dimethylthiazol-2-yl)-5(3-carboxymethonyphenol)-2-(4-sulfophenyl)-2H-tetrazolium (MTS) assay due to evaporation in wells on the edge of a 96 well plate. *Biotechnology Letters, 27*, 805–808.

Hanging a Painting

Chenghao Ding and Lavani Samantray

Abstract Usually, we hang paintings on two nails so that if one nail drops, the painting will not drop. However, is it possible to hang a painting on two nails, such that if either nail drops, the painting will drop? In this report, we investigate this problem and extend it to n nails: Is it possible to hang a painting on n nails such that the painting drops if any one of the nails drops? Through abstraction using algebra and illustration with diagrams, we will show that this is to be possible and there also happens to be more than one solution. We extend further to show that it is possible to make the painting drop only if k of the nails drop. Finally, we give some examples to extend our method to similar problems.

Keywords Algebraic representation · Visualisation · Puzzles

1 Introduction

Usually, we hang paintings on two nails so that even if one nail drops, the painting will remain hanging. However, is it possible to hang a painting on two nails such that if either nail drops, the painting will drop? This is an interesting problem which we will explore in this report, through which we would like to show how (not) to hang a painting using a string and nail(s). We find this problem very intriguing. There are some possible real-life applications, such as for use in museums to prevent people from stealing paintings (If someone tries to remove any one nail, the painting will fall). It can be used to hang a long string on several nails and allows one to quickly take the string off by removing any one nail. Furthermore, it can be used in entertainment (such as magic shows / carnival games) to amaze people. A similar problem has appeared in the Euler Math Competition 2015 [1].

C. Ding (✉) · L. Samantray
NUS High School of Mathematics and Science , Singapore, Singapore
e-mail: h2010032@nushigh.edu.sg

L. Samantray
e-mail: h2010117@nushigh.edu.sg

H. Guo et al. (eds.), *IRC-SET 2022*,
https://doi.org/10.1007/978-981-19-7222-5_4

2 Problem

2.1 Definitions

A painting is said to be hung on n nails by a string if the string wraps around every nail at least once. For example, Fig. 1 below shows a painting hung on 1 nail.

We can also hang a painting on 2 nails. We know that there are many ways of hanging a painting on 2 nails, as shown in Fig. 2.

2.2 Problem Description

We would like to investigate if it is possible to hang the painting on 2 nails such that the painting would drop if either of the nails drops.

As an extension, we would also like to investigate if it is possible to hang the painting on many nails, say n nails, such that the painting drops if any one of the nails drops.

Fig. 1 Hanging a painting on 1 nail

(a) Hanging a painting on two nails

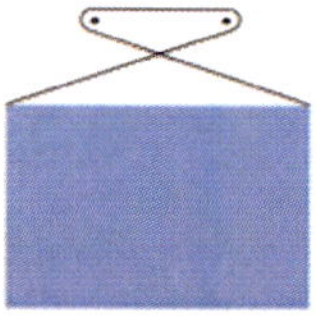

(b) Another way to hang the painting on two nails

Fig. 2 Hanging a painting on two nails

3 Removing 1 Out of n Nails

3.1 Basic Observations

For the case of 1 nail, clearly removing the only nail would make the painting fall, as shown in Fig. 3a.

For the case of 2 nails, after some trying, we may find out that the following arrangement in Fig. 3b allows the painting to fall if either nail is removed.

3.2 Notations

Now, we will consider the general case with n nails. Let us denote the n nails as $n_1, n_2, \ldots, n_n$.

We define the function x_i as a clockwise movement of the string about n_i and $x_i{}^{-1}$ as an anticlockwise movement of the string about n_i, with x_i^{-1} being the inverse of x_i. Note that $(x_i^{-1})^{-1} = x_i$ because wrapping the string clockwise around a nail undoes the anticlockwise movement of the string around the nail. For example, the movement of the string in Fig. 1, a painting on one nail, can be written as x_1, whilst the movement of the string in Fig. 2a, a painting on two nails, can be written as x_1x_2.

Note that a movement is only considered a movement when the string is supported by the nail. In particular, the movement of the string in Fig. 4 is $x_1x_2x_3x_1^{-1}$ and not $x_1x_2x_3x_2x_1^{-1}$ since the string is not supported by n_2 when it passes under the nail.

When a movement and its inverse are placed next to each other, the movement of the string can be simplified by omitting it, as when the movement and its inverse are adjacent, they cancel each other out, and that is equivalent to the string not being wrapped around the nail. For example, $x_1x_2x_2^{-1}$ can be simplified to x_1 as shown in Fig. 5, but $x_1x_2x_1^{-1}$ cannot be simplified as x_1 and x_1^{-1} are not next to each other.

When the movement of the string is simplified to where there are no functions in the movement, we will define the movement of the string to be e.

The inverse of a sequence of movements $x_{i_1}x_{i_2}\ldots x_{i_n}$ is equal to $x_{i_n}^{-1}x_{i_{n-1}}^{-1}\ldots x_{i_1}^{-1}$ as the last movement done has to be undone, followed by the next one and so on. For example, $(x_1x_2)^{-1}$ is equal to $x_2^{-1}x_1^{-1}$, as shown in Fig. 6.

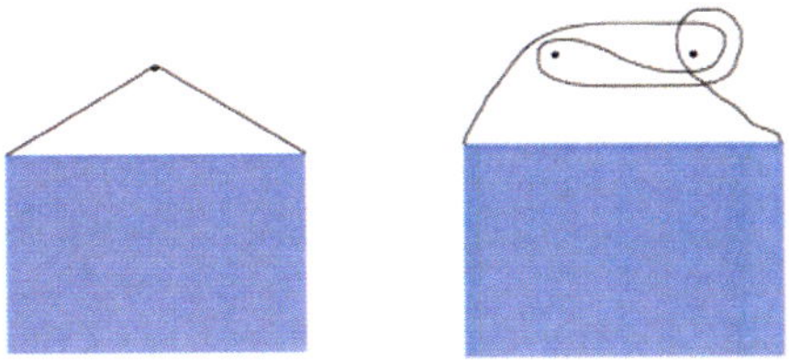

(a) Solution for 1 nail (b) Solution for 2 nails

Fig. 3 Solution for basic cases

Fig. 4 $x_1x_2x_3x_1^{-1}$, not $x_1x_2x_3x_2x_1^{-1}$

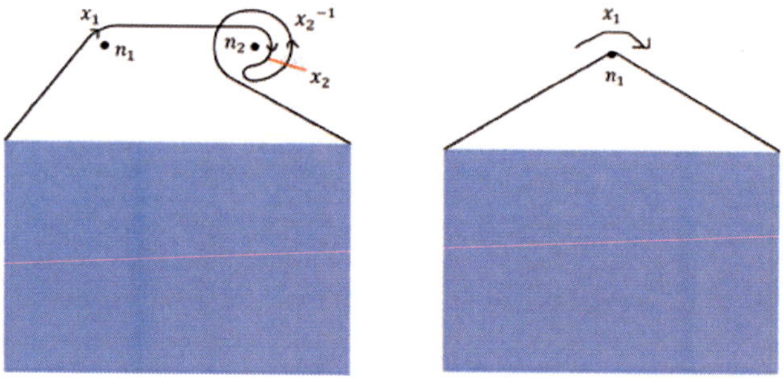

(a) $x_1x_2x_2^{-1}$ (b) This is equivalent to Figure 5a

Fig. 5 Inverse of 1 nail

Fig. 6 Inverse of multiple nails

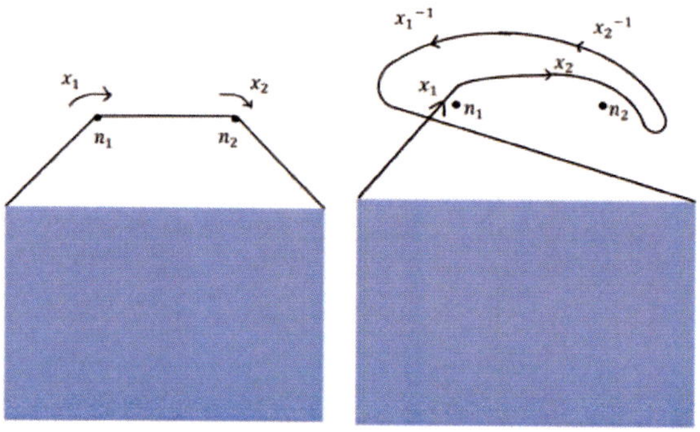

(a) x_1x_2 (b) $x_1x_2x_2^{-1}x_1^{-1}$

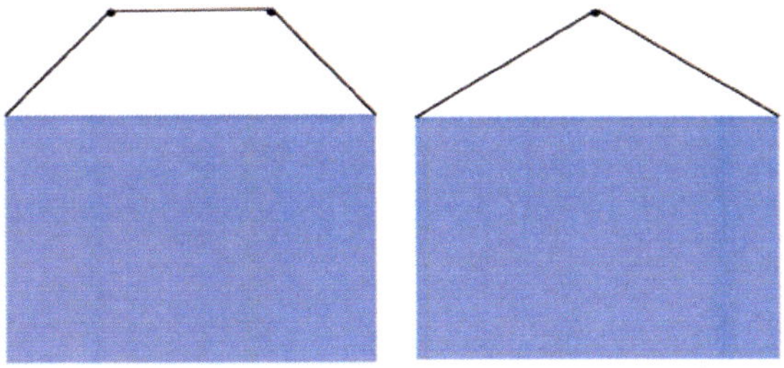

(a) x_1x_2 (b) x_1

Fig. 7 Removing nail n_2 from x_1x_2 will result in x_1

We denote a nail n_i dropping as removing all instances of x_i and x_i^{-1}. For example, removing nail n_2 from x_1x_2 in Fig. 7 will result in x_1 as all instances of x_2 and x_2^{-1} will be removed. A painting drops when the movement of the string is reduced to e.

3.3 Proving Our Basic Observations with These Notations

Our solution for the 1 nail puzzle in Fig. 3a can be written as x_1. It is easy to see that when nail n_1 is removed, the movement of the string will be reduced to e.

Our solution for the 2 nail puzzle in Fig. 3b can be written as $x_1x_2x_1^{-1}x_2^{-1}$.

When nail n_1 is removed, the movement of the string will be reduced to $x_2x_2^{-1}$, which will be further reduced to e, as shown in Fig. 8a. When nail n_2 is removed, the movement of the string will be reduced to $x_1x_1^{-1}$, which will be further reduced to e, as shown in Fig. 8b.

We can also find many more solutions to the 2 nail puzzle, such as $x_1x_2^{-1}x_1^{-1}x_2$, shown in Fig. 9.

(a) If the first nail n_1 is removed (b) If the second nail n_2 is removed

Fig. 8 Explanation for $n = 2$

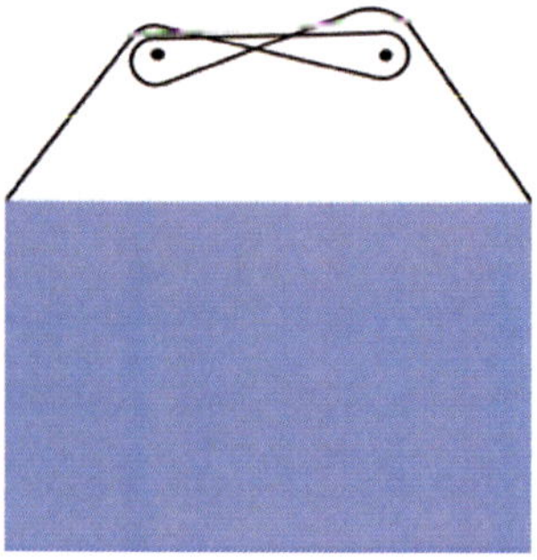

Fig. 9 Another solution to the 2 nail puzzle

3.4 Solution for Any n

3.4.1 Simple Solution

We can generalise our solution to the 2 nail puzzle by replacing each x_i with a solution from a smaller problem. For example, the solution for the 3 nail puzzle is $(x_1x_2x_1^{-1}x_2^{-1})x_3(x_1x_2x_1^{-1}x_2^{-1})^{-1}x_3^{-1}$ (note that $x_1x_2x_1^{-1}x_2^{-1}$ is a solution to the 2 nail puzzle).

Let us define S_i to be the solution for i nails, and $|S_i|$ to be the length of S_i. For S_2, we have found that the solution is $x_1x_2x_1^{-1}x_2^{-1}$. Let us define

$$f(x_i, x_j) = x_ix_jx_i^{-1}x_j^{-1}.$$

Now by using S_2, we can build a solution for the 3 nail puzzle by using the same pattern. Hence

$$\begin{aligned}
S_3 &= f(S_2, x_3) \\
&= S_2x_3S_2^{-1}x_3^{-1} \\
&= (x_1x_2x_1^{-1}x_2^{-1})x_3(x_1x_2x_1^{-1}x_2^{-1})^{-1}x_3^{-1} \\
&= x_1x_2x_1^{-1}x_2^{-1}x_3x_2x_1x_2^{-1}x_1^{-1}x_3^{-1}
\end{aligned}$$

We can verify that this solution works as firstly any value and its inverse are not next to each other so this solution can hold the painting without removing any nails. By removing n_1, all x_1 and x_1^{-1} disappear so we are left with just $x_2x_2^{-1}x_3x_2x_2^{-1}x_3^{-1}$. Since the x_2 and x_2^{-1} are next to each other, they cancel each other out and we are left with $x_3x_3^{-1}$, which will now cancel each other out too, so the painting will fall. It is a similar case when we remove n_2 and n_3 (Fig. 10).

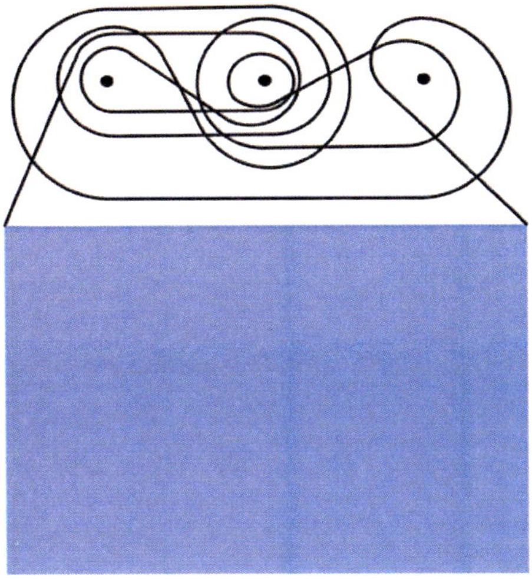

Fig. 10 Solution for the 3 nail puzzle

Now by using S_3, we can build a solution for the 4 nail puzzle.

$$\begin{aligned} S_4 &= f(S_3, x_4) \\ &= S_3 x_4 S_3^{-1} x_4^{-1} \\ &= (x_1x_2x_1^{-1}x_2^{-1}x_3x_2x_1x_2^{-1}x_1^{-1}x_3^{-1})x_4(x_1x_2x_1^{-1}x_2^{-1}x_3x_2x_1x_2^{-1}x_1^{-1}x_3^{-1})^{-1}x_4^{-1} \\ &= x_1x_2x_1^{-1}x_2^{-1}x_3x_2x_1x_2^{-1}x_1^{-1}x_3^{-1}x_4x_3x_1x_2x_1^{-1}x_2^{-1}x_3^{-1}x_2x_1x_2^{-1}x_1^{-1}x_4^{-1} \end{aligned}$$

Similarly, this solution works as any value and its inverse are not next to each other so this solution can hold the painting without removing any nails. By removing each nail, the painting will drop as the values are next to their inverse, causing them to cancel each other out.

Hence, the solution to the case with n nails is $S_n = f(S_{n-1}, x_n)$, with $S_1 = x_1$.

3.4.2 More Efficient Solution

Now, we can replace the 1 nail on the right with multiple nails for a more balanced solution. For example, we can write the solution to the 4 nail problem as $f(x_1x_2x_1^{-1}x_2^{-1}, x_3x_4x_3^{-1}x_4^{-1})$, where $x_1x_2x_1^{-1}x_2^{-1}$ is the solution to the 2 nail problem, instead of $f(S_3, x_4)$.

Formally, define $S_{i,j}$ to be the solution to problem with nails $n_i, n_{i+1}, \ldots, n_j$. Clearly, $S_{i,i} = x_i$ and as we found, $S_{i,i+1} = x_i x_{i+1} x_i^{-1} x_{i+1}^{-1}$. Let

$$S_{i,j} = f(S_{i,k}, S_{k+1,j}), \quad i \le k < j.$$

For the solution to be most balanced, we can let k be in the middle of i and j, so $k = \left\lfloor \frac{i+j}{2} \right\rfloor$.

Thus, we can write the solution for the i nail problem as

$$S_{1,i} = f\left(S_{1,\lfloor i/2\rfloor}, S_{\lfloor i/2\rfloor+1,i}\right).$$

For example, we can write the solution for the 4 nail problem as

$$\begin{aligned} S_{1,4} &= f(S_{1,2}, S_{3,4}) \\ &= f(x_1x_2x_1^{-1}x_2^{-1}, x_3x_4x_3^{-1}x_4^{-1}) \\ &= x_1x_2x_1^{-1}x_2^{-1}x_3x_4x_3^{-1}x_4^{-1}x_2x_1x_2^{-1}x_1^{-1}x_4x_3x_4^{-1}x_3^{-1} \end{aligned}$$

This solution only has 16 moves in comparison to the 22 move solution from S_4 above. As n gets larger, the difference becomes much greater too. We will explain this difference below.

Remark A code to generate solutions for both the simple and more efficient solution can be found in Appendix A.

3.4.3 Quantifying the Difference

We first prove the following properties of $|S|$.

Lemma 3.1 $|S_{i,j}| = |S_{i+k,j+k}|$ *for any integer* k.

Proof The number of nails in $[i, j]$ is the same as that in $[i+k, j+k]$, so we are just changing the labels of the nails, and thus there is no change in the number of moves.

Lemma 3.2 $|S_{i_1}S_{i_2}\dots S_{i_j}S_{i_{j+1}}\dots S_{i_k}| = |S_{i_1}S_{i_2}\dots S_{i_j}| + |S_{i_{j+1}}\dots S_{i_k}|$ *for all* $1 \leq j < k$.

Proof When joining 2 sets of moves together, the length of the combined moves is the sum of the length of the 2 sets of moves.

Lemma 3.3 $|S_i| = |S_i^{-1}|$

Proof By definition of S^{-1}.

Lemma 3.4 $|S_{1,n}| < |S_{1,n+1}|$ *for all positive integer* n.

Proof We use strong induction on n. For the base case, $1 = |S_{1,1}| < |S_{1,2}| = 4$. Now suppose it is true for all integer $n \in [1, k-1]$. We form two cases: k is even and k is odd.

Case 1: k is even. We can write $k = 2m$ for some positive integer m. Then

$$
\begin{aligned}
& |S_{1,k}| < |S_{1,k+1}| \\
\Longleftrightarrow\quad & |S_{1,2m}| < |S_{1,2m+1}| \\
\Longleftrightarrow\quad & |f(S_{1,m}, S_{m+1,2m})| < |f(S_{1,m+1}, S_{m+2,2m+1})| \\
\Longleftrightarrow\quad & |S_{1,m}S_{m+1,2m}S_{1,m}{}^{-1}S_{m+1,2m}{}^{-1}| < |S_{1,m+1}S_{m+2,2m+1}S_{1,m+1}{}^{-1}S_{m+2,2m+1}{}^{-1}| \\
\Longleftrightarrow\quad & |S_{1,m}S_{m+1,2m}S_{1,m}{}^{-1}S_{m+1,2m}{}^{-1}| < |S_{1,m+1}S_{m+1,2m}S_{1,m+1}{}^{-1}S_{m+1,2m}{}^{-1}| \\
\Longleftrightarrow\quad & |S_{1,m}| + |S_{1,m}{}^{-1}| < |S_{1,m+1}| + |S_{1,m+1}{}^{-1}| \\
\Longleftrightarrow\quad & |S_{1,m}| < |S_{1,m+1}|
\end{aligned}
$$

where the last equation is true by the induction hypothesis.
Case 2: k is odd. We can write $k = 2m - 1$ for some positive integer m. Then

$$
\begin{aligned}
& |S_{1,k}| < |S_{1,k+1}| \\
\Longleftrightarrow\quad & |S_{1,2m-1}| < |S_{1,2m}| \\
\Longleftrightarrow\quad & |f(S_{1,m}, S_{m+1,2m-1})| < |f(S_{1,m}, S_{m+1,2m})| \\
\Longleftrightarrow\quad & |S_{1,m}S_{m+1,2m-1}S_{1,m}{}^{-1}S_{m+1,2m-1}{}^{-1}| < |S_{1,m}S_{m+1,2m}S_{1,m}{}^{-1}S_{m+1,2m}{}^{-1}| \\
\Longleftrightarrow\quad & |S_{m+1,2m-1}| + |S_{m+1,2m-1}{}^{-1}| < |S_{m+1,2m}| + |S_{m+1,2m}{}^{-1}| \\
\Longleftrightarrow\quad & |S_{m+1,2m-1}| < |S_{m+1,2m}| \\
\Longleftrightarrow\quad & |S_{1,m-1}| < |S_{1,m}|
\end{aligned}
$$

where the last equation is true by the induction hypothesis.
Therefore $|S_{1,n}| < |S_{1,n+1}|$ by mathematical induction.

Now, we will find the number of moves for the solution in Sect. 3.4.1. From the recursion, we can see that the solution has an exponential number of moves from the recursion. In fact, we shall prove that the exact number of moves is $2^n + 2^{n-1} - 2$.

Theorem 3.5 *The solution in Sect. 3.4.1 has* $2^n + 2^{n-1} - 2$ *moves.*

Proof For the base case, $S_1 = x_1$ has $1 = 2^1 + 2^0 - 2$ moves, which is correct. Now suppose that S_n has $2^n + 2^{n-1} - 2$ moves. Then

$$
\begin{aligned}
S_{n+1} &= f(S_n, n_n) \\
&= S_n n_n S_n^{-1} n_n^{-1}
\end{aligned}
$$

So the length of S_{n+1} is $2(2^n + 2^{n-1} - 2) + 2 = 2^{n+1} + 2^n - 2$.

For the solution in Sect. 3.4.2, we can see that it is clearly more efficient that the one in Sect. 3.4.1, but we can prove that it in fact has at most $4n^2$ moves (which is quadratic), a significant improvement from the exponential solution.

Theorem 3.6 *The solution in Sect. 3.4.2 has at most* $4n^2$ *moves.*

Proof We can split into 2 cases.
Case 1: n is a power of 2. In this case, the number of moves is n^2 and we can prove this by induction.

For the base case $1 = 2^0$, $S_{1,1} = x_1$ has $1 = 1^2$ moves, which is correct. Now suppose $S_{1,2^n}$ has $(2^n)^2$ moves. Then

$$\begin{aligned} S_{1,2^{n+1}} &= f(S_{1,2^n}, S_{2^n+1,2^{n+1}}) \\ &= S_{1,2^n} S_{2^n+1,2^{n+1}} S_{1,2^n}{}^{-1} S_{2^n+1,2^{n+1}}{}^{-1} \end{aligned}$$

So $S_{2^{n+1}}$ has $4 \times 2^{2n} = 2^{2n+2} = (2^{n+1})^2$ moves.

Case 2: n is not a power of 2. From Lemma 3.4, the number of moves is less than the number of moves if n is rounded up to a power of 2, and the number of moves for that is at most $(2n)^2 = 4n^2$. Note that this is a loose bound.

4 Removing k Out of n Nails

Now that we have solved the 1 out of n nails problem, we can consider the k out of n nails problem, as follows:

Is it possible to hang a painting on n nails, such that painting drops if any k of the nails drop ($1 \leq k \leq n$), but removing fewer nails does not?

4.1 Special Cases

We begin by first analysing some special cases.

4.1.1 $k = n$

If we are to remove all the nails, clearly the painting will fall. Thus we only need to consider how to ensure that the painting does not fall when a fewer number of nails are removed.

If a painting falls when less than n nails are removed, some moves must be reduced by its inverse. A simple solution to avoid this is to not have any inverses in the move.

So for 1 nail, a possible solution is x_1; for two nails a possible solution is x_1x_2, and for n nails, a possible solution is $x_1x_2 \ldots x_{n-1}x_n$ (Fig. 11).

4.1.2 $k = n - 1$

For n nails, how may we remove nails such that the painting will drop only if $n-1$ nails are removed? For that to be possible, we must have an inverse for every movement so that they can be reduced. Furthermore, the movement and its inverse

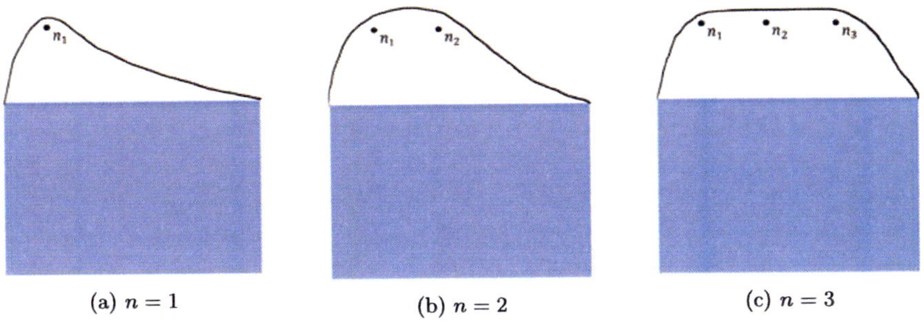

Fig. 11 Basic solutions to the $k = n$ case

Fig. 12 Basic solutions to the $k = n - 1$ case

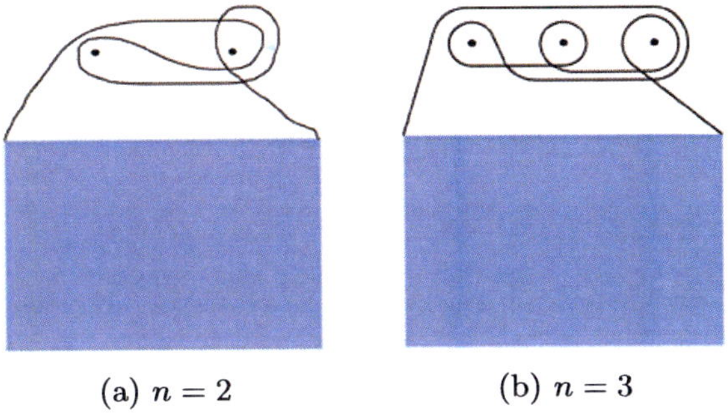

must not be too close to each another, otherwise they would be next to each another before $n - 1$ nails are removed, and the painting would drop sooner.

For the 2 nail puzzle, we know that for the painting to drop if only $n - 1$ nails, in this case 1 nail, is removed, the solution is $x_1 x_2 x_1^{-1} x_2^{-1}$.

We can build on this for the 3 nail puzzle. So one solution for the painting to drop if only 2 nails are removed is $x_1 x_2 x_3 x_1^{-1} x_2^{-1} x_3^{-1}$. It is easy to see that this works as well.

We can generalise the solution for removing $n - 1$ out of n nails to be

$$x_1 x_2 \ldots x_{n-1} x_n x_1^{-1} x_2^{-1} \ldots x_{n-1}^{-1} x_n^{-1}.$$

We can verify that this works as there are $n - 1$ moves between any move and its inverse, so by removing less than $n - 1$ nails, there will still be moves between any nail and its inverse, so they will not be reduced. There will also not be a case of multiple nails reducing each other since the movements and their inverses are arranged in the same order (Fig. 12)

4.1.3 $k = 2$

We can generalise this from the 1 out of n nail problem by treating every possible combination of 2 nails $x_i x_j$ $(i < j)$ as 1 nail, then solving the $k = 1$ problem.

Formally, let N_i be the group consisting of the i-th combination of 2 nails. For example, for the problem of removing 2 out of 4 nails, $N_1 = n_1n_2$, $N_2 = n_1n_3$, $N_3 = n_1n_4$, $N_4 = n_2n_3$, $N_5 = n_2n_4$, $N_6 = n_3n_4$. For convenience, let $y_i = x_{N_i}$ and we can solve this from our solution to the 1 nail problem:

$$\begin{aligned}
S_{1,6} &= f(S_{1,3}, S_{4,6}) \\
&= f(y_1y_2y_1^{-1}y_2^{-1}y_3y_2y_1y_2^{-1}y_1^{-1}y_3^{-1}, y_4y_5y_4^{-1}y_5^{-1}y_6y_5y_4y_5^{-1}y_4^{-1}y_6^{-1}) \\
&= y_1y_2y_1^{-1}y_2^{-1}y_3y_2y_1y_2^{-1}y_1^{-1}y_3^{-1}y_4y_5y_4^{-1}y_5^{-1}y_6y_5y_4y_5^{-1}y_4^{-1}y_6^{-1}y_1^{-1}y_2^{-1}y_1 \\
&\quad {}^{-1}y_2^{-1}y_1^{-1}y_2y_1y_3y_4^{-1}y_5^{-1}y_4y_5y_6^{-1}y_5^{-1}y_4^{-1}y_5y_4y_6
\end{aligned}$$

Substituting back, we have the solution

$$\begin{aligned}
&x_1x_2x_1x_3x_2^{-1}x_1^{-1}x_3^{-1}x_1^{-1}x_1x_4x_1x_3x_1x_2x_3^{-1}x_1^{-1}x_2^{-1}x_1^{-1}x_4^{-1}x_1^{-1} \\
&x_2x_3x_2x_4x_3^{-1}x_2^{-1}x_4^{-1}x_2^{-1}x_3x_4x_2x_4x_2x_3x_4^{-1} \\
&x_2^{-1}x_3^{-1}x_2^{-1}x_4^{-1}x_3^{-1}x_2^{-1}x_1^{-1}x_3^{-1}x_1^{-1} \\
&x_1x_2x_1x_3x_4^{-1}x_1^{-1}x_3^{-1}x_1^{-1}x_2^{-1}x_1^{-1}x_1x_3x_1x_2x_1x_4 \\
&x_3^{-1}x_2^{-1}x_4^{-1}x_2^{-1}x_2x_3x_2x_4x_4^{-1}x_3^{-1}x_4^{-1}x_2^{-1}x_3^{-1}x_2^{-1}x_2x_4x_2x_3x_3x_4.
\end{aligned}$$

Note that the solution above can be simplified by omitting moves adjacent to their inverse as follows.

$$\begin{aligned}
&x_1x_2x_1x_3x_2^{-1}x_1^{-1}x_3^{-1}x_4x_1x_3x_1x_2x_3^{-1}x_1^{-1}x_2^{-1}x_1^{-1} \\
&x_4^{-1}x_1^{-1}x_2x_3x_2x_4x_3^{-1}x_2^{-1}x_4^{-1}x_2^{-1}x_3x_4x_2x_4x_2x_3x_4^{-1} \\
&x_2^{-1}x_3^{-1}x_2^{-1}x_4^{-1}x_3^{-1}x_2^{-1}x_1^{-1}x_3^{-1}x_2x_1x_3x_4^{-1} \\
&x_1^{-1}x_3^{-1}x_1^{-1}x_2^{-1}x_3x_1x_2x_1x_4x_3^{-1}x_2^{-1}x_4^{-1}x_3x_2x_3^{-1}x_4^{-1}x_2^{-1} \\
&x_3^{-1}x_4x_2x_3x_3x_4.
\end{aligned}$$

In general, for n nails, we can reduce it to the following problem: Remove 1 of the $\binom{n}{2}$ combinations of 2 nails such that the painting will fall. We can solve this with our solution to the removing 1 of n nails problem.

This generalises directly.

4.2 General Solution

For general k, we can generalise from the $k = 2$ solution and reduce it to the following problem: Remove 1 of the $\binom{n}{k}$ combinations of k nails such that the painting will fall. Similarly, we can solve this with our solution to the removing 1 of n nails problem.

Lastly, we will show that the painting would not fall when less than k nails are removed. When less than k nails are removed, there must still be at least 1 nail in every group N_i. By our solution in Sect. 3, the painting would not drop as no N_i has been completely removed yet.

5 Similar Problems

With a similar method as that in Sect. 4, we can solve problems of the form "Wrap the string around n nails in m groups such that removing all the nails in a group makes the painting drop, but otherwise the painting would not drop".

Consider the following problem: Hang a painting on two red nails and two blue nails such that removing both nails of the same colour makes the painting fall, but otherwise the painting would not fall (Fig. 13)

To solve this problem, we can take inspiration from Sect. 4. Let the 2 red nails be n_1 and n_2, and the 2 blue nails be n_3 and n_4. We can treat the nails n_1n_2 and n_3n_4 each as 1 nail since removing any of them makes the painting fall, then solve the $k = 1$ problem for $n = 2$.

Let $n_1' = n_1n_2$, $n_2' = n_3n_4$ and $y_i = x_i'$. Then

$$\begin{aligned} S_{1,2} &= y_1 y_2 y_1^{-1} y_2^{-1} \\ &= x_1 x_2 x_3 x_4 x_2^{-1} x_1^{-1} x_4^{-1} x_3^{-1} \end{aligned}$$

Fig. 13 Removing both nails of the same colour makes the painting fall

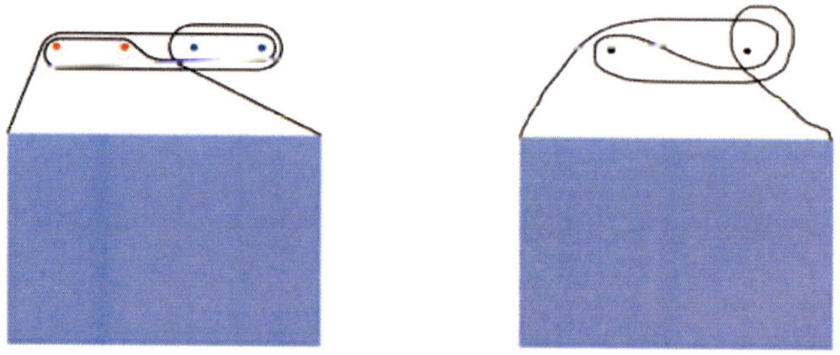

(a) Solution for the similar problem (b) Solution for $k = 1$, $n = 2$

Fig. 14 Comparison of solutions

so we have the solution $x_1x_2x_3x_4x_2^{-1}x_1^{-1}x_4^{-1}x_3^{-1}$ (Fig. 14).

We may also observe that this looks very similar to the $k = 1$ problem for $n = 2$. In fact, the only thing we changed was to replace 1 nail with 2 nails of the same colour.

6 Further Investigations

We hope to investigate the following questions:

1. Is there a more efficient (shorter) solution to the 1 out of n nail problem?
2. Is there a more efficient (shorter) solution to the k out of n nail problem?
3. What if we have more than 1 string to hang the painting with?

Acknowledgements We would like to thank our school, our teacher-in-charge Mr Chai Ming Huang, and our teacher-mentor Ms Ooi Liu Qi for their guidance and support.

7 Code for Generating Solutions

```
def inv(s):  # returns the inverse of a movement s
    ans = ""
    movements = s.split("x")[1:]
    movements.reverse()
    for i in movements:
        if len(i) >= 2 and i[-3:] == "^-1":
            ans += "x" + i[:-3]
        else:
            ans += "x" + i + "^-1"
    return ans

```

```
def simple(n):  # returns a simple solution for n nails
    if n == 1: return "x1"
    return simple(n - 1) + "x" + str(n) + inv(simple(n - 1)) + "x" + str(n) + "^-1"

def efficient(i, j):  # returns an efficient solution for nails n_i, n_i+1, ..., n_j
    if i == j:
        return "x" + str(i)
    k = (i + j) // 2
    return efficient(i, k) + efficient(k + 1, j) + inv(efficient(i, k)) + inv(efficient(k + 1, j))

# examples
for i in range(1, 5):
    print(simple(i))

for i in range(1, 5):
    print(efficient(1, i))
```

Reference

1. Euler Math Competition 2015. Available at https://sites.google.com/site/ntumathcompetitions/euler-2015

EEG-Based Assessment of the Influence of Music-Induced Emotions on Cognitive Task Performance

Joel Yap Ray Kai, Aloysius Khoo Kai Jun, Evan Poh, Sandra Tan Seok Hwee, and Aung Aung Phyo Wai

Abstract Many youths show a preference for listening to music in the background while doing work. In this study, we examined the effect of the emotions, induced by different moods of background music, on the cognitive functions of an individual. Participants were asked to perform a test battery of a total of three cognitive tasks, each focussing on a different cognitive function: attention, working memory, and task-switching abilities. The set of tasks was repeated three times, each with different background music and its attendant emotions: happy, peaceful, and no music. Surveys were also conducted at regular intervals to assess emotional state and stress levels. EEG data collected was analysed by using various software, including BrainVision LSL Analyser, Matlab and JASP, to demonstrate the effect of music on attention, memory, and task-switching capabilities. We have found that, contrary to popular belief, peaceful music stimulates greater EEG variation. This leads to greater distractibility, affecting task performance. The application of our results has the potential to improve currently misinformed practises in educational pedagogy and classroom practise.

Keywords EEG · Music · Emotions · Cognitive performance

J. Y. R. Kai (✉) · A. K. K. Jun · E. Poh · S. T. S. Hwee
Hwa Chong Institution, Singapore, Singapore
e-mail: 171367H@hci.edu.sg

A. K. K. Jun
e-mail: 171345R@hci.edu.sg

E. Poh
e-mail: 171339R@hci.edu.sg

S. T. S. Hwee
e-mail: sandra@hci.edu.sg

A. A. P. Wai
Nanyang Technological University, Singapore, Singapore
e-mail: apwaung@ntu.edu.sg

H. Guo et al. (eds.), *IRC-SET 2022*,
https://doi.org/10.1007/978-981-19-7222-5_5

1 Introduction

Music is a therapeutic tool widely used amongst youth, who mainly listen to music for self-awareness, social relatedness, arousal and mood regulation (Schäfer et al. [17]), or to invoke nostalgia or solace (Travis [18]).

Studying with music has become a habit amongst many students (Cockrill et al. [4]), despite many being vaguely aware that their performance in cognitive tasks may be impaired (Kotsopoulou and Hallam [10]). Music has been shown to interfere with cognitively-demanding tasks, such as arithmetic and reading comprehension (Chew et al. [3]). On the other hand, the habituation hypothesis (Wolf and Weiner [20]) suggests that music with which one is familiar may have a positive effect on performance.

There has been much research on the usefulness of music in positively influencing one's performance in different types of work (Rickard et al. [15]; Angel et al. [1]; Dolegui [5]), and the effect of emotions on cognitive performance (Papousek et al. [13]; Eysenck [6]). Unfortunately, most studies focus on the link between music and cognitive performance, or emotions and performance. There seems as yet not to be a study specifically investigating the effect of emotions, induced by different background music, on the cognitive functions of an individual. In addition, there have not been any studies in the local context examining the effect of music-induced emotions on task performance in Asian students.

The present study hence assesses the extent of benefit, and impacts of various music-induced emotions on cognitive task performance, as well as determines which emotions are more favourable for positive neural stimulation. We also sought to determine if one's familiarity with the audio stimuli affected their response and performance.

2 Objective and Hypotheses

The objectives of this study are to determine whether happy music or peaceful music can better aid in optimising one's cognitive performance and to ascertain the effect of background music on emotion.

Hence, we hypothesise that music-induced happiness and peacefulness will be positively correlated with cognitive task performance, with peaceful music being more conducive than happy music in enhancing cognitive performance; subjects who listen to or play classical music will also perform better with music playing in the background.

3 Methodology

We selected participants aged between 16 to 18 and comprised a mix of both male and female students. We collected data from 23 subjects in total. A Muse 2.0 headband (InteraXon Inc., Toronto ON, Canada) was used to record the EEG waves of participants. Tasks were built using the PsychoPy software package and presented on an external monitor (Fig. 1).

During the experiment, participants were asked to perform a total of three cognitive tasks, each focussing on a different aspect: attention, working memory, and task-switching abilities. Attention (Gabrieli and Norton [7]), working memory (Pugin et al. [14]) and task-switching (Monsell [11]) are important areas of human awareness that are used daily. The set of tasks was repeated three times, each with a piece of different background music: happy, peaceful, and quiet (no music). The order of tasks and music was randomised for each subject.

For attention, we adapted the experimental design from Sauseng et al. [16]. This began with a fixation cross on a black screen. At the start of each trial (10 s), an arrow would appear (340 ms), indicating the position (left or right) of the subsequently-presented visual stimulus (200 ms). Either an upright or inverted triangle was shown, and subjects were instructed to respond with a button press corresponding to its direction, as seen in Fig. 2. 75% of trials had the cue arrow correspond with the target direction, whilst 25% were incongruent, with a total of 32 trials being conducted.

The memory task was an N-back test (Kirchner [9]), where $N = 2$, as illustrated by Fig. 3. A random sequence of letters was presented on-screen (2 s), with an interstimulus interval of 1 s. Subjects were instructed to indicate whether the current letter matched the one shown 2 positions prior in the sequence.

The task-switching task was modified from the Brain@Work multitask study (Haavisto et al. [8]). The task involved arithmetic and visual segments. The arithmetic

Fig. 1 Experimental setup, which included a monitor, a keyboard, and a mouse

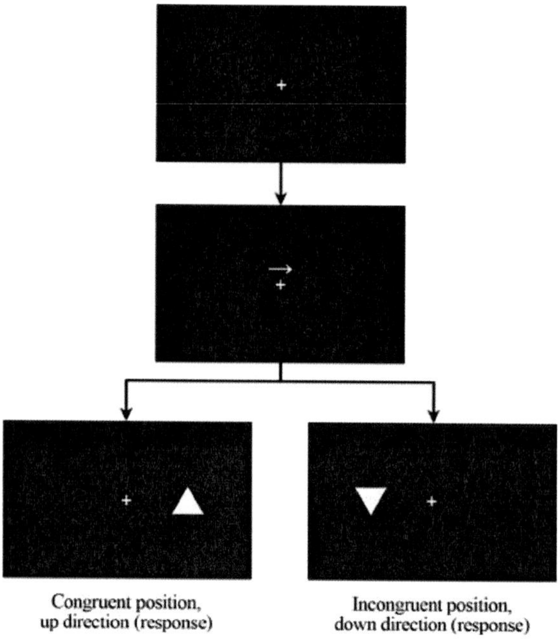

Fig. 2 Brief outline of a single trial, attention sub-task

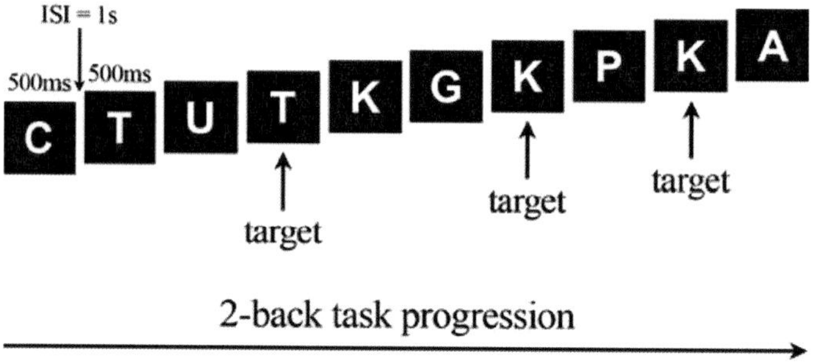

Fig. 3 Outline of the N-back memory task, $N = 2$

portion involved subjects indicating the sum of two 3-digit numbers, using a mouse to click the digits of their answer. The visual portion required subjects to press a spacebar when a dot in the centre of the screen reaches as close to the outer edge of a set of concentric circles as possible. Figure 4 shows the screen which participants would see during this task. Both subtasks had to be completed within the time limit of 7 s per trial, with a total of 42 trials being run.

Surveys were conducted before and after the experiment and between each complete set of tasks. The pre-experiment survey assessed participants' fatigue and emotional state; the survey between sets assessed participants' attention level, as well as

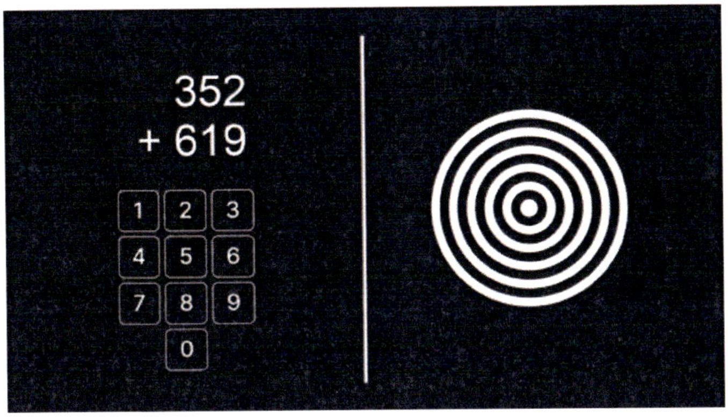

Fig. 4 Task-switching task, including arithmetic (*L*) and visual (*R*) segments

whether they felt the music was enjoyable or distracting; the post-experiment survey assessed their overall sentiments towards the whole experiment.

The data collected in this study were analysed using Tableau and JASP. The collected EEG data were processed using MATLAB©to filter artefacts, segment tasks, extract features and classify. We applied band-pass filters (0.5–45 Hz) to raw EEG signals and removed artefacts using a set of moving average filters. Then, we segmented the filtered data by synchronising timestamps of different unique tasks carried out during the experiment. With individual segmented data, we extracted 6 band-power features (delta, theta, alpha, low beta, high beta, gamma) for attention classification and statistical analysis.

4 Results and Discussion

4.1 Participants' Perceptions of Task Efficacy are Inaccurate

During surveys conducted throughout the experiment, subjects reported that attention, memory and task-switching tasks were more challenging, on average, when they were exposed to happy music (Fig. 5). This perception was particularly significant for task-switching. This could be due to pieces of music associated with happiness being more lively and having greater variation in terms of tempo, dynamic range and rhythmic complexity (Wang [19]), hence stimulating more intense emotions which could have interfered with task completion. This likely led to participants preferring peaceful music (71% accuracy) instead of happy music (61%), given that it was more conducive to dealing with the tasks at hand. Subjects also reflected being able to better follow the tasks while listening to peaceful music (70% ease) than when listening to happy music (60%).

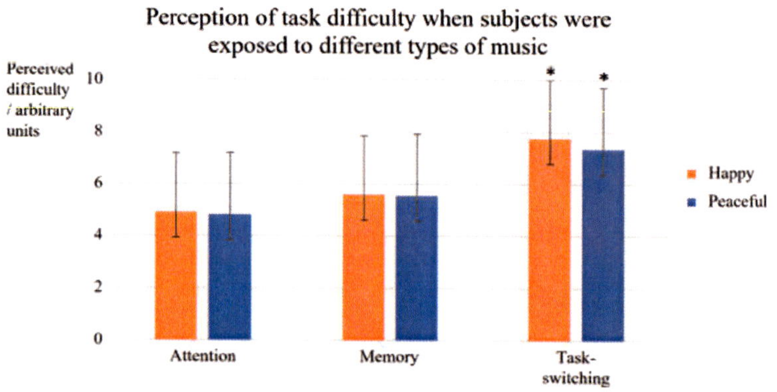

Fig. 5 Participants generally found that task difficulty increased when happy music was played

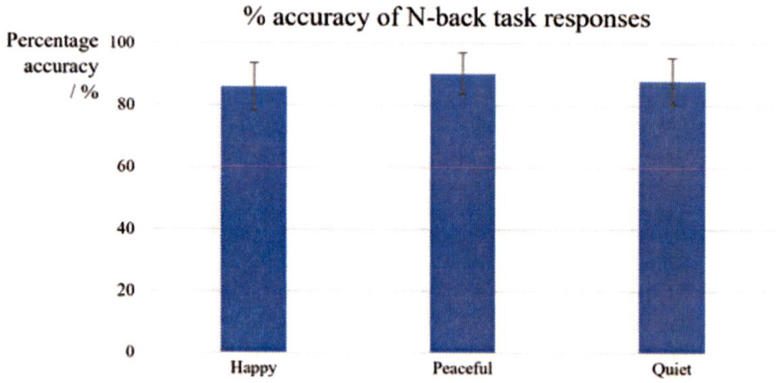

Fig. 6 The percentage of correct responses during the attention task was highest when no music was played

Contradictorily, the average accuracy of correct responses for attention tasks was highest when no music was played (Fig. 6) and the best response times for task-switching tasks occurred when happy music was played (Fig. 7). Thus, user perception of how music affects their work is warped and possibly unreliable.

4.2 *Peaceful Music Induces Greater Variation in EEG Patterns*

Since user perception was contradictory and the results of different types of music on various tasks were unclear, we turned to the EEG patterns of participants to gauge brain activity, and, by proxy, the level of distractibility caused by each type of music (Figs. 8, 9 and 10). Surprisingly, in all three tasks, peaceful music generated

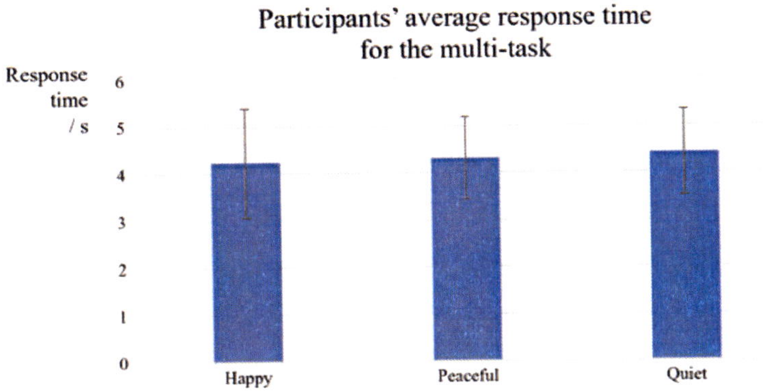

Fig. 7 Response time for the task-switching task was fastest when happy music was played

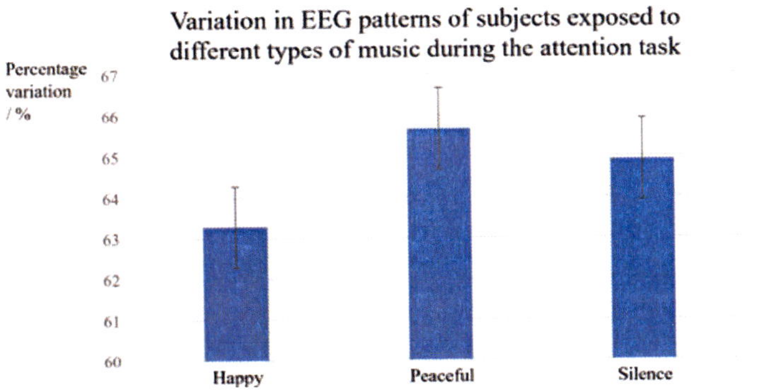

Fig. 8 Peaceful music generated the greatest variation in EEG patterns in the attention task

the greatest variation in EEG patterns generated, with differences most significantly obvious during attention tasks.

Our findings are contrary to studies by Cahn and Polich [2]. Peaceful music theoretically generates theta and alpha waves in the brain, resulting in mindfulness and intense focus that contributes to "flow" (Nakamura and Csikszentmihalyi [12]). We believe that such effects might only be applicable in adult brains; adolescent brains still undergoing development, however, respond differently to peaceful music. This finding has implications for classroom practise, where adult teachers may use calming or peaceful music to guide students to better learning outcomes, but these may, in fact, backfire.

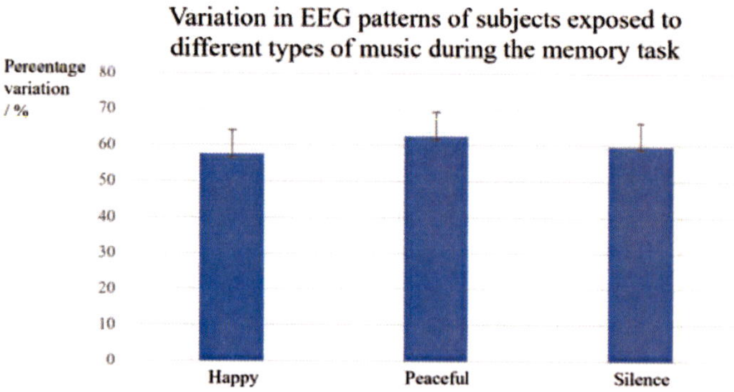

Fig. 9 During the memory task, the difference in percentage variation between music types is much smaller

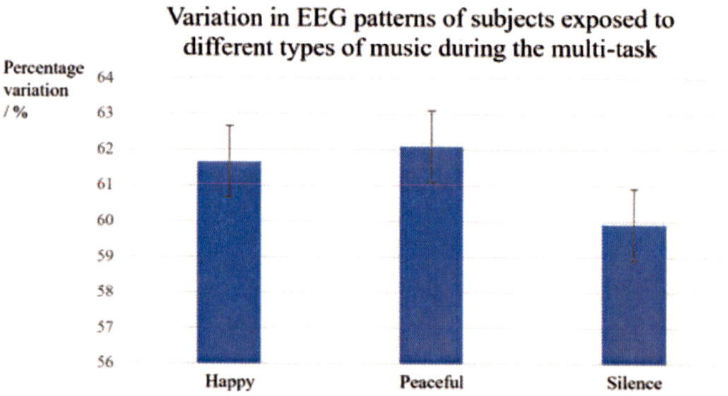

Fig. 10 When engaging in task-switching, both music types produced greater variation than silence

5 Conclusion

Through our initial research, we anticipated that a peaceful emotional state would enhance cognitive performance by allowing for a greater level of focus and minimising distractions. However, participants consistently responded faster with happy music and gave more accurate responses when the room was silent. Unlike conventional wisdom, we found that peaceful music results in a greater level of variation of EEG patterns, which indicates a greater level of cognitive activity. Therefore, our findings should inform classroom practise for better teaching and learning outcomes in adolescent students.

Acknowledgements We would like to thank our teacher-mentor, Dr. Sandra Tan, for her invaluable guidance and feedback, as well as assistance on administrative matters, throughout the duration of our research project. We would also like to thank the participants of our study for their active enthusiasm, and taking time off their busy schedules to help us in our data collections.

References

1. Angel, L. A., Polzella, D. J., & Elvers, G. C. (2010). Background music and cognitive performance. *Perceptual and Motor Skills, 110*(3), 1059–1064. https://doi.org/10.2466/pms.110.C.1059-1064
2. Cahn, B. R., & Polich, J. (2006). Meditation states and traits: EEG, ERP, and neuroimaging studies. *Psychological Bulletin, 132*(2), 180–211. https://doi.org/10.1037/0033-2909.132.2.180
3. Chew, A. S. Q., Yu, Y. T., Chua, S. W., & Gan, S. K. E. (2016). The effects of familiarity and language of background music on working memory and language tasks in Singapore. *Psychology of Music, 44*(6), 1431–1438. https://doi.org/10.1177/0305735616636209
4. Cockrill, A., Sullivan, M., & Norbury, H. (2011). Music consumption: Lifestyle choice or addiction. *Journal of Retailing and Consumer Services, 18*, 160–166. https://doi.org/10.1016/j.jretconser.2010.12.003
5. Dolegui, A. S. (2013). The impact of listening to music on cognitive performance. *Inquiries Journal, 5*(9)
6. Eysenck, M. W., Derakshan, N., Santos, R., & Calvo, M. G.: Anxiety and cognitive performance: Attentional control theory. *Emotion, 7*(2), 336–353. https://doi.org/10.1037/1528-3542.7.2.336
7. Gabrieli, J. D. E., & Norton, E. S. (2012). Reading abilities: Importance of visual-spatial attention. *Current Biology, 22*(9), R298–R299. https://doi.org/10.1016/j.cub.2012.03.041
8. Haavisto, M. L., Porkka-Heiskanen, T., Hublin, C., Härmä, M., Mutanen, P., Müller, K., Virkkala, J., & Sallinen, M. (2010). Sleep restriction for the duration of a work week impairs multitasking performance. *Journal of Sleep Research, 19*(3), 444–454. https://doi.org/10.1111/j.1365-2869.2010.00823.x
9. Kirchner, W. K. (1958). Age differences in short-term retention of rapidly changing information. *Journal of Experimental Psychology, 55*(4), 352–358. https://doi.org/10.1037/h0043688
10. Kotsopoulou, A., & Hallam, S. (2010). The perceived impact of playing music while studying: age and cultural differences. *Educational Studies, 36*(4), 431–440. https://doi.org/10.1080/03055690903424774
11. Monsell, S. (2003). Task switching. *Trends in Cognitive Sciences, 7*(3), 134–140. https://doi.org/10.1016/S1364-6613(03)00028-7
12. Nakamura, J., & Csikszentmihalyi, M. (2014). The concept of flow. In M. Csikszentmihalyi (Ed.), *Flow and the foundations of positive psychology: the collected works of Mihaly Csikszentmihalyi* (pp. 239–263). Springer. https://doi.org/10.1007/978-94-017-9088-8_16
13. Papousek, I., Schulter, G., & Lang, B. (2009). Effects of emotionally contagious films on changes in hemisphere-specific cognitive performance. *Emotion, 9*(4), 510–519. https://doi.org/10.1037/a0016299
14. Pugin, F., Metz, A. J., Stauffer, M., Wolf, M., Jenni, O. G., & Huber, R. (2014). Working memory training shows immediate and long-term effects on cognitive performance in children. *F1000Research, 3*, 82. https://doi.org/10.12688/f1000research.3665.2
15. Rickard, N. S., Toukhsati, S. R., & Field, S. E. (2005). The effect of music on cognitive performance: Insight from neurobiological and animal studies. *Behavioral and Cognitive Neuroscience Reviews, 4*(4), 235–261. https://doi.org/10.1177/1534582305285869
16. Sauseng, P., Klimesch, W., Stadler, W., Schabus, M., Doppelmayr, M., Hanslmayr, S., Gruber, W. R., & Birbaumer, N. (2005). A shift of visual spatial attention is selectively associated with human EEG alpha activity. *European Journal of Neuroscience, 22*(11), 2917–2926. https://doi.org/10.1111/j.1460-9568.2005.04482.x
17. Schäfer, T., Sedlmeier, P., Städtler, C., & Huron, D. (2013). The psychological functions of music listening. *Frontiers in Psychology*. https://doi.org/10.3389/fpsyg.2013.00511
18. Travis, R. (2013). Rap music and the empowerment of today's youth: Evidence in everyday music listening, music therapy, and commercial rap music. *Child and Adolescent Social Work Journal, 30*(2), 139–167. https://doi.org/10.1007/s10560-012-0285-x

19. Wang, W. C. (2014). A study of the type and characteristics of relaxing music for college students. *Proceedings of Meetings on Acoustics, 21*(1), 035001. https://doi.org/10.1121/1.4902001
20. Wolf, R. H., & Weiner, F. F. (1972). Effects of four noise conditions on arithmetic performance. *Perceptual and Motor Skills, 35*(3), 928–930. https://doi.org/10.2466/pms.1972.35.3.928

Manga Layout Analysis via Deep Learning

Nyx Audrey Angelo Iskandar

Abstract Manga Layout Analysis is an end-to-end system integration of instance segmentation models, novel algorithms, and optical character recognisers as a complete solution for the adaptation of manga into other media. As an innovative and unified approach, this system combines the aforementioned deep learning tools into one process to contribute to document layout analysis concerning manga on which only limited research has been conducted and aid Japanese companies adapt manga into anime and other media. This integrated system hence consists of instance segmentation models trained with the Manga109 dataset, algorithms leveraging the results of the instance segmentation models to analyse and evaluate the structure and semantics of manga, and Japanese hiragana and kanji optical character recognisers trained with Kuzushiji-MNIST datasets. The instance segmentation models for text and frames built using Mask R-CNN and PointRend achieved mAP values of 0.93 and 0.95, respectively (IoU = 0.5); for text, frames, and faces, the Mask R-CNN model achieved an mAP value of 0.90 (IoU = 0.5). The ResNet-50 and ResNet-101 optical character recognisers achieved high F1 scores of 0.89 and 0.90, respectively.

Keywords Document layout analysis · Manga · Deep learning · Instance segmentation · Algorithms · Optical character recognition

1 Background and Purpose

The Japanese animation industry responsible for adapting manga into anime, movies, and other media is infamous for its heavy workload and low pay, with animators working 12 to 18 hours daily for only an average of USD 200 of pay every month. Manga Layout Analysis (MLA) hence aims to assist the Japanese animation industry by automating the process of adapting manga, increasing its efficiency and productivity. MLA consists of instance segmentation models for manga (IS), algorithms to perform face-text association (FTA) and text order determination (TOD), and manga

N. A. A. Iskandar (✉)
Raffles Institution, Singapore, Singapore
e-mail: xyntechx@gmail.com

H. Guo et al. (eds.), *IRC-SET 2022*,
https://doi.org/10.1007/978-981-19-7222-5_6

optical character recognisers for Japanese hiragana and kanji (M-OCR). An image of a manga page would thus be passed into the MLA system to be automatically processed, analysed, and evaluated by IS, FTA, TOD, and M-OCR, in that order. All in all, this research aims to create the best possible deep learning integrated MLA system in terms of accuracy, precision, as well as training and execution times.

2 Literature Review

2.1 Instance Segmentation

Dubray and Laubrock implemented a solution to detect and segment speech balloons (text) using deep convolutional neural network (Deep CNN) [1]. Even though their model achieved a high F1 score of over 0.94, it fails to replicate such a performance on manga pages. This could be due to the difference in how English alphabets and Japanese characters appear; as the model was trained solely on English comic books, it is unaccustomed to segmenting Japanese speech balloons. Hence, it is imperative that MLA is trained with Japanese manga to achieve the maximum possible performance.

Dutta and Biswas also proposed a solution to extract panels (frames) and characters (faces) from Bengali comic books using CNN and You Only Look Once (YOLO) [2]. The model in said paper achieved near perfect F1 scores for both frames and faces. However, even though faces are curved with nonlinear outlines, the segmentation for them is all rectangular. Although this is likely due to the ground-truth annotations being rectangular, it is still a downside of the model as the high F1 scores could be attributed to the relatively easier task of segmenting rectangular shapes than nonlinear shapes. This means that ground-truth annotations of MLA, to ensure accurate and precise segmentation of nonlinear shapes, must follow the nonlinear boundaries of the shapes as well. As for the model type, Mask R-CNN and PointRend are better options to segment nonlinear shapes than plain CNN and YOLO due to their ability to segment finer details given their more developed mechanisms.

2.2 Optical Character Recognition

One of the most well-known OCRs today is the Tesseract OCR developed chiefly by Smith [3]. Although older models simply leveraged on character pattern recognition, Tesseract's latest releases include an LSTM (which is focused on line recognition). According to third-party analyses, the F1 scores of Tesseract English OCRs averaged at approximately 0.9. This high score can perhaps be attributed to the use of LSTM, but the M-OCRs of MLA have achieved similar performances without the use of said architecture for the identification of Japanese hiragana and kanji characters. Hence, simple convolutional neural network-based multi-class classification models seem to suffice for the M-OCRs, maintaining high performance and training and execution speeds.

2.3 Manga Layout Analysis

MLA is concerned with segmenting text, frames, and faces via a single unified model and identifying Japanese hiragana and kanji characters using deep learning. It also employs algorithms to perform face-text association and text order determination by leveraging on the vertices returned by instance segmentation. The challenge is thus how to perform these tasks for manga as accurately and efficiently as possible.

For the instance segmentation function of MLA, Mask R-CNN [4] and PointRend [5] IS models are employed. Mask R-CNN is a framework built on Faster R-CNN and fully convolutional network (FCN) to perform object detection while concurrently generating a high-quality pixel-to-pixel segmentation mask. Mask R-CNN has many benefits, including relatively fast training and execution times as well as a greater segmentation accuracy and precision compared to older models. MLA also uses PointRend (Point-based Rendering), which views the challenge of instance segmentation as a rendering problem, performing point-based segmentation at adaptively selected locations according to an iterative subdivision algorithm. Compared to Mask R-CNN, PointRend has faster training and execution times as well as a more accurate and precise segmentation, especially for curved boundaries.

The IS models are all trained using the Manga109 dataset [6]. This is to ensure the models are accustomed to the structure of Japanese manga, further ensuring accurate and precise segmentation of regions of interest specifically for manga.

For the optical character recognition function of MLA, convolutional neural network-based multi-class classification M-OCR models are used to identify Japanese hiragana and kanji characters. One model is built on ResNet-50, and the other on ResNet-101 [7]. After undergoing the same training process in terms of the dataset used, number of epochs, learning rate, and other parameters, M-OCR-ResNet-101 expectedly achieved a higher F1 score than M-OCR-ResNet-50. Having said that, the difference in their performances is slight, and M-OCR-ResNet-50 has higher training and execution speeds.

Both M-OCRs are trained using the Kuzushiji-49 dataset and a subset of the Kuzushiji-Kanji dataset of Kuzushiji-MNIST [8] for hiragana and kanji characters, respectively. These datasets were chosen as they provide a comprehensive collection of hiragana and kanji characters commonly found in manga.

3 Methodology

3.1 Instance Segmentation

There are four MLA IS models: IS[Mask R-CNN]-3 (Manga109), IS[Mask R-CNN]-2 (Manga109), IS[PointRend]-2 (manual), and IS[Mask R-CNN]-2 (manual).

IS[Mask R-CNN]-3 (Manga109) is a Mask R-CNN model performing instance segmentation for text, frames, and faces. This model was trained using the original annotations of the Manga109 dataset whose format was modified slightly for training compatibility; originally, one XML file contained annotations for one book, but after modification, one XML file contained annotations for one image (two pages of one book). It uses only the annotations for text, frame, and face. This model, whose backbone is ResNet-101, underwent four rounds of training with one CUDA-enabled GPU. Table 1 shows the configurations.

For all training rounds, the list of backbone strides for each FPN Pyramid layer is [4, 8, 16, 32, 64], the batch size is 2, and the number of images per GPU is 2. The learning momentum is 0.9, the weight decay is 0.0001, and the pool size is 7. The loss weights implemented are rpn_class_loss, rpn_bbox_loss, mrcnn_class_loss, mrcnn_bbox_loss, and mrcnn_mask_loss, all of which have a value of 1.0. Each image used for training is a square of shape [1024, 1024, 3] (the minimum and maximum image dimensions are 800 and 1024, respectively), and the mask shape is [28, 28]. The number of classes is 4 (background + text + frame + face), and the image meta size is 16.

IS[Mask R-CNN]-2 (Manga109) is a Mask R-CNN model performing instance segmentation for text and frames. This model was trained using the dataset for IS[Mask R-CNN]-3 (Manga109) but without the annotations for face. This model has a ResNet- 101 backbone and underwent four rounds of training with the configurations in Table 1. However, the number of classes is 3 (background + text + frame), and the image meta size is 15.

IS[PointRend]-2 (manual) is a PointRend model performing instance segmentation for text and frames. This model was trained using a COCO-like dataset containing Manga109 images which were manually annotated. This model, whose backbone is ResNet-101, only underwent one round of training using one CUDA-enabled GPU. It segments 2 classes (text + frame) with a learning rate of 0.0005, a learning

Table 1 Training configurations for Mask R-CNN IS models

Round	Learning rate	Number of epochs	Layers	Steps per epoch
1	0.001	10	Heads	131
2	0.0001	40	All	131
3	0.001	1	All	1000
4	0.001	1	All	2524

momentum of 0.9, and a maximum iteration count of 2500. Its weight decay is 0.0001, number of workers is 2, and images per batch is 2.

IS[Mask R-CNN]-2 (manual) is a Mask R-CNN model performing instance segmentation for text and frames. This model was trained using the same dataset used to train IS[PointRend]-2 (manual) and underwent four rounds of training with the configurations in Table 1.

3.2 *Face-Text Association*

FTA uses the output vertices of the segmentation results of IS[Mask R-CNN]-3 (Manga109). FTA then calculates the centres of text and faces using each of their four vertices, identifying which text and faces lie inside each frame by determining whether the centres of the text and faces lie within the bounding box of the frame. FTA thus associates each face to the text nearest to it in the same frame by comparing the distances between the centres of each face and those of each text in a particular frame.

3.3 *Text Order Determination*

TOD also uses the output vertices of the segmentation results of IS[Mask R-CNN]-3 (Manga109). TOD first orders the frames; the topmost and rightmost frame is the first frame. TOD then traverses the image half chiefly in the x-direction to find the next frame, moving down in the y-direction and repeating the process upon exhausting a row. Once all frames have been ordered, TOD orders text. Once it has ordered all the text in one frame, TOD orders text in the subsequent frame.

3.4 *Optical Character Recognition*

M-OCR-ResNet-50 and M-OCR-ResNet-101 are convolutional neural network-based optical character recognition models built on ResNet-50 and ResNet-101, respectively. The models each underwent five rounds of training with one GPU. The configurations are as shown in Table 2. For all training rounds, the batch size is 128, the number of classes to classify is 99 (hiragana + kanji), and the number of steps per epoch is 1839.

Both models are trained using the Kuzushiji-49 dataset which contains 48 hiragana characters and 1 hiragana iteration mark and a subset of the Kuzushiji-Kanji dataset which contains 50 kanji characters. The images underwent several image augmentation steps, including rotation of range 10, zoom of range 0.05, width-shift

Table 2 Training configurations for M-OCRs

Round	Learning rate	Number of epochs
1	0.01	30
2	0.01	50
3	0.001	50
4	0.005	100
5	0.005	50

of range 0.1, height-shift of range 0.1, and shearing of range 0.15. The images were also resized to have a shape of (32, 32, 1). The image labels were binarised for multi-class classification.

4 Results and Discussion

4.1 Instance Segmentation

Table 3 shows the mAP values obtained by each of the ISs on different IoU thresholds. Sample outputs for the ISs are found in Fig. 1.

IS[Mask R-CNN]-3 (Manga109) performed slightly worse than IS[Mask R-CNN]-2 (Manga109) despite being trained using the same dataset and configurations. This difference could be attributed to IS[Mask R-CNN]-3 (Manga109) having to segment faces which are inherently difficult to segment. This is because there are

Table 3 mAP values of ISs

Name	mAP (IoU = 0.5)	mAP (IoU = 0.75)
IS[Mask R-CNN]-3 (Manga109)	0.90	0.80
IS[Mask R-CNN]-2 (Manga109)	0.93	0.86
IS[PointRend]-2 (manual)	0.95	0.92
IS[Mask R-CNN]-2 (manual)	0.93	0.85

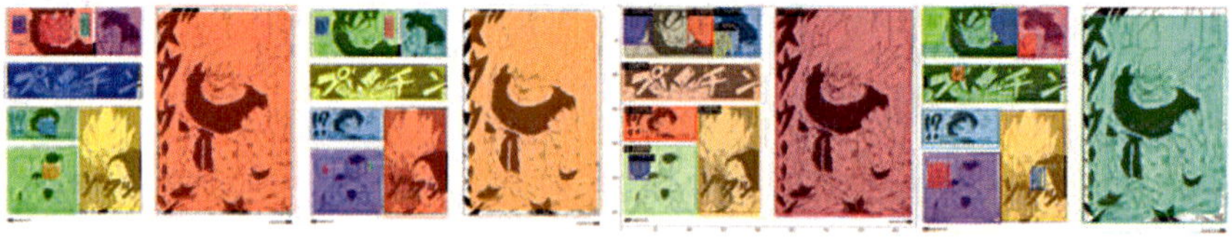

Fig. 1 From left to right: sample outputs of IS[Mask R-CNN]-3 (Manga109), IS[Mask R-CNN]-2 (Manga109), IS[PointRend]-2 (manual), and IS[Mask R-CNN]-2 (manual)

many different faces in each manga page and book, thus the model must adjust to this difference to ensure a high performance.

IS[Mask R-CNN]-2 (Manga109) performed only slightly better than IS[Mask R-CNN]-2 (manual). While their mAP values for IoU 0.5 are identical, their mAP values for IoU 0.75 differ by 0.01. This could be due to the more difficult task of segmenting nonlinear shapes faced by IS[Mask R-CNN]-2 (manual); the original annotations in the Manga109 dataset are all rectangular, while the manual annotations follow the boundaries of the text, frames, and faces exactly.

IS[PointRend]-2 (manual) performed better than IS[Mask R-CNN]-2 (manual), especially for IoU 0.75. This is due to the PointRend architecture being more suitable for segmenting shapes with rounded boundaries. Training and executing IS[PointRend]-2 (manual) are also faster than doing so for IS[Mask R-CNN]-2 (manual), proving that the PointRend architecture is more exact and efficient than Mask R-CNN. Hence, these features enable IS[PointRend]-2 (manual) to more readily segment text and frames than IS[Mask R-CNN]-2 (manual).

Based on these findings, IS[PointRend]-2 (manual) is the best model both in terms of performance and speed. Performance-wise, the PointRend architecture yields more detailed masks of higher resolution than Mask R-CNN, making IS[PointRend]-2 (manual) a more precise model than its Mask R-CNN counterparts especially for rounded objects such as faces which require finer segmentation of curved boundaries. Despite this more detailed segmentation, IS[PointRend]-2 (manual) is approximately 200% faster than the three Mask R-CNN models during training and slightly faster during execution (the execution times for all models do not exceed 5 s; hence, the differences in execution times are not significant).

However, the FTA algorithm requires the segmentation of faces; thus, IS[Mask R-CNN]-3 (Manga109) is used to develop it. TOD, although not requiring the segmentation of faces, uses IS[Mask R-CNN]-3 (Manga109) for consistency with FTA. The use of IS[Mask R-CNN]-3 (Manga109) does not pose a significant issue because its accuracy and precision are comparable to those of the most performant model, IS[PointRend]-2 (manual), and its execution time is similar to that of the other three models.

The ISs have also been tested on manga with English text instead of Japanese. The results for English manga are comparable to those for Japanese manga as shown in Fig. 2. The comparable results, even though the ISs were trained on solely Japanese manga in the Manga109 dataset, are possible due to the ISs learning that the shape of the speech bubble in which the text is located holds a greater importance than what is written inside. Hence, ISs do not discriminate between English and Japanese characters.

4.2 Face-Text Association

As seen in Fig. 3, FTA successfully associated each segmented face and text, even in a frame where there are multiple faces and text. However, even though FTA alone

Fig. 2 IS[PointRend]-2 (manual) on English (left) and Japanese (right) manga

is accurate and completely functional, it is heavily dependent on the results of the segmentation of text, frames, and faces by IS[Mask R-CNN]-3 (Manga109). Hence, there are instances where FTA fails to associate faces to text correctly due to the model failing to detect either the faces, text, or both mainly due to the presence of bleeding frames and extremely small speech bubbles. Moreover, FTA is not optimised for interframe association (associating a face in one frame to text in other frames), which can occur in cases of bleeding frames (frames with images overflowing out of them usually to convey a climactic moment in a story). To address this, FTA can be further developed to consider the direction each speech bubble tail is pointing towards and assess whether the tail's "line of action" intersects a face's area obtained via segmentation. Overall, FTA is still a successfully implemented algorithm leveraging the results of the aforementioned IS model.

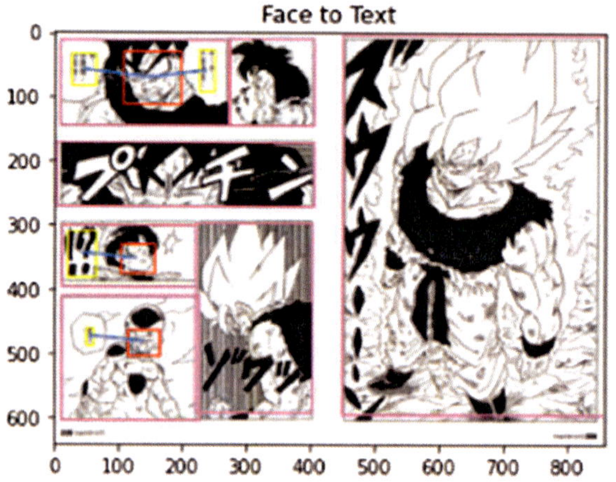

Fig. 3 Sample output of FTA

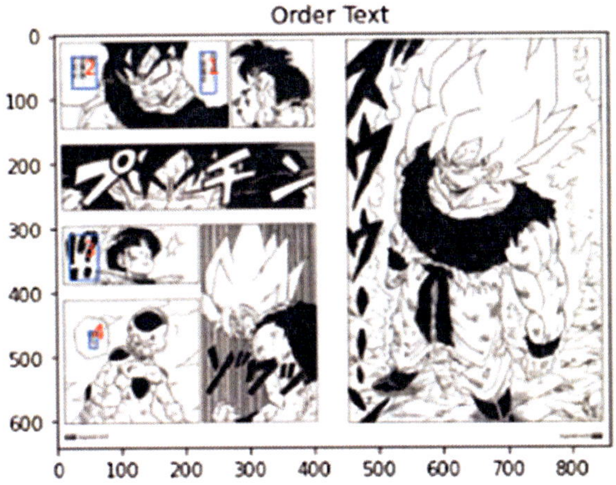

Fig. 4 Sample output of TOD

4.3 *Text Order Determination*

Figure 4 shows sample output images of TOD. As seen, TOD successfully orders text as per the reading order of manga (from top to bottom and from right to left); even when the Japanese characters in manga are translated to English, the reading order remains the same. Furthermore, TOD is able to distinguish between text in speech bubbles and onomatopoeia and text outside of frames, allowing them to only order text within speech bubbles. However, similar to FTA, TOD is dependent on the performance of IS[Mask R-CNN]-3 (Manga109), failing to identify and thus order the text when the IS model fails to segment the text correctly. Furthermore, TOD does not understand the text it orders; hence, TOD will not adapt to any unconventional manga reading order. Despite such drawbacks, the mechanism of TOD is capable of ordering text according to the Japanese reading sequence accurately.

4.4 *Optical Character Recognition*

Table 4 shows the F1 scores obtained by each of the M-OCRs. Sample outputs of M-OCR-ResNet-50 and M-OCR-ResNet-101 are found in Fig. 5, where each hiragana or kanji character is paired with its numerical class label. A green number signifies a correct classification, while a red number signifies an incorrect classification.

It is observed that there are 9 times more green than red numbers for both images in Fig. 5; hence, it can be concluded that both M-OCRs manage to correctly identify 90% of the hiragana and kanji characters accurately. Comparing the two models, M-OCR-ResNet-101 is slightly more accurate in classifying the Japanese hiragana and kanji characters, but M-OCR-ResNet-50 has slightly higher training and execution

Table 4 F1 scores of M-OCRs

Name	F1 score (accuracy)	F1 score (macro-average)	F1 score (weighted average)
M-OCR-ResNet-50	0.87	0.79	0.89
M-OCR-ResNet-101	0.89	0.81	0.90

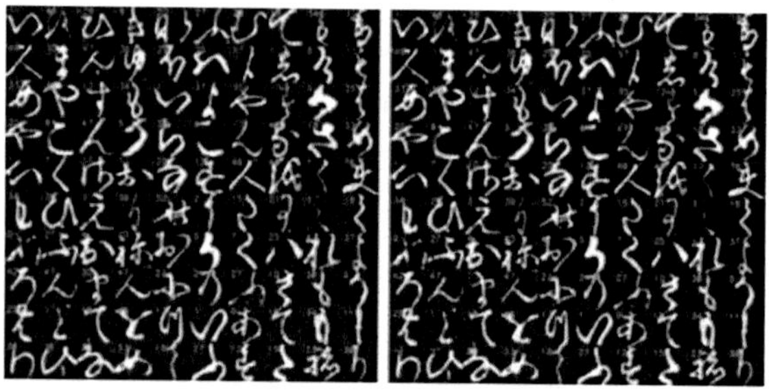

Fig. 5 Sample outputs of M-OCR-ResNet-50 (left) and M-OCR-ResNet-101 (right)

speeds due to its more lightweight architecture. Hence, MLA has no preference over the two models as both produce similar results and take similar periods of time for training and execution.

5 Conclusion

All in all, Manga Layout Analysis via deep learning has pushed the boundaries of document layout analysis. The ISs have achieved comparable performance to existing instance segmentation models on related documents (Western comic books) with the addition of simultaneous segmentation of multiple classes (text, frame, face) and adaptation to the structure of manga. FTA and TOD have also successfully performed their tasks according to the semantics of manga. The M-OCRs have achieved performances similar to or even better than state-of-the-art OCRs such as the Tesseract English OCR despite not implementing additional architectures such as LSTMs. Hence, a highly performant integrated MLA system has been successfully innovated through IS, FTA, TOD, and M-OCR. In conclusion, Manga Layout Analysis has succeeded overall as an accurate and efficient end-to-end system integration of instance segmentation models, novel algorithms, and optical character recognisers as a complete solution for the adaptation of manga into other media.

References

1. Dubray, D., & Laubrock, J.: Deep CNN-based speech balloon detection and segmentation for comic books. In *2019 International conference on document layout analysis and recognition (ICDAR)*.
2. Dutta, A., & Biswas, S. (2019). CNN based extraction of panels/characters from Bengali comic book Page images. In *2019 International conference on document layout analysis and recognition workshops (ICDARW)*.
3. Smith, R. (2007). An overview of the tesseract OCR engine. In *Ninth international conference on document analysis and recognition (ICDAR 2007)*, vol 2.
4. He, K., Gkioxari, G., Dollar, P., & Girshick, R.: Mask R-CNN. In *2017 IEEE international conference on computer vision (ICCV)*.
5. Kirillov, A., Wu, Y., He, K., & Girshick, R.: Pointrend: Image segmentation as rendering. In *2020 IEEE/CVF conference on computer vision and pattern recognition (CVPR)*.
6. Aizawa, K., Fujimoto, A., Otsubo, A., Ogawa, T., Matsui, Y., Tsubota, K., & Ikuta, H. (2020). Building a manga dataset 'manga109' with annotations for multimedia applications. *IEEE Multimedia, 27*(2), 8–18.
7. He, K., Zhang, X., Ren, S., & Sun, J. (2016). Deep residual learning for image recognition. In *2016 IEEE conference on computer vision and pattern recognition (CVPR)*.
8. Clanuwat, T., Bober-Irizar, M., Kitamoto, A., Lamb, A., Yamamoto, K., & Ha, D. (2018). Deep learning for classical Japanese literature.

A Comparative Study on the Characteristics of the Inflow Path of Fine Dust and the Trend of Climate Change in Korea and Singapore Using Satellite Data

Sean Shi-An Lim and Harini Manivannan

Abstract Fine dust is very small, less than 10 μm in diameter, so it cannot be filtered out of the lungs and absorbed into the body, causing respiratory diseases. It also disturbs the earth's heat system and affects climate change. Therefore, the importance of predicting fine dust in Korea is increasing. We checked the trajectory of the dust that moved for 72 h before they arrive in Korea and Singapore using the HYSPLIT model (NOAA). Also, the analysis model GIOVANNI, which provides geophysical data, is used. We got data of aerosol optical depth, angstrom exponent, and black carbon, which are related to features of total suspended particles (TSP), from GIOVANNI. The final goal of this research is to create a database of characteristics by inflow trajectories, and therefore we will be able to decide the dangerousness of aerosol which inflows to Korea and Singapore.

Keywords Fine dust · Carcinogenic · Satellite · Global warming · Climate change

1 Introduction

Fine dust not only gives negative effects on human health, but also distorts the heat system and influences climate change. It has the second highest influence on climate change after CO_2 [1]. Also, the metallic substances absorbed by fine dust cause climate change due to their strong thermal conductivity. In this study, we used three data for finding out the features of suspended particles in each region. Fine dust has diameters of less than 10 μm and is very hard to filter out. It can cause respiratory diseases and is classified as a group 1 carcinogenic agent by the international agency for research on cancer.

Aerosol optical depth shows how much light is scattered or absorbed after passing aerosol. It is composed of various ingredients such as see salts, fine dusts, black

L. Sean Shi-An (✉) · H. Manivannan
NUS High School, Clementi 129957, Singapore, Singapore
e-mail: H1710121@nushigh.edu.sg

H. Manivannan
e-mail: H1710042@nushigh.edu.sg

H. Guo et al. (eds.), *IRC-SET 2022*,
https://doi.org/10.1007/978-981-19-7222-5_7

carbon, sulfates, and so on. AOD increases as total suspended particles (TSP) increases, so we can analyze the movement or distribution of fine dust.

Angstrom exponent is used as qualitative indicators for sizes of aerosol particles because it is inversely related to the average size of the particles in the aerosol. Value under 1 usually shows that aerosol is mainly composed of see salt, and value over 2 usually shows that aerosol is mainly composed of fine dust.

Black carbon is closely related to PM 2.5 because of its small size, and it also has negative effects on human health. It is formed by the incomplete combustion of fossil fuels. It can make fine dusts gathered, and therefore is important to study.

The trajectory of the 1000 m height air mass shows the movement of air flow from various regions to Singapore and Korea during the different seasons. It aids in tracking the concentration of the fine dust in the air using the meteorological data of mixed layer depth. The relative humidity also tracks the speed of air flow. Primarily, the air flow trajectories show us the origin and flow of fine dust and how it affects the surrounding areas.

1.1 Aim and Hypothesis

The aim of this project is to track the movement and geophysical data of dust over two countries with differing characteristics such as the presence of seasons so as to better understand and create a database of characteristics of fine dust.

2 Materials and Methods

2.1 HYSPLIT

The Hybrid Single-Particle Lagrangian Integrated Trajectory model (HYSPLIT) [2] is a computer model that is used to compute air parcel trajectories to determine how far and in what direction a parcel of air, and subsequently air pollutants, will travel. HYSPLIT is also capable of calculating air pollutant dispersion, chemical transformation, and deposition. The HYSPLIT model was developed by the National Oceanic and Atmospheric Administration (NOAA) Air Resources Laboratory and the Australian Bureau of Meteorology Research Center in 1998.

In this study, HYSPLIT was used to track the trajectory of air mass at various heights. 72 h trajectory of 500, 1000 and 6000 m height air mass at 3/22, 6/22, 9/22, and 12/22 was extracted for every year from 2015 to 2020. Meteorological data of mixed layer depth and relative humidity was thus plotted.

Mixed layer depth data showed various air concentrations at different heights of air mass. As most of the fine dust was concentrated at the 1000 m height air mass, the 1000 m high air trajectory was used to monitor the fine dust flow and its effects on the neighboring areas.

Relative humidity was tracked in order to measure the speed of the air flow as it is directly proportional to the speed of air. As relative humidity increases, the concentration of water vapor in the air increases causing air mass to become denser. This would in turn result in an increase of air mass, causing air flow to be slower.

2.2 Giovanni

Giovanni is a tool that provides geophysical data from observations by NASA satellites and Earth surface. In addition to providing data on wind direction, pressure, and temperature, which are basic weather elements, specific and diverse geophysical data such as black carbon concentration, aerosol optical depth, evaporation rate, and sea surface temperature can be analyzed using the Grid Analysis and Display System (GrADS). This tool provides convenient geophysical data, allowing the user to select the desired resolution, region, and timing according to the purpose of the research. The main data from Giovanni in this study are aerosol optical depth and angstrom exponent, which are related to the absolute amount of total suspended particles (TSP).

Giovanni was used to supply data for three main items.

Firstly, aerosol optical depth (AOD). AOD is a measure of light extinction by dust and haze. It shows how much light is scattered or absorbed after passing aerosol. AOD is calculated as $-\ln(I/I_0)$, where I is intensity of radiation after passing aerosol and I_0 is intensity of radiation before passing aerosol. Aerosol is composed of various ingredients such as see salts, fine dusts, black carbon, sulfates from factories, and some other organic materials.

Because aerosol affects the radiative equilibrium, distribution of it will affect climate change. AOD is closely related to concentration of fine dusts, and therefore we can gain insights about movement of fine dust by investigating AOD.

Secondly, angstrom exponent. The angstrom exponent is generally used as qualitative indicators for sizes of aerosol particles because it is inversely related to the average size of the particles in the aerosol. If the exponent is less than 1 than it indicated that the aerosol is mainly composed of sea salt. If the exponent is greater than 2, than it indicates that the aerosol is mainly composed of fine dust which usually comes about as a byproduct of urban pollution and biomass burning.

Lastly, black carbon. Black carbon is formed from the incomplete combustion of fossil fuels such as gasoline. It negatively impacts human health and can cause fine dust to gather, clump, and aggregate. Hence, regions with high black carbons normally also have high levels of fine dust.

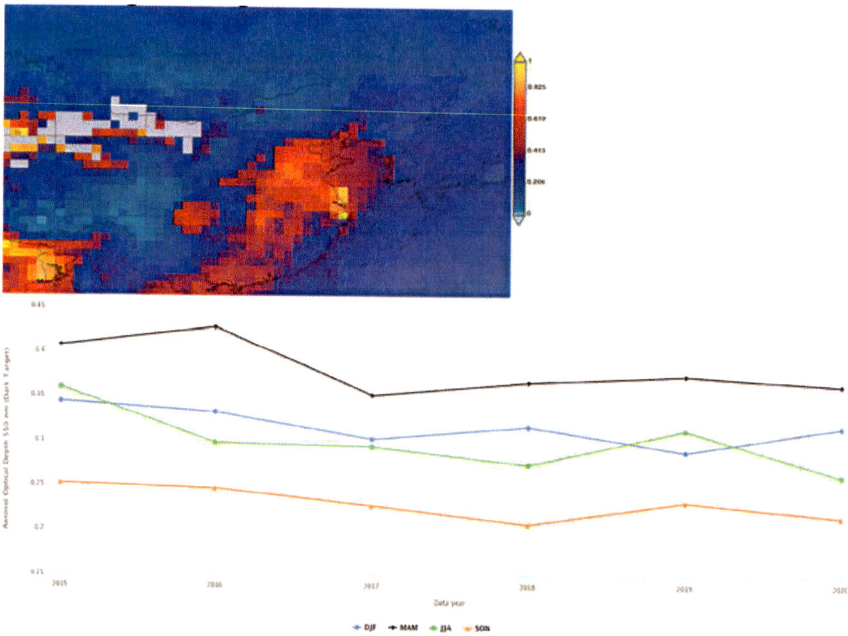

Fig. 1 Map and interannual time series of AOD from 2015 to 2020 of Korea

3 Results

3.1 Giovanni

Aerosol Optical Depth

See Figs. 1 and 2.

Angstrom Exponent

See Figs. 3 and 4.

Black Carbon

See Figs. 5 and 6.

3.2 HYSPLIT

See Figs. 7 and 8.

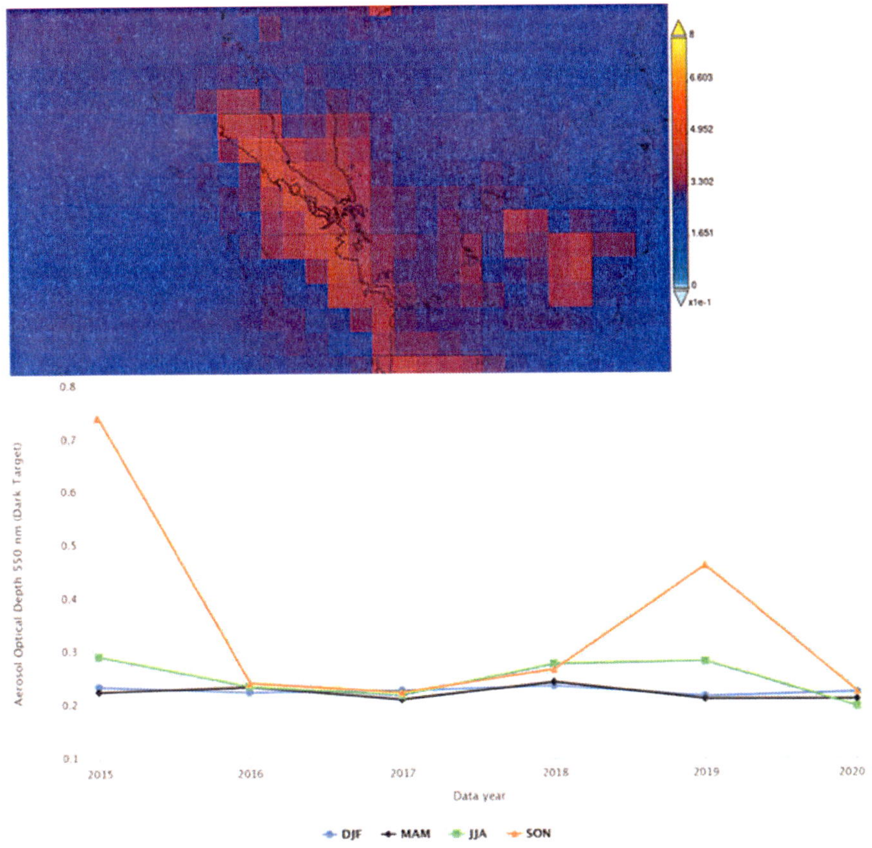

Fig. 2 Map and interannual time series of AOD from 2015 to 2020 of Singapore

4 Discussion

4.1 Giovanni

AOD

From the time averaged map of AOD from 2015 to 2020 near Korea, we can see that AOD is very high at the East side of China. We estimate it as an effect of industrial complex, thermal power station and aerosol resulting from heating.

The graph showing the seasonal time series AOD graph of this area from 2015 to 2020 provides us with further insights of AOD changes with seasons. In spring, it shows the highest value, and we estimate this because of yellow dust from west. The next highest value is observed in winter, and it is estimated as the effect of heating

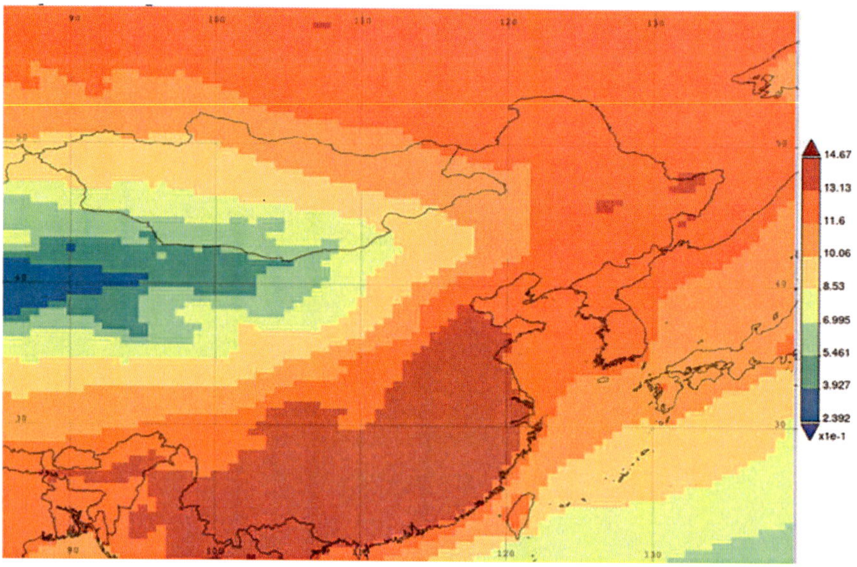

Fig. 3 Map of Angstrom exponent of Korea

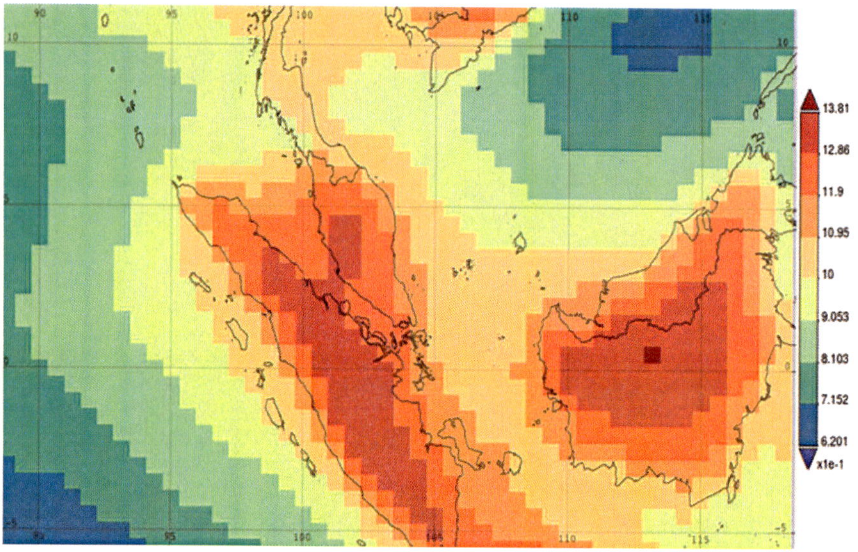

Fig. 4 Map of Angstrom exponent of Singapore

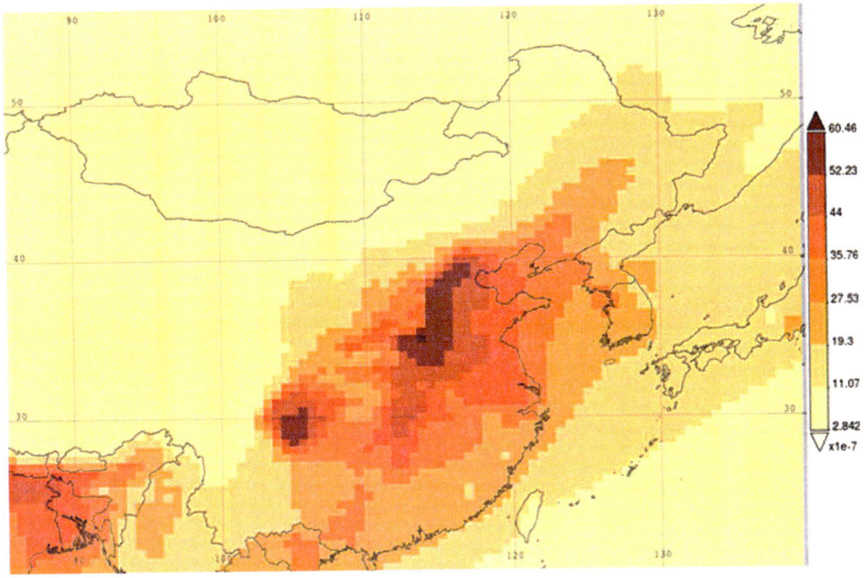

Fig. 5 Map of black carbon of Korea

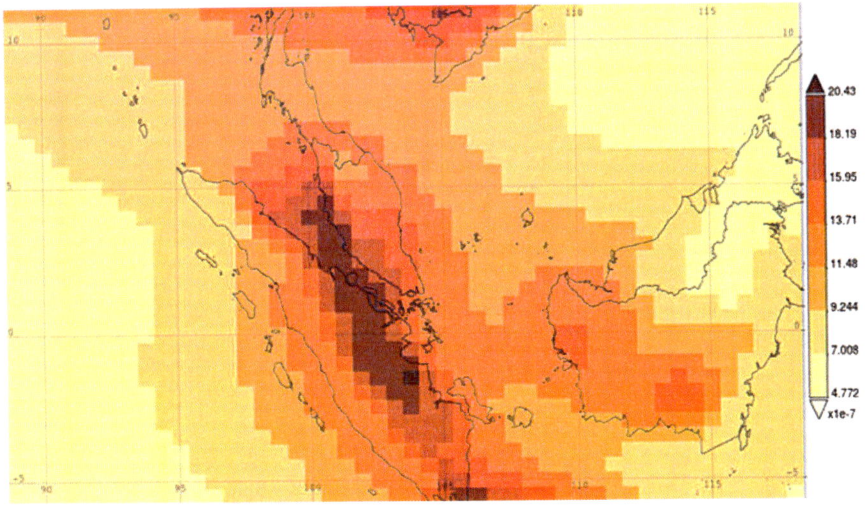

Fig. 6 Map of black carbon of Singapore

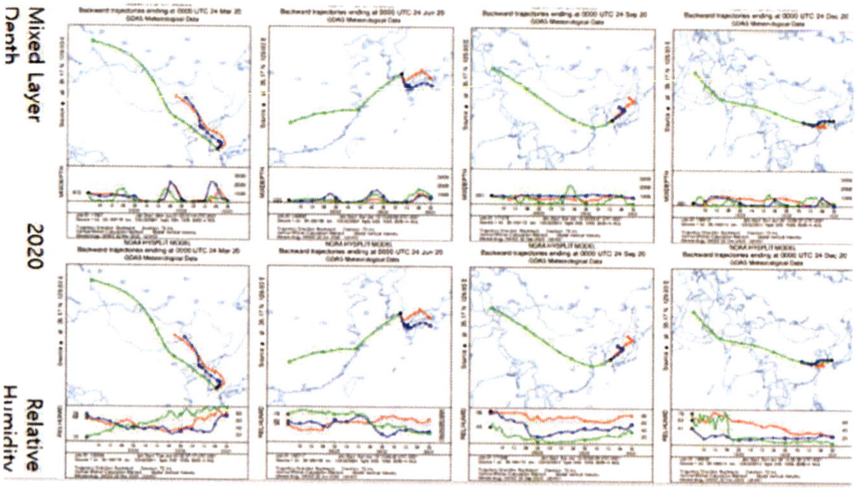

Fig. 7 Map showing flow path of dust in Korea

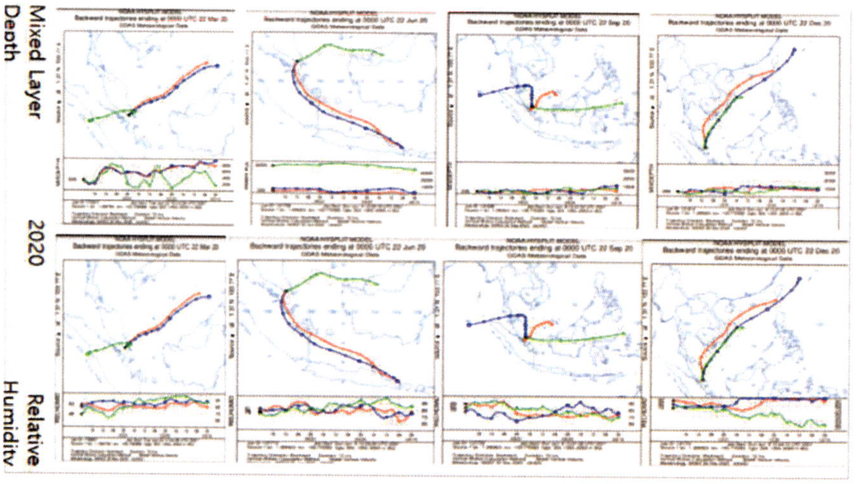

Fig. 8 Map showing flow path of dust in Singapore

in households. As we can see in the graph, AOD value of Korea and nearby regions shows changes through the year and is affected by the change in season.

This image is the time averaged map of AOD from 2015 to 2020 near Singapore, and we can see that the AOD value is very high at the section of Indonesia (from Medan to Riau) that is near to sea and facing Malaysia. We estimate it as an effect of burning of forest which causes haze and smoke.

The graph showing AOD changes through the years provide us with additional insights on how AOD value near Singapore is influenced by external sources.

The peaks in AOD can be seen in summer and autumn in 2015. This correlates with the 2015 southeast Asian haze which lasted from June to October. A smaller peak reappears in 2019 showing a downplayed version of AOD levels in 2015. This correlates with a less severe haze that spanned February to September in 2015.

Apart from these two events, there is little correlation between AOD levels and seasons showing that human activities have the greatest impact on AOD levels in the area surrounding Singapore.

Angstrom Exponent

In the southwest region from Mongol, angstrom parameter value appears low, indicating natural dusts with bigger particles. In the southeast region of China, angstrom parameter value appears higher, and it shows that artificial dusts with smaller particles are more present there.

Black Carbon

The time averaged map of black carbon of region near Korea is coded such that darker color shows higher concentration of black carbon. We can see that the east region of China shows high black carbon concentration and estimate it as an effect of industrial complex and thermal powers stations. Because black carbon plays like important rule on the formation of fine dust, its distribution should be studied.

4.2 *HYSPLIT*

Korea

The graphs show the air trajectory of dust particles around Korea during the four seasons. From the graphs in March, we can derive that air flows from the region of Mongol to Korea. This allows us to conclude that natural dust with larger particles are being pushed over onto Korea during spring.

The graphs from the 4th quarter of the year show the airflow from middle and east China where pollution is in abundance. We can hence, conclude that artificial dust with smaller particles flow from China to Korea in the winter.

The air trajectory in September fluctuates irregularly over the years. We have concluded that anticyclones, which involves mass circulation of air may have resulted in the varying trajectories during autumn.

Singapore

The graphs show the air trajectory and the movement of dust particles around the region of Singapore during the four seasons. From the graphs of March, June, and September, we can derive that the air is directed from the region around Indonesia (Medan to Balikpapan) during spring, summer, and autumn. Artificial and smaller particles flows from the Indonesian region. This is likely due to the heaps of forest fires around those regions that produces haze which is directed toward Singapore, especially during the South West Monsoon season. Haze consists largely of fine dust, commonly known as particulate matter which are smaller than 2.5 μm. Continuous exposure over several years to high ambient pollution from these fine particles may lead to a higher risk of cardiovascular effects, such as heart attacks, reduced lung development, as well as the development of chronic respiratory diseases, such as asthma, in children [3].

In the winter, the direction of air flow is from the north of Singapore, around the Malaysian and Cambodian regions. Hence, we can conclude that natural dust with larger particles flows in during the winter season.

4.3 Improvements and Future Work

We plan to further investigate all of the components and their concentration in fine dust particles using sampling and test kits as this would help nations and organizations to take better preventative measures against fine dust. We also hope to track the air trajectories of fine dust flow throughout the entire Southeast Asian region as this would give us a greater insight into the origin and pathways taken by fine dust.

References

1. Carmichael, G. R., Adhikary, B., Kulkarni, S., D'Allura, A., Tang, Y., Streets, D., ... & Marrapu, P. (2009). Asian aerosols: current and year 2030 distributions and implications to human health and regional climate change. *Environmental science & technology, 43*(15), 5811–5817.
2. Hysplit. Air Resources Laboratory. (n.d.). Retrieved January 27, 2022.
3. Dai, B. T., Jayarajah, K., Lim, E.-P., Misra, A., & Nayak, S. (2016, January 1). A study on Singapore Haze. A study on Singapore haze. In *Proceedings of the 17th international conference on distributed computing and networking*. Retrieved January 27, 2022.

Analysis of Public Transport Accessibility of Different Areas in Singapore

Samuel Chan Guan Xu

Abstract Singapore has a robust public transport network, and it is important to ensure its accessibility, so that residents can travel around Singapore with ease in an environmentally friendly manner, regardless of financial background. We look primarily into the accessibility of bus stops in Singapore, we split the Singapore map into subzones following the "Master Plan 2019 Subzone Boundary (No Sea)" and measure subzone accessibility in terms of distance to the bus stops in each subzone and the number of bus services per bus stop. Python is used to analysis the datasets obtained through public sources (e.g. LTA DataMall in Dynamic datasets: bus arrival, bus routes, bus services, bus stops [1]) and the results are choropleths showing the accessibility of each subzone of Singapore. The results show that the Central Business District (CBD) is extremely accessible business rich area. Town centrals of public housing also have high accessibility ratings. Major main roads in Singapore like Commonwealth Avenue which stretches until Boon Lay and have HDBs on both sides of the road are highly accessible by bus. Conversely, private housing estates along Bukit Timah Road have extremely low accessibility to buses as the people living there are likely to own vehicle and have no need for buses. The most inaccessible land types are "Open Space", "Port/Airport", and "Special Use", which have no need for bus accessibility due to them not being public or residential places. The results from this study can be used to further develop the bus transport system in Singapore.

Keywords Public transport · Urban planning · Geometry

1 Introduction

Singapore is a well-developed country with a dense transport network due to our small size and relatively large population. We have a tight road network and high traffic volume despite private vehicle ownership being extremely expensive. However, there

S. C. G. Xu (✉)
NUS High School of Math and Science, 20 Clementi Ave 1, Singapore 129957, Singapore
e-mail: samuelchangx@gmail.com

H. Guo et al. (eds.), *IRC-SET 2022*,
https://doi.org/10.1007/978-981-19-7222-5_8

is still a significant portion of people in Singapore that take public transport daily, which is cheaper than private transport. With the increasing standard of living in Singapore, the number of people taking public transport in Singapore is likely to increase. Furthermore, Singapore is moving towards being an environmentally friendly country and with numerous government campaigns on going green, Singaporeans, regardless of financial background, are likely to take public transport. As Singapore is a fast-paced society, it is essential that public transport is easily accessible to all so people can keep up with their busy lives. It is reasonable to assume that people living in private housing are more likely to own vehicles, as they have the means, and are less likely to take public transport. Therefore, accessibility to public transport in those areas is hypothesised to be low as it is unnecessary to develop public transport there. On the other hand, the town central of public housing, such as The Clementi Mall at Clementi or Compass One at Sengkang, is expected to have high accessibility to public transport. This is with the assumption that people staying in public housing are of lower to middle income and are more likely to take public transport and hence it is logical for the government to develop public transport in those areas. This is evident as town centrals often have a bus interchange in proximity. However, we might see a different trend than people with more wealth having a higher usage of private vehicles as there are other factors involved, such as the decision to be eco-friendly. We also hypothesise that land use type will influence the accessibility of public transport. Parks, places of nature and reservoirs are unlikely to have roads passing through them and hence public transport accessibility should be low there. Places that have a lot of business such as our Central Business District should have a high accessibility of public transport. The results from this study could be used to understand the profile of different areas, the reasoning behind their different accessibility of public transport, as well as the efficiency of land use in general. The study could also reveal some unexpected information of the Singapore public transport system and areas where they can improve on in a resource efficient manner. The result of this study would be a choropleth map of Singapore, highlighting the different accessibilities of different areas in Singapore using colour codes.

In this study, we will explore the accessibility of different areas in Singapore mainly in terms of their accessibility to bus stops. The "different areas" in Singapore will be defined based on the "Master Plan 2019 Subzone Boundary (No Sea)" [2] and the accessibility of each area will be studied. Each of the subzones in the "Master Plan 2019 Subzone Boundary (No Sea)" will be considered an area. The accessibility will be analysed through a range of factors, including the distance to the nearest bus stop and their connectivity within the wider network of bus stops. Data on the buses in Singapore will be obtained from publicly available sources such as LTA DataMall [1] and analysis will be performed on the data to aid in the research of this study. Areas, or subzones, not suitable for transport development will be excluded from the analysis and measures will be developed to quantify the public transport accessibility of the other subzones, through a few rounds of analysis. Various analysis techniques based on elementary geometry will be employed to characterise spatial patterns in relation with public transport accessibility. The characteristics of the areas include the centrality or location within the larger urban system, allowing comparison with other urban attributes, such as the land use type.

2 Methodology

Analysis of data and research in this study is done on Python using Jupyter Notebook and with necessary Python Libraries. Datasets are obtained from LTA DataMall [1] and subzone data is obtained from "Master Plan 2019 Subzone Boundary (No Sea)" [2] Each subzone has a code and a name (e.g. MESZ01: MARINA EAST) The subzone data consist of the coordinates of the outline of each subzone.

To start off the analysis, a shapefile called "BusStopLocation" was downloaded from LTA DataMall [1]. Using python, the file was read, and its content was converted into a Python dictionary for easier manipulation of the data. This python dictionary, titled "bus_dict" used the standard 5-digit bus stop number code (e.g. 11,401) that all bus stops in Singapore have, as the key and returned the coordinates of the corresponding bus stop in SVY21, the coordinate system that Singapore uses.

With the coordinates of all the bus stops and the coordinate outline of each subzone, we can find all the bus stops in each subzone. 300 m buffer was given to each subzone from their perimeter and all bus stops that existed within each subzone and their 300 m were found. A new dictionary was then written to store this new information, with the key being the subzone code and the information returned being the subzone name and all the bus stops within that subzone and its 300 m buffer. All these was accomplished with the help of the Shapely library in Python. The reasoning behind that 300 m buffer is the accessibility of a subzone should not be limited the public transport within that subzone as people might choose to walk out of the subzone to take public transport and that would add towards the accessibility of the subzone. 300 m was chosen as we felt it was a reasonable distance that people would walk.

With the knowledge of all the bus stops in each subzone, we now begin our bus stop accessibility analysis on each subzone. Equal circles are drawn around each bus stop, which is taken as a node, and all the circles expand equally until the whole subzone is filled. The area of the polygon that is not covered by the union of the circles is also recorded for every subzone.

Figure 1 on the left is an example of the analysis. The black outlined rectangle is the subzone boundary, and the red dots are the bus stops within 300 m of the subzone's buffer. The circles (in blue) are drawn around them. The polygon, which area is recorded, is the white space in the middle of the circles, which is the difference between union of the circles and the area of the subzone.

The radius of the circles increases in 10 m intervals, from 0 m. Each time, the area of the polygon formed by the circles and the boundary of the subzone is measured and recorded as a percentage of the total subzone area. This continues for each subzone until the percentage of the polygon to the total subzone area is 0%, meaning the subzone is completely covered.

Fig. 1 Example the analysis. Red dots are the bus stop, and the blue circles are the increasing radius

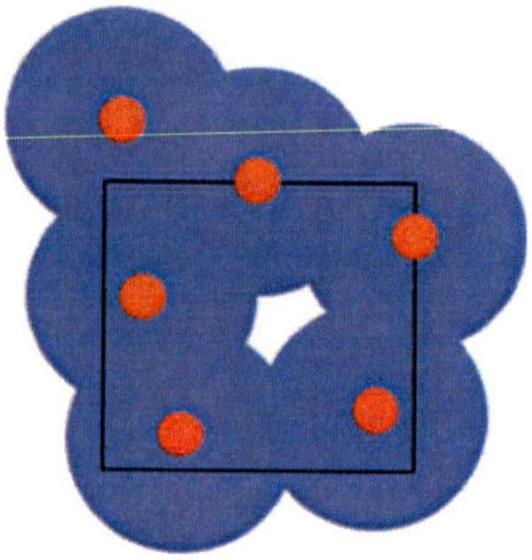

The radius of the circles when the subzones were first fully covered were recorded and put into a histogram (Fig. 2) with the x-axis being the radius mentioned above and the y-axis being the number of subzones that were fully covered when the circles were at the corresponding radius. The histogram shows that most subzones are completely covered at around 300 m. We can infer that in most subzones, you are around 300 m away from the nearest bus stops. Hence, we can conclude and justify that the accessibility radius of a bus stop is approximately 300 m.

If all area within 300 m radius of a bus stop is labelled the accessible area, then we label the area in the subzone that is not covered by all the bus stop's 300 m radius as the inaccessible area. The area of all the inaccessible areas were recorded and compared with the "Master Plan 2019 Land Use Layer" [3] to determine the type of land that are usually inaccessible. The result of this analysis is in the Results section.

The percentage of the total accessible area against the total subzone area was calculated for all subzones. A choropleth was made using "Master Plan 2019 Subzone Boundary (No Sea)" [2] to show accessibility of each subzone. The colours represent the percentage accessible area of each subzone in 10% intervals. Subzones that are 0% or 100% accessible have their own unique colours.

Finally, further research was done after evaluation of the preliminary results. Bus stops that serve more services are no doubt more "accessible" as they provide access to more places. Hence, the analysis was redone using bus stops that serve at least

Fig. 2 Histogram of the number of subzones that are fully covered when the radius around each bus stop is a certain distance

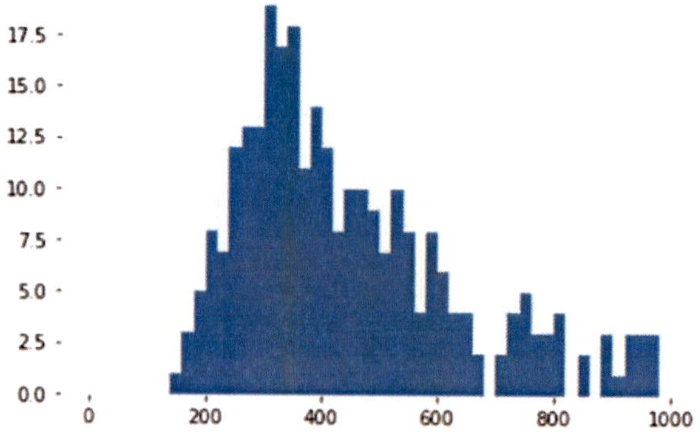

2 services, 3 services, 4 services, and 5 services. Buses that served night services, express services and similar services were removed from the service count as they travel the same route as the main services and hence do not increase accessibility. The choropleths for this repeated analysis were all created too and will be in the Results. An overlay of the MRT map using "Master Plan 2019 Rail Line Layer" [4] was placed on the maps for evaluation.

3 Results and Discussion

Starting from the histogram (Fig. 2.) obtained in the early stages of the research, we concluded that most subzones were 100% accessible when the circles of each bus stop in the subzone and its 300 m buffer had a 300 m radius. The value of 300 m makes a lot of sense as it is around the maximum distance anyone would walk for a bus stop and hence it makes sense that in most subzones, you are at most 300 m away from the nearest bus stop. This might even be by the design of LTA of the relevant authorities. This is evidence that the authorities plan out our bus transport system meticulously.

During the experiment, a total of 10 subzones had to be removed from the analysis as they contained no bus stops within them and their 300 m buffer. The table (Fig. 3) shows the subzone code and subzones that have no bus stops.

These subzones are usually offshore islands that are not connected to the mainland. Some of these areas also have restricted access. Hence, accessibility at these areas is 0% and their no bus stops there.

The inaccessible areas and their land use types were also obtained from the research. Figure 4 shows the land use type of the inaccessible areas and the amount of area of that type that is inaccessible in mm^2. The list shown is ordered from the most inaccessible area type to the least. We see that "OPEN SPACE" is the most inaccessible land use type. We can infer that because it is just open space and nothing else, there is no point building public transport for that. What is interesting is that "MASS RAPID TRANSIT", which is a public transport is in the inaccessible list. This is very unusual as most MRT have bus stops near them. I would conclude that maybe these areas are the places that the trains are stored and not MRT stations (Fig. 5).

WISZ01	Jurong Island and Bukom	NESZ01	Pulau Tekong, Pulau Tekong Kechil and Pulau Ubin
WISZ02	Semakau	PGSZ07	Coney Island
WISZ03	Sudong	SLSZ02	Pulau Punggol Barat
SISZ01	Sentosa	SMSZ04	Pulau Seletar
SISZ02	Southern group, Tuas	SMSZ02	Tanjong Irau

Fig. 3 List of subzones, and their codes, with no bus stops in them

```
'OPEN SPACE', 63261889.869602665],
'PORT / AIRPORT', 38984395.89214721],
'SPECIAL USE', 36671471.611524396],
'RESERVE SITE', 34173691.31112244],
'WATERBODY', 31591843.04209843],
'BUSINESS 2', 27340322.419456962],
'RESIDENTIAL', 19793837.833072744],
'ROAD', 15904459.159668885],
'PARK', 10012546.766815279],
'SPORTS & RECREATION', 8864575.520449614],
'UTILITY', 4942664.919506878],
'AGRICULTURE', 4720183.587031959],
'TRANSPORT FACILITIES', 2707834.3798769363],
'CEMETERY', 2021722.7823290878],
'EDUCATIONAL INSTITUTION', 1960059.701638134],
'CIVIC & COMMUNITY INSTITUTION', 1602988.0513605392],
'BUSINESS 1', 667628.5915946268],
'BUSINESS PARK', 558598.5921131952],
'WHITE', 530731.5061607675],
'HEALTH & MEDICAL CARE', 263188.6320597409],
'BEACH AREA', 254111.5256650579],
'COMMERCIAL & RESIDENTIAL', 250073.22360347517],
'COMMERCIAL', 182712.37331109564],
'PLACE OF WORSHIP', 112232.13509744962],
'RESIDENTIAL WITH COMMERCIAL AT 1ST STOREY', 91728.61640447073],
'HOTEL', 82405.83905815122],
'BUSINESS 2 - WHITE', 63215.78429890139],
'BUSINESS 1 - WHITE', 30945.52065102433],
'MASS RAPID TRANSIT', 19488.662710523877]]
```

Fig. 4 List of land use type of inaccessible places

Choropleth 1 is the result of the analysis including bus stops serving at least 1 service. The scale is provided above and just overlaps offshore islands which are irrelevant to the research. The yellow dots are Mass Rapid Transit (MRT) and Light Rapid Transit (LRT) stations. Immediately, we notice that the Central Business District (CBD) of Singapore has a high accessibility. This coincides with our hypothesis that areas of business are likely to be more accessible as people need to go there to generate income. The more populated town centrals such as Clementi MRT, Jurong East MRT, and Sengkang MRT are also highly accessible as seen from the warmer colours of their subzones. This is also in line with our hypothesis. Jurong East has many malls serving the whole Jurong population, most of which live in public housing, and so bus accessibility there must be high to meet the demand. Clementi has many schools nearby, and Jurong East and Sengkang both have public hospitals in proximity. Therefore, accessibility at those areas must be high. Public housing clusters in Yishun, Serangoon and Bishan are also highly accessible. Major roads

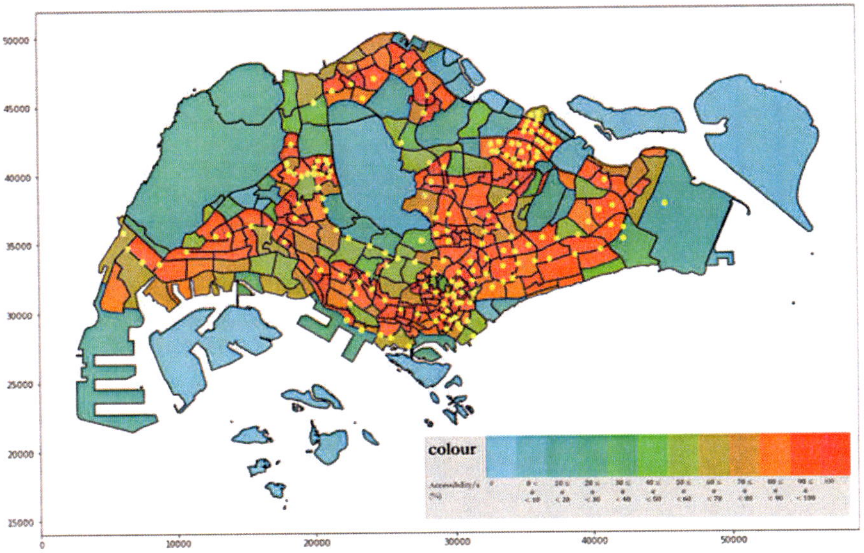

Fig. 5 Choropleth [5] of the results of the analysis considering all bus stops

like Commonwealth Avenue which stretches to Boon Lay Way also have a high bus stop accessibility. These roads have the East West Line of the MRT running above them. There are also many HDBs neighbourhoods along those stretches of roads. In contrast, Bukit Timah Road, which has the Downtown Line running below it, has a low accessibility to bus stops, displayed by the cooler colours of their subzones. This is because that stretch of roads are filled with private housing and condominiums. The owners of these properties are very likely to be of higher income and own vehicles. Hence, there is less demand for public transport there (Fig. 6).

Choropleth 2 is the result of the analysis only including bus stops serving at least 5 services. This choropleth hence only considers higher value bus stops (bus stops that serve more services provide more accessibility value than bus stops that serve less services). The general trends Choropleth 1 follows remains. The CBD, Commonwealth Avenue to Boon Lay stretch and the major town centrals have high accessibility to bus stops despite only considering bus stops that serve at least 5 bus services. Unexpectedly, there are interesting occurrences in our analysis. Despite having more private housing and being a significant distance away from the MRT, the Marine Parade subzone remains highly accessible by bus. This is possibly due to there being popular schools such as Tao Nan School and the popular malls such as Parkway Parade. The demand to go to those areas are very high, especially Tao Nan School where parents would buy property near the school just to get their child a place in the school. As such, the bus stops are more frequently placed there to cope with the dense population. Alternatively, the Singapore Government might be encouraging people with the means to purchase a car to take public transport instead as it is more

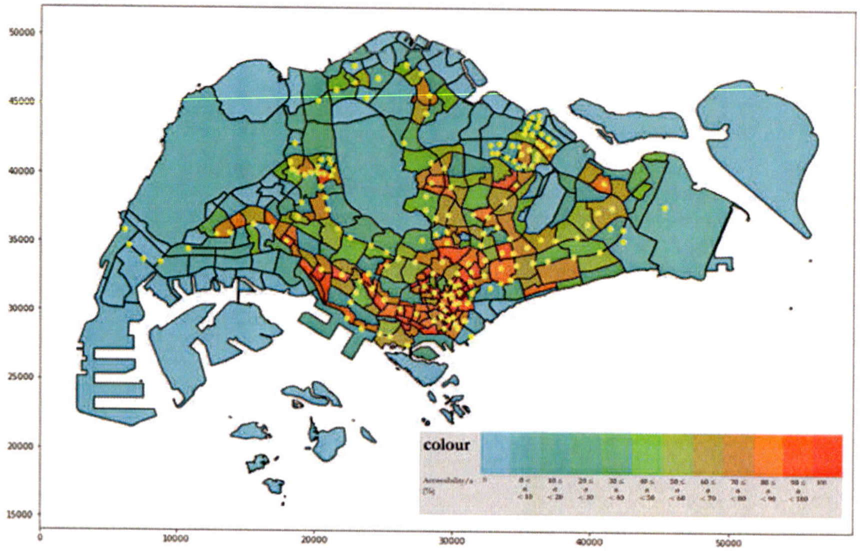

Fig. 6 Choropleth [5] of the results of the analysis considering bus stops with at least 5 services

environmentally friendly and doing so helps Singapore take a step towards being a green nation.

4 Conclusion

The bus transport system in Singapore is well-developed and well-planned out. The relevant authorities have carefully allocated their resources to develop the public transport in areas that need it most to suit the demands of the people to the best that they can. If we have more resources, Singapore can consider building more public transport network in the private housing areas. Although they would most probably own vehicles, it is still helpful to place bus stop at those areas to encourage the people there to take public transport. Public transport is more environmentally friendly than private transport. Furthermore, Singapore is very small and public transport should be sufficient to get around smoothly. The data obtained from this research can be used by the authorities to further improve the bus transport network. Looking at the choropleths in Appendix 2, we can see that the north side of Singapore contain bus stops that serve very few services (the accessibility of the subzones decreases as we consider bus stops with less and less services). Improvements can be made there.

Appendix 1—Master Plan 2019 Subzone Boundary (No Sea)

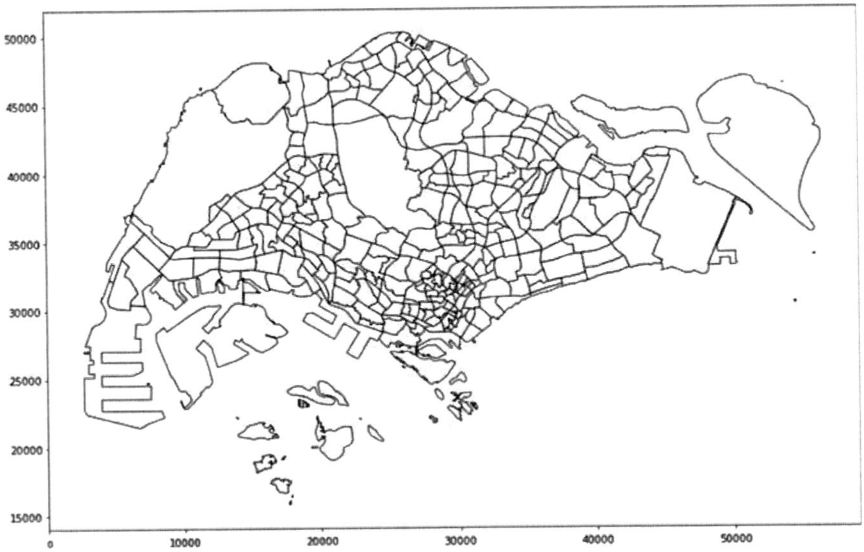

Plain choropleth

Appendix 2—More Choropleths

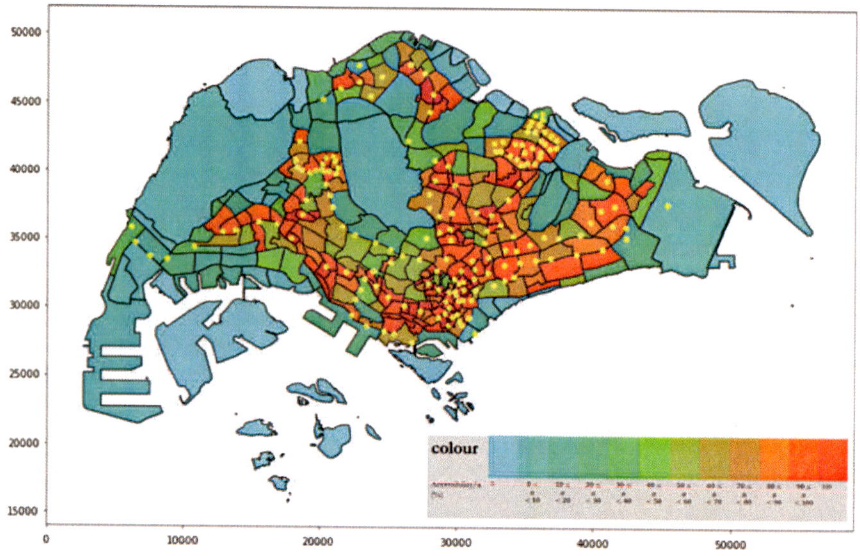

Same analysis but only considering bus stops with at least 2 services

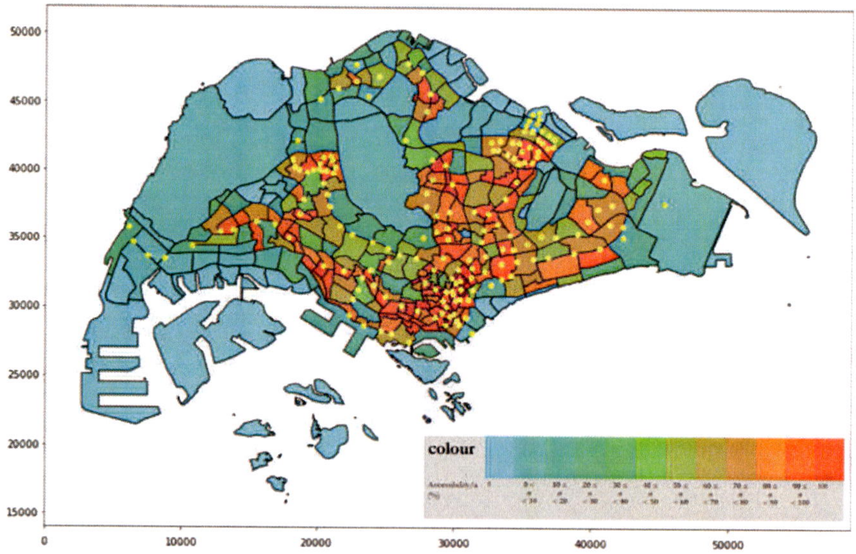

Same analysis but only considering bus stops with at least 3 services

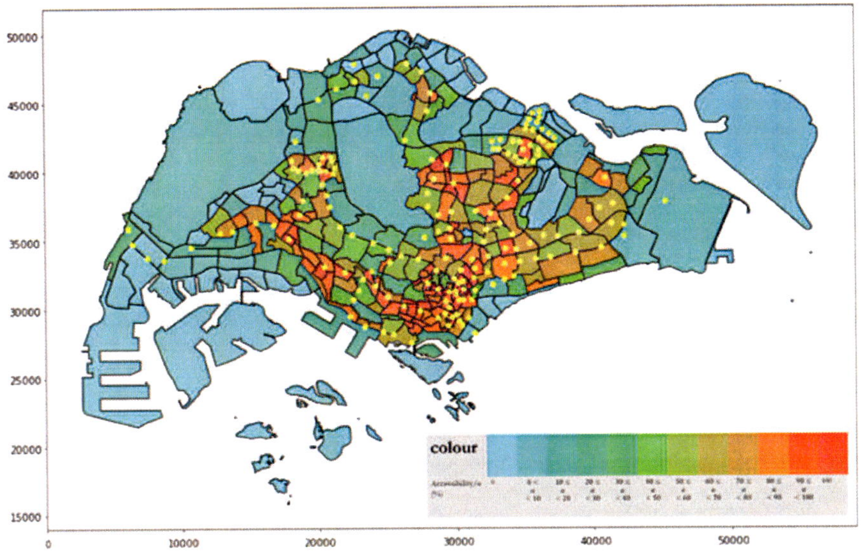

Same analysis but only considering bus stops with at least 4 services

References

1. Dynamic datasets: Bus arrival, bus routes, bus services, bus stops. https://datamall.lta.gov.sg/content/datamall/en.html. Last accessed December 31, 2021.
2. Master plan 2019 subzone boundary (no sea). https://data.gov.sg/dataset/master-plan-2019-subzone-boundary-no-sea. Last accessed December 31, 2021.
3. Master plan 2019 land use layer. https://data.gov.sg/dataset/master-plan-2019-land-use-layer. Last accessed December 31, 2021.
4. Master plan 2019 rail line layer. https://data.gov.sg/dataset/master-plan-2019-rail-line-layer
5. Colours for choropleth. https://colordesigner.io/gradient-generator. Last accessed December 31, 2021.

Carbon Monoxide Concentration in Atmosphere—A Review

Anmol Kumar, Shubham Kumar, and Annu Kumari

Abstract Carbon monoxide (CO) is colorless, odorless, and tasteless gas formed by the incomplete combustion of fuels. Due to order less and tasteless nature of CO, it doesn't sense to our organs. It poses a serious threat to our health. Clothes dryers, water heaters, furnaces or boilers, and fireplaces, both gas and wood burning, are some frequent sources of CO in our houses: automobiles, grills, generators, power tools, lawn mowers, wood stoves, tobacco smoke, gas stoves, and ovens. In this article, we will discuss the concentration of carbon monoxide gas. Nowadays, machine learning is playing vital role for future prediction. Many scientists and researchers have done commendable work by implementing different machine learning algorithms for the future prediction of CO concentration in environment. In this article, we will review the different research paper related to the prediction of CO concentration. In different research paper, many different algorithms are used and we review these papers properly to better prediction of CO.

Keywords Time series forecasting · ARIMA · Auto-regression · LSTM · Holt Winter

1 Introduction

Polluted air is hazardous to living organism, and it must be controlled quickly in both city and village areas, and therefore reliable air quality prediction is required. Different types of pollution exist in environment like soil pollution, water pollution, sound pollution, air pollution, and noise pollution but among them the most serious is air pollution, which must be addressed promptly since people consume oxygen via the air. In the air pollution, there are many pollutant present in the air such as CO_2, SO_2, NO_2, and CO. Among this pollutant, CO is hazardous. In this paper, we review the different papers related to the prediction of carbon monoxide concentration in the atmosphere.

A. Kumar · S. Kumar (✉) · A. Kumari
Mahatma Gandhi Central University, Motihari, India
e-mail: shubhamkumar@mgcub.ac.in

H. Guo et al. (eds.), *IRC-SET 2022*,
https://doi.org/10.1007/978-981-19-7222-5_9

Carbon monoxide gas is dangerous for our health, and it may cause death. It is colorless, odourless, and tasteless gas which doesn't senses our organ. This gas is spread by the incomplete combustion of hydrocarbon fuels. It is also produced by burning of charcoal, wood, propane, and other gases. CO affects directly to the human being. The CO poisoning may cause some issues like headache, weakness, dizziness, vomiting, and chest pain. People who suffers from chronic heart disease, anemia or people who are getting problem while breathing. If the carbon monoxide is too much in air, then the oxygen present in red blood cell replace with poisonous gas carbon monoxide. It may damage the tissue, and it might be possible that a person gets death.

Carbon monoxide poisoning is especially harmful for:

Unborn babies such that the baby who are ready to born, i.e., babies inside the stomach of his/her mother, old people, and youngster too.

With increased population, city expansion, and industrial development, the consequences of air pollution are spreading at a quicker rate and with a changing composition. While air pollutants produced locally have local effects, rising energy use, fossil fuel burning, and motor vehicle use all contribute to the worsening of air quality in urban regions. In India, the Central Pollution Control Board (CPCB) is responsible for monitoring the level of pollutant in the air. As, we know that carbon monoxide is one of the main pollutants in residential area and the main source of CO are motor vehicles, and incomplete combustion of hydrocarbon carboxyhemoglobin levels in the blood are increased by high CO concentrations. Blood loses its capacity to absorb oxygen due to excessive levels of carboxyhemoglobin. It reduces the amount of oxygen transported by the blood to the various areas of the body.

2 Related Work

A lot of work have been done in this area and many researchers already discussed the prediction of CO concentration in the atmosphere and also discussed the idea to control the emission of CO in the atmosphere by using many machine learning algorithm or model, and in this paper [1], the author deals with the forecasting of carbon monoxide concentration using fuzzy logic for Konya (a city in Turkey). Three separate fuzzy logic algorithms were employed in the modeling. These were the algorithms FCM, GP, and SC. The best results are obtained using FCM 4 and 5 input models, as well as SC 5 input models. In the paper [2], three statistical models—ARIMA, SARIMAX, and Holt Winter model—two machine learning models—linear regression and random forest model—and one deep learning model—long short-term memory (LSTM)—were used by the author to determine the CO2 emission effect in India for the next 10 years [3]. In this paper, author used artificial neural network (ANN) and adaptive neuro-fuzzy inference system (ANFIS) models to calculate the daily carbon monoxide (CO) content in the atmosphere of Tehran City. In this research, the author concluded that FS-ANN provides more appropriate result than FS-ANFIS for predict daily CO concentration [4]. The author investigates the organization of IoT-produced sensor data from diverse sites in the National Capital Region (NCR). Using a linear regression model, this data predicts the air quality for the following day. This research discusses AQI prediction based on data provided by IoT systems.

In this paper, it is based on the concept of time series analysis (TSA) models [5]. The primary goals of these models are to recognize the phenomena under research using data trend and periodicity reconstruction, as well as to count the next value of time series model. This study proposes a hybrid multi-step-ahead forecasting model. This framework may be used with a variety of prediction techniques, including LSSVR and LSTM, which are explored in this article [6]. In this research to analyze better carbon monoxide content in Bay of Algeciras (a city in Spain) author used two-stage procedure. The experiment demonstrates that two-stage models beat single models when it comes to estimating CO concentrations [7]. Deep learning model to forecast air pollution. ML was used in this study to identify the best model features and to assess the appropriateness of the employed delayed values [8]. Author used machine learning models to identify and predict PM2.5 levels. The results suggest that machine learning models (logistic regression and auto-regression) may be utilized effectively to identify air quality and forecast PM2.5 levels in the future [9].

Review paper on air quality prediction using machine learning [10]. They read and study the different algorithms and concluded that the neural network and boosting model outperform all other methods [11]. Author compares the different machine learning algorithm. For the prediction of CO concentrations, a total of 11 predictive models were created utilizing both single-based learning algorithms and meta-learning ensemble approaches using a well-known toolkit called Waikato Environment for Knowledge Analysis (WEKA) [12]. In this research, author used machine learning model, i.e., support vector machine, which accounts for the content of carbon monoxide in the atmosphere. To predict CO concentrations, in this investigation, 11 input vectors were loaded into the SVR model. As a result, the SVR model's tuning parameters C and g, as well as the RBF kernel function's tuning parameters g, must be modified.

In their research, author used one machine learning model support vector machine to predict carbon monoxide concentration in the atmosphere of Tehran [13]. To forecast average daily CO concentrations, two types of SVM regression models, namely - ε-SVM and ν-SVM approaches, were applied. In this research, nonporous gold (NPG) performs exceptionally well in CO oxidation. To handle high-dimensional inputs, neural networks are a frequent and efficient deep learning approach. This may express complex functions with basic feature engineering inputs, eliminating the laborious development of higher-order features [14].

In this research, the proposed model's the primary purpose is to simulate and estimate the amount of CO emitted by autos at a given time and location [15]. In this work, a hybrid model was created by combining three models, i.e., CFS, SVR, and GIS. The hybrid model predicted vehicular CO correctly (80.6) and had the lowest RMSE [16]. In this research, author talked about the short-term evaluation of carbon monoxide. The static feedforward network and the NARX model, two different types of neural networks, were used. In this result, the NARX performs better than static feedforward network. In this research, author discussed the online SVM and conventional SVM. In this research, the author presents an online SVM model whose fundamental difference from the standard SVM model is the method in which the data is delivered, i.e., the data are supplied in sequence for the proposed online SVM model, whereas they are supplied in batch for the traditional SVM model. Author declares that online SVM is much better than conventional SVM [17].

In their research, author employs a multivariate time series model to analyze and forecast carbon monoxide air concentrations [18]. Bayesian multivariate time series model is used for better performance. By using Gibbs sampling, Bayesian analysis is performed [19]. In their research, the author reflects the idea of nonlinear time series analysis technique mainly to predict the ambient carbon monoxide content in the environment. ARIMA, ANN, and NRMSE models were used to predict the concentration of CO. In this paper [20], author discussed the carbon monoxide evaluation in the atmosphere using an algorithm based on artificial neural network and multiple regression. In this research, gray relationship analysis is used as input variable in artificial neural network. After investigation in this paper, it is concluded that neural network model is best-fitted model as compared to regression model [21]. In their journal, author discusses how the CO is dangerous for health? How it is killing the live. Carbon monoxide is harmful because it has a higher affinity for hemoglobin than oxygen, resulting in considerable oxygen displacement from the hemoglobin molecule and a decrease in levels of blood oxygen [22]. In their research, author reviews about the carbon monoxide intoxication. CO quickly attaches to Hb, resulting in the production of carboxyhemoglobin (COHb). The capacity of carrying oxygen of blood get declines, which turned in to result tissue hypoxia. Because the clinical indications and symptoms of CO poisoning are unclear, all prospective victims should be treated with O_2 inhalation as soon as their blood COHb content is determined [23]. In their research, author deals with impact of carbon monoxide on public health. CO's health effects vary depending on the concentration and length of exposure. The treatment of CO-poisoned patients begins with removing the patient from the poisonous environment and providing pure oxygen to expedite CO elimination and increase tissue oxygenation [24]. In their research, author reviews the paper about carbon monoxide sensing technology for next-generation cyber and physical system. The cyber-physical system (CPS) is a novel theoretical concept that combines the cyber world (computation) and the physical world into a single modeling framework.

3 Methodology

Auto-Regressive Integrated Moving Average (ARIMA): ARIMA models are the model in which the value of a variable in current period is related to its own value in the previous as well as value of the residual in the previous period, i.e., ARIMA is combination of both auto-regressive terms (p) and moving average (q) term [2]. An ARIMA model is defined by three terms: p, d, and q. P denotes the order of the

AR term, q the order of the MA term, and d the number of differencing necessary to make the time series stable [19].

Auto-Regression (AR): Auto-regression is defined as it is a time series model that uses previous observation time steps as input for a regression equation to predict the value in the following time steps. This model assumes that previous time step results may be used to forecast the value at the following time step. It is very simple and easy model to forecast on a time series problem.

Moving Average (MA): It represents that the current values depend on some other factor of previous values.

$$\text{Target Output} = \text{Input}_1 * \text{Weight}_1 + \text{Input}_2 * \text{Weight}_2 + \cdots + \text{Input}_n * \text{Weight}_n + \text{Bias}$$

Here,

n = total number of input feature.
Weights = learned regression coefficient during training the model based on train data.
Bias = intercept of linear regression. Input is previous data.

Long Short-Term Memory (LSTM): This model belongs to deep learning technique which obeys the rule of recurrent neural network (RNN). This model was chosen because it is one of the most effective models for analyzing and predicting time series data. LSTM was created to prevent the issue of long-term reliance. It has capacity to remember information for long term. This is characterized as default behavior of LSTM. It is not a hard work to learn for this model. In LSTM, the key cell state is very much popular. It is like a conveyor belt. LSTM has ability to add or remove information to the cell state which is carefully regulated by the structures called gates. It remembers the previous information and uses it for processing the current input [2, 6].

Holt Winter Model: It is very easy model that deals with the behavior of time series data. It has three components, i.e., level, trend (growing or decreasing), and seasonal (repetitive pattern). It doesn't apply on random data points. It is defined by its three order parameters, alpha, beta, and gamma.

Alpha—coefficient of level smoothing
Beta—coefficient for the trend smoothing
Gamma—coefficient for the seasonal smoothing.

This model includes two variations based on the seasonal component of data. One among them is additive Holt Winter model, while another is the multiplicative Holt Winter model.

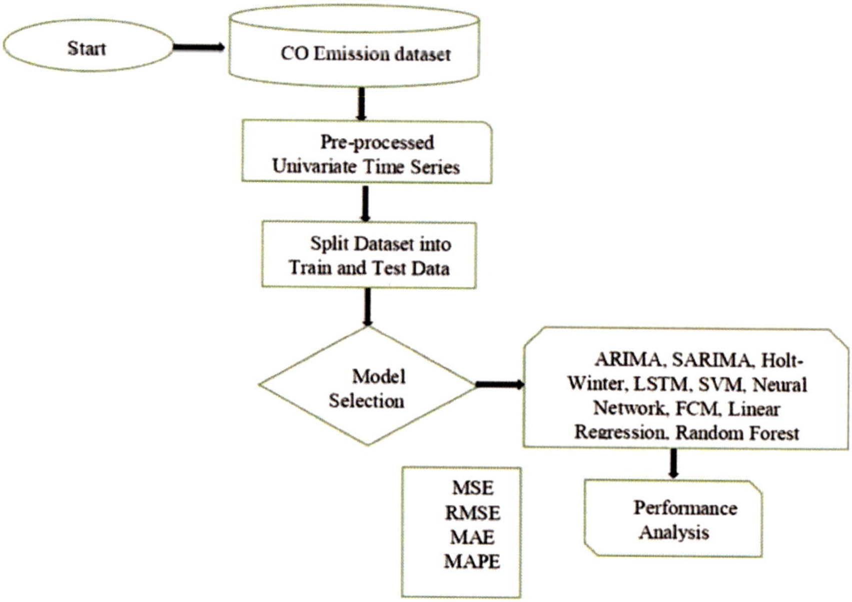

Fig. 1 Flowchart of methodology

Fuzzy Cognitive Maps (FCMs): Fuzzy cognitive maps (FCMs) are a modeling technique which is based on knowledge and experience. It condenses the greatest characteristics of fuzzy logic and neural networks into a graphical model made of nodes-concepts (system-descriptive components) linked by weighted edges. (representing the cause-and-effect relationships among the concepts) [1].

Support Vector Machine (SVM): SVMs are supervised learning algorithms for classification, regression, and outlier identification. Both linear and nonlinear classification can be performed by the SVM. The Kernel function is used to perform nonlinear classification. The SVM performs admirably thanks to the careful selection of kernel. The SVM method is widely regarded as the most effective method for diagnosing coronary diseases. The SVM approach is largely recognized as the most effective way for detecting coronary artery disease [13] (Fig. 1).

4 Performance Matrices

In this part, we discussed the different performance metrics used to analyze the effective scenario of the models. This technic predicts CO emissions, which is a type of regression issue. There are various evaluation techniques present, and here we use some important among them. To calculate the different error for performance analysis, we must define residual error. Residual error is equal to $(y - y')$, where y indicates actual value and y' indicates predicted value.

Mean Square Error (MSE): By using mean square error, it is easy to calculate the square of the residual error which is defined above for each data point and then calculate the average.

$$\text{MSE} = \frac{1}{n}\sum_{i=1}^{n}(Y_i - \widehat{Y}_i)^2 \tag{1}$$

where

n = number of data points
Y_i = observed value
$\widehat{Y}_i$ = predicted value.

Root Mean Square Error (RMSE): It is same as to mean square error but only difference is that it is a square root of MSE.

$$\text{RMSE} = \sqrt{\frac{1}{n}\sum_{i=1}^{n}(Y_i - \widehat{Y}_i)^2} \tag{2}$$

Mean Absolute Error (MAE): It is defined as the average of all absolute error.

$$\text{MAE} = \frac{1}{n}\sum_{i=1}^{n}|X_i - X| \tag{3}$$

where

Xi = observed value
X = predicted value.

Mean Absolute Percentage Error (MAPE): It is defined as the mean or average of absolute percentage error of forecast.

$$\text{MAPE} = \frac{1}{n}\sum_{i=1}^{n}\left|\frac{A_t - F_t}{A_t}\right| \tag{4}$$

Here,

n = the number of fitted point
A_t = the actual value
F_t = the forecast value.

5 Analysis of Result Obtained by Various Researcher with Different Algorithms

See Figs. 2 and 3 and Table 1.

6 Conclusion

Lots of work has been done and different algorithms used to predict the future concentration of carbon monoxide. Different algorithm provided different outcomes with respect to their criterion. As we discussed above that CO is dangerous and poisonous gas for living organism so it becomes more practical to know about the concentration of CO in our environment and try to control it. In this paper, effect of carbon monoxide on different area is also discussed. Not only the prediction of CO but also some other effective role of CO in the environment. In future, many more works related to the prediction of CO concentration should be enhanced. In this paper, there are some important models/algorithms and are used which gives the different result and helps to predict the concentration of CO in atmosphere. Here different dataset is used for different algorithm, and hence, we got different result. That model has been declared best fitted which has less value of MSE and RMSE. The basic structure of this review paper is based on the comparison of different algorithm result. Some important algorithm result has been discussed and compared it with another algorithm result [2]. In this, the author used 40 years data of CO_2 and greenhouse gas emission and try to predict the concentration of next 10 year. In their research, MSE and RMSE values are 1099.065 and 1286.177, respectively. In most cases, author used root mean square error (RMSE) and mean absolute error (MAE) to predict the accuracy of model. Less the error rate, the prediction will be more accurate. Similarly, some other researchers used different algorithms like FCM, SC, neural network, SVM, and SVR. Their results are discussed in the above table. Till now, I am able to justify my review on the CO concentration, i.e., among the different algorithms like time series forecasting algorithm, deep learning, and many more, and the more suitable models are LSTM and Holt Winter.

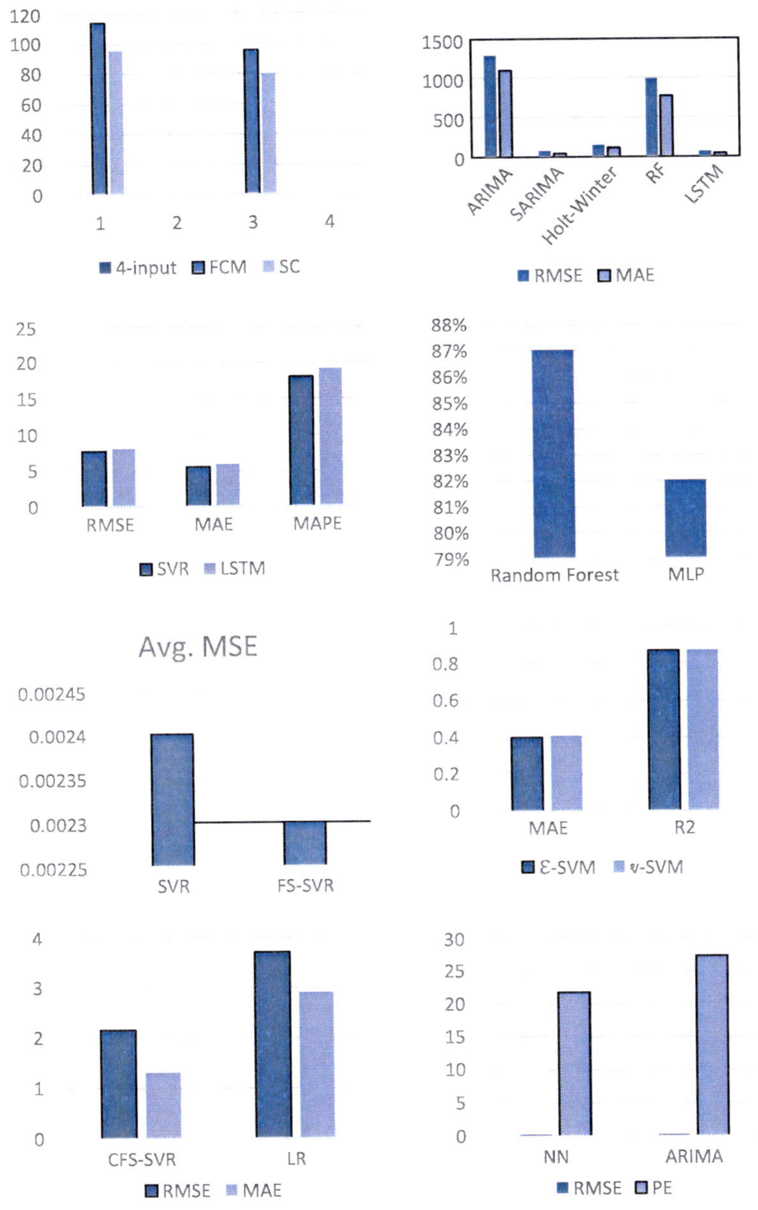

Fig. 2 Bar graph of above eight comparative analysis

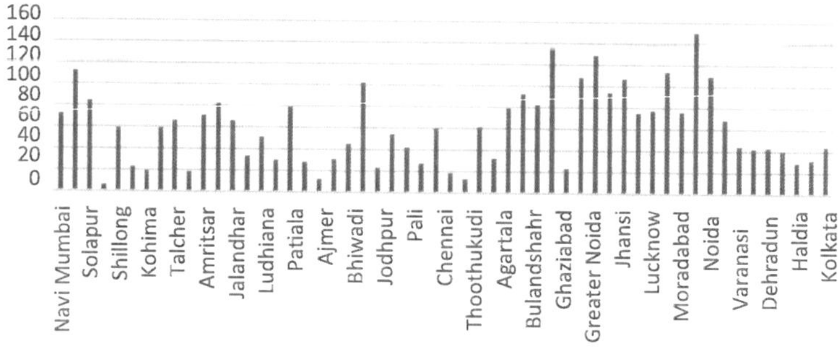

Fig. 3 Carbon monoxide concentration (avg.) in Indian cities

Table 1 Comparative analysis of some important model/algorithm

Ref. No.	Title	Model/algorithm	Result/accuracy			
[1]	Carbon monoxide prediction with parameters of air quality and fuzzy logic for Konya: A case study of Meram	FCM, and SC	**4-input**	**FCM**	**SC**	
			RMSE	114.078	95.357	
			MAE	95.924	79.622	
[2]	Time Series Models based on Machine learning for Effective emission of CO_2 prediction in India	ARIMA, SARIMAX, Holt Winter model, Linear Regression, Random Forest Long Short-Term Memory (LSTM)		**RMSE**	**MAE**	
			ARIM A	1286.177	1099.065	
			SARI MA	68.221	44.155	
			Holt Winter	144.12	113.441	
			RF	987.755	771.631	
			LSTM	60.635	45.524	
[6]	An ensemble multi-step- ahead forecasting system for fine particulate matter in urban areas	Neural network, Least Square SVR, Long short-term memory (LSTM)		**RMSE**	**MA E**	**MAP E**
			SVR	7.81	5.562	18.075
			LST M	8.118	5.903	19.075

(continued)

Table 1 (continued)

Ref. No.	Title	Model/algorithm	Result/accuracy		
[11]	Modeling the atmospheric concentration of carbon monoxide by using ensemble learning algorithm	Random forest, Multi- layer Perceptron	**Accuracy**		
			Random Forest	87%	
			MLP	82%	
[12]	Reliable prediction of carbon monoxide using developed support vector machine	SVR, and FS-SVR		**SVR**	**FS-SVR**
			Avg. MSE	0.0024	0.0023
[13]	Prediction of CO in the Tehran City with the help of support vector machine	ε-SVM, ν-SVM		**ε-SVM**	**ν-SVM**
			MAE	0.40	0.41
			R^2	0.87	0.87
[15]	CO emission prediction produced by the vehicles by using support vector regression	CFS-SVR, LR		**CFS-SVR**	**LR**
			RMSE	2.156	3.7032
			MAE	1.317	2.9067
[19]	Using nonlinear time series analysis technique Prediction of ambient carbon monoxide concentration	NN, ARIMA		**NN**	**ARIMA**
			RMSE	0.27	0.31
			PE	21.82	27.28

References

1. Demir, E. A. Ç. V., & Sevimli, M. F. (2022). Carbon monoxide forecasting with air quality parameters and fuzzy logic for Konya: A case study of Meram. March, 2020–2023.
2. Singh, S. K., & Kumari, S. (2022). Machine learning-based time series models for effective CO_2 emission prediction in India.
3. Noori, R., Hoshyaripour, G., Ashrafi, K., & Araabi, B. N. (2010). Uncertainty analysis of developed ANN and ANFIS models in prediction of carbon monoxide daily concentration. *Atmospheric Environment, 44*(4), 476–482. https://doi.org/10.1016/j.atmosenv.2009.11.005
4. Kumar, R., Kumar, P., & Kumar, Y. (2020). Time series data prediction using IoT and machine learning technique. *Procedia Computer Science, 167*(2019), 373–381. https://doi.org/10.1016/j.procs.2020.03.240

5. Guarnaccia, C., Bretón, J. G. C., Quartieri, J., Tepedino, C., & Bretón, R. M. C. (2014). An application of time series analysis for predicting and management of carbon monoxide concentrations. *International Journal of Mathematical Models and Methods in Applied Sciences, 8*(1), 505–515.
6. Ahani, I. K., Salari, M., & Shadman, A. (2020). An ensemble multi-step-ahead forecasting system for fine particulate matter in urban areas. *Journal of Cleaner Production, 263*, 120983. https://doi.org/10.1016/j.jclepro.2020.120983
7. Turias, I. J., et al. (2017). Prediction of carbon monoxide (CO) atmospheric pollution concentrations using meterological variables. *WIT Transactions on Ecology and the Environment, 211*, 137–145. https://doi.org/10.2495/AIR170141
8. Heydari, A., Majidi Nezhad, M., Astiaso Garcia, D., Keynia, F., & De Santoli, L. (2022). Air pollution forecasting application based on deep learning model and optimization algorithm. *Clean Technologies and Environmental Policy, 24*(2), 607–621. https://doi.org/10.1007/s10098-021-02080-5
9. Aditya, C. R., Deshmukh, C. R., Nayana, D. K., Gandhi, P., & Astu, V. (2018). Detection and prediction of air pollution using machine learning models. *International Journal of Engineering Trends and Technology, 59*(4), 204–207. https://doi.org/10.14445/22315381/ijett-v59p238
10. Madan, T., Sagar, S., & Virmani, D. (2020). Air quality prediction using machine learning algorithms-A review," In *Proceedings—IEEE 2020 2nd international conference on advances in computing, communication control and networking, ICACCCN 2020*, pp. 140–145. https://doi.org/10.1109/ICACCCN51052.2020.9362912
11. Masih, A. (2018). Modelling the atmospheric concentration of carbon monoxide by using ensemble learning algorithms. *CEUR Workshop Proceedings, 2298*(February).
12. Moazami, S., Noori, R., Amiri, B. J., Yeganeh, B., Partani, S., & Safavi, S. (2016). Reliable prediction of carbon monoxide using developed support vector machine. *Atmospheric Pollution Research, 7*(3), 412–418. https://doi.org/10.1016/j.apr.2015.10.022
13. Akbarzadeh, A., Vesali Naseh, M., & Node Farahani, M. (2020). Carbon monoxide prediction in the atmosphere of tehran using developed support vector machine. *Pollution, 6*(1), 43–57.
14. Zhou, Y., Deng, H., Huang, X., Hu, Y., Ye, B., & Lu, L. (2021). Predicting the oxidation of carbon monoxide on nanoporous gold by a deep-learning method. *Chemical Engineering Journal, 427*(May 2021), 131747. https://doi.org/10.1016/j.cej.2021.131747
15. Azeez, O. S., Pradhan, B., & Shafri, H. Z. M. (2018). Vehicular CO emission prediction using support vector regression model and GIS. *Sustainability, 10*(10). https://doi.org/10.3390/su10103434
16. Mohebbi, M. R., Karimi Jashni, A., Dehghani, M., & Hadad, K. (2019). Short-term prediction of carbon monoxide concentration using artificial neural network (NARX) without traffic data: Case study: Shiraz City. *Iranian Journal of Science and Technology, Transactions of Civil Engineering, 43*(3), 533–540. https://doi.org/10.1007/s40996-018-0210-4
17. Wang, W., Men, C., & Lu, W. (2008). Online prediction model based on support vector machine. *Neurocomputing, 71*, 550–558. https://doi.org/10.1016/j.neucom.2007.07.020
18. Tonellato, S. F. (2001), A multivariate time series model for the analysis and prediction of carbon monoxide atmospheric concentrations. *Journal of the Royal Statistical Society: Series C (Applied Statistics), 50*(2), 187–200. https://doi.org/10.1111/1467-9876.00228
19. Chelani, A. B., & Devotta, S. (2007). Prediction of ambient carbon monoxide concentration using nonlinear time series analysis technique. *Transportation Research Part D: Transport and Environment, 12*, 596–600. https://doi.org/10.1016/j.trd.2007.07.006
20. Shams, S. R., Jahani, A., Moeinaddini, M., & Khorasani, N. (2020). Air carbon monoxide forecasting using an artificial neural network in comparison with multiple regression. *Modeling Earth Systems and Environment, 6*(3), 1467–1475. https://doi.org/10.1007/s40808-020-00762-5
21. Byard, R. W. (2019). Carbon monoxide—the silent killer. *Forensic Science, Medicine and Pathology, 15*(1), 1–2. https://doi.org/10.1007/s12024-018-0040-5
22. Prockop, L. D., & Chichkova, R. I. (2007). Carbon monoxide intoxication: An updated review. *Journal of the Neurological Sciences, 262*(1–2), 122–130. https://doi.org/10.1016/j.jns.2007.06.037

23. Raub, J. A., Mathieu-Nolf, M., Hampson, N. B., & Thom, S. R. (2000). Carbon monoxide poisoning—A public health perspective. *Toxicology, 145*(1), 1–14. https://doi.org/10.1016/S0300-483X(99)00217-6
24. Nandy, T., Coutu, R. A., Ababei, C. (2018). Carbon monoxide sensing technologies for next-generation cyber-physical systems. *Sensors (Switzerland), 18*(10). https://doi.org/10.3390/s18103443

Investigating Effects of Concentration, Angle, Diameter and Impact Velocity of Soap Film and Soap Droplet on Trajectory of Droplet

Xinxuan Hou and Yuzhe Jiang

Abstract Soap is an essential item in our lives for housekeeping and has a variety of applications, including the making of soap films. Soap films are known for their ability to filter particles by size, but little research is done about soap films in the shape of a bubble (soap bubbles), and how soap droplets interact with it. A 0.05 mL soap droplet dropped onto the soap bubble under different sets of conditions. Our experiment aims to investigate the factors that determine whether a soap droplet dropped onto the soap bubble passes through or is deflected off the soap bubble. After collecting our experimental data, we used machine learning for modelling. Predictions for points in the range of experimental conditions (width 2–12 cm, distance 1–19 cm, angle 45°, 60°, 90° and concentration of soap droplet and bubble 4% and 6%) were generated. Our experiment demonstrates that independent factors such as greater height width, impact velocity and angle have a positive correlation with penetration. Our findings also show that a concentration of 6% w/w droplet and 4% w/w bubble provided the most significant probability of penetration. Such findings are all aligned with our hypothesis, based on our knowledge on the matter. Soap films have a hydrophilic head and a hydrophobic tail, just like our cell membranes. By diving deeper into fluid mechanics and providing a model of this experimental design, our work may provide insights on the movement of molecules through our biological cells, which may be useful for future generations of engineers and biologists.

Keywords Soap bubble · Soap film · Soap droplet · Fluid mechanics

1 Introduction

In our daily lives, we see common examples of raindrops striking rain bubbles and beer being poured on beer bubbles. Surprisingly, most of these bubbles do not break because of the interactions between the falling liquid and the film/bubble [1].

Understanding soap films is useful for a variety of applications. In the context of fluid mechanics, soap films provide the closest substitutes for a two-dimensional

X. Hou · Y. Jiang (✉)
NUS High School of Mathematics and Science, Singapore, Singapore

H. Guo et al. (eds.), *IRC-SET 2022*,
https://doi.org/10.1007/978-981-19-7222-5_10

environment, enabling scientists to test features of the Navier–Stokes equations. An industrial application of soap films is curtain coating, a process that creates an uninterrupted curtain of fluid that falls onto an object. Another is foam, a system of connected spherical films that one is often trying to minimize or generate in a chemical process. A deeper analysis of the dynamics of soap films is crucial for the understanding of foam. Soap films are also relevant to lipid membranes [2]. Both soap and lipid molecules contain hydrophilic heads and hydrophilic tails. Soap films can be imagined as inverted larger versions of the cell plasma membrane. The inner and outer surfaces of the soap film are hydrophobic. In comparison, the extracellular matrix in the cell in contact with the outer membrane and the inner contents of liposomes are hydrophilic [3]. Liquid droplets falling onto soap films are like endocytosis [4] (specifically pinocytosis). The droplets are coated with the soap film, similar to large particles entering the cell being surrounded by the plasma membrane, forming a vesicle in the cell. Thus, such experiments may provide insights on how molecules enter the biological cell.

This research will study whether concentration of soap films and soap droplets, size of soap bubble, angle of incidence and impact velocity of droplet affect whether the droplet passes through or bounces off the soap bubble. These factors are chosen as they displayed both penetration and deflection possibilities when a discovery experiment was carried out on multiple factors to determine their suitability for more experimentation.

2 Related Work

In this section, we will discuss existing literature that model interactions between soap film and soap droplets.

The issue of grouping the various variables that lead to the outcome of interactions between soap film and soap droplet has been partly addressed by introducing Weber number, which describes the ratio of kinetic energy to surface tension. Such a model aims to directly correlate different Weber numbers to different outcomes such as packing and collapsing [5]. Our study aims to use a neural network approach instead.

During droplet penetration, the translational kinetic energy of the droplet is converted into deformational potential energy of the surface tension for the soap bubble. The energy is partly restored to the translational motion following impact. The remaining energy is converted into internal drop motion, including waves and oscillations.

When droplets hit surfaces at an angle, the contact time is reduced by modifying the spreading and retraction process of impacting drops. During impact, the drop promotes a larger spreading in the tangential direction than in the lateral direction, followed by a fast retraction in the lateral direction and finally leaves the surface in an elongated shape. This results in droplet asymmetric bouncing at an angle [6]. In this paper, a movable soap film is used instead of a rigid surface, providing more possibilities.

3 Methodology

The soap used to make bubbles is composed of glycerine, sodium dodecyl sulphate (SDS) and deionized water (DI water). For 4% concentration, a mixture of 4% w/w glycerine, 15% w/w SDS and 81% w/w DI water was used. Similarly, for 6% concentration. A mixture of 6% w/w glycerine, 15% w/w SDS and 79% w/w DI water was used. The SDS used was in 500 g packaging manufactured by Kimic Chemitech. Using an electronic weighing scale and weighing boat, we weighed the correct mass of the materials needed and mixed them on a plastic plate. After the SDS was completely dissolved in DI water with no visible signs of SDS residue remaining, the first step of the experiment is carried out by dipping a circular plastic ring onto the plastic plate. It is carefully lifted out of the plastic plate. Following this, a sweeping upwards motion is carried out to detach the soap film from the plastic ring to form a circular soap film (soap bubble). The soap bubble is caught by the plastic ring and held by hand. The hand is rested on the table to ensure the stability of the bubble. Next, a soap droplet is dropped onto the soap bubble using a syringe filled with soap water fixed onto the retort stand. The observations are recorded by a phone with a slow-motion filter.

The video recording of the process was analysed with Tracker. The data was collected and analysed using Python with the scikit-learn module via neural networks. Firstly, the data collected was randomly divided into two lists of 127 data points (80%) for training data and 32 data points (20%) for test data. Next, the training data was used to train the computer to produce 10,000 neural networks with different configurations consisting of two hidden layers, with nodes ranging from 1 to 100 for each layer.

Lastly, they were tested with the 32 data points and given an accuracy score based on the number of data points they could accurately predict. The neural network with the highest accuracy score of 87.5%, with the first hidden layer having four nodes and the second hidden layer with three nodes were chosen. It was then used to generate predictions for any point in the range of experimental conditions, providing the graphs in the results.

The variable distance and width are the distance from the tip of the syringe to the point where the droplet strikes the bubble and the width of the bubble, respectively. The variable angle is defined by the altitude of the trajectory of the droplet with respect to a tangent plane of the spherical bubble at the point the droplet strikes the bubble, as shown in Fig. 1.

4 Results

To analyse the graphs in Figs. 2 and 3, we used Excel and plotted 4 graphs. They are the number of data points with droplet penetration for distances 1–19 cm, widths 6–12 cm, angles 45°, 60° and 90° and the concentration (of droplet and bubble). As

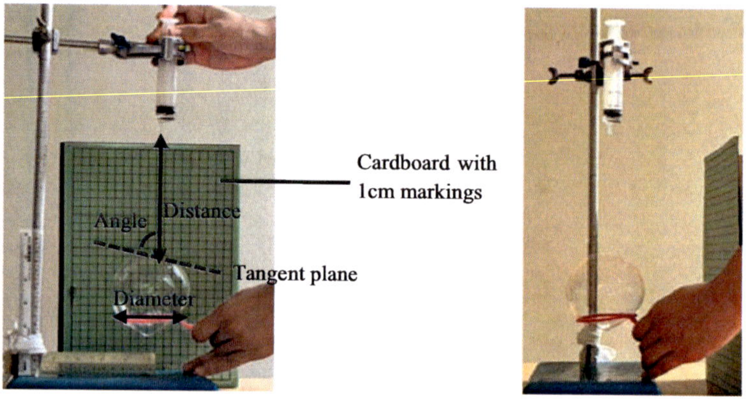

Front view of set-up

Side view of set-up

Fig. 1 Experimental set-up

shown in Fig. 4a, when the width increases from 7 to 11 cm, the total number of instances where the droplet penetrates through the bubble increases from 120 to 135.

In Fig. 4b, when distance increases, the total number of instances of droplet penetration increases linearly from 17 to 113.

For angle, as shown in Fig. 4c, 45° gives the lowest penetration of 266, followed by 335 for 60°, and 792 for 90°.

For concentration, when droplet concentration is lower than bubble concentration, in the case of 4% droplet concentration and 6% bubble concentration, it results in the lowest number of penetrations, 206. When droplet concentration and bubble concentration are equal, 4% droplet concentration and 4% bubble concentration results in a value of 360, comparable to 371 from 6% droplet and bubble concentration. When the droplet concentration is greater than the bubble concentration, 456 penetrations result from 6% droplet concentration and 4% bubble concentration.

4.1 Key Findings

- Higher distance is equivalent to a higher likelihood of droplet penetration.
- Larger angle is equivalent to a higher likelihood of droplet penetration.
- Higher width is equivalent to a higher likelihood of droplet penetration.
- Droplets tend to penetrate when the droplet concentration is greater than the bubble concentration.

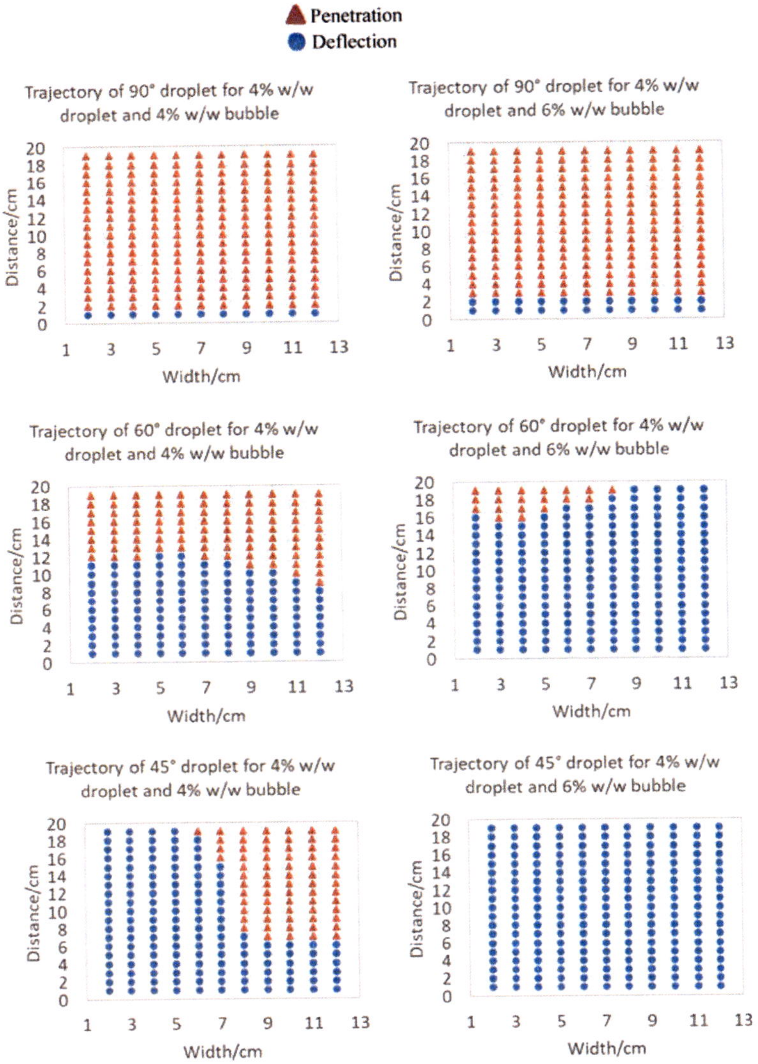

(a) 4% conc. droplet, 4% conc. bubble (b) 4% conc. droplet, 4% conc. bubble

Fig. 2 Results for 4% droplet concentration (conc.)

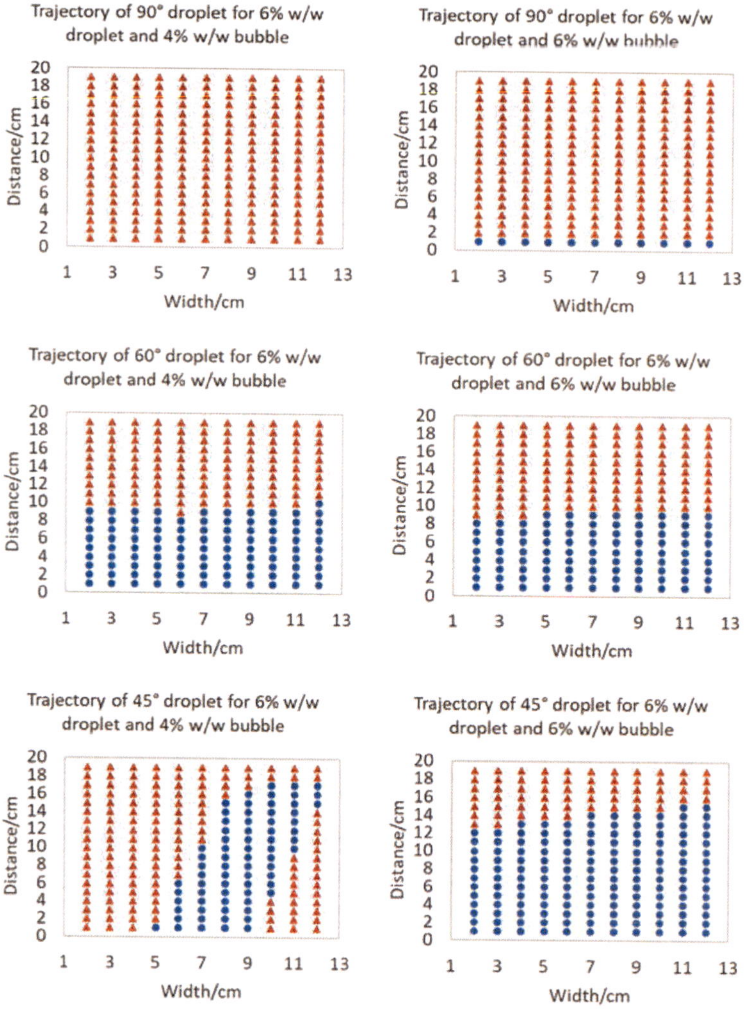

(a) 6% conc. droplet, 4% conc. bubble (b) 6% conc. droplet, 6% conc. bubble

Fig. 3 Results for 6% droplet concentration (conc.)

5 Discussion

5.1 Higher Distance is Equivalent to a Higher Likelihood of Droplet Penetration

The droplet deforms the soap film when it hits the soap bubble. When the height of the fall of the droplet is small, the impact velocity of the droplet is low. The initial kinetic

Fig. 4 Number of data points with droplet penetration for various factors

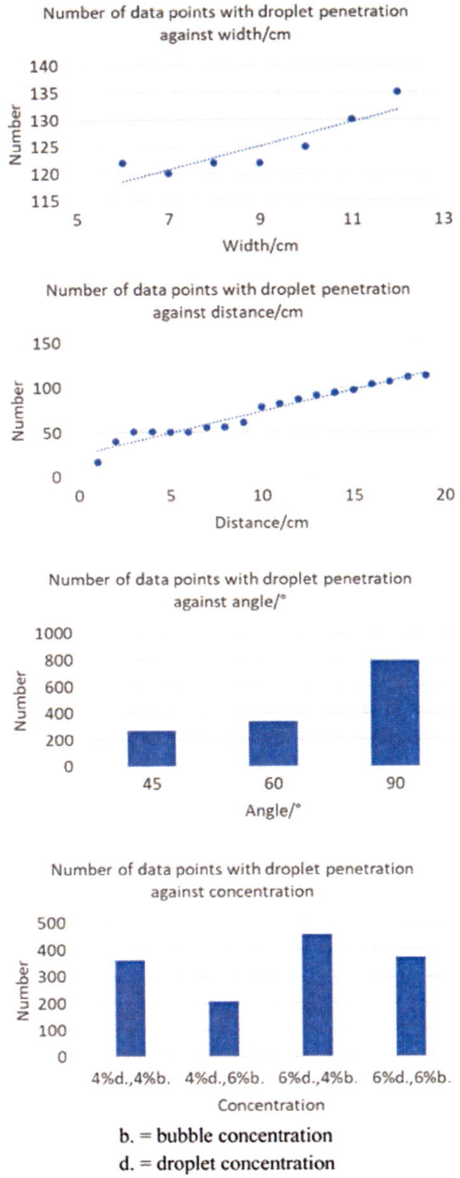

energy of the droplet is partially converted into potential energy of deformation of the film, which acts like a liquid trampoline: the droplet bounces off the film. However, at greater distances, the high initial kinetic energy causes large deformation of the film such that the bubble's shape holding the droplet becomes unstable [7]. The droplet's leading-edge coalesces with the film and the droplet is embedded in the film. It exits

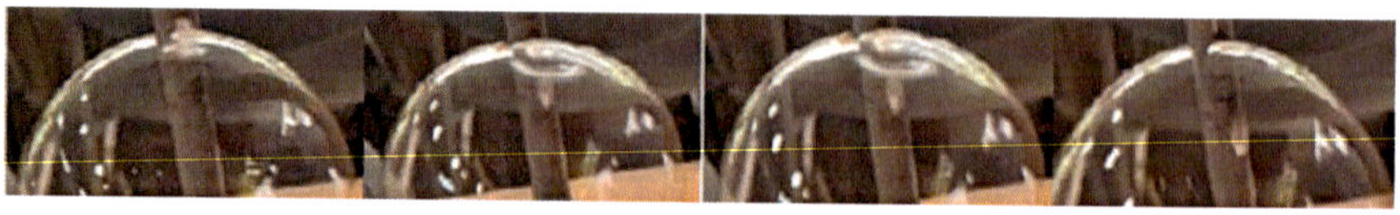

Fig. 5 4 Snapshots of soap droplet passing through soap bubble

the bubble and its trailing-edge emerges without breaking the film, which is said to 'self-heal', resulting in droplet penetration.

For a droplet with width *W* and impact velocity V striking a soap bubble with surface tension σ, the Weber number, We = $\rho V2W/\sigma$, prescribes the relative magnitudes of the initial kinetic energy of the droplet and the interfacial energy of the system. At low We, the droplet bounces off the soap bubble, while for high We, it passes through without bursting the bubble. At higher distances, *V* is larger, causing We to be high. This is consistent with the figures, where droplets tend to penetrate at higher distances.

5.2 *Larger Angle is Equivalent to a Higher Likelihood of Droplet Penetration*

As the angle increases, the droplet is in contact with a smaller surface area of the bubble. The kinetic energy of the droplet remains the same as the angle increases while the surface area of the bubble in contact with the droplet decreases during penetration. The force of the droplet on the surface of the bubble increases, increasing the impulse acting on the bubble surface. This increase in impulse causes the structure of the bubble to become unstable and deforms to allow the droplet to penetrate the bubble more easily (Fig. 5).

5.3 *Higher Width is Equivalent to a Higher Likelihood of Droplet Penetration*

The gas pressure inside a small bubble is higher than in a large bubble; therefore, the surface tension of a small bubble will be higher. Surface tension is the property of the surface of a liquid that allows it to resist an external force, due to the cohesive nature of its molecules [5]. Liquids with high surface tension exhibit significant resistance to penetration. Smaller-sized bubbles are more stable and droplets are less likely to penetrate when dropped onto a smaller-sized soap bubble. This is consistent with higher Weber numbers allowing droplets to pass through the bubble [8]. As Weber number represents the ratio of kinetic energy to surface tension, a greater surface tension results in a lower weber number, and therefore, the droplet is less likely to penetrate the smaller bubble.

5.4 *Droplets Tend to Penetrate When the Droplet Concentration is Greater than the Bubble Concentration*

This is an unexpected result. One possible explanation for this would be that a higher droplet concentration causes the droplet to become elongated vertically and shortened horizontally, causing the droplet to become needle like shaped, reducing the surface area of the droplet in contact with the bubble during collision; therefore, greater force is applied to deform the shape of the bubble. When the bubble concentration is low, the thickness of the bubble is greater and therefore deforms more easily as the upper layer of the bubble exerts force on the lower layer.

Our study has reported experimental results of droplets colliding with soap bubbles. The four factors explained in the results play a huge role in determining the two outcomes of the experiment—deflection or penetration. In particular, a distance of 19 cm; an angle of 90°; a width of 12 cm; and 6% conc. droplet with 4% conc. bubble will result in greatest probability of penetration.

6 Limitations and Further Study

Due to wind and the fact that we used our hand to hold on to the ring, the soap bubble may move slightly, causing it to be deformed and not perfectly spherical. This may result in a difference in results compared to when the experiment is repeated under the standard conditions with perfectly spherical bubbles. Our future work would be to have more duplicates to ensure the reliability of the data, especially since the variable air pressure and air may have an impact on the results of the experiment. In addition, the ring can be secured by a retort stand instead of held by hand to reduce the movement of the soap bubble. The experiment can also be conducted in the laboratory with closed windows to ensure less variation in air pressure and air flow. The concentration of SDS that we used is only 4% and 6% for low and high concentrations, respectively. For future work, we can use a wider range of concentrations (less than 4% and more than 6%) to analyse the impacts of concentration on the trajectory of the droplet.

7 Conclusion

Many factors affect whether the droplet bounces or penetrates when it hits a soap bubble. The ones we covered in our research are size of bubble, impact velocity of droplet, concentration of droplet and bubble and angle of impact of droplet. Our findings may allow biologists to have a better understanding on plasma membranes and how they may be punctured. (For example, when the velocity of the particle colliding with the plasma membrane is high and the spherical cells are small.) Such knowledge may be useful in determining how molecules enter our biological cells. Membrane transport is essential for cellular life. As cells go through their lifecycle, the incorporation of biological molecules and the discharge of waste products is necessary for normal cellular function.

Appendix

Soap making steps

For 4% concentration (20 g):

1. Weigh 3 g of glycerine, 0.8 g of SDS and 16.2 g of water using a weighing machine.
2. Transfer them to a bowl and mix evenly using a glass rod.

For 6% concentration (20 g):

3. Weigh 3 g of glycerine, 1.2 g of SDS and 15.8 g of water using a weighing machine.
4. Transfer them to a bowl and mix evenly using a glass rod.

Bubble making process:

Step 1: Dip the bubble wand into the soap solution.

Step 2: Lift the bubble wand from the solution such that a bubble is formed.

Step 3: After the bubble is formed, let it fall and catch it using the wand.

References

1. Yang, Y.-J., Mei, C.-X., Zhang, X.-D., Wei, Y.-J., & Liu, S.-H. (2019). Kinematics and passing modes of a droplet impacting on a soap film. *Acta Physica Sinica, 68*, 156101.
2. Brasz, F. Soap films: Statics and Dynamics—Princeton University. Retrieved June 22, 2022, from https://www.princeton.edu/~stonelab/Teaching/FredBraszFinalPaper.pdf
3. Muller, P., Rudin, D. O., Tien, H. T., & Wescott, W. C. (1962). Reconstitution of excitable cell membrane structure in vitro. *Circulation, 26*, 1167–1171.
4. Stogin, B. B., Gockowski, L., Feldstein, H., Claure, H., Wang, J., & Wong, T. -S. (2018). Free-standing liquid membranes as unusual particle separators. *Science Advances, 4*.
5. Surface Tension and Water. Retrieved June 22, 2022, from https://www.usgs.gov/special-topics/water-science-school/science/surface-tension-and-water
6. Wang, H., Liu, C., Zhan, H., Liu, Y. (2019). Droplet asymmetric bouncing on inclined superhydrophobic surfaces. *ACS Omega* (pp. 12238–12243)
7. Courbin, L., & Stone, H. A. (2006). Impact, puncturing, and the self-healing of soap films. *Physics of Fluids, 18*, 091105.
8. Zou, J., Wang, W., Ji, C., & Pan, M. (2017). Droplets passing through a soap film. *Physics of Fluids, 29*, 062110.

Using IMU to Determine Velocity of Falling Object and Hence Develop Framework to Optimize Parachute Size for Falling UAV

Anhe Xin

Abstract With the ever-increasing popularity of using unmanned aerial vehicles (UAVs) as a means of delivery, inspection, videography and cinematography, the chances of them falling out of the sky and damaging property and causing injuries has never been greater. To solve this problem, one drone service company has experimented with parachutes. Hence, there is a need for a method to find the optimum size of parachute drones depending on their variables.

Keywords Inertial Measurement Device · IMU · Unmanned Aerial Vehicle · UAV · Parachute

1 Introduction

The popularity of unmanned aerial vehicles (UAVs), which are machines capable of flight without any human operators onboard [1], is ever increasing [2, 3]. Under the Department of Defence, UAVs which have a maximum take-off mass of under 20 pounds, an operational ceiling of less than 1200 ft, and a maximum operational speed of below 100 km are classified as small UAVs (SUAVs) [4–6]. A SUAV can take the form of a Quadcopter [7], which is known as a Quadcopter SUAV (QSUAV). The majority of registered UAVs in the United States are recreational [8] UAVs [9]. As recreational QSUAVs are likely flown by inexperienced operators, the chances of a QSUAV entering a free fall state [10] is greater than ever, meaning that the importance of safety has never been greater. One way to slow the descend of a QSUAV in free fall is through the use of a parachute [11–13]. This has been experimented with by the Swiss Post and Matternet using the Matternet M2V9 [14]. However, drones cannot come equipped with parachutes straight from the factory, as under different use cases the variables of the drone will differ, causing a different size of parachute to be the most optimal.

A. Xin (✉)
NUS High School of Mathematics and Science (NUS High), Singapore, Singapore
e-mail: anhe.xin@gmail.com

H. Guo et al. (eds.), *IRC-SET 2022*,
https://doi.org/10.1007/978-981-19-7222-5_11

The contribution of this paper is to use an inertial measurement device (IMU) to determine the speed of a falling object and hence calculates the optimal size of parachute for a falling object through the terminal velocity (V_t) equation.

This paper is organized as follows: Firstly, Sect. 2 explores the background of this research and the state of the art. Then, Sect. 3 proposes a solution with the help of the related work. Afterwards, Sect. 4 carries out an experiment according to the proposed solution. Finally, Sect. 5 discusses the limitations, assumptions and extensions of this paper.

2 Related Work

It is difficult to recover a drone when a loss of control is experienced in a limited space [15]. Hence, some method is needed to quickly slow a drone down without much input from the operator. One such way is by an airbag with drag flaps proposed in [16] by Cawthorn. However, that comes with the drawback that the system is not aerodynamically self-righting, which may lead to problems with the orientation of the drone being undesired when landing and bouncing in a seemingly random direction. The quick deployment time of the airbag may also cause injuries if improperly deployed. In [11], Al-Madani et al. designed a system that can automatically deploy a parachute from a drone when falling is detected. However, the article does not do any work on how to determine the dimensions of a parachute needed under different circumstances. Most attempts at an automated detection method to determine if a drone are in trouble comes with drawbacks. Kwong et al. proposed using fuzzy logic to determine if the sensors of a drone are operating normally in [17]. But, the responsiveness of a drone differs with the load it is carrying, and hence, the fuzzy logic may be mislead when the drone changes or releases its payload. Rehage et al. aimed to use hardware redundancy to reduce the chance that the drone will experience issues due to sensor failure in [18], at the cost of increased weight. A different solution was explored by Wiseman who used a camera mounted on the drone to detect possible abnormalities with the drone in [19] with the drawback that this might be mislead by the swinging and changes of payload and added weight. Hence, having the parachute deployed by the drone operator would be better as it does not add too much weight to the drone and will not be tricked by changes in mass or the natural effects of the payload. Given that, there has been work done to create a mathematical model of the IMU and parachute (assembly) such as the one by Shao et al. in [20]. Marcović et al. managed to use an IMU to quantify vertical jump height with the means of vertical velocity in [21]. This is limited by the fact that Marcović et al. used airtime to find out the velocity and distance, which would not work in the case of a falling object as the initial altitude of the object is not known and may vary. Obtaining translations from an IMU have already been done by Rahni and Yahya in [22] with the help of Bar-Itzhack and Fegley in [23]. With these translations, it would be possible to easily measure global acceleration values and find out if they are abnormal.

3 Proposed Solution

3.1 How to Obtain Data

To measure the horizontal, vertical, and rotational speed of the assembly a HW579 10 degree of freedom (DOF) IMU. The HW579 is comprised of a HMC5833 3DOF magnetometer, an ADXL345 3DOF accelerometer and a ITG3200 4DOF gyroscopic and thermometric sensor.

A LOLIN D32 Pro [24] microcontroller that can act as a WiFi and Bluetooth transceiver is used to obtain readings from the IMU. The microcontroller sends the readings through the MQTT protocol, where they are received by a ThingsBoard database run on a Raspberry Pi. The readings can then be retrieved from the database and analyzed.

The parachute was made using a superior brand plastic table cloth that is made from high density polyethylene [25]. One sheet of the tablecloth was cut into a square with sides of 70 cm. Then, the parachute was made according to method 2 of [26], with each string being made of nylon fishing line and having a length of 2 times the edge length of the parachute.

The microcontroller and IMU were attached to a 3D printed housing with double-sided tape (sensor). The nylon string was then cut in half, and the ends of each half were tied to the holes of the sensor. The assembly was then dropped 5 times from a height of approximately 5 m to obtain data that allow for calculation of the V_t of the parachute. This assembly amounted to a mass of 44g in total.

3.2 Maximal Impact Force

Humans can survive up to a g-force of 20 for time periods below 10 s according to [27]. The mean mass of the humans surrogates used by the NHTSA are 70 kg [28], which when experiencing a g-force of 60 will have been displaced by a force of approximately 12000N. A QSUAV is defined to have a mass of below 20lb or approximately below 9kg. Assuming that the mass of the QSUAV is 9 kg, and in a perfectly inelastic collision with a human, it takes 20 ms for the QSUAV to decelerate to a stop completely, the maximal velocity at which the QSUAV be travelling at when it hits a human is 31 $\mathrm{m\,s^{-1}}$. This means that the V_t of the assembly needs to be below 31 m/s.

3.3 Terminal Velocity

The equation for V_t is $V_t = \sqrt{\frac{2mg}{\rho A C_d}}$, where m is the mass of the assembly, g is the gravitational field strength acting on the assembly, ρ is the density of the medium the assembly is travelling through, A is the projected area of the assembly, and C_d is the drag coefficient of the assembly. By substituting in known values, Eq. 1 is formed.

$$V_t = \sqrt{\frac{2m \cdot 9.81}{1.23A \cdot C_d}} \tag{1}$$

To find the C_d of the assembly, an experiment was carried out to find the V_t of a square parachute with an area of 4900 cm^2.

4 Experiment

4.1 Experimental Design

The assembly would be dropped from the top of a playground with a height of approximately 5 m, during which the IMU would be collecting readings. However, due to the chance of unpredictable and uncontrollable events occurring such as changes in wind direction in the middle of a drop, the assembly would be dropped 5 times in total.

4.2 IMU Coordinate System to Global Coordinate System

To convert the IMU readings from the assembly coordinate system (S) to the global coordinate system (G), the method as mentioned in [22] was used. Firstly, the readings obtained by the accelerometer are expressed in an acceleration vector $\mathbf{a_S}$ as in Eq. 2.

$$\mathbf{a_S} = \begin{bmatrix} A_x \\ A_y \\ A_z \end{bmatrix} \tag{2}$$

where A_x, A_y and A_z are the acceleration in the x, y and z axis in S respectively. Then, the readings obtained by the gyroscope are expressed in the form of direction cosine matrix $R_{GS}(\theta)$ in Eq. 3. θ is defined in Eq. 4.

$$R_{GS}(\theta) = \begin{bmatrix} \cos R_y \cos R_z & \sin R_x \sin R_y \cos R_z - \cos R_x \sin R_z & \cos R_x \sin R_y \cos R_z + \sin R_x \sin R_z \\ \cos R_y \sin R_z & \sin R_x \sin R_y \sin R_z + \cos R_x \cos R_z & \cos R_x \sin R_y \sin R_z - \sin R_x \cos R_z \\ -\sin R_y & \sin R_x \cos R_y & \cos R_x \cos R_y \end{bmatrix} \quad (3)$$

$$\theta = \begin{bmatrix} R_x & R_y & R_z \end{bmatrix} \quad (4)$$

where R_x, R_y and R_z are the difference between the angle of the x, y and z axis in S and G, respectively. They can be calculated from the readings of the gyroscopic sensor and the magnetometer. Using Eq. 5, the acceleration in G can be determined and represented as the vector $\mathbf{a_G}$.

$$\mathbf{a_G} = R_{GS}(\theta)\mathbf{a_S} = \begin{bmatrix} G_x \\ G_y \\ G_z \end{bmatrix} \quad (5)$$

where G_x, G_y and G_z are the acceleration in the x, y and z axis in G respectively.

4.3 Data Processing

The data collected were read from the Thingsboard database as a csv file.[1] The csv file was split into the 5 different drops performed, and then, all the readings within each drop were smoothed with a method that was proposed by Savitzky and Golay in [29], the *Savitzky-Golay Filter*. The csv files were cropped to show only readings from the actual drop. The starting and ending indexes of the drops are shown in Table 1. Thereafter, the drift and zero errors in all readings were corrected by subtracting the average of all data points before the drop started. This also prevented Earth's gravitational field from interfering with the readings. Afterwards, the readings were converted to G using the method detailed in Sect. 4.2, and the vertical acceleration in G was integrated to obtain the velocity of the assembly, of which the greatest reading was taken to be the V_t.

Figures 1, 2, 3, 4 and 5 show the graphs of the vertical acceleration in red and vertical velocity readings obtained by the accelerometer in blue.

4.4 Results

Due to the diverse nature of the V_t readings, the mean of the readings is taken to be the V_t of the assembly, 8.24 m s^{-1}. With the assumption that the ratio of the projected area of the parachute when deployed to the area of the HDPE square that the parachute

[1] All Python programs (py files) and comma delimited files (csv files) mentioned in this section can be found in https://gitlab.com/XinAnhe/parachute/.

Fig. 1 The first drop

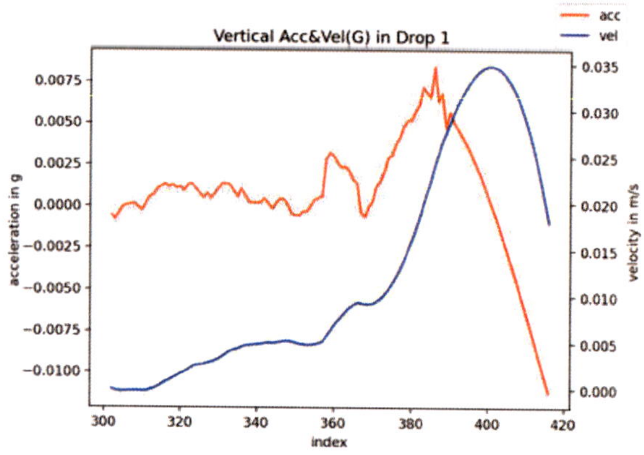

Fig. 2 The second drop

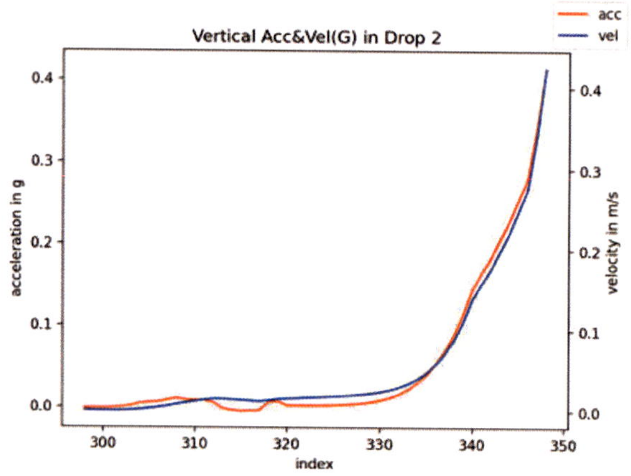

Fig. 3 The third drop

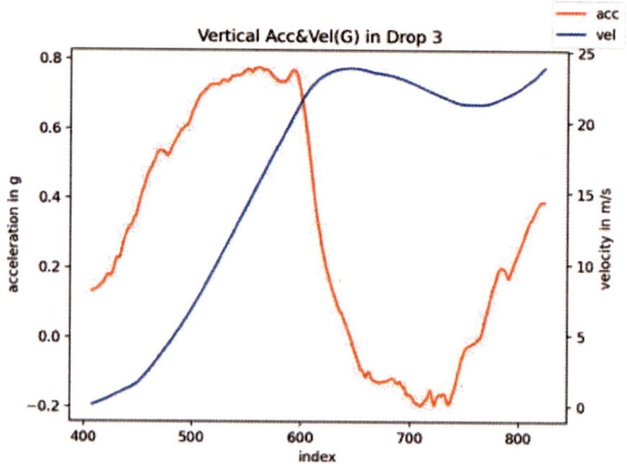

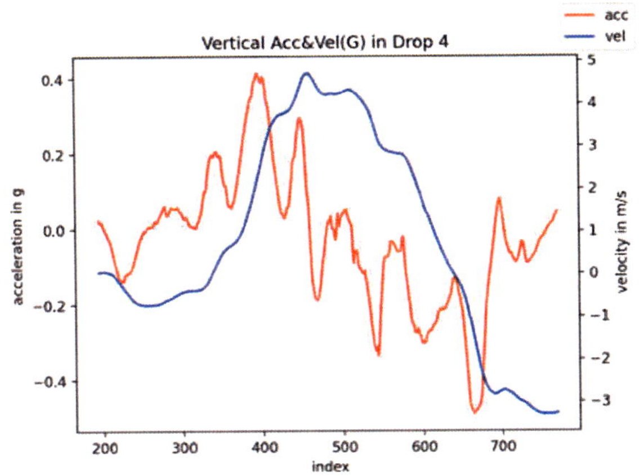

Fig. 4 The fourth drop

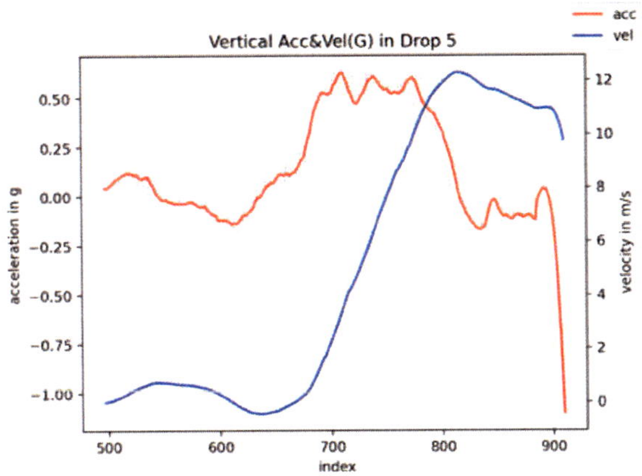

Fig. 5 The fifth drop

was made of is constant and that all the drag is coming from the parachute, we can redefine the projected area in Eq. 1 to be the area of the original HDPE square, which will be referred to as A' in future, and still scale it without any problems. However, this comes at the cost that the C_d calculated does not represent the coefficient of drag of the parachute, but a different measurement of drag that only applies to square parachutes constructed using the same method, which will be referred to as C_{ds} in future. With some variables redefined, we can now substitute in $m = 0.044\,\text{kg}$ (44 g) as mentioned in Sect. 3.1, $A' = 0.0049\,\text{m}^2$ ($4900\,\text{cm}^2$) as mentioned in Sect. 3.3 and $V_t = 8.24\,\text{m}\,\text{s}^{-1}$ as mentioned in Sect. 4.3 into Eq. 1 with C_d replaced by C_{ds} and A replace by A' to obtain $C_d s = 2.11$.

Table 1 Start and end index and terminal velocity of each drop

Drop No.	Start index	V_t in m s^{-1}
1	300	0.03
2	295	0.42
3	400	23.86
4	315	4.72
5	490	12.29

5 Discussion and Conclusion

In Sect. 4.4, the C_{ds} of the assembly was found to be 2.11. By rearranging Eq. 1 to make A the subject and replacing A with A' and C_d with C_{ds} and substituting in $V_t = 31\,\text{m s}^{-1}$ as calculated in Sect. 3.2 and $C_{ds} = 2.11$, we get Eq. 6.

$$A' = \frac{963}{125{,}000} \cdot m \tag{6}$$

Equation 6 shows the minimal A' needed for a HDPE square parachute such that when the parachute is deployed in the case that a QSUAV of mass m enters free fall, the impact of the QSUAV hitting a person is in compliance with [27].

5.1 Extension to UAVs with Special Requirements

However, in the case that the QSUAV is delivering a package which can only withstand a force of less that what a human can withstand, Eq. 1 can be modified similarly to Eq. 6 with the exception that the variable V_t is not replaced, which will give Eq. 7.

$$A' = \frac{981}{130} \cdot m V_t^{-2} \tag{7}$$

$$V_t = \frac{Ft}{m} \tag{8}$$

$$A' = \frac{981}{130} Ft \tag{9}$$

Equation 8 is a rearrangement of the physics equation $F = m \cdot \frac{v}{t}$ with v replaced by V_t. This allows for calculations of the force exerted when the UAV has a collision with the variable F. By substituting Eq. 8 into Eq. 7, we get Eq. 9.

Equation 9 calculates the minimum A' of a parachute needed such that upon a perfectly inelastic collision at V_t the force that is released would be less than F.

5.2 Limitations

One major limitation of this paper is that the framework provided does not work for parachutes that are not square in shape. Another limitation is that the framework does not apply to parachutes of other materials, especially if there is a large difference between the density of that material and HDPE. As the mass ratio between the parachute and the sensor and housing are different, the characteristics of the parachute falling will be greatly different. These limitations lead to there being a possibility that the actual minimum A' needed for a parachute may be vastly different from what is calculated using the framework.

5.3 Future Work

5.3.1 More Intelligent Offline Calibration

In future, a calibration algorithm that can calibrate data differently across time should be implemented. This is due to the fact that sometimes the accelerometer and gyro loses their zero in the middle of the drop, and by using the 'average' method of calibration used in this paper, the readings will not be accurate.

5.3.2 Increase Stability

If the sensor and housing were to be rotated in a way to lower its centre of gravity, the stability of the parachute will be greatly improved. This will lead to a lower chance that there will be rotation and acceleration that happens between the readings that the IMU takes, leading to a lower chance that there will be a loss of calibration of the gyro and accelerometer.

Acknowledgements I would like to express my gratitude for my project mentor Ms Lim Li Chen from *NUS High* for her supervision and for proofreading this paper. I would also like to thank my friends, Ryan Chan Chee How and Gao Ziyao, also from *NUS High*, for their inspirational discussions and their assistance in experimental planning and building the parachutes. I would also like to appreciate my teacher, Mrs Teo Huan Xin, for her help and advice in this project.

References

1. Hu, J., Niu, H., Carrasco, J., Lennox, B., & Arvin, F. (2022). Fault-tolerant cooperative navigation of networked UAV swarms for forest fire monitoring. *Aerospace Science and Technology, 123,* 107494. https://doi.org/10.1016/j.ast.2022.107494.
2. Global Drone Service Market Report 2019: Market is Expected to Grow from USD 4.4 Billion in 2018 to USD 63.6 Billion by 2025, at a CAGR of 55.9%. https://markets.businessinsider.

com/news/stocks/global-drone-service-market-report-2019-market-is-expected-to-grow-from-usd-4-4 billion in-2018-to-usd-63-6-billion-by-2025-at-a-cagr-of-55-9-1028147695
3. Intelligence, I.: Drone market outlook in 2022: Industry growth trends, market stats and forecast. https://www.insiderintelligence.com/insights/drone-industry-analysis-market-trends-growth-forecasts/
4. Classification of the Unmanned Aerial Systems | GEOG 892: Unmanned Aerial Systems. https://www.e-education.psu.edu/geog892/node/5
5. Federation of American Scientists: Intelligence Resource Program. https://irp.fas.org/
6. United States Army (2010) US Army Unmanned Aircraft Systems Roadmap
7. Hoffmann, G., Huang, H., Waslander, S., & Tomlin, C. (2007). Quadrotor helicopter flight dynamics and control: Theory and experiment. In AIAA Guidance, Navigation and Control Conference and Exhibit. American Institute of Aeronautics and Astronautics, Hilton Head. https://doi.org/10.2514/6.2007-6461
8. Recreational. https://www.collinsdictionary.com/dictionary/english/recreational
9. Drones by the Numbers. https://www.faa.gov/uas/resources/by_the_numbers/. Modified 2022-05-31T15:01:26-0400
10. Thornton, S. T., & Marion, J. B. (2004). Classical dynamics of particles and systems (5th ed.). Brooks/Cole.
11. AL-Madani, B., Svirskis, M., Narvydas, G., Maskeliūnas, R., Damaševi & čius, R. (2018). Design of fully automatic drone parachute system with temperature compensation mechanism for civilian and military applications. *Journal of Advanced Transportation,2018*, 1–11. https://doi.org/10.1155/2018/2964583
12. Everett, W. (1973). Development of an improved midair-retrieval parachute system for drone/RPV aircraft. In 4th Aerodynamic Deceleration Systems Conference. American Institute of Aeronautics and Astronautics. https://doi.org/10.2514/6.1973-469
13. Wang, Z. (1999). Investigation of parachute deployment assembly for drones. In 15th Aerodynamic Decelerator Systems Technology Conference. American Institute of Aeronautics and Astronautics, Toulouse, France. https://doi.org/10.2514/6.1999-1722
14. Swiss Post (2020) Drone Operations—Report of the expert board to Swiss Post (p. 7)
15. Khantsis, S. (2006). Control system design using evolutionary algorithms for autonomous shipboard recovery of unmanned aerial vehicles.
16. Cawthorn, D. (2016). Development of a multirotor drone airbag. Flexible Master, University of Southern Denmark.
17. Kwong, W., Passino, K., Laukonen, E., & Yurkovich, S. (1995). Expert supervision of fuzzy learning systems for fault tolerant aircraft control. *Proceedings of the IEEE, 83*(3), 466–483. https://doi.org/10.1109/5.364491.
18. Rehage, D., Carl, U. B., & Vahl, A. (2005). Redundancy management of fault tolerant aircraft system architectures–Reliability synthesis and analysis of degraded system states. *Aerospace Science and Technology, 9*(4), 337–347. https://doi.org/10.1016/j.ast.2005.02.002.
19. Wiseman, Y. (2014). Device for detection of fuselage defective parts. *Information (Japan), 17,* 4189–4194.
20. Shao, P., Wu, C., & Ma, S. (2013). Research on key problems in assigned-point recovery of UAV using parachute. In 2013 IEEE International Conference of IEEE Region 10 (TENCON 2013) (pp. 1–4). IEEE. https://doi.org/10.1109/TENCON.2013.6719061
21. Marković, S., Dopsaj, M., Tomažič, S., Kos, A., Nedeljković, A., & Umek, A. (2021). Can IMU provide an accurate vertical jump height estimate? *Applied Sciences, 11*(24), 12025. https://doi.org/10.3390/app112412025.
22. Rahni, A. A. A., & Yahya, I. (2007). Obtaining translation from a 6-DOF MEMS IMU—An overview. In 2007 Asia-Pacific Conference on Applied Electromagnetics (pp. 1–5). IEEE. https://doi.org/10.1109/APACE.2007.4603861
23. Bar-itzhack, I., & Fegley, K. (1969). Orthogonalization techniques of a direction cosine matrix. IEEE Transactions on Aerospace and Electronic Systems, AES-5(5), 798–804. https://doi.org/10.1109/TAES.1969.309878. http://ieeexplore.ieee.org/document/4103397/
24. D32 Pro. https://www.wemos.cc/en/latest/d32/d32_pro.html

25. Our Products—Superior Link Marketing Pte Ltd. https://www.superiorlinkmarketing.com/mobile/our-products.html
26. How to Make a Toy Parachute. https://www.wikihow.com/Make-a-Toy-Parachute
27. Creer, B. Y., Smedal, H. A., & Wingrove, R. C. (1960). NASA Technical Note D-337. https://ntrs.nasa.gov/citations/19980223621
28. Kleinberger, M., Sun, E., Eppinger, R., Kuppa, S., & Saul, R. (1998). Development of improved injury criteria for the assessment of advanced automotive restraint systems (p. 120).
29. Savitzky, A., & Golay, M. J. E. (1964). Smoothing and differentiation of data by simplified least squares procedures. *Analytical Chemistry, 36*(8), 1627–1639. https://doi.org/10.1021/ac60214a047.

Deceiving Traffic Sign Recognition with Physical One-Pixel Attacks

Juncheng Huang, Xinnan Jiang, and Yi Fei Xia

Abstract Traffic sign recognition (TSR) is an important component of advanced driver-assistance systems, which are increasingly prevalent to minimise human error. Most TSR use convolutional neural networks (CNNs) due to their fast execution and superhuman accuracy. However, CNNs are difficult to analyse and sometimes exhibit counterintuitive behaviour, making them vulnerable to adversarial attacks. One-pixel attacks do not require information from the target model and change merely one-pixel. In this study, we examine three methods of one-pixel attacks—brute-force, differential evolution and a hybrid algorithm—and evaluate their speed and effectiveness in generating adversarial attacks against TSR. We demonstrated that one-pixel attacks remain effective against TSR. We related an attack's effectiveness to the position of the changed pixel and proposed possible explanations to better understand this attack—the first step in defending against it. We also successfully replicated one-pixel attacks in a physical setting using printed modified road signs. This proves that TSR in vehicles can be attacked by simply adding a small, coloured spot to an existing road sign. The position and colour of this spot are easily generated without access to the target model's inner information. Thus, we demonstrated a novel method of physical adversarial attack against TSR which is more inconspicuous and easily executed than existing ones.

Keywords Traffic Sign Recognition · Convolutional Neural Network · Autonomous vehicle · Black-box attack

1 Background

Every year, 1.35 million people worldwide are killed by traffic accidents [1], mostly due to human factors [2, 3]. Advanced driver-assistance systems (ADAS) can greatly reduce human error [4, 5] and are now present in 30% of new cars in the United States [6]. This shows that automation is rapidly becoming the norm in modern vehicles.

J. Huang · X. Jiang (✉) · Y. F. Xia
Hwa Chong Institution, Singapore, Singapore

H. Guo et al. (eds.), *IRC-SET 2022*,
https://doi.org/10.1007/978-981-19-7222-5_12

ADAS comprises many systems, of which traffic sign recognition (TSR) is a crucial component [7]. Other systems relying on TSR will become mandatory in many regions by 2022 [8], increasing the need to prevent attacks against TSR.

Most current research on TSR focuses on convolutional neural networks (CNNs) [9, 10], with their fast execution and superhuman accuracy [11]. CNNs assign importance to various aspects of an input image to differentiate them. However, CNNs learn automatically, making them difficult to interpret and sometimes display counterintuitive behaviour [12]. Therefore, adversarial examples with almost undetectable differences have successfully deceived CNNs meant for various purposes into misclassifying images [13].

Components of ADAS including TSR [16] are also vulnerable to such attacks [14, 15], causing traffic signs to be misrecognised. Existing adversarial attacks on CNNs algorithmically generate and inject imperceptible noise into the input image to deceive the CNN [13]. However, they require control over the computer vision pipeline. As such, it is simpler to manipulate external factors, including the physical environment [16]. Most modern TSR use the camera as its sole input source [9, 10] due to its unique ability to detect colours and textures, which are important information in traffic signs, as well as its low cost and high availability. However, when compared to other sensors that collect 3D data, only being able to capture 2D image data makes cameras more susceptible to external disturbances such as weather. More importantly, it offers opportunities for adversarial attacks on TSR.

Existing physical adversarial attacks against TSR require complex changes [16], which are difficult to execute on existing traffic signs inconspicuously.

Other attacks which modify the input image digitally could be adapted to develop more feasible physical adversarial attacks. Whilst many generate imperceptible noise which are not sufficiently robust to real-world distortions [13], one-pixel attacks modify a single pixel to deceive deep neural networks (DNNs) [17]. Adding patches to an existing traffic sign could be more unnoticeable and easily replicated than current physical adversarial attacks. One-pixel attacks also do not require information of the target model being attacked [17]. With no current research on the vulnerabilities of TSR towards one-pixel attacks, let alone physical one-pixel attacks specifically, it is pertinent to investigate them.

Su et al. [17] expressed the generation of adversarial images in one-pixel attacks as an optimisation problem and used differential evolution (DE) as the optimisation algorithm. DE uses Darwinian principles of evolution to tap on the parallels between biological evolution and optimisation problems [18, 19]. However, no other methods to generate adversarial images for one-pixel attacks have been attempted. Considering that most TSR models resize the input image to 48×48 pixels or smaller [9], a brute-force algorithm may be feasible and should be considered, in addition to other optimisation algorithms.

In this study, we propose performing one-pixel attacks on TSR and evaluating different optimisation algorithms used to perform one-pixel attacks. We aim to demonstrate the viability of utilising one-pixel attacks for more feasible, inconspicuous, and thus dangerous physical adversarial attacks on TSR.

2 Experimental Goals

We evaluated three methods of one-pixel attacks—brute-force, DE [17] and a hybrid of both—based on their speed and effectiveness at deceiving TSR. We would then replicate successful one-pixel attacks in a physical setting to test the potential of utilising one-pixel attacks for physical adversarial attacks on TSR. In particular, we aim to answer the following questions in our experiments:

1. How fast are the different methods of one-pixel attacks?
2. How successful are the different methods of one-pixel attacks at deceiving TSR models?
3. Are one-pixel attacks successful at deceiving TSR models in a physical setting?

3 Methods

3.1 German Traffic Sign Recognition Benchmark (GTSRB)

There currently exist numerous traffic sign benchmarks, including the Belgian Traffic Sign Dataset [20] and GTSRB [11]. We chose GTSRB since it is one of the most commonly used benchmarks amongst existing TSR models [9–11]. It consists of over 50,000 images classified into 43 classes, which are representative of traffic signs captured in real life.

3.2 Traffic Sign Recognition Model

We used TrafficSignNet, an existing CNN which achieved 95% accuracy on GTSRB [21], as the target of the adversarial attacks. During pre-processing, all input images were down- or up-sampled to 32×32 pixels, similarly to other TSR models [9]. The minimum confidence score threshold for a traffic sign to be detected was set at 50%, the default value in the TensorFlow Object Detection API [16].

3.3 Experimental Setup

We ran all operations on a device with an Intel Core i5-8259U CPU, an Intel Iris Plus Graphics 655 GPU and 8 GB of RAM using the PyTorch framework.

We chose 100 images randomly from images that achieved a confidence score of at least 99.9% in the validation dataset of GTSRB as sample input to attack. They include at least one image from each class.

We recorded the total time taken to generate adversarial attacks for all 100 images for each method.

A successful adversarial attack occurs when a traffic sign is detected and the class with the highest confidence score is not the true class of the image, meaning that the sign is misclassified.

For each generated adversarial attack, we recorded the class with the highest confidence score post-attack, its confidence score and the confidence score of the true class post-attack.

3.4 Brute-Force Algorithm

Given the limited number of pixels in the input of TSR models [9], a brute-force approach to one-pixel attacks could be used to test modifying every single pixel to a finite range of colours.

However, the RGB colour space contains 2563 colours, rendering it too time-consuming to iterate through every colour for every pixel. In order to reduce the number of iterations whilst maintaining an adequate range of colours, we restricted the colours tested to black, white, the primary colours (red, green, blue) and the secondary colours (cyan, magenta, yellow)—which have the greatest contrast against the primary colours.

For each input image, our brute-force algorithm iterated through every pixel. For each pixel, it would be modified to the eight colours tested and all pixel-colour combinations were input into the TSR model. The pixel-colour combination which decreased the confidence score of the true class by the largest extent, and when there exists other successful attacks, was successful itself, was output as the generated adversarial attack.

3.5 Differential Evolution

DE is an optimisation algorithm which does not use gradient information, making it suitable for one-pixel attacks where the gradient information of the model is not easily obtainable. It could run faster than the brute-force algorithm for one-pixel attacks. In general, an initial population of candidate solutions is randomly generated, which then combine to generate offspring which compete with the rest of the population, with fitter candidates replacing less fit ones. For one-pixel attacks, each candidate is represented by

$$\boldsymbol{v} = (x, y, r, g, b)$$

where $\boldsymbol{v}$ is a vector, x and y are the coordinates of the changed pixel, and r, g and b are its RGB colour values after the change. Each candidate's fitness is negatively correlated to the confidence score of the input image's true class when the candidate image with the modified pixel is passed through the TSR model. In other words, DE searches for the minimum point of the confidence score of the image's true class.

We used SciPy's implementation of DE [22]. Due to speed constraints, we set the maximum number of iterations to 10. The final candidate solution was taken as the generated adversarial attack.

3.6 Brute-Force Algorithm Enhanced with Differential Evolution

We limited the range of colours in the brute-force algorithm and the number of iterations in DE to achieve reasonable computational speeds, negatively affecting the quality of adversarial examples generated. To mitigate these shortcomings whilst maintaining computational efficiency, we propose a hybrid algorithm.

We first ran the brute-force algorithm using only two colours—black and white. After the best pixel position was generated, we generated the optimal colour change using a modified DE algorithm with 3-dimensional vectors representing the RGB values. With lower-dimensional vectors, DE could run for 20 iterations without affecting computational speed. This avoids arbitrarily limiting the number of colours possible. The best pixel-colour combination was chosen with the same criteria as the brute-force algorithm.

3.7 Physical Adversarial Attack

To generate physical adversarial examples, we gave a total of 43 original traffic sign templates, one from each class in GTSRB, as input to the brute-force algorithm. We resized 15 successful adversarial examples for 11 classes and their corresponding original templates to fit within a 3.5×3.5 cm box and printed them. We placed physical traffic signs 15.0 cm from the ground. We took photos from a fixed distance of 50.0 cm at a height of 10.0 cm from 0° and 45° with respect to the perpendicular to the sign (Fig. 1) using an iPhone 11, with the camera always pointed at the sign, to demonstrate attacks within a controlled setting.

We cropped captured photos to focus on the traffic sign (Fig. 2). We then fed cropped photos of adversarial examples and their original signs into the TSR model as input. We only evaluated classes where the original sign was classified correctly. We evaluated physical attacks similarly to virtual one-pixel attacks.

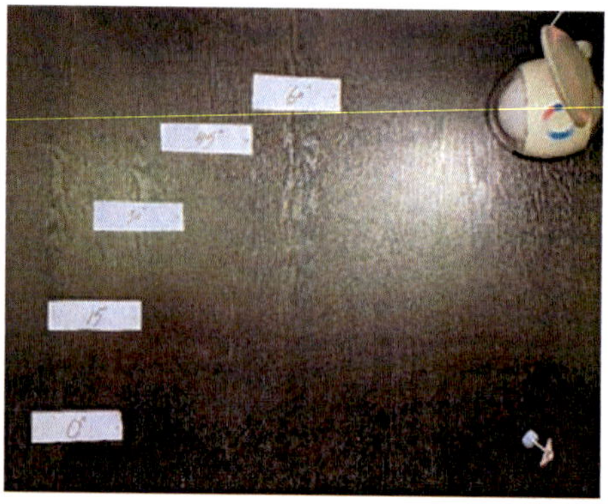

Fig. 1 Setup for physical adversarial attacks

Fig. 2 Images from physical adversarial attacks

4 Results

4.1 Evaluation of Attack Algorithms

Numerous successful one-pixel attacks were generated with the three algorithms. We further evaluated these three methods of one-pixel attacks on both the time taken and their success rates. Initial average confidence of the true class is 99.9% (Table 1).

DE took an average of 130 s to generate an adversarial attack for one image, which is 59.6% less than the brute-force algorithm and 69.0% less than the hybrid

Table 1 Time taken and effectiveness of one-pixel attack algorithms

Algorithm	Time per iteration (s)	Success rate (%)	Average confidence of true class (%)
Brute-Force	322	18.0	81.2
DE	130	8.0	90.2
Brute-Force + DE	420	26.0	70.8

algorithm. However, it had the lowest success rate and decrease in confidence of the true class, so DE is the least successful method for one-pixel attacks.

We limited DE to a maximum of 10 iterations, which prevented it from reaching a more optimal solution. Thus, DE should not be discounted as a method of one-pixel attack and further studies can be conducted to evaluate the effectiveness of DE compared to other methods for one-pixel attacks by using better hardware and allowing DE to take the same amount of time as other methods.

The hybrid algorithm took 30.4% more time than the brute-force algorithm to generate an adversarial attack, because DE needed more iterations than iterating through a limited range of eight colours to find the optimal colour. The hybrid algorithm was 44.4% more likely to succeed at deceiving the TSR model and decreased the average confidence of the true class by an additional 10.4 percentage points. This could show the importance of finding the optimal colour for the pixel change in order to attack TSR successfully.

The success rate achieved by our hybrid algorithm is on par with similar untargeted one-pixel attacks on DNNs which used the original CIFAR-10 test dataset [17]. This proves that one-pixel attacks are equally applicable to the CNNs used by TSR models. Su et al. [17] obtained much higher success rates when a perturbed CIFAR-10 dataset was used, where original images underwent modifications such as duplication, rotation, clipping, blurring and adding a few random bad pixels. Thus, a higher success rate in attacking TSR models could be obtained by using a similarly modified GTSRB dataset, and the models' actual vulnerability can be further demonstrated by attacking other real-life images of traffic signs.

Analysis. One-pixel attacks are difficult to understand, as is expected with the unpredictability of DNNs [12]. Vargas and Su [23] proposed the use of propagation maps to analyse one-pixel attacks, which show how minor local perturbations can propagate into much larger changes across layers of the network.

Since black-box attacks, such as one-pixel attacks, do not have information of the target model, we analysed the relationship between the confidence score of the true class and pixel position (Fig. 3). In successful attacks, pixel changes which led to a large decrease in the confidence score of the true class tended to cluster and form a peak on the graph (Fig. 3), showing that attacks on nearby pixels would also be effective and that pixel position is an important factor in one-pixel attacks. This is consistent with Vargas and Su, who proved this to be true for many different neural networks [23]. This demonstrates that the receptive field, which is the region of the input space that affects a particular unit of the network, is the vulnerable component [23] rather than the neurons in the model.

Vargas and Su [23] showed that failed attacks propagated changes similarly to successful attacks and hypothesised that one-pixel attacks are closely related to disturbances in saliency maps, which highlight the parts of the input image that have a big impact on the prediction. This means that the pixel change misleads the model to focus more on the change rather than the actual identifying features of the object to be classified, without directly modifying the original identifying features [23].

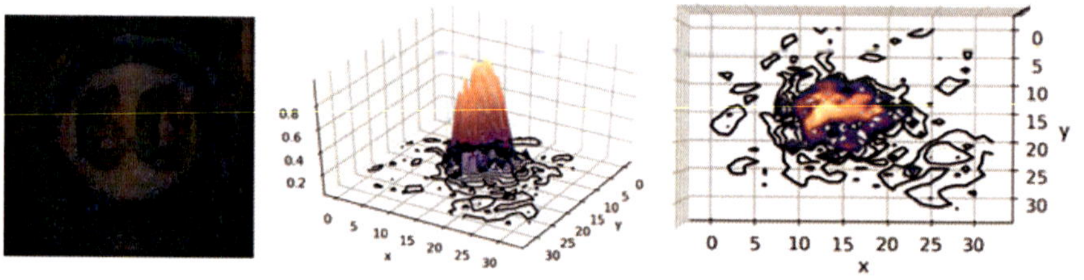

Fig. 3 Analysis of successful adversarial attack

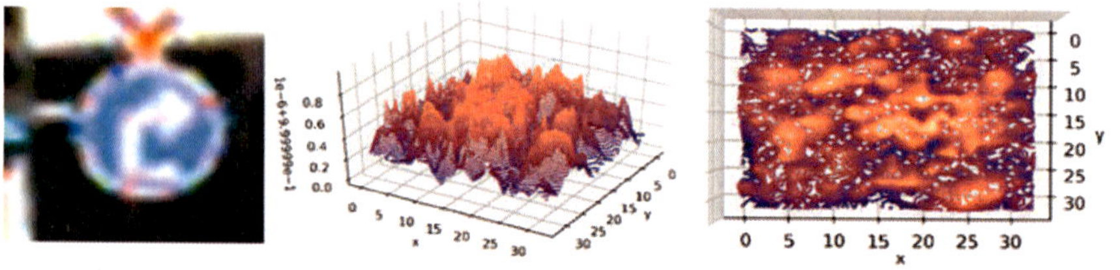

Fig. 4 Analysis of unsuccessful adversarial attack

In our study, failed one-pixel attacks displayed many much smaller peaks in the decrease in confidence score of the true class (Fig. 4). Thus, should the conflicting salience hypothesis be true [23], it could be possible that the saliency of the identifying features in the original image is strong enough such that no pixel change can mislead the model to misclassify the image. It is also possible that pixel changes directly modified the parts of the image which already had high saliency originally. In this case, one-pixel attacks would be less successful when the saliency of the original image is more evenly distributed, since the neural network would have more features to focus on and a single pixel change would be insufficient to disrupt all of them.

Further studies could be done to further analyse one-pixel attacks, given the widespread use of computer vision DNNs, including for TSR. Deeper analysis whilst considering past research would better serve the understanding of such inconspicuous adversarial attacks and aid the development of defence mechanisms against them. It could also shed insight on how attackers could potentially analyse the target model for black-box attacks, such as through analysing the model's prediction behaviour when individual pixels are changed.

4.2 Physical Adversarial Attacks

We replicated one-pixel attacks in a physical setting. Out of 11 classes, 10 adversarial examples from eight classes, which had their original signs classified correctly, are

Table 2 Effectiveness of physical one-pixel attacks

Angle (°)	Average original confidence of true class (%)	Success rate (%)	Average post-attack confidence of true class (%)
0	81.9	20.0	71.0
45	71.8	30.0	62.5
Overall	76.8	25.0	66.8

evaluated. We achieved a 25% success rate and the average confidence of the true class decreased by 9.2 to 10.9 percentage points. Therefore, it is shown that such minor perturbations can remain effective even in a physical setting (Table 2).

One possible explanation is that localised vulnerable regions, represented by peaks (Fig. 4), are similar across images with similar features, such as those of the same class. However, due to the small scale of the perturbations made in one-pixel attacks, they are less robust to physical distortions.

Whilst the similar results between photos taken from both 0° and 45° demonstrate robustness against different viewing angles, further studies can be conducted with larger, multi-pixel attacks to improve robustness against other physical distortions such as brightness and distance, and the ability to evade human observation can be tested to evaluate the feasibility of using these more noticeable but effective modifications.

5 Conclusion and Future Work

In this study, we used three algorithms to generate one-pixel adversarial attacks on TSR models and evaluated them on their speed and effectiveness. Whilst DE had the fastest performance, we achieved the highest success rate by enhancing the basic brute-force algorithm with DE to optimise the colour of the pixel change. Our success rate of 26.0% is on par with existing one-pixel attacks against general image classification models [17], demonstrating that TSR is equally vulnerable to one-pixel attacks.

We analysed one-pixel attacks through both a successful example and a failed example. Based on existing research on the analysis of one-pixel attacks by Vargas and Su [23], we suggested two potential explanations. A successful pixel change could need to gain a higher saliency than the original identifying features, such that it can "distract" the model from the original image, or it could require the original defining features to be concentrated within a smaller region in order to directly modify them to deceive the model.

We also successfully replicated one-pixel attacks in a physical setting, which is more inconspicuous and accessible for attackers, thus posing greater danger to TSR systems.

To better prove the feasibility of one-pixel attacks on TSR, both virtual and physical one pixel attacks could potentially be tested under actual traffic conditions and against other state-of-the-art commercially used TSR models, instead of relying on GTSRB and a controlled indoor setting. Multi-pixel attacks could be used to strengthen the robustness of physical adversarial attacks to physical distortions, which would be more dangerous to TSR. More importantly, deeper analysis should be done on the nature of one-pixel attacks, especially physical ones. Although existing research has proposed defences against virtual one-pixel attacks [24], our novel physical one-pixel attacks have exposed further vulnerabilities, which would need to be defended against, especially for TSR models, to ensure safety.

References

1. Global status report on road safety 2018. World Health Organization, Geneva, Switzerland, 2018 [Online]. https://www.who.int/publications/i/item/9789241565684
2. Singh, S. (2015). Critical reasons for crashes investigated in the National Motor Vehicle Crash Causation Survey. National Highway Traffic Safety Administration, Washington, DC, USA, DOT HS 812 115, February 2015 [Online]. https://crashstats.nhtsa.dot.gov/Api/Public/ViewPublication/812115
3. National Motor Vehicle Crash Causation Survey: Report to Congress. National Highway Traffic Safety Administration, Washington, DC, USA, DOT HS 811 059, July 2008 [Online]. https://crashstats.nhtsa.dot.gov/Api/Public/ViewPublication/811059
4. Crayton, T. J., & Meier, B. M. (2017). Autonomous Vehicles: Developing a public health research agenda to frame the future of transportation policy. *Journal of Transport & Health, 6*, 245–252. https://doi.org/10.1016/j.jth.2017.04.004
5. Brookhuis, K. A., de Waard, D., & Janssen, W. H. (2001). Behavioural impacts of advanced driver assistance systems–an overview. *EJTIR, 1*(3). https://doi.org/10.18757/ejtir.2001.1.3.3667
6. Level 2 autonomous driving Q4 2020 and full year 2020. Canalys. https://www.canalys.com/newsroom/canalys-autonomous-driving-starts-to-hit-mainstream-as-35-million-new-cars-had-level-2-features-in-q4-2020
7. Intelligent Speed Assistance (ISA). European Transport Safety Council. https://etsc.eu/intelligent-speed-assistance-isa
8. Document 32019R2144, Regulation (EU) 2019/2144, Nov. 2019. [Online]. https://eur-lex.europa.eu/eli/reg/2019/2144/oj
9. Cireşan, D., Meier, U., Masci, J., & Schmidhuber, J. (2011). A committee of neural networks for traffic sign classification. In *The 2011 International Joint Conference on Neural Networks* (pp. 1918–1921). https://doi.org/10.1109/IJCNN.2011.6033458
10. Shustanov, A., & Yakimov, P. (2017). CNN Design for Real-Time Traffic Sign Recognition. *Procedia Engineering, 201*, 718–725. https://doi.org/10.1016/j.proeng.2017.09.594
11. Stallkamp, J., Schlipsing, M., Salmen, J., & Igel, C. (2012). Man vs. computer: Benchmarking machine learning algorithms for traffic sign recognition. *Neural Networks, 32*, 323–332. https://doi.org/10.1016/j.neunet.2012.02.016
12. Szegedy, C., Zaremba, W., Sutskever, I., Bruna, J., Erhan, D., Goodfellow, I., & Fergus, R. (2013). *Intriguing properties of neural networks*. arXiv:1312.6199
13. Gragnaniello, D., Marra, F., Poggi, G., & Verdoliva, L. (2018). Analysis of adversarial attacks against CNN-based image forgery detectors. In *2018 26th European Signal Processing Conference (EUSIPCO)* (pp. 967–971). https://doi.org/10.23919/EUSIPCO.2018.8553560

14. Petit, J., Stottelaar, B., Feiri, M., & Kargl, F. (2015). Remote attacks on automated vehicles sensors: Experiments on camera and LiDAR. In *Proceedings of Black Hat Europe* (pp. 1–13) [Online]. https://www.blackhat.com/docs/eu-15/materials/eu-15-Petit-Self-Driving-And-Connected-Cars-Fooling-Sensors-And-Tracking-Drivers-wp1.pdf
15. Jing, P., Tang, Q., Du, Y., Xue, L., Luo, X., Wang, T., Nie, S., & Wu, S. (2021). Too good to be safe: Tricking lane detection in autonomous driving with crafted perturbations. In *Proceedings of 30th USENIX Security Symposium*, August 11–13, 2021 (pp. 3237–3254) [Online]. https://www.usenix.org/system/files/sec21-jing.pdf
16. Chen, S.-T., Cornelius, C., Martin, J., Chau, D. H. (2018). ShapeShifter: robust physical adversarial attack on faster R-CNN Object Detector. In M. Berlingerio, F. Bonchi, T. Gärtner, N. Hurley, & G. Ifrim, (Eds.), *Machine learning and knowledge discovery in databases*, Dublin, Ireland, September 10–14, 2018 (pp. 52–68). https://doi.org/10.1007/978-3-030-10925-7_4
17. Su, J., Vargas, D. V., & Sakurai, K. (2019). One pixel attack for fooling deep neural networks. *IEEE Transactions on Evolutionary Computation, 23*(5), 828–841. https://doi.org/10.1109/TEVC.2019.2890858
18. Storn, R., & Price, K. (1997). Differential evolution—A simple and efficient heuristic for global optimization over continuous spaces. *Journal of Global Optimization, 11*, 341–359. https://doi.org/10.1023/A:1008202821328
19. Das, S., & Suganthan, P. N. (2011). Differential evolution: A survey of the state-of-the-art. *IEEE Transactions on Evolutionary Computation, 15*(1), 4–31. https://doi.org/10.1109/TEVC.2010.2059031
20. Timofte, R., Zimmermann, K., & Van Gool, L. (2009). Multi-view traffic sign detection, recognition, and 3D localisation. *Workshop on Applications of Computer Vision (WACV), 2009*, 1–8. https://doi.org/10.1109/WACV.2009.5403121
21. Sahay, R. (2019). TrafficSignNet [Source code]. https://github.com/rajatsahay/TrafficSignNet
22. Virtanen, P., Gommers, R., Oliphant, T. E., Haberland, M., Reddy, T., Cournapeau, D., et al. (2020). SciPy 1.0: Fundamental algorithms for scientific computing in Python. *Nature Methods, 17*, 261–272. https://doi.org/10.1038/s41592-019-0686-2
23. Vargas, D. V., & Su, J. (2020). Understanding the one-pixel attack: Propagation maps and locality analysis. In *CEUR Workshop Proceedings*, Vol. 2640 [Online]. https://kyushu-u.pure.elsevier.com/en/publications/understanding-the-one-pixel-attack-propagation-maps-and-locality-
24. Chen, D., Xu, R., & Han, B. (2019). Patch selection Denoiser: An effective approach defending against one-pixel attacks. In: T. Gedeon, K. Wong, & M. Lee (Eds.), *International Conference on Neural Information Processing*, Sydney, Australia, December 12–15, 2019 (pp. 286–296). https://doi.org/10.1007/978-3-030-36802-9_31

Automated Categorization of Mathematics Questions

Peiran Qiu and Yao Yao

Abstract Sorting out mathematic questions into respective topics manually is highly time-consuming but necessary. A reliable automated categorization algorithm to categorize mathematic questions is needed to speed up the process. In this paper, we present an accurate text categorization computer algorithm that categorizes junior college pure mathematics questions into corresponding chapters. This algorithm extracts keywords from the questions in the dataset and calculates the Jaccard similarity coefficient between questions based on the keywords extracted. Different LaTeX formulas were taken into consideration during keyword extraction. Upon close examination of the algorithm and by testing with sample questions, we concluded that the algorithm efficiently achieves high accuracy in classifying pure mathematics questions. Furthermore, a program with a graphic user interface was developed to provide a potential practical use of the algorithm.

Keywords Computer algorithm · Categorization · Keywords · Similarity · Jaccard Index

1 Background and Purpose

As technology develops, more people have adapted to the new lifestyle: study and work online. Through the increasing use of digital documents, many have found their life more convenient.

Since 2002, computer scientists have started to work on automated text categorization using machine learning, for which English texts were categorized into predefined categories [1]. This technology has improved efficiency of work and saved huge labor power. In addition, keywords extraction has been researched for many years and keywords provide an overview of a document's content [2].

Since 2019, online learning and remote working escalated during the outbreak of COVID-19 [3]. Many schools hence brought education online. Teachers provide

P. Qiu (✉) · Y. Yao
Hwa Chong Institution, 661 Bukit Timah Rd, Singapore 269734, Singapore
e-mail: robin.qiu@foxmail.com

H. Guo et al. (eds.), *IRC-SET 2022*,
https://doi.org/10.1007/978-981-19-7222-5_13

digital notes for students to help them study more efficiently. Through conversations with teachers of Hwa Chong Institution (HCI) in Singapore, it was discovered that these teachers would manually sort questions after receiving massive digital papers, by copying and pasting each question into different folders. One paper, with around 12–14 questions, would usually take these teachers one hour to sort, which took much time and effort.

Therefore, to tackle the urgent need of teachers, a computer algorithm, which contains a set of steps to solve a specific task, is successfully figured out to meet the demand.

2 Material and Method

2.1 Material

The Mathematics Department of HCI kindly provides 274 questions (given by chapters) from 11 topics, including Arithmetic Progression and Geometric Progression, Complex Numbers, Differentiation Equation, Differentiation and Applications, Equations and Inequalities, Functions, Integration Techniques, Integration Applications, Maclaurin and Binomial Series, Sigma Notation and MOD and Vectors. These 274 questions are the sample dataset and arranged by chapters. Two sets of modified Preliminary Examination papers (with questions from 11 chapters only) are also provided to test the accuracy of the algorithm.

Therefore, with the provided samples, a computer algorithm for automated categorization of junior college mathematics questions was developed.

2.2 Method

Question Input. Python-docx [4] library is used in the algorithm to input the question from the docx file. Questions are pre-prepared and formatted in a table form (see Fig. 1) with an additional empty row or a row of labeled difficulty of the question. The LaTeX formula needs to start and end with a '$' sign. Diagrams and graphs are omitted by the algorithm, which will be discussed later.

Removal of Stop Words (Abundant Words). Natural Language Toolkit (aka. nltk) library [5] and gensim.parsing.preprocessing library [6] are used to remove stop words. The purpose of this step is to minimize the impact of abundant words since they do not contribute to explaining the question.

Extract Keywords and Key LaTeX Words and Pre-process the Sample Set. As there is no general method to extract keywords in different applications, a new suitable method extracting keywords must be carefully designed considering the application

Fig. 1 Prelim paper 1 question 5 from HCI

5	`The function $\text{f}$ is defined by $\text{f}:x\mapsto \ln {{\left( 2-x \right)}^{2}}-2,\text{ }x\in \mathbb{R},\text{ }x\ne 2.$`
(i)	`By considering the graph of $y=\text{f}\left( x \right),$ give a reason why ${{\operatorname{f}}^{-1}}$ does not exist.` [2]
	`The function $\text{h}$ is defined by $\text{h}:x\mapsto \text{f}\left( x \right),\text{ }x\ge a,\text{ }a\in \mathbb{Z}.$`
(ii)	`Find the largest possible domain of $\text{h}$ such that ${{\text{h}}^{-1}}$ exists.` [1]
(iii)	`Define ${{\text{h}}^{-1}}$ in a similar form.` [3]
(iv)	`Find the set of values of $x$ which satisfies the equation $\text{h}{{\text{h}}^{-1}}\left( x \right)={{\text{h}}^{-1}}\text{h}\left( x \right)$.` [2]
	1 1 2 2

and dataset [7]. As LaTeX formulas are largely used in our sample set and they largely contribute to understanding the meaning of questions, we proposed to consider key LaTeX words in our extraction process. A key LaTeX word is defined as the word "xx" in "\xx" format (e.g., "frac", "theta", "pi"). If a word or LaTeX word appears in at least six different questions from the sample set, it will be identified as a keyword or LaTeX word. Words in different tenses are considered as different keywords (such as "shows" and "shown") as they have different meanings in a question. In total, 231 keywords and 37 key LaTeX words are identified from the 11 chapters. These keywords of samples are pre-processed and stored in a.csv file for use in the final application.

After that, a binary list is created for each of the 274 sample questions. The keyword and key LaTeX word list for a question is binary (i.e., contains only 1 and 0). One means the word is in the text of the question at least once. Zero means this word is not in the text of the question. The length of each list is 268. This binary list is pre-processed for all the samples and stored in a.csv file for use in next steps.

Similarity Calculation using Jaccard Index. To categorize a new question, keywords of the question will be extracted after the removal of stop words. Then a binary list like the one used in the previous part will be created for the new question. Similarities between this question and each question in the sample set will then be calculated based on the two sets of keywords identified in the previous part using binary lists. When calculating the similarity, Jaccard Index (see Fig. 2), developed by Paul Jaccard [8], is used to measure the similarity between two sets of data. The formula is calculated as the intersection size divided by the union size of two sample sets, which provides a straightforward measure of similarity between datasets [9]. Scikit-learn provides a function to easily calculate Jaccard similarity coefficient score between two one-dimensional array-like matrices [10].

Fig. 2 Formula or Jaccard Index between set A and set B

$$J(A, B) = \frac{|A \cap B|}{|A \cup B|}$$

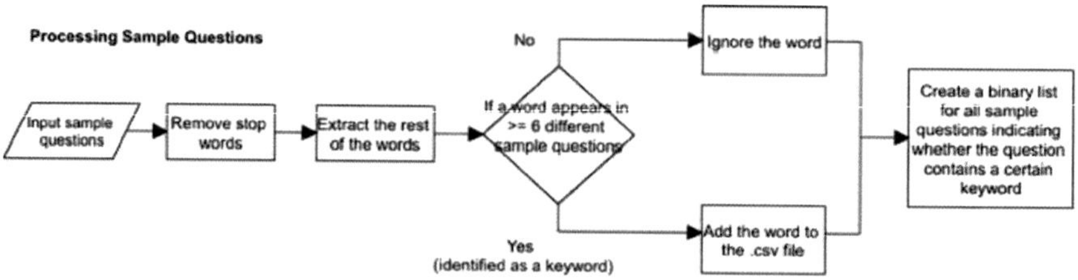

Fig. 3 Flowchart of the algorithm on processing sample questions

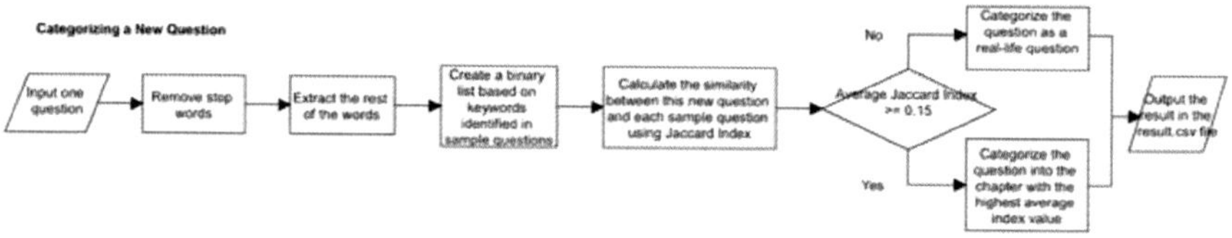

Fig. 4 Flowchart of the algorithm on categorizing a new question

Results. In the application, the algorithm will calculate the Jaccard Index between each new question and 274 samples. Average indexes between all questions from each chapter and the new question are calculated and sorted. The chapter with the highest average index value is the result. However, if the highest index value is below the preset threshold value ($\theta = 0.15$), the question will be categorized as a question set in real-life context.

Flowchart of the Algorithm. Figures 3 and 4 are the flowcharts of the algorithm. Figure 3 shows how the algorithm processes the 274 sample questions. Figure 4 shows how to categorize a new question.

3 Accuracy Analysis

3.1 Similarities Between Sample Questions

The similarities between one question and another is calculated using Jaccard Index for all the 274 sample questions of the 11 chapters, with 1.0 meaning all keywords extracted from each question are the same, and 0.0 meaning none of the keywords is the same.

Figure 5 is a visualization of the similarities within the 274 sample questions using Jaccard Index. On the x-axis and y-axis of the heatmap, the scales refer to indexes of the 274 sample questions, starting from Question 0. Each point on the heatmap refers to the similarity between two sample questions, for which the indexes of the

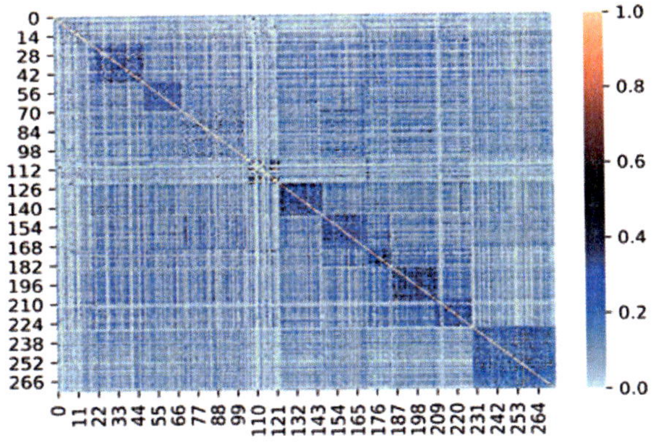

Fig. 5 Visualization of similarities between 274 questions from the 11 chapters

two questions are presented by the coordinates. For example, the point (88, 70) refers to the similarity between Question 88 and Question 70 of the 274 sample questions.

A diagonal line of light red color is observed from the heatmap (see Fig. 6). Each point on this line has the same x-coordinate and y-coordinate. Hence, the similarities calculated are between the same question. Therefore, each point has the highest similarity of 1.0, which presents a light red color.

Patterns of dark blue rectangles can also be observed on the heatmap (Fig. 7). Since the 274 sample questions are arranged by chapters, each rectangle contains the similarities between two sample questions from the same chapter. As questions from the same chapter contain more similar keywords than other chapters, the similarities within chapters are higher, as proved from the dark blue color on the heatmap. It can also be proved by these dark blue rectangles that for sample questions from different chapters, the similarities are lower, as other areas on the heatmap show a light blue color. These patterns show clearly that questions in the same chapter are found to be highly similar by our algorithm.

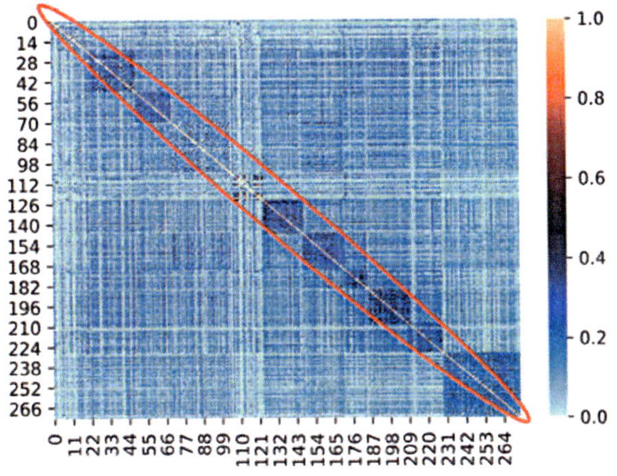

Fig. 6 Diagonal line of light red color on heatmap

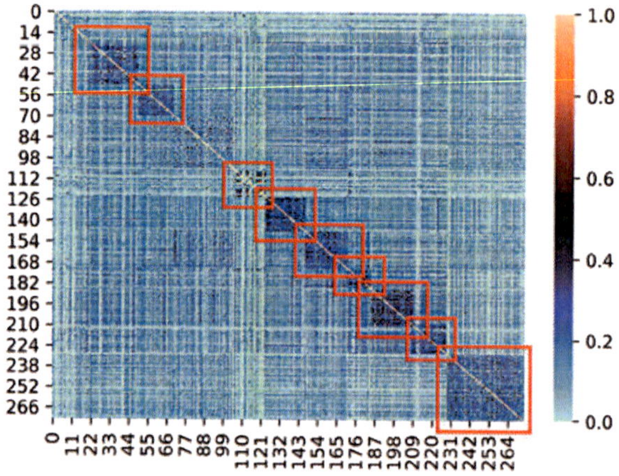

Fig. 7 Rectangles of dark blue color on the heatmap

As it shows a clear pattern for which those questions from the same chapter are highly similar after counting the similarities, this program is likely to categorize junior college mathematics questions into corresponding chapters by calculating the similarity between the keywords extracted from targeted questions and from all the questions from the 11 chapters in the database. The chapter with the highest similarity will be the wanted chapter.

3.2 *Testing with Past Year Papers*

The algorithm is tested with two modified past year math prelim papers from HCI.

Figures 8 and 9 further prove that the program is exceedingly accurate, for which all the questions from past year papers are categorized into correct chapters. This algorithm can also differentiate those questions of real-life context from other questions which are purely mathematical.

3.3 *Final Application*

A graphic user interface (see Fig. 10) is developed using Python Tkinter [11].

The total number of the questions identified and the progress bar will be displayed while the application is running (see Fig. 11).

A file check procedure is implemented since it is critical to have NLTK package and three.csv files to be under the same file path of the executable file.

Failure to select a specific file of questions before clicking 'Start' will lead to an error warning (see Fig. 12). This can be solved by selecting a specific file at

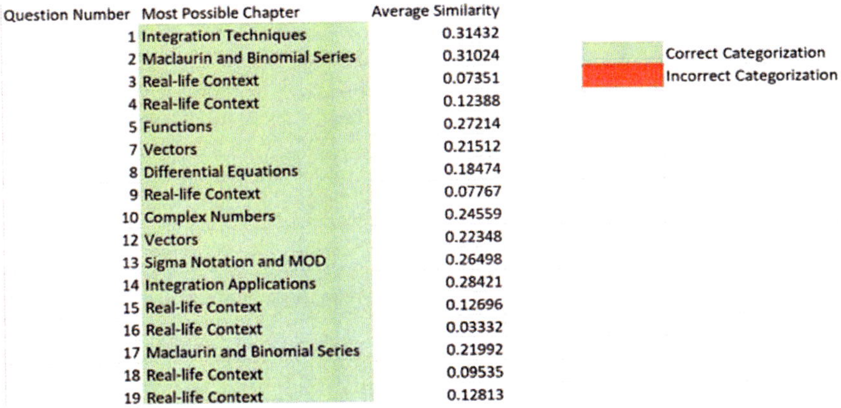

Question Number	Most Possible Chapter	Average Similarity
1	Integration Techniques	0.31432
2	Maclaurin and Binomial Series	0.31024
3	Real-life Context	0.07351
4	Real-life Context	0.12388
5	Functions	0.27214
7	Vectors	0.21512
8	Differential Equations	0.18474
9	Real-life Context	0.07767
10	Complex Numbers	0.24559
12	Vectors	0.22348
13	Sigma Notation and MOD	0.26498
14	Integration Applications	0.28421
15	Real-life Context	0.12696
16	Real-life Context	0.03332
17	Maclaurin and Binomial Series	0.21992
18	Real-life Context	0.09535
19	Real-life Context	0.12813

Fig. 8 Test results of prelim 2015 paper from HCI

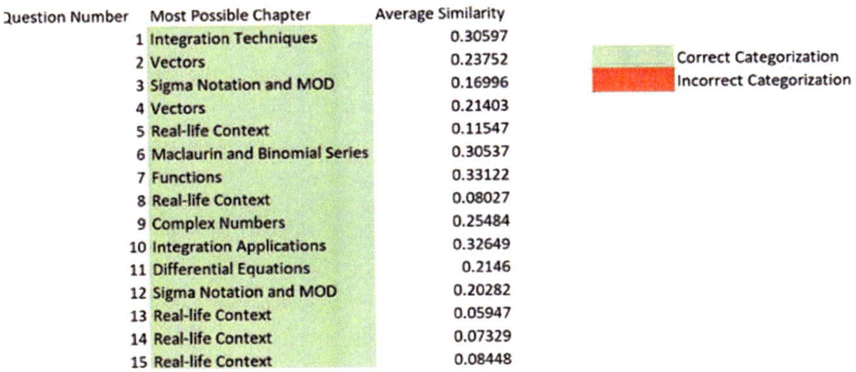

Question Number	Most Possible Chapter	Average Similarity
1	Integration Techniques	0.30597
2	Vectors	0.23752
3	Sigma Notation and MOD	0.16996
4	Vectors	0.21403
5	Real-life Context	0.11547
6	Maclaurin and Binomial Series	0.30537
7	Functions	0.33122
8	Real-life Context	0.08027
9	Complex Numbers	0.25484
10	Integration Applications	0.32649
11	Differential Equations	0.2146
12	Sigma Notation and MOD	0.20282
13	Real-life Context	0.05947
14	Real-life Context	0.07329
15	Real-life Context	0.08448

Fig. 9 Test results of prelim 2016 paper from HCI

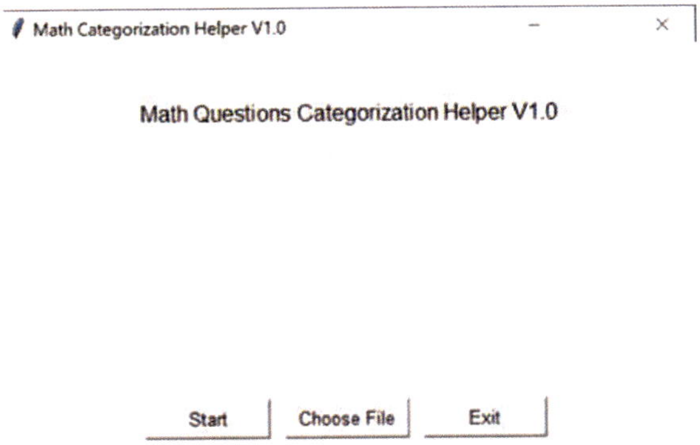

Fig. 10 User interface of math categorization helper V1.0

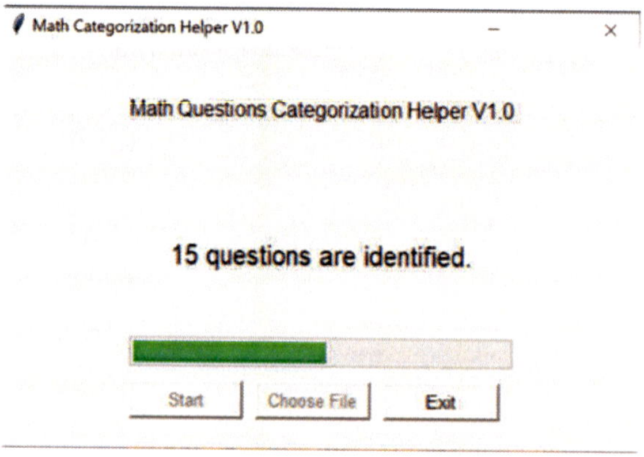

Fig. 11 Math categorization helper V1.0 categorizing questions from selected file

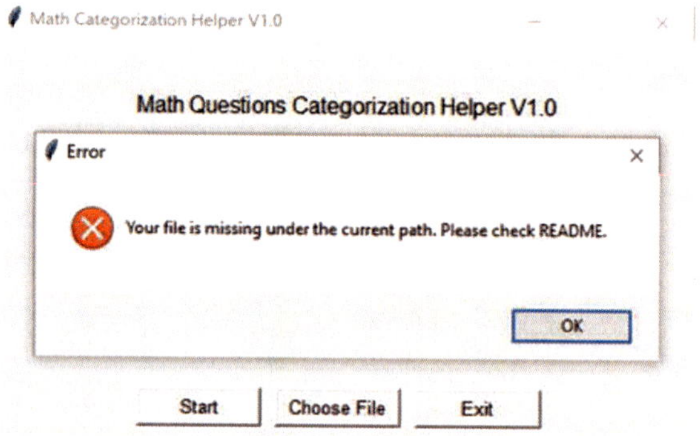

Fig. 12 Error warning of failing to select a specific file at the user interface

the interface. Missing original files in the folder of the final application will lead to another error warning (see Fig. 13). This can be solved by putting all files in the same folder.

A message box (see Fig. 14) will be prompted once the application finishes categorization and outputs the results.

The application is packed by PyInstaller [12] (with command line –onefile and –noconsole) to an executable file. Three.csv files and two packets from NLTK are included under the same file path as the executable file. Then the folder together with the README.txt file is packed to a zip file.

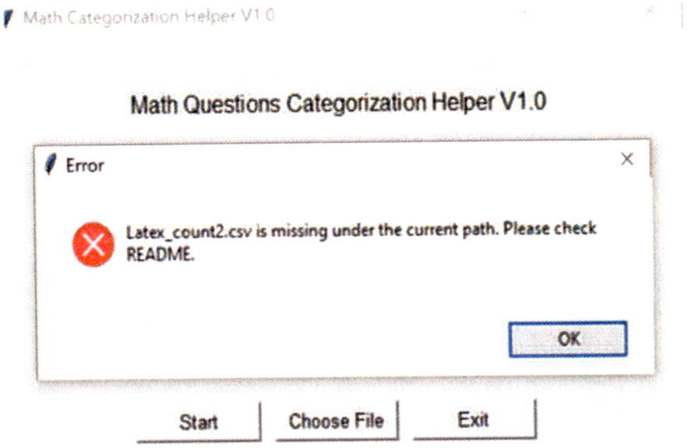

Fig. 13 Error warning of missing original files in the folder of final application

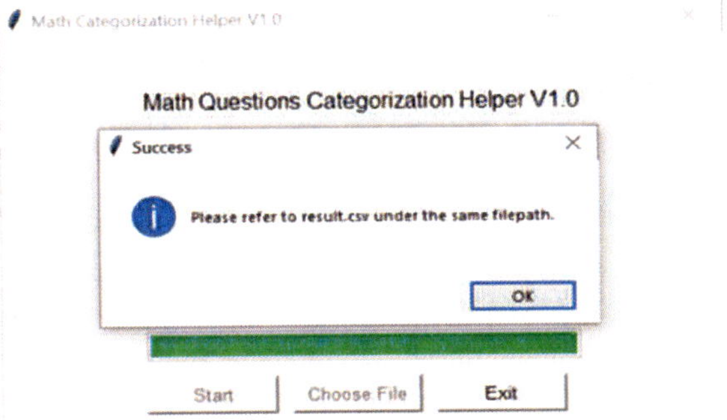

Fig. 14 Message box prompted to users after successfully outputting

4 Discussion

4.1 Small Sample Set

In this case, only a small, labeled sample set can be provided. Therefore, traditional machine learning models may not be well trained due to this constraint. Our algorithm requires a smaller sample set (in average 25 samples per chapter) while achieving high accuracy during the test.

4.2 Time Complexity

On a 3.80 GHz CPU frequency, without user interface, the algorithm takes averagely 5.83 s in total to output the results for categorizing a preliminary examination paper with 15 questions with about 1850 words (including LaTeX expressions). In addition, the initialization takes about 1.33 s.

4.3 Limitation Analysis

Since the format of diagrams and graphs are very different in the given sample set and optical character recognition (OCR) or image classification may be required to understand the picture or extract keywords from a picture, information in diagrams or graphs is omitted in our current algorithm. Therefore, the application cannot deal with questions from chapters where diagrams and graphs are critical in understanding a question (e.g., Functions). OCR and image classification may be used if there are large enough sample sets in future.

It should also be noted that some keywords or questions in the samples may not be useful or sometimes a bit misleading. The LaTeX keywords used in different formulas in different questions are also not very consistent. Although the program still gives out results with high accuracy, the questions in the samples, keywords, and key LaTeX words can be manually filtered to ensure the accuracy of categorization.

4.4 Possible Extension

In future, more topics (e.g., Statistics) in the junior college curriculum can be covered in this program. Furthermore, like mathematics problems, Physics, Chemistry, and Biology problems contain unique terminologies. Thus, the algorithm may also be able to categorize problems in these subjects into corresponding topics since the unique terminology is likely to increase the accuracy of categorization.

4.5 Other Possible Applications of the Algorithm

In addition, this algorithm could be used in recommendation systems for students to find suitable practice questions. Students could input the contexts of questions they have. Then, the algorithm would automatically categorize the question. The system could next display practice questions of the same category to the students. This will be helpful for students to practice or prepare for exams more efficiently.

5 Conclusion

In this study, a computer algorithm and an application to categorize junior college mathematics questions into corresponding chapters are developed. LaTeX keywords were taken into consideration during extraction process of keywords. For this algorithm, Jaccard Index is firstly used to categorize mathematics questions by calculating similarities between the input questions and the 274 sample questions. The application can reduce human work and it has been proved to have a high accuracy. Despite some minor limitations, we believe this categorization algorithm will have more applications in future.

Acknowledgements We would like to express our deep and sincere gratitude to our project mentor, Dr. Kwan Wei Lek, Senior Lecturer, Associate Head of Cluster for Undergraduate Programme and Outreach (Science, Mathematics and Technology), Singapore University of Technology and Design, for providing invaluable guidance and encouragement throughout this project.

We are also extremely grateful to Hwa Chong Institution Mathematics Department, especially to Mr. Ng Say Tiong and Mr. Yee Weng Hong, for their suggestions about this project and help in providing us sample and test sets.

We would also like to thank Ms. Yu Qian, Ms. Goh Keng Wah, and Mr. Loh Tze Wei, teachers in charge of the Computing Talented Programme, for their genuine support throughout this project.

Lastly, we thank all the people who have supported us to complete this project directly or indirectly.

References

1. Sebastiani, F. (2002). Machine learning in automated text categorization. *ACM Computing Surveys, 34*(1), 1–47. https://doi.org/10.1145/505282.505283
2. Beliga, S. (2014). *Keyword extraction: A review of methods and approaches*. University of Rijeka. 10.1.1.704.9230
3. Favale, T., Soro, F., Trevisan, M., Drago, I., & Mellia, M. (2020). Campus traffic and e-learning during COVID-19 pandemic. *Computer Networks, 176*, 107290. https://doi.org/10.1016/j.comnet.2020.107290
4. Canny, S. (2013). *Python-docx 0.8.11 documentation*. https://python-docx.readthedocs.io/en/latest/
5. Bird, S., Klein, E., & Loper, E. (2009). *Natural language processing with python*. O'Reilly Media Inc. https://www.nltk.org/book/
6. Řehůřek, R. (2019). *Gensim Documentation*. https://radimrehurek.com/gensim_3.8.3/auto_examples/index.html
7. Firoozeh, N., Nazarenko, A., Alizon, F., & Daille, B. (2019). Keyword extraction: Issues and methods. *Natural Language Engineering* (pp. 1–33). https://doi.org/10.1017/s1351324919000457
8. Jaccard, P. (1901). Étude comparative de la distribution florale dans une portion des Alpes et du Jura. https://doi.org/10.5169/seals-266450
9. Verma, V., & Aggarwal, R. K. (2020). A comparative analysis of similarity measures akin to the jaccard index in collaborative recommendations: Empirical and theoretical perspective. *Social Network Analysis and Mining, 10*(1). https://doi.org/10.1007/s13278-020-00660-9
10. Pedregosa, F., Varoquaux, G., Gramfort, A., Michel, V., Thirion, B., Grisel, O., Blondel, M., Prettenhofer, P., Weiss, R., Dubourg, V., & Vanderplas, J. (2011). Scikit-learn: Machine learning in Python. *JMLR, 12*, 2825–2830
11. Van Rossum, G. (2020). The Python library reference, release 3.8.2. Python Software Foundation.
12. PyInstaller Development Team. (2021). PyInstaller Documentation. https://pyinstaller.org/en/stable/index.html

Design of Cost-Effective Metasurface for Efficient Broadband Polarisation Conversion

Gareth Chua, Joel Au, and Tse Tong Chia

Abstract The conversion of polarisation of electromagnetic waves has been widely studied due to its applications in antennas and radars. Metasurfaces are a popular method to convert polarisation as they can achieve high polarisation conversion ratio (PCR). Although metasurfaces that can achieve PCR > 0.9 have been designed, many of these designs have other limitations such as narrow bandwidth or thick and expensive substrate. In this paper, we propose two novel cost-effective metasurface designs which can achieve efficient broadband polarisation conversion. We optimise our design parameters using parameter sweep in CST Microwave Studio to obtain the set of parameters which gives the highest PCR for the greatest frequency range. We also test the performance of these designs experimentally using a 19 × 26 array of unit cells. The first design achieves a PCR > 0.9 between 8.00 and 17.85 GHz, covering X-band and K_u-band. The second design achieves a PCR > 0.9 between 8.90 GHz and 12.20 GHz, covering X-band. The results remain similar for wave incident angles of up to 12.5°. Our designs open up many opportunities in cloaking technology and use a cheaper substrate (FR-4), which has not been widely used in other research papers dealing with polarisation converting metasurfaces.

Keywords Metasurface · Broadband · Polarisation conversion · Radar cross-section (RCS) · Electromagnetic (EM) waves · FR-4 · Thin substrate · Air gap · Polarisation conversion ratio (PCR) · X-band · K_u-band

G. Chua (✉) · J. Au
NUS High School of Mathematics and Science, Singapore, Singapore
e-mail: H1710039@nushigh.edu.sg

J. Au
e-mail: H1710010@nushigh.edu.sg

T. T. Chia
DSO National Laboratories, Singapore, Singapore

H. Guo et al. (eds.), *IRC-SET 2022*,
https://doi.org/10.1007/978-981-19-7222-5_14

1 Introduction

Metasurfaces consist of sub-wavelength resonant unit cells that can change the properties of electromagnetic waves upon reflection, such as its polarisation. Polarisation refers to the direction of the electric field vector of an electromagnetic wave, and it is an important property of electromagnetic waves as seen in polarisation-sensitive antennas and in radar cross-section. Therefore, the manipulation of the polarisation of electromagnetic waves is often desirable, and several methods have been developed to achieve this. Metasurfaces in particular are able to achieve high PCR, making the polarisation conversion process much more efficient.

Many different metasurfaces have been developed, using different designs such as L-shaped resonators [1–3], split ring resonators [4–9] and double-headed arrows [1, 10]. However, the designs which were able to achieve PCR > 0.9 had other limitations such as narrow bandwidth or a thick substrate, which is more expensive. Some papers have shown that a thin substrate with an air gap can be used instead of a thick substrate [8]. However, most of these designs were unable to achieve both high PCR and large operating bandwidth. In this paper, we present two new metasurface designs which can achieve efficient broadband polarisation conversion using a thin substrate with air gap. Both designs can achieve PCR > 0.9 for large bandwidth whilst using less than 0.7 mm thick FR-4, which makes the metasurface more cost-effective whilst still being highly efficient.

2 Materials and Methods

We arranged our design patterns to be diagonally symmetric so that the polarisation conversion behaviour is the same for x-polarised and y-polarised waves.

2.1 *Designs*

Figure 1 shows the 3D view of the two metasurface designs. The first design consists of a pair of L-shaped patterns [2] and a circular split ring [6] in the middle. The second design consists of a pair of L-shaped metallic patterns [2] and a double-slit circular split ring resonator [4]. Both designs use FR-4 as the dielectric substrate with 0.035 mm thick copper for the metal patterns and the bottom metal plate. There is an air gap between the substrate and the bottom metal layer for both designs. The optimised unit cell dimensions are shown in the tables in Fig. 1.

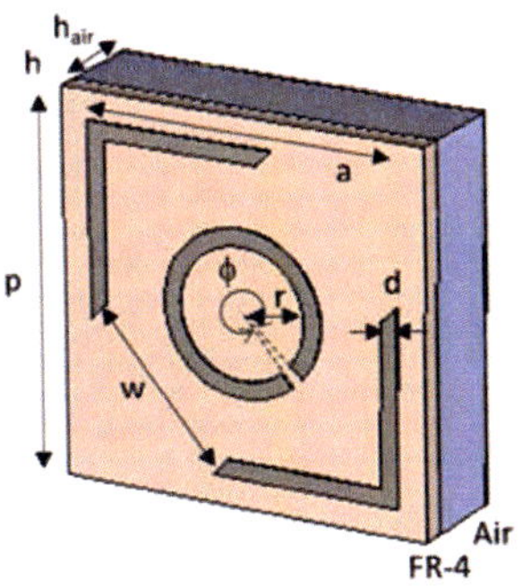

p	10 mm
a	8.5 mm
w	4.8 mm
r	1.7 mm
ϕ	350°
d	0.45 mm
h	0.63 mm
h_{air}	2.37 mm

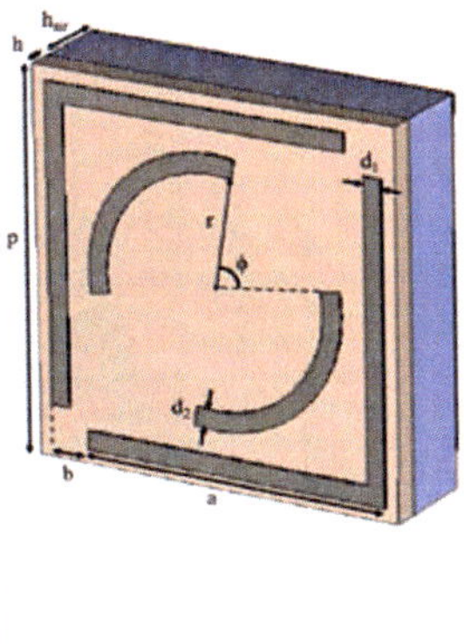

p	10 mm
a	8.27 mm
b	1 mm
r	3 mm
ϕ	70°
d_1	0.5 mm
d_2	0.5 mm
h	0.68 mm
h_{air}	2.12 mm

Fig. 1 3D view of unit cell for (left) Design 1 and (right) Design 2

2.2 *Optimisation Process*

We used CST Microwave Studio to simulate and test our designs. To maximise the frequency range that our metasurfaces could achieve PCR > 0.9,[1] we ran hundreds of simulations on CST for each design to get the best possible combination of parameters. The thickness of the substrate (h) was not varied as we had to use the FR-4 available. The thickness of the air gap (h_{air}) was also constrained by the thickness of the Styrofoam boards we had. We also avoided using a thick air gap to reduce the total thickness of the metasurface. Lastly, we decided on a fixed value of p for both designs prior to experimentation.

Design 1

We noticed that a larger frequency range could be achieved when ϕ is large regardless of the values of the other parameters. When ϕ is large, the length of the ring is increased, and the metasurface resonates at a lower frequency. Therefore, we fixed the value of ϕ to 350°. For the rest of the simulation, we ran a parameter sweep to optimise the values of a, w, r and d. This process eventually gave us the optimised set of parameters shown in Fig. 1.

Design 2

From early simulations, we noticed that d_1 and d_2 had no significant impact on the results, and so, we fixed their values. We noticed that the angle, ϕ, had a large impact on the PCR, and there was a small range of ϕ that we could provide PCR > 0.9 for the desired frequency range of 8.5–11.5 GHz. For the rest of the simulation, we only used three values of ϕ: 60°, 70° and 80°. For all 3 angles, we ran a parameter sweep for each of the three variables r, a and b in increasing intervals of 0.2 mm, which

[1] $PCR = \frac{R_{yx}^2}{R_{xx}^2 + R_{yx}^2}$

made it fine enough to optimise the parameter values. This process eventually gave us the optimised set of parameters shown in Fig. 1.

2.3 Finite Array Experimentation

After designing the unit metasurface cells, we simulated an array of 19 × 26 cells for each design in CST. Figure 2a and b show the 19 × 26 array of cells for Designs 1 and 2, respectively. The air gap for the 19 × 26 arrays was created by gluing Styrofoam between the bottom metal plate and the base of the FR-4 substrate. Figure 3 shows the fabricated 19 × 26 array for Design 1.

We also fabricated the two arrays and measured their reflection coefficients for the *x*- and *y*-polarisations using a bistatic measurement system. The system works by emitting a wave of either *x*- or *y*-polarisation using a transmit horn. The wave

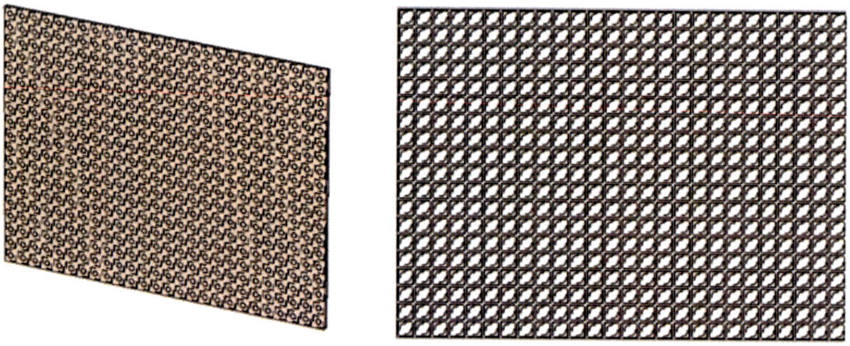

Fig. 2. 19 × 26 array of metasurface cells for (left) **a** Design 1 and (right) **b** Design 2

Fig. 3 Fabricated 19 × 26 array of metasurface cells for Design 1

reflected by the metasurface is detected by the receiving horn for both polarisations. Both the transmit and receive horns are positioned at a bistatic angle of 25°. The bistatic measurement system allows the possibility to measure the performance of our metasurface at various bistatic angles, which we can compare with the simulated results in CST.

3 Results

3.1 *Simulated Unit Cell Performance*

We used the frequency domain solver with unit cell boundary conditions in CST to simulate our unit cell, which forms an infinite array of cells with no edge effects. Figure 4 shows the reflection coefficients for Designs 1 and 2 for the normal incidence case. The 2 subscript letters of R represent the polarisation of the wave: R_{xy} denotes the reflection coefficient for incident y-polarisation and reflected x-polarisation.

As shown in Fig. 4, R_{xx} and R_{yy} fall below -10 dB for a significant frequency range, which means that the magnitude of the reflected co-polarised wave is reduced by at least 3 times. This gives a rough gauge of the frequency range that our metasurfaces can achieve for PCR > 0.9 since an ideal polarisation converting surface maximises the cross-polarisation (reflected polarisation orthogonal to incident polarisation) and minimises the co-polarisation. Figure 5 shows the PCR for Designs 1 and 2. The frequency range for a PCR > 0.9 is 8.00–17.85 GHz for Design 1 and 8.90–12.20 GHz for Design 2. Due to the diagonal symmetry of both Designs 1 and 2, R_{xx} and R_{yy} are expected to be identical, similarly for the R_{xy} and R_{yx} curves. Whilst this is observed for the simulated results in Design 2, there are slight discrepancies for Design 1. The cause is most likely due to the meshing and smallest divisions that we ran the simulation at due to time constraints.

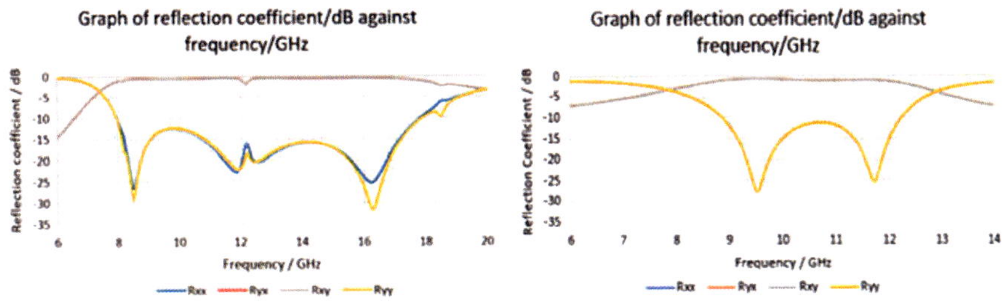

Fig. 4 Reflection coefficient at normal incidence for (left) **Design 1** and (right) **Design 2**

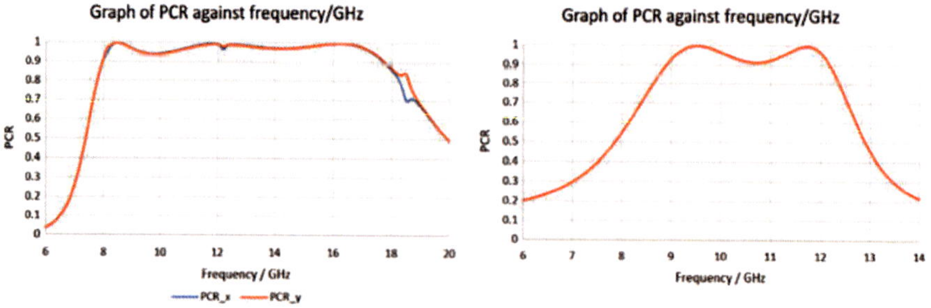

Fig. 5 PCR at normal incidence for (left) Design 1 and (right) Design 2

3.2 *Performance of Finite Array of Cells*

Normal Incidence

We used the time domain solver in CST to obtain the PCR for the finite 19 × 26 array of cells. We note that the simulated frequency range for PCR > 0.9 for the finite array is narrower than that of the unit cell. The frequency range where PCR > 0.9 in Design 1 is 8.00–17.85 GHz for the unit cell (refer to Sect. 3.1) whilst it is 8.30–14.10 GHz for the finite array. Similarly, the frequency range is 8.90–12.20 GHz (refer to Sect. 3.1) in Design 2 for the unit cell whilst it is 9.15–12.00 GHz for the finite array. We attribute the discrepancy to the non-negligible edge effects in the 19 × 26 panels.

As can be seen from Fig. 6, the trend of the experimental results seems to follow closely to the simulated results. However, the upper frequency where the PCR > 0.9 for Design 1 extends to 18 GHz for the measured case, which is beyond the simulated frequency of 14.1 GHz. The discrepancy could be due to the high sensitivity of the metasurface performance to the air gap thickness. When creating the air gap for the prototype, the styrofoam board was not big enough to fit the entire metasurface of Design 1. As such the edges of the panel were flimsy and the air thickness was not uniform across the surface, which could lead to the extended frequency range of the PCR.

For Design 2, Fig. 6 shows that the experimental results give a good fit with the simulated data, thus verifying the design. Compared to Design 1, the styrofoam used for Design 2 is more uniform and larger in size. Hence, the uniform air gap thickness across the panel produces results that are similar to the simulation.

Non-normal Incidence

Using the same time domain solver in CST, we obtained the PCR of both designs for a wave with incidence angle of 12.5°, as shown by the blue lines in Fig. 7. The experimental bistatic results are shown by the red lines in Fig. 7. The experimental PCR for Design 1 follows a relatively similar shape as that of the simulation, except for its extended range, which has been explained by the uneven air gap thickness in

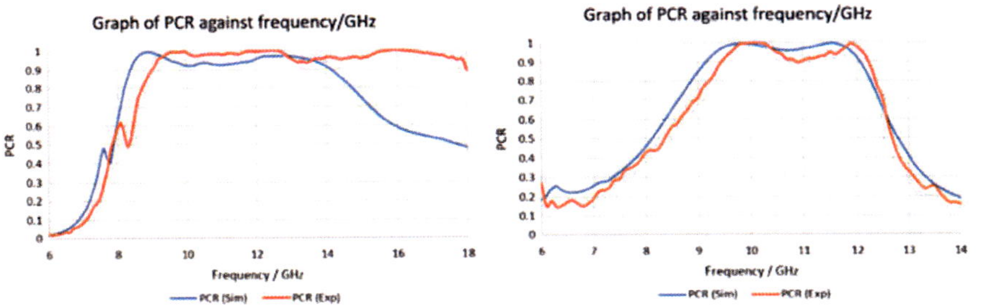

Fig. 6 Comparison of simulated and measured PCR at normal incidence for (left) **Design 1** and (right) **Design 2**

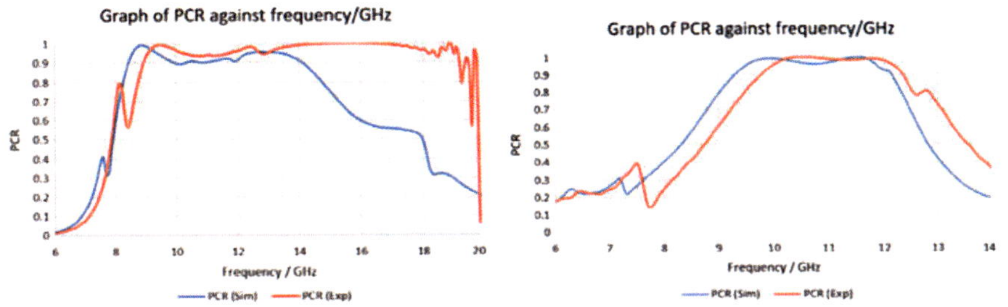

Fig. 7 Comparison of simulated and measured *bistatic* PCR at 12.5° incidence for (left) **Design 1** and (right) **Design 2**

this section—"Normal incidence." Additionally, the experimental PCR for Design 2 follows a relatively similar shape as that of the simulation, besides being shifted to the right by about 0.5 GHz.

4 Discussion

The diagonally symmetric arrangement of the metasurface element will allow equal interaction for both *x* and *y*-polarised waves, making it suitable for polarisation conversion. Many papers employ designs like a double-slit circular split ring resonator [4], split ring resonators [6] and metallic arms [2] to achieve polarisation conversion.

Another reason for the special property of metasurfaces to convert polarisation is that at particular frequencies, certain elements in the design resonates with the wave. Figure 8 shows the surface current of the unit cell of Design 2 for a wave at normal incidence for two different frequencies. Figure 8a illustrates denser greener arrows at the L-shaped arms, indicating their resonance, whilst Fig. 8b shows the resonance of the double-slit circular split ring. The surface current analysis of Design

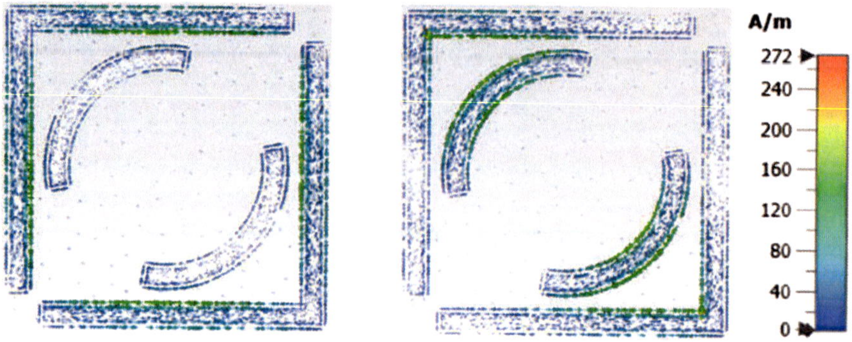

Fig. 8 Diagram of surface current for normal incidence at (left) **a** 9.5 GHz and (right) **b** 11.5 GHz

2 shows how multiple elements can be meticulously designed to allow for multiple resonant frequencies, producing unique behaviours of each metasurface. At resonant frequencies, the direction of induced current determines the type of resonance. If the direction of the surface currents on the top and bottom layers is antiparallel, a magnetic dipole is induced, and it creates a magnetic resonance. If the surface currents are parallel, an electric dipole is induced, and it creates an electric resonance. Both resonances will result in a polarisation conversion as demonstrated by our design performance.

5 Conclusion

In summary, our research has overcome many limitations that previous metasurface designs had. Both our designs have a wide bandwidth. Specifically, Design 1 can achieve PCR > 0.9 for 8.00–17.85 GHz. Design 2 can achieve conversion for 8.90–12.20 GHz, which some designs failed to achieve [6, 7]. We also tested the performance of our metasurfaces at various incidence angles and conclude that both Designs 1 and 2 operate optimally from 0° to 12.5°. Experimental results collected for the two finite 19 × 26 arrays at normal incidence and 12.5° incidence also generally agree with the simulated results, which verify the two designs' functionality. Finally, both the designs used a FR-4 thickness of less than 0.7 mm, which is relatively inexpensive and affordable, as compared to thicker substrates made of materials like Rogers [4, 5, 11]. The two designs presented in this paper open up many opportunities in cloaking technology and use a cheaper substrate (FR-4), not seen in most other research papers dealing with polarisation converting metasurfaces.

Acknowledgements We would like to thank Dr Chia Tse Tong of DSO National Laboratories and Dr Chiam Sher Yi of NUS High School of Mathematics and Science for their guidance and support throughout the entire project. We would also like to thank 3DS for providing the CST Microwave Studio licences for this work as well as for their technical support.

References

1. Sun, H., Gu, C., Chen, X., Li, Z., Liu, L., Xu, B., & Zhou, Z. (2016). Broadband and broad-angle polarization-independent metasurface for radar cross section reduction. *Scientific Reports, 7*. https://doi.org/10.1038/srep40782
2. Jia, Y., Liu, Y., Zhang, W., & Gong, S. (2016). Ultra-wideband and high-efficiency polarization rotator based on metasurface. *Applied Physics Letters, 109*, 051901. https://doi.org/10.1063/1.4960355
3. Kamal, B., Chen, J., Yin, Y., Ren, J., & Ullah, S. (2020). *A broad band polarization converting metasurface for C and X-band applications* (pp. 1–4). https://doi.org/10.1109/ICSPCC50002.2020.9259460
4. Kamal, B., & Chen, J., Yin, Y., Ren, J., & Ullah, S. (2020). *Wide band cross polarization converting metasurface based on circular split rings resonators* (pp. 223–225). https://doi.org/10.1109/ICCEM47450.2020.9219496
5. Guo, W.-L., Wang, G.-M., Chen, K., & Li, H.-P., Xu, H.-X., & Feng, Y. (2019). Broadband polarization-conversion metasurface for a cassegrain antenna with high polarization purity. *Physical Review Applied, 12*. https://doi.org/10.1103/PhysRevApplied.12.014009
6. Mustafa, M. E., Izhar, R., Wahidi, S., Tahir, F., & Abbasi, Q. (2019). *A Broadband 90° Polarization Rotator Metasurface* (pp. 1–2). https://doi.org/10.1109/COMPEM.2019.8778944
7. Mustafa, M. E., & Tahir, F. (2019). *A broadband metasurface for cross polarization conversion applications* (pp. 1–2). https://doi.org/10.1109/COMPEM.2019.8779218
8. Fang, W., Fan, D., Xie, X. Y., Liu, X., Sun, S., & Chen, P. (2019). *A broadband radar cross section reduction metasurface based on polarization conversion and scattering cancellation* (pp. 1–3). https://doi.org/10.1109/CSQRWC.2019.8799173
9. Lin, B.-q., & Guo, J., Lv, L., Liu, Z., Ji, X., & Wu, J. (2019). An ultra-wideband reflective phase gradient metasurface using pancharatnam-berry phase. *IEEE Access* (pp. 1–1). https://doi.org/10.1109/ACCESS.2019.2894133
10. Ameri, E., Esmaeli, S. H., Shakouri, A. H., Sedighy S. H., & Karimian, S. (2020). Ultra-wideband RCS reduction using polarization conversion metasurfaces. In *2020 28th Iranian Conference on Electrical Engineering (ICEE)* (pp. 1–4). https://doi.org/10.1109/ICEE50131.2020.9261054
11. Yang, J., Cheng, Y., Ge, C., & Gong, R. (2018). Broadband polarization conversion metasurface based on metal cut-wire structure for radar cross section reduction. *Materials, 11*, 626. https://doi.org/10.3390/ma11040626

Modelling of Taxi Drivers' Decision-Making and Its Application on Optimising Changi Airport Taxi Management System

Zeyu Cui and Xinrui Shi

Abstract Nowadays, major airports all over the world, including Singapore's world-renowned Changi Airport, often have a large inflow and outflow of passengers due to frequent airline travels. In many circumstances, a comprehensive understanding of the airport's taxi system, which entails modelling and optimising taxi driver's decision-making process, as well as implementing more efficient taxi arrangement corresponding to relevant parameters, is necessary for optimising the performance of airport operations and at the same time, increasing the profit of taxi drivers. As such, the problem is addressed in the perspectives of both airport managers and taxi drivers, making the interests of all stakeholders well-considered.

Keywords Monte Carlo method · Decision-making model · Queuing theory · Airport taxi system

1 Introduction

The ongoing pandemic has a profound impact on Singapore's tourism industry, with Changi Airport being a prominent example. Despite the hard hit by Covid-19 pandemic recently, as one of the world's major aviation hubs, it is arguably true that in the foreseeable future, with the expanding vaccine access on a global scale and increased easing of COVID-19 related restrictions, the passenger volume will recover to its standard level within a time span of two to three years.

In this light, it is imperative for the authority of Changi Airport to devise a comprehensive approach to tackle the huge volume of tourists in the most efficient manner, as well as for the taxi drivers to optimise their strategies to increase their revenue so as to recover from the loss due to the pandemic.

Z. Cui (✉) · X. Shi
Hwa Chong Institution, Singapore 269734, Singapore
e-mail: cuizeyuresearch@gmail.com

H. Guo et al. (eds.), *IRC-SET 2022*,
https://doi.org/10.1007/978-981-19-7222-5_15

Hence, we introduced the fundamental problem of our paper: given the influence of various observable factors such as flight arrival, number of existing taxis, passenger inflow, etc., are we able to establish a viable mathematical model to offer the best strategies to satisfy both the profit-maximising objective of taxi drivers as well as the efficiency-maximising objective of airport managers?

2 Objective

(1) Providing an optimal decision-making model for taxi drivers arriving at the airport.
(2) Establishing a comprehensive model to suggest optimisation of the taxi management system for the airport managers.
(3) Providing a returning decision-making model to help drivers decide whether it is beneficial to return to the airport to pick up the next passenger.

3 Assumptions and Notations

3.1 Assumptions

(1) The impact of speed difference on fuel consumption rate is ignored, i.e. fuel consumption per kilometre is a constant value.
(2) Cost of wear and tear of the vehicle is considered constant.
(3) All statistics are true, accurate and credible.

3.2 Notations

See Table 1.

4 Methodology

4.1 Decision-Making Model for Taxi Drivers Arriving at the Airports

Taxi drivers face two options after drop-off at the airport's departure area:

Table 1 Symbols and notations used in this paper

Notation	Meaning	Unit
φ	Fuel cost coefficient	\$/km
S_0	Distance between the airport and the downtown	km
C'	Miscellaneous costs	\$
$P(s)$	Taxi fare of a single trip	\$
$N(t)$	Number of taxi passengers	passenger/h
$K(t)$	Ratio of passengers leaving the airport by taxi	
$f(t)$	Number of flight arrivals	flight/h
$\rho_f(t)$	Number of passengers per incoming flight	passenger/flight
t_{per}	Time taken between the first taxi entering and last taxi leaving	h
E	Average time spent by taxis picking up one passenger	h
L	The length of one parking lot	m
v	Average taxi speed at the boarding area	m/s
t_r	Driver's reaction time	s
R_i	Revenue under option i	\$
ΔP_i	Profit under option i per unit time	\$/h
γ	Ratio between the distance travelled and S_0	km
T_x	Time taken by short-distance drivers to finish another ride	h
T_y	Time taken under option y	h

(A) Proceed to the arrival area and wait to pick up new passengers [1]. They will have to queue, which incurs an amount of waiting time and then sending passengers to their destination. The profit under this option can be expressed as:

$$P_A = R_A - C_A \tag{1}$$

(B) Leave the airport immediately without passengers and continue operation afterwards. The profit under this option can be expressed as:

$$P_B = nP_i(t) - \left(\varphi \sum_{i=0}^{n} X_i + C'\right) \tag{2}$$

The Singapore taxi fare consists of a basic charge p_0 and various kilometre prices. The kilometre price is calculated by the product of distance travelled S and the unit fare per kilometre p_s. Therefore, total fare of a single trip $P(s)$ is as follows:

$$P(s) = p_0 + [p_s \cdot S] \tag{3}$$

The number of passengers who intend to take taxis to leave airport at time t, $N(t)$ is determined by the product of proportion of passengers who choose to transit by taxi and the total number of arriving flight passengers. Therefore, $N(t)$ can be expressed as:

$$N(t) = K(t)\, f(t)\, \rho_f(t) \tag{4}$$

where $\rho_f(t)$ is the number of passengers per incoming flight at time t; $f(t)$ is the number of flight arrivals at time t; $K(t)$ is the ratio of passengers leaving the airport by taxi.

A decision-making model for taxi drivers is devised in order to find the optimal decision strategy after arriving at the airport. The group base the decision between options A and B on profits per unit time. The waiting time is approximated using queuing model:

$$M/M/c/\infty/\infty/FCFS \tag{5}$$

4.2 Model of Pick-Up Points

Based on the number of passenger arrivals at the airport, structure of taxi pick-up points at different airports varies. To determine the structure of pick-up points in Changi Airport, the group paid an in-person visit to Changi Airport. The layout of Changi Airport taxi boarding area consists of two taxi lanes. The waiting area is labelled as area A, and the pick-up area is labelled as area B (see Fig. 1).

Consider the "pick-up" process as a Poisson process, the time taken between the first taxi entering and last taxi leaving, in the same round can be expressed as:

$$t_{\text{per}} = E + t_{\text{in}} + t_{\text{out}} = \int_0^\infty \lambda e^{-\lambda t} \left(1 - e^{-\lambda t}\right)^{n-1} t\, \mathrm{d}t + \frac{2}{k}\left[\frac{nL}{v} + (n-1)\, t_r\right] \tag{6}$$

where k is taken as 2 in Changi Airport's context. Since our desired outcome is to optimise the efficiency of this "pick-up process," we set up a model taking the number of taxis leaving area B per hour as our objective function: $\exists \frac{n}{k} \in N^*$

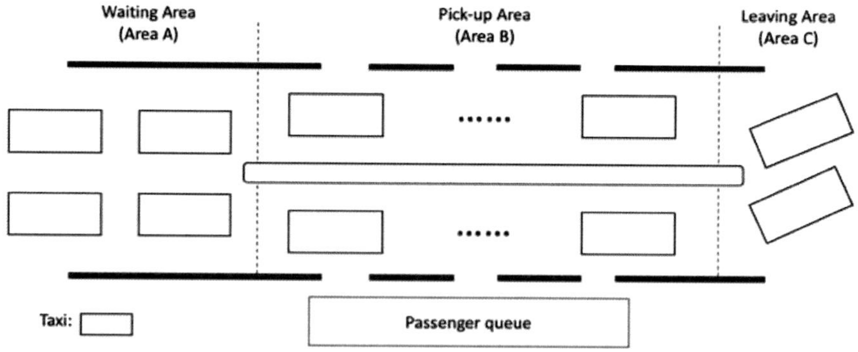

Fig. 1 Structure of pick-up points

$$\text{s.t.}\max\left(\frac{n}{t_{\text{per}}}\right) \tag{7}$$

4.3 *Returning Model for Taxi Drivers*

It can be deduced that after the drivers have picked up the passengers at the arrival area, the taxi drivers carrying passengers for only a short distance are likely to gain less profits than long-distance drivers as they earn less from the metered system due to shorter distance travelled. Therefore, a returning decision-making model should be designed to help drivers make wiser decisions to minimise the gap in profits amongst all drivers. For taxi drivers carrying passengers for only a short distance, they have two options to choose between:

X. Return to the airport to pick up passengers.
Y. Empty cruise to the urban area and provide taxi service there afterwards.

It is observed that taxi drivers travelling a short distance have a higher tendency to return to the airport, due to relatively less time costs for travelling back and the lucrative surcharge given at the airport. Let the ratio between the distance travelled by them and the distance between the airport and the urban area be γ, which will be used in the following discussions.

4.3.1 Option X

If drivers decide to return to the airport to pick up passengers, the total time taken can be expressed as:

$$T_x = \gamma T + T_q(t) + T \tag{8}$$

where T is the time taken for a ride from the airport to the urban area. $T_q(t)$ is the queuing time at the airport.

As such, the profit earned under option X, P_x, is

$$P_x = R_x - C_x = R_x - \varphi(\gamma S) - \varphi S - C' = R_x - \varphi(1+\gamma)S - C' \tag{9}$$

where R_x is the revenue earned from the second ride, S_0 is the distance between the airport and the urban area, and C' is all miscellaneous costs.

4.3.2 Option Y

If drivers decide to continue heading towards the urban area without passengers, the composition of the total time T_y is

$$T_y = (1 - \gamma)\,T + \sum_{i=1}^{n} T_{\mathrm{di}} \tag{10}$$

As such, the profit earned under option y, P_y, is

$$P_y = nP_i(t) - \varphi(n\overline{X}_i) - \varphi X_0 - C' \tag{11}$$

In this case,

$$\varphi X_0 = \varphi\,(1 - \gamma)\,S \tag{12}$$

To find the profit per unit time under the two options, we can equalise time to solve for n.

$$T_x = T_y \tag{13}$$

$$\gamma T + T_q\,(t) + T = (1 - \gamma)\,T + \sum_{i=1}^{n} T_{\mathrm{di}} \tag{14}$$

To know whether it is beneficial for short-distance drivers to return to the airport after dropping off the first group of passengers, the returning decision-making model can be constructed using profit per unit time.

5 Data Analysis and Evaluation

5.1 Profit-Maximisation for Taxi Drivers

The result is derived by using government database [2–4] and following the procedures below:

(1) Find $T_q(t)$ using state transition equation based on the concept of queuing theory.
(2) Derive ΔP_A from P_A and $T_q(t)$.
(3) Similarly, derive ΔP_B from P_B and T'.
(4) Plot ΔP_A and ΔP_B onto the same time graph for comparison.

In Fig. 2, the orange line represents the calculated profit earned per unit time under option A. The blue line represents the calculated profit earned per unit time under option B. In the perspective of taxi drivers, option B is preferred when the blue line is above the orange line; as in this case, it is more profitable to leave the airport straight away instead of waiting at the airport for a ride. During other times when the red line is above the blue line, option A is preferred due to short queuing time.

From the graph, we can determine 6 points of intersection. As points of intersection represent the equilibrium state, 6 equilibrium states are located at 02:07, 03:00, 06:18, 11:32, 15:53 and 19:10. Therefore, strategy for taxi drivers at Changi Airport to maximise their profit is (see Fig. 2):

Fig. 2 Graph of option A and option B

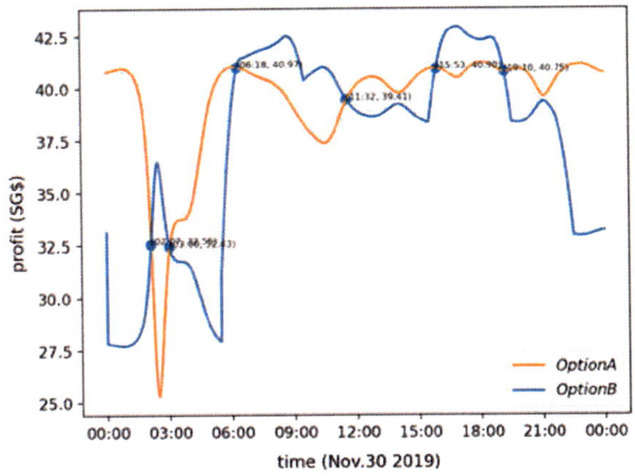

Fig. 3 Profits when $\gamma = 0.2$

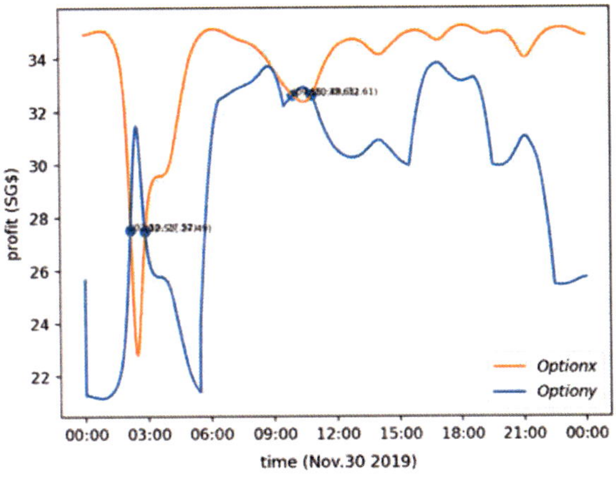

(1) During 03:00–06:18, 11:32–15:53 and 19:10–02:07, it is better for taxi drivers to choose option A (i.e. queuing up for passengers)
(2) During 02:07–03:00, 06:18–11:32 and 15:53–19:10, it is better for taxi driver to choose option B (i.e. leaving airport without passengers).

With reference to the criteria for short-distance taxi drivers in other regions, the group stipulates that the distance travelled must be no more than 10 km. Taking γ as 0.2, 0.3 and 0.4 separately, we obtain the result as shown in Figs. 3 and 4.

In the graphs below (see Figs. 3 and 4), the orange line represents the profit per unit time earned under option x, and the blue line represents the profit per unit time earned under option y, when $\lambda = 0.2$ and 0.3, respectively. Intersection of the two lines represents the equilibrium points where choosing option x and option y yield equal profits. At times when the orange line lies above the blue line, it is more profitable to choose option x, in other words, to return to the airport for another ride. Therefore, we can conclude from the two graphs that:

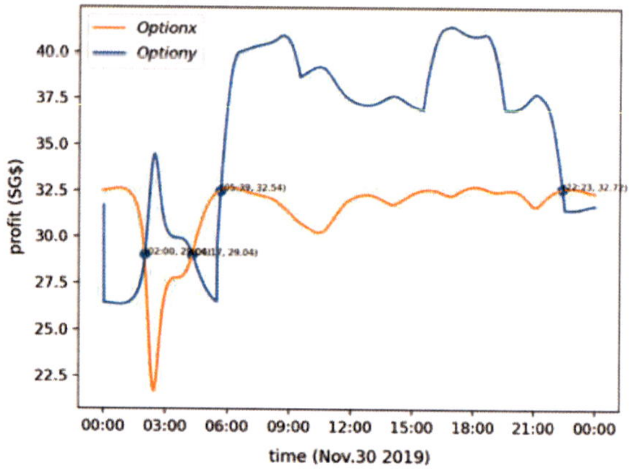

Fig. 4 Profits when $\gamma = 0.3$

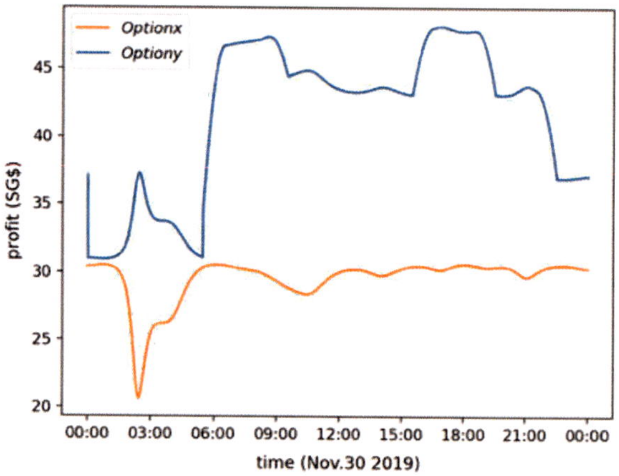

Fig. 5 Profits under two options when $\gamma = 0.4$

(1) When $\gamma = 0.2$, during 02:52–09:55 and 10:42–02:02, it is better for taxi drivers to choose option x.
(2) When $\gamma = 0.3$, during 04:17–05:39 and 22:23–2:00, it is better for taxi drivers to choose option x.

By calculation, the total duration in a day to choose option x is 1343 min min and 299 min when $\gamma = 0.2$ and 0.3, respectively. It shows that the duration in a day where choosing x is more profitable falls drastically when value of γ increases, meaning that with the increase in the distance of each ride, it is less profitable for taxi drivers to return to the airport for another ride, which indirectly demonstrates the reliability of our model.

When $\gamma = 0.4$ (see Fig. 5), it is worth noticing that the blue line is always above the orange line, meaning that it is always more profitable for taxi driver to choose option y throughout the day. Following the trends discussed above, it can be safely concluded that any value of $\gamma \geq 0.4$ will result in option y being the more profitable choice at all time. In other words, when the distance of the first ride exceeds 6.86 km, it is always better for taxi drivers to continue driving to the urban area rather than returning to the airport.

5.2 Optimal Distribution of Pick-Up Points

Here, we use the Monte Carlo method to determine the optimal solution. For Changi Airport, the length of one parking lot is 5 m, and the speed limit is 5 m/s. Usually, the human reaction time for a driver is 2 s. The result is derived by following the procedures below:

(1) Set the initial value $n = 2$;
(2) If $n > 100$, jump to **step 8**.
(3) Generate n random variates of an exponential distribution with parameter $\lambda = \frac{1}{\bar{t}_{\text{board}}}$, where $\bar{t}_{\text{board}}$ is the average boarding time of passengers, taken to be 20 s;
(4) Set E to be the maximum of the n random numbers and calculate the value of t_{per};
(5) Repeat **step 3** and **step 4** for 100 times and store each value of t_{per} as $t_{\text{per}}(i)$;
(6) Set t_{per} to be mean$\{t_{\text{per}}(i)\}$;
(7) Calculate the value of the objective function, set $n = n + 2$, return to **step 2**;
(8) Plot a graph of the objective function against the variable n.

Following the above procedure, the result graph is obtained (see Fig. 6). According to the graph, the taxi flow per hour increases sharply at first, then stabilises to a maximum. As it is impractical to add parking lots in an unrestricted manner, coupled by the increased management difficulty and probability of accidents, we base on our choice of the number of parking lots on 80% of maximum efficiency. As such, the result of our model would be setting 22 boarding points, or 44 parking lots in total.

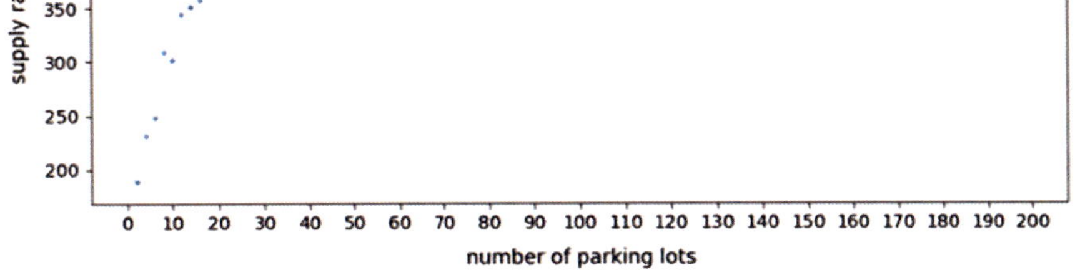

Fig. 6 Relationship of taxi efficiency with number of parking lots

Table 2 Robustness test against fluctuation in λ, L and v

Variable	Relative change (%)	Boarding points	Taxi flow per hour	Change of taxi flow (%)
λ	−10	66	499.2	4.52
	−5	50	474.4	−0.67
	0	44	477.6	0
	5	44	480.8	0.67
	10	40	481.6	0.84
L	−10	48	509.6	6.7
	−5	46	492.8	3.18
	0	44	477.6	0
	5	42	464	−2.85
	10	40	452	−5.36
v	−10	44	448.8	−6.03
	−5	44	463.2	−3.02
	0	44	477.6	0
	5	46	492	3.02
	10	46	506.4	6.03

5.2.1 Evaluation

To test the robustness of the model against changes to its dependent constants, we modified λ, L and v to observe changes in the outputs.

According to Table 2, the target output, taxi flow per hour, fluctuates up and down with at most 6.7%, showing a strong level of robustness of our model.

6 Discussion

6.1 Limitation

1. Some of the data required by the model are not directly available. Therefore, these data are obtained by estimation and processing of other data indirectly. This results in some errors.
2. There is no quantitative analysis for some factors that influence decision-making (e.g. weather, season), which are thus not considered in this paper.

6.2 Future Extension

1. Investigate and gather more detailed taxi operation data (e.g. GPS data) to make some of the parameters in the models more precise.
2. Include other relevant parameters (such as weather and season) into the models to improve accuracy.

7 Conclusion

In conclusion, the group has successfully established a robust model for Changi Airport's taxi management which considers both taxi drivers' profit-maximising objective as well as the airport management's efficiency-maximising objective. For the former, the model is useful in deciding the more profitable choice to make during different time period of a day, which could facilitate taxi drivers in their decision-making when put into application. For the latter, based on calculated results from the model, as well as in consideration of practicality and increased management difficulty, the group propose an optimum number of 22 boarding points, i.e. 44 parking lots in total.

Further work may be conducted to factor in more complex parameters to improve the accuracy of our model. For practical use, the model can be incorporated into mobile applications, allowing taxi drivers to have easy access to its suggestions calculated based on real-time data, thus helping them to better gauge the circumstance and make beneficial decisions.

References

1. Changi Airport Transfer. (2014, March 2). Singapore Airport Arrival and Departure Transfer: Taxi Singapore. Taxi Singapore I Taxi Cab in Singapore. https://www.taxisingapore.com/changi-airport/. Accessed January 19, 2022
2. Civil aircraft arrivals, departures, passengers and mail, Changi Airport, Monthly. Data.gov.sg. (n.d.). https://data.gov.sg/dataset/civil-aircraft-arrivals-departuassengers-and-mail-changi-airport-monthly. Accessed January 19, 2022
3. Public transport utilisation—Average trip distance. Data.gov.sg. (n.d.). https://data.gov.sg/dataset/public-transport-utilisation-average-trip-distance. Accessed January 19, 2022
4. Taxi fare in Singapore. (n.d.). Taxi Fare in Singapore. Taxi Prices in Singapore. Taxi Price Calculator. https://www.numbeo.com/taxi-fare/in/Singapore. Accessed January 19, 2022

Identification of Mosquito Larvae in Drains Using Deep Learning

Eungi Hong and Mirdhini Shri Rajaram

Abstract This study assesses the feasibility of using a deep learning model to alleviate the high rates of dengue infections in Singapore, which has reached as high as 2700 cases in the first half of 2021 (as of 1 June). The high number of cases can be partially attributed to the ineffectiveness of the current measures put in place. In public spaces, the NEA places Gravitraps to trap female Aedes mosquitoes. In homes, NEA officers make visits to flats and inspect homes one by one. Apart from being a slow and tedious process, there are also limitations to inspections. It is risky and dangerous to check places like roof gutters, and inconvenient to check small crevices such as in flowerpots and corners. In addition, NEA officers apply Temephos, the sand granular insecticide, to drains, the top breeding habitat for mosquitos. However, it is simply impossible to monitor and get rid of all potential breeding sites present in over 7000 km of drains manually. Therefore, we propose a more efficient measure—a deep learning model able to identify the presence of mosquitos when attached to autonomous robots traveling in drains and alert authorities of the location of identified breeding sites. This model could also be implemented into other tools such as extendable sticks for NEA officers to use during home inspections.

Keywords Dengue · Autonomous robot · Deep learning model

1 Introduction

In order to limit the high rates of dengue infections in Singapore, inspectors investigate households one by one, and apply sand granular insecticide to drains [1]. As these solutions to remove mosquito breeding sites are extremely manpower-intensive and time-consuming, a remote autonomous robot to replace the human resources needed would be extremely useful. Hence, this research will build and assess the feasibility of a binary image classifier that can differentiate between mosquito larvae-infested waters and clearwater* using convolutional neural networks. It can be used by a

E. Hong · M. S. Rajaram (✉)
Raffles Institution, Singapore, Singapore
e-mail: mirdhini.r05@rafflesgirlssch.edu.sg

H. Guo et al. (eds.), *IRC-SET 2022*,
https://doi.org/10.1007/978-981-19-7222-5_16

remote autonomous robot that would travel in drains and alert authorities of the exact identified locations. Alternatively, the software could also be incorporated into long extendable sticks for inspectors to use to check in places they are unable to manually, inside small cracks and crevices, or on top of roofs, for instance. This will reduce the over-reliance on manpower for monitoring drains, and home inspections can be done more easily and quickly, consistent with Singapore's objective of achieving manpower sustainability.

Therefore, this study may serve as an effective and practical solution to the rampant dengue infections in Singapore.

The contributions of this paper are twofold:

- A trained binary classifier model that can distinguish mosquito larvae-infested waters from clearwaters*.
- Optimal conditions that enhance the model's performance (such as the number of epochs).

This study aims to investigate the feasibility of a deep learning model that can identify the presence of mosquito larvae in drains and other common breeding sites. "Clearwater" refers to water absent of mosquito larvae.

The remainder of this paper is structured as follows. Section 2 reviews related literature. In Sect. 3, the methodology is proposed. In Sect. 4, experimental settings and results are reported and analyzed. Section 5 illustrates the conclusion and future work.

2 Literature Review

The purpose of this section is to present a review of convolutional neural networks and possible factors that affect the performance levels of image recognition when used in the context of identifying mosquito larvae. We viewed Udemi's "Convolutional Neural Networks with Tensorflow in Python" course to understand basic concepts of Deep Learning, CNN, and image classification.

2.1 Convolutional Neural Network (CNN)

Convolutional Neural Network (CNN) is a Deep Learning algorithm that when given an image, appoints importance to various aspects in it and differentiates them from one another. A CNN is able to capture spatial dependencies in an image. Different layers exist within a CNN [2].

Firstly, in the convolutional layer, the Kernel (K) carries out the convolution operation. It shifts at its specified stride length, every time performing a matrix multiplication operation between K and the portion P of the image over which the kernel is hovering. It moves over the complete height and width of the image. For colored images with multiple channels (RGB), the kernel has the same depth of 3. Traditionally, the first convolutional layer is responsible for capturing the low-level features such as edges and colors [2].

Secondly, the pooling layer is used to reduce the spatial size of the convolved feature. This helps reduce the computational power required to process the data via dimensionality reduction. It is also useful for extracting dominant features, thus ensuring effective training of the model. Max Pooling, the more common type of pooling, returns the maximum value from the portion of the image covered by the Kernel. Max Pooling serves as a noise suppressant [2].

Depending on the image complexities, the number of these layers are altered for capturing details of different levels. However, the addition of layers also means more computational power is needed [2].

2.2 *Factors Affecting CNN Model*

There are many factors in the model that will affect its performance. Examples of factors are number of epochs, batch size, learning rate, etc.

Epochs. The number of epochs is a hyperparameter that defines the number of times the algorithm goes through the entire training dataset. One epoch means that each sample in the training dataset updates the model once [3].

With a high number of epochs, "overfitting" might occur, which means that the model is fitted too perfectly to the training examples, and hence has limited capability for generalization. This can be seen from Fig. 1 where each increase in epoch (x axis) significantly affects the accuracy of the model.

On the other hand, with a low number of epochs, "underfitting" might occur, which means that the model has failed to learn the data thoroughly enough. This can be seen from Fig. 2 where there is no impact of the training data on the overall accuracy of the model.

Model convergence occurs when both loss and accuracy stabilize after a number of epochs. This means that it reaches a state where the loss is very close to the global/local minimum and it remains within the specific loss range regardless of any more epochs. Additional training will not improve the model [3].

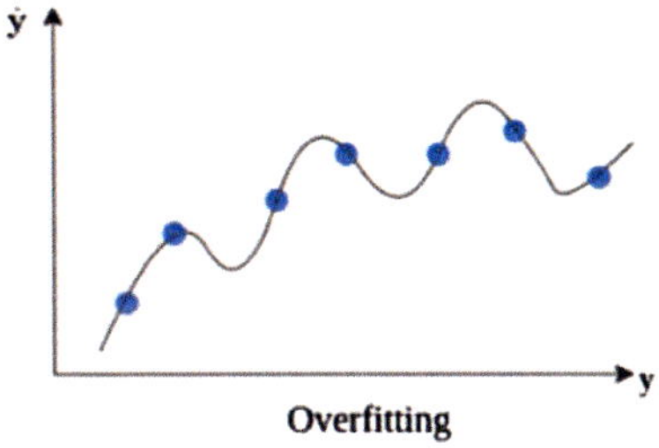

Fig. 1 Aylin Tokuç's overfitting graph (2022) [4]

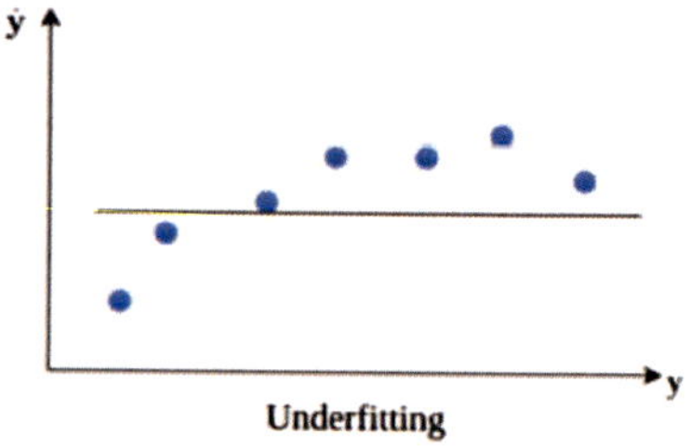

Fig. 2 Aylin Tokuç's underfitting graph (2022) [5]

The learning curve graphs are examined to decide on model convergence. Generally, "loss versus epoch" or "accuracy versus epoch" graphs are plotted. During the training, the loss decreases, and accuracy increases with the number of epochs.

Both loss and accuracy stabilize when the neural network eventually converges [3]. When building the model, if the number of epochs is set too low, the training stops before the model converges. On the other hand, if the number of epochs is set too high, overfitting may occur and computing power and time are wasted [3].

A widely adopted solution to this is early stopping. The main idea in "early stopping" is to stop when model loss starts to increase, or when the accuracy starts to decrease. To decide on the change in these values, the model is tested with the validation set after every epoch [3].

Batch size. Larger batch sizes are used to allow computational speedups. However, too large of a batch size will lead to poor generalization [6].

Both smaller and bigger batch sizes have their benefits. On one extreme, using a batch equal to the entire dataset guarantees convergence to the global optima of the objective function. However, this is at the cost of slower, empirical convergence to that optima. On the other hand, smaller batch sizes have been empirically shown to have faster convergence. This is because smaller batch sizes let the model "start learning before having to see all the data." However, the model is not guaranteed to converge to the global optima. It will bounce around the global optima. Therefore, with no computational constraints, it is often advised to start with a small batch size then increase it [6].

It has been observed that smaller batch sizes not only have faster training dynamics but also generalization to the test dataset versus larger batch sizes. It is commonly known that there is a "sweet spot" for batch size between 1 and the number of training images that provides the best generalization. This "sweet spot" usually depends on the dataset and the model in question. The reason for better generalization is attributed to the existence of "noise" in small batch size training. Since neural network systems are extremely prone to overfitting, seeing many small batch sizes, with each batch being a "noisy" representation of the entire dataset, will cause a sort of "tug-and-pull" dynamic. This "tug-and-pull" dynamic prevents the neural network from overfitting and hence performing badly on the test set [6].

2.3 *Previous Studies*

A previous study has shown that identification of mosquitoes themselves is possible, though with images that were highly enlarged used as dataset. While this reflects potential in developing deep learning models that can be used for identification of mosquito larvae as well, it would be highly likely that images from real-world situations would be of smaller size and lower resolution [7].

3 Methodology

This section will discuss the data collection as well as model building, training, and optimization process (Fig. 3).

The above image shows the rough outline of the robot and the arrow points to the location of the camera where images will be captured. The camera will be angled within the range of 0 to 45° to the ground surface. This angle was taken into consideration when collecting images.

3.1 *Data Collection*

At the start of our data collection process, a simulation to depict drain water was created using common household items. However, this method of image collection was unsuccessful as accurately depicting the conditions in drains with household items was difficult. For example, when obtaining images of clearwater, the bottom of the container (which was different from that of a drain) was distinctly visible in the image, contributing only to misleading noise that confused the machine. Therefore, other alternatives were explored. A variety of Youtube videos of site visits of dengue breeding sites, drains and rainwater were collected, downloaded and converted to JPG images using FreeStudio's "DVDVideoSoft Free Video to JPG Converter." They best represented the type of images a moving robot would capture. This is because the

Fig. 3 A robot prototype with a camera [8]

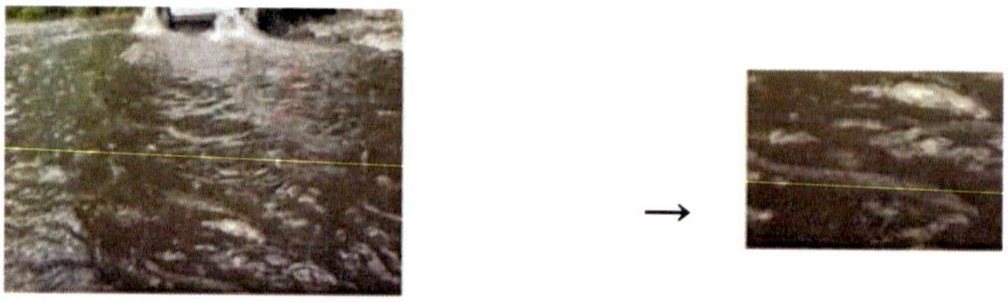

Fig. 4 Modifications to image [9]

camera angle, as well as the objects captured, were in motion, exactly replicating the types of images the robot will capture while moving along drains.

Photos were carefully selected by taking into account different variables such as the angle of the camera, the approximate distance of the camera from the water level, and how zoomed in the images were.

The images gathered were cropped to remove irrelevant parts that would not be present in drains. Below is an example of an image of clearwater, where trees and cars are removed (Fig. 4).

3.2 Parameters of the Dataset

The images collected were then preprocessed in the code so that they could be used to train the model. The images were then standardized for the loader, at the height and width of 180 pixels. The dataset was split such that 80% of the images were used for the training dataset and 20% of the images were used for the validation dataset. Using the seed variable, the order of the images was also shuffled so that the machine would not learn from images from the same video back-to-back and pick up on irrelevant similarities between them. Additionally, by using tf.keras.layers.Rescaling, the RGB channel values that originally had a range of [0, 255] were rescaled to [0, 1] so that the values were compatible for a neural network. The two classes were labeled as "larvawater" and "clearwater" for training.

3.3 The Model

There are 3 blocks of CONV2D+Max Pooling layers + dropout in the model. The output of the CNN will be the input for the 3 fully connected layers. There are 2 Dense layers. The last layer then has 2 outputs for its 2 classes: larvawater and clearwater (Fig. 5).

Initial Run. As seen in the plateaus of the Fig. 6, the model converges at approximately 60 epochs. Convergence means that the model has reached its optimal stage.

Layer (type)	Output Shape	Param #
sequential_1 (Sequential)	(None, 180, 180, 3)	0
rescaling_3 (Rescaling)	(None, 180, 180, 3)	0
conv2d_7 (Conv2D)	(None, 180, 180, 16)	448
max_pooling2d_7 (MaxPooling2D)	(None, 90, 90, 16)	0
conv2d_8 (Conv2D)	(None, 90, 90, 32)	4640
max_pooling2d_8 (MaxPooling2D)	(None, 45, 45, 32)	0
conv2d_9 (Conv2D)	(None, 45, 45, 64)	18496
max_pooling2d_9 (MaxPooling2D)	(None, 22, 22, 64)	0
dropout_4 (Dropout)	(None, 22, 22, 64)	0
flatten_2 (Flatten)	(None, 30976)	0
dense_5 (Dense)	(None, 128)	3965056
dense_6 (Dense)	(None, 2)	258

Total params: 3,988,898
Trainable params: 3,988,898
Non-trainable params: 0

Fig. 5 The summary of the model

However, when images from a different video were uploaded for validation, the model was unable to identify the presence of mosquito larvae accurately. This could have been due to the lack of variety in the dataset. Since numerous images were obtained from each video, the images in the validation set had similarities to those in the training set, irrelevant to the presence of mosquito larvae. The model could have picked up on these details instead, resulting in (falsely) high performance levels.

Data augmentation. Following this run, images were gathered from a wider variety of videos, and the time interval between each frame in the video was increased so that no two images would look similar (Fig. 7).

In total, 750 images for larvae water and 818 images for clearwater were gathered.

Additional training data was also generated through the random transformation of the images, using "layers.RandomRotation," "layers.RandomZoom" set to 0.1, and "layers.RandomFlip". It is known that 1000 images per class is the minimum requirement for deep learning models. However, taking into consideration the 20% of the images used as "test set," the total number of images used for training per class was only approximately 600. Therefore, as a more efficient method to gather more data, data augmentation was used, allowing the model to be exposed to a more varied dataset.

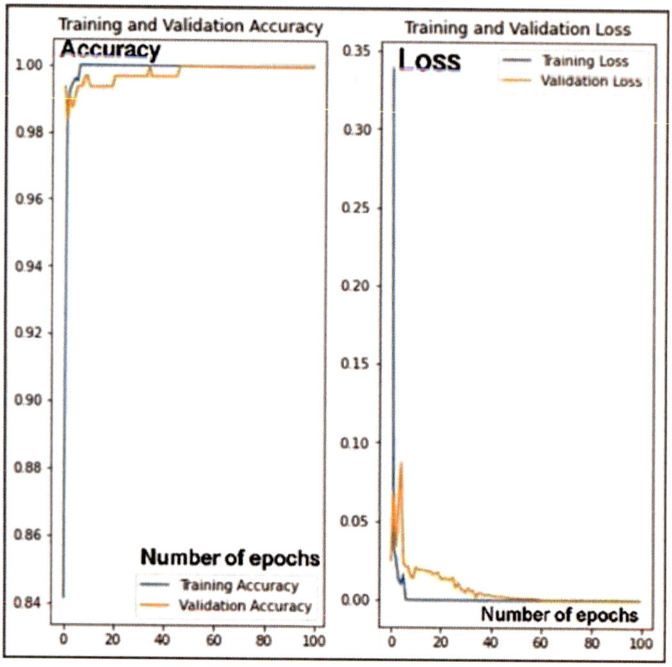

Fig. 6 Graph of initial run

Furthermore, the Dropout function was added. Dropout is a technique where randomly selected neurons are ignored during training. They are "dropped-out" randomly. This means that their contribution to the neural network is temporarily removed. This is believed to result in multiple independent internal representations being learned by the network [10].

The effect is that the network becomes less sensitive to the specific weights of neurons. This in turn results in a network that is capable of better generalization and is less likely to overfit the training data [10].

Figure 8 shows the results after the improvement of images and data augmentation.

3.4 Optimization of Epochs

```
earlystopping=tf.keras.callbacks.EarlyStopping(
    monitor="val_accuracy", patience=5, verbose=1,
    mode='max', restore_best_weights=True)
```

Using the function tf.keras.callbacks.EarlyStopping, it was ensured that the model would stop training after the validation accuracy hits the maximum possible value of 1.00. The value of the patience function was set to 5, which means that the model

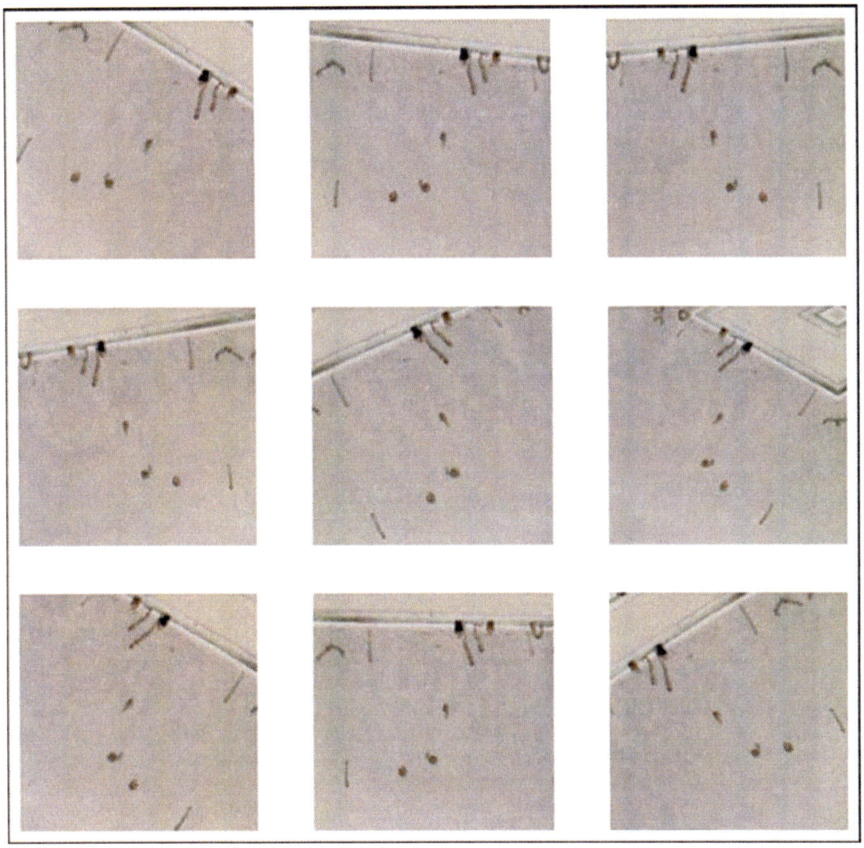

Fig. 7 Augmented examples of same image

would stop training at 5 epochs after it reaches the validation accuracy of 1.00. As seen from graph 3, the model stopped training at 9 epochs.

However, this function was ineffective. This is because it stopped the training without considering whether the graph plateaued after the point where it reached maximum accuracy. As seen in Fig. 9, the training stops at epoch 9 since the model hit accuracy of 1.00 at 4 epochs. However, there is a drop to 0.9936 at 6 epochs instead of plateauing. Therefore, the training stopped before convergence. Hence, it can be concluded this function is not effective in determining the optimal number of epochs.

Our next alternative was to analyze the graph to find out the optimal number of epochs. The graph reaches the accuracy of 1.00 and loss of 0.00 at the 63rd epoch and remains relatively consistent from then on. Therefore, since model convergence requires the value to be constant for a few epochs, it can be said that the optimal number of epochs is approximately 68.

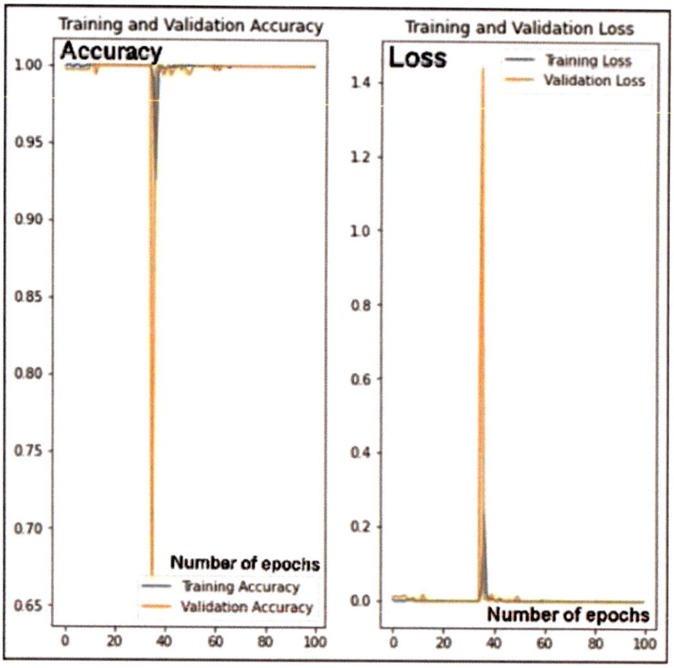

Fig. 8 Graph after data augmentation

4 Experimental Settings and Results

The software used to create and train the deep learning model is Google Colab. As stated in Sect. 3.4, by analyzing the graph, the optimum number of epochs is 68. The resulting final validation accuracy of the model is 1.00, as shown in graph 4 (Fig. 10).

3 images of varying difficulties from each class were then used as inputs in the model and the average confidence of the model was obtained (Tables 1 and 2).

To answer our aims and objectives, it can be concluded that it is feasible with an accuracy of over 96% to detect the presence of mosquito larvae in waters using deep learning. The percentage confidence of the clearwater image 1 and larvawater image 3 is significantly lower than the average. For clearwater image 1, this could be due to pattens of the ripples of the clearwater that resemble the shape of mosquito larvae. For clearwater image 3, the confidence could be lower due to the larvae being present in only corners of the image. Therefore, further fine tuning would be required to increase the confidence of the model under all circumstances.

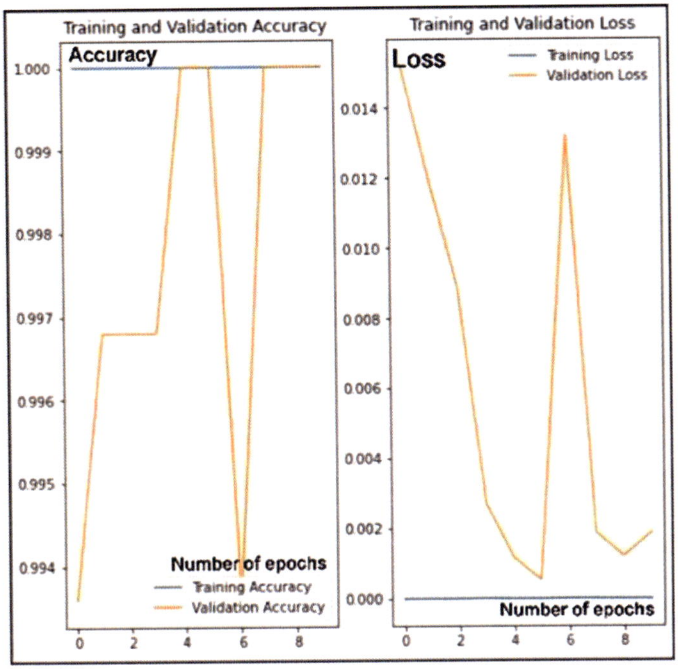

Fig. 9 Graph with early stopping

5 Conclusion

In the future, we plan to do the following works:

1) Use an amphibious robot to collect images from drains, which is the exact location where it will be deployed, forming a more accurate dataset.
2) Install this image recognition model in the software of an amphibious robot so that it is able to perform its function in real-time while traveling through drains.
3) Allow the robot to send real-time images to authorities every time it concludes the presence of mosquito larvae in the drains. This will allow authorities to confirm with their own eyes and come to the location to eliminate the potential Aedes mosquito breeding site.

In addition, our model can also be developed for other uses. In the case of manual home inspection done by NEA officials, they have to be able to look into deep crevices between flower pots and roof gutters that are difficult to carry out manually or pose a safety risk. Hence, this model can also be developed into an app so that it can act as the "third eye" and added assistance for the dengue inspection officers, as seen in Fig. 11.

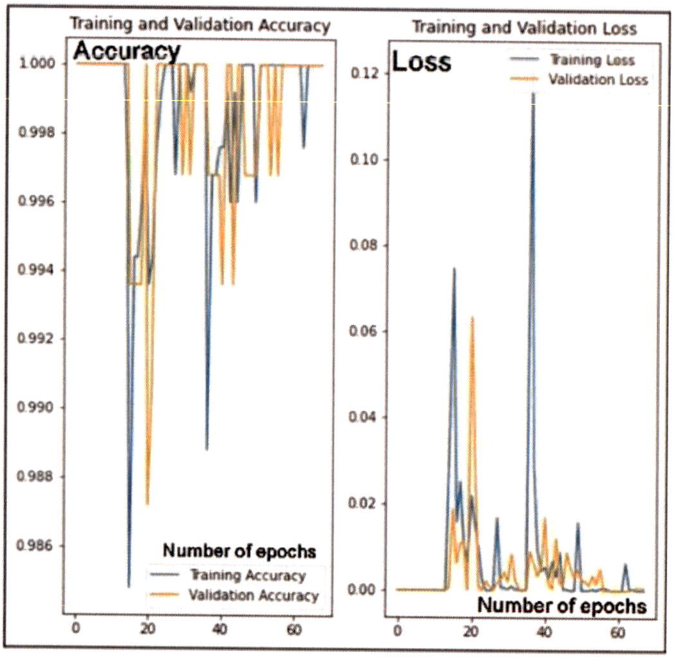

Fig. 10 Final graph

Table 1 The test images used to evaluate the confidence of model

larvawater image 1	larvawater image 2	larvawater image 3
		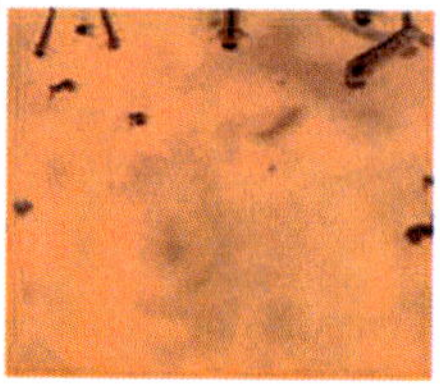
clearwater image 1	clearwater image 2	clearwater image 3

Table 2 The confidence of model for each test image

Class	Confidence of model/%			
	Image 1	Image 2	Image 3	Average
larvawater	99.85	100.00	89.00	96.28
clearwater	90.73	100.00	99.70	96.81

Fig. 11 Proposed application of model [11, 12]

References

1. National Environment Agency. (2020, June 18). *Why is NEA placing mosquito traps outside my house? Will this increase my chances of being bitten by mosquitoes?* Retrieved January 14, 2022, from https://www.nea.gov.sg/media/nea-vox/index/why-are-nea-officers-scattering-chemicals-into-drains-and-are-they-safe
2. Saha, S. (2018, December 16). *A comprehensive guide to convolutional neural networks—the ELI5 way*. Retrieved January 11, 2022, from https://towardsdatascience.com/a-comprehensive-guide-to-convolutional-neural-networks-the-eli5-way-3bd2b1164a53
3. Baeldung (2021, February 27) *Epoch in Neural Networks*. Retrieved January 12, 2022, from https://www.baeldung.com/cs/epoch-neural-networks
4. Aylin Tokuç, A. (2022, January 1). Overfitting Retrieved January 15, 2022, from https://www.baeldung.com/cs/ml-underfitting-overfitting
5. Aylin Tokuç, A. (2022, January 1) Underfitting Retrieved January 15, 2022, from https://www.baeldung.com/cs/ml-underfitting-overfitting
6. Shen, K. (2018, June 20) Effect on batch size on training dynamics. Retrieved January 12, 2022, from https://medium.com/mini-distill/effect-of-batch-size-on-training-dynamics-21c14f7a716e
7. Kittichai, V., Pengsakul, T., Chumchuen, K., Samung, Y., Sriwichai, P., Phatthamolrat, N., Tongloy, T., Jaksukam, K., Chuwongin, S., & Boonsang, S. (2021, March 1). Deep learning approaches for challenging species and gender identification of mosquito vectors. Retrieved August 2, 2022, from https://www.nature.com/articles/s41598-021-84219-4
8. FAULHABER Asia Pacific Pte Ltd. *Different versions of robots, depending on the application*. Retrieved January 16, 2022, from https://www.faulhaber.com/en/markets/factory-automation-robotics/sewer-robot/

9. Post 10 (2020, July 17) *Draining massive flood by unclogging drains, big whirlpools.* Retrieved January 16, 2022, from https://www.youtube.com/watch?v=MV9dOHRDb3c&t=157s
10. Brownlee, J. (2016, June 20) *Dropout regularization in deep learning models with Keras.* Retrieved January 12, 2022, from https://machinelearningmastery.com/dropout-regularization-deep-learning-models-keras/
11. Bilanol/Getty. Retrieved January 16, 2022, from https://www.houselogic.com/organize-maintain/home-maintenance-tips/fast-fixes-common-gutter-problems
12. sbs. Retrieved January 16, 2022, from https://www.sbsmobile.com/fra/en/photography/6330-telescopic-selfie-8018417210525.html

Mobile Computer Vision for Scout Teams

Chun Fong Matthew Aaron Chan, Shan Fei Faith Goh, Lai Hui Charmaine Lee, Rui Ying Caitlin Mok, and Kin Loong Bryan Ng

Abstract This project uses new advancements in computer vision to effectively identify threats in windows for use during urban operation. To make an objective decision, comparisons were made between R-FCN (with ResNet101), Faster R-CNN (with ResNet50), ResNet50SSD and Faster R-CNN (with ResNet50), based on the criteria of Ease of Training, Ease of Implementation, Accuracy, Latency and Computational Resources, to discern the most effective neural network available for threat identification with the use of mobile devices. Based on statistics, available literature and further evaluation, EfficientDet-LiteD3 was selected as the model architecture for training. The model was developed using CVAT, Roboflow and Colab Pro Notebook. While training the model, additional data was gathered on quantisation options and optimal training parameters and the correlation between annotated images and accuracy was analysed. The trained EfficientDet-LiteD3 model achieved an accuracy rate of 75.9% on the test set, maintaining a latency between frames of 167 ms on a Samsung S20.

Keywords Object detection · Deep learning · Convolutional neural network · Mobile device

---

C. F. M. A. Chan (✉)
St. Joseph Institution, Singapore, Singapore
e-mail: matthew.aaron.chan.chun.fong@sji.edu.sg

S. F. F. Goh
Raffles Institution, Singapore, Singapore
e-mail: 23YGOHS125A@student.ri.edu.sg

L. H. C. Lee
Dunman High School, Singapore, Singapore
e-mail: lee.laihui.charmaine@dhs.sg

R. Y. C. Mok
Eunoia Junior College, Singapore, Singapore
e-mail: 2022.caitlin.mok@ejc.edu.sg

K. L. B. Ng
Defence Science and Technology Agency, Singapore, Singapore
e-mail: NKINLOON@dsta.gov.sg

H. Guo et al. (eds.), *IRC-SET 2022*,
https://doi.org/10.1007/978-981-19-7222-5_17

1 Introduction

Given the age of technological advancement, specifically in computer vision (CV), it is imperative that military equipment begins exploiting modern technology to increase efficiency. This project aims to assist in this transition, as we assimilate neural networks for more accurate threat identification in urban areas.

With modernisation, a greater proportion of buildings in urbanised areas have been converted into high-rise structures. This poses a greater challenge when detecting threats, as the numerous windows present on the side of these buildings must be constantly monitored for the presence of combatants. This leads to vulnerability when key threats are missed amid concrete jungles, where scores of high-rise buildings are in view at once. Our application addresses this problem by providing a solution for threat detection through windows that can be deployed on scout teams and on autonomous robots. This enables scout teams on urban missions to detect threats in windows more easily, drastically reducing the risk associated with advancing into unfamiliar territories, and equipping forces with means to scout such territories out accurately and effectively without additional risks to human life.

2 Materials and Methods

2.1 *Initial Review*

Based on this implementation, we conducted an initial review to understand the current capabilities and limitations of CV, research prior breakthroughs in its lengthy history and discover resources crucial to our workflow.

Image classification, the mainstream form of computer vision, is unsuitable for this project, due to the high level of uncertainty in threat detection. We opted to focus on the field of object detection instead, which can localise multiple threats simultaneously. After narrowing down the most promising model infrastructures for object detection, a thorough analysis for each infrastructure was conducted.

2.2 *Establishing Criteria for Analysis*

Through extensive research [1], four models were shortlisted for in-depth analysis.

1. **R-FCN (with ResNet101).** It uses position-sensitive score maps to address a dilemma between translation invariance in image classification and translation variance in object detection [2].

2. **Faster R-CNN (with ResNet50).** An adapted version of Fast R-CNN which uses feature maps generated by ResNet50 for a Regional Proposal Network (RPN), which subsequently generates Regions of Interest (ROI) for object localisation [3].

   We also considered models employing Single Shot Detector (SSD), which uses the output from image classification 'backbones' to predict the locations of bounding boxes and classify them in one pass of the network.
3. **ResNet50SSD.** A model employing skip connections and an SSD, simplifying the network while tackling many challenges plaguing cutting-edge models [4].
4. **EfficientDetLite.** An upgraded version of MobileNetSSD which utilises ingenious scaling algorithms and innovations to FPNs to improve model performance [5].

We decided on 4 criteria to assess these shortlisted models: **Ease of Training, Ease of Implementation, Accuracy**, and **Latency and Computational Resources**.

Ease of training was a concern due to time constraints. Models which required a large number of annotated images or long periods of time to train could jeopardise our ability to fine-tune parameters and improve the final accuracy of our model.

Ease of implementation took into consideration the availability of Android Libraries, the effort required to convert respective models to TFLite and the adaptability of each model onto a smartphone. Greater ease of implementation would shorten the time needed to implement our custom model while making it easier for teams with little to no prior coding experience to use the model as well.

These two criteria had less significance in our analysis, since they had no direct impact on our model performance and were not relevant once we successfully converted the models and found a suitable demo application.

Accuracy was defined as the percentage of predicted bounding boxes with an IoU (Intersection over Union) of over 50% with our ground truth boxes (Ref. to Fig. 1). Bounding boxes with this IoU can reliably pinpoint threats. Accuracy is imperative to the success of our model in field conditions, as small differences in accuracy may become the difference between life and death. A higher accuracy also reduces the likelihood of false positives which may mislead or distract soldiers.

Lastly, **latency** was considered, as a faster model can provide more predictions every second. These models recognise and identify threats in a short time, allowing for detection even when threats appear for a split second.

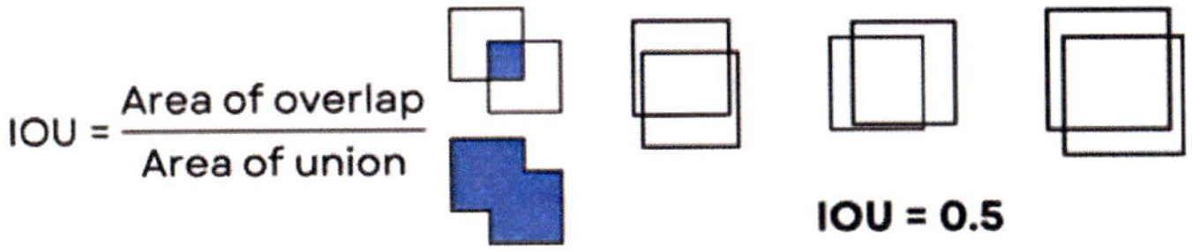

Fig. 1 A diagram showing intersection over union and boxes with IOU 50%

Given the correlation between the **amount of computations required** and **latency**, we combined these into 1 criterion. This is paramount, as an excessive number of FLOP/s (floating point operations per second) would overload the CPU/GPU, resulting in our application hanging or crashing constantly [6]. A lower CPU/GPU usage would also extend battery life, especially important given the inability to charge the phone during operations and the application requiring prolonged use.

Accuracy and latency are inversely related, as models with greater accuracy need more time to analyse each frame. We concluded that latency had higher priority, as accuracy could be improved without affecting latency by using better cameras and higher resolution images. Increased latency also has other effects like high computational costs and excessive battery usage. Thus, we decided on a 200 ms limit for latency, within which we would choose the model with the greatest accuracy.

2.3 Analysing the Models

For the **ease of training** criterion, SSD models require more images to train due to their inability to accurately differentiate objects with similar categories, such as windows with and without humans. However, they require less time to train given only one pass of the network is required to provide a prediction. EfficientDetLite also has a drastically reduced parameter count compared to competitors, resulting in faster predictions from it. On the other hand, Faster R-CNN takes the most time to train as its anchor boxes take a long time to converge on optimal values and its multiple constituent parts must be trained individually. R-FCN also performs worse than SSD models when comparing training time but is still 2.5–3× faster than Faster R-CNN.

For **ease of implementation**, ResNet50SSD and EfficientDetLite can be easily converted for mobile use due to support in the TensorFlow2 (TF2) Model Zoo [2], TFLite Converter and TF Model Maker Library. However, Faster R-CNN is unsupported by the TFLite Converter and requires extra time, effort and coding expertise to convert. Likewise, R-FCN is only available on the TF1 Model Zoo, an outdated resource, which would result in suboptimal model performance [7]. TF1 models were also saved in a different format from TF2 models, and thus required additional time and effort during TFLite conversion.

For **accuracy**, we used literature values from Common Objects in Context (COCO) to assess the models (Ref. to Table 1). COCO is an extensive and challenging object detection database on which models are evaluated for their ability to detect everyday objects, such as bicycles, backpacks and kites [8]. We felt that COCO metrics could predict a model's accuracy on windows, given that they are a common occurrence in daily life, much like the items in the database.

For **latency**, we converted models into TFLite format and tested them on the TFLite Android Object Detection Demo, which allowed us to quickly swap in models

Table 1 Table of COCO accuracy analysis data

Model name	COCO accuracy
Faster R-CNN (with ResNet50)	29.3% [2]
R-FCN (with ResNet100)	30.4% [9]
EfficientDetLite	26.4–43.18%* [10]
ResNet50SSD	34.3% [2]

* due to scalable nature, see Fig. 6 or 'Final Decision'

with beginner coding experience, while its simplistic nature allowed us to troubleshoot bugs as they arose. We then ran the application on a Samsung S20 phone, obtaining latency metrics for respective models.

We plotted our findings on the accuracy and latency of several models, allowing us to visualise the different latency-accuracy trade offs each model brought to the table. To ensure we had a wide array of models to compare, we also incorporated several non-shortlisted models into our analysis to increase our sample size (Ref. to Fig. 2).

Final Decision: EfficientDetLite. While EfficientDetLite may require a greater excess of images, it makes up for this with its more favourable training time and ease of implementation. In addition, EfficientDetLite has an exceptional accuracy-latency trade off, with superior accuracy even when disregarding the 200 ms latency threshold. Out of the 5 scalable models included in EfficientDetLite, we specifically chose the D3 variant, as it was the best model below the latency threshold.

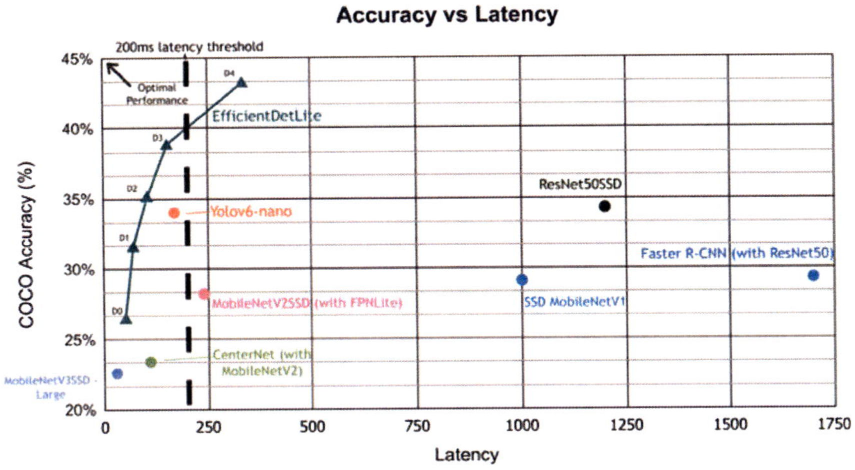

Fig. 2 Graph of accuracy over latency for the models

2.4 Annotations

For reliable and accurate results, a large dataset was needed to identify key characteristics consistently present in windows with vastly different appearances. The images the model was trained on were available online and taken under various lighting conditions to increase accuracy (Ref. Figs. 3 and 4). Google Lens functions were also used to increase the number of images. Visipics was then used to identify and delete duplicate images. Annotating on Computer Vision Annotation Tool (CVAT) requires a strict set of guidelines to ensure consistency and optimal model performance: images labelled must have no diagonal objects, no overlapping bounding boxes and extremely tight bounding boxes [11]. This is crucial, given that bad input data would result in bad output data from the model.

Fig. 3 Identification of windows with humans

Fig. 4 Identification of windows without humans

2.5 Model Learning

We used our custom training pipeline to allow the model to identify windows with and without humans. First, we transferred the annotations from CVAT to Roboflow, enabling us to quickly modify the train-valid-test split and the image augmentation techniques employed (see Sect. 3.1 for further details). Following augmentation, we imported the modified dataset into a Colab Pro Notebook, where we made use of two P100 GPUs to train our models at notable speeds. Colab Notebooks also ensured that model training was consistent, easily reproducible and semi-automated, allowing us to obtain results quickly during the fine-tuning process where many different training runs had to be performed with slightly different parameters and input data. After exporting and quantising the model, it was tested on the demo application, providing us with our final product and its corresponding latency metrics.

3 Results

Prior to model training, we conducted various experiments to optimise our final model.

3.1 Overfitting Prevention

Overfitting happens when a model learns the details in the training data to the extent that it negatively impacts the performance of the model on unseen data, as the noise or random fluctuations in the training data is learned as concepts by the model.

We came up with multiple solutions to prevent this. For example, we tested each augmentation individually to prevent overfitting and increase model accuracy, repeating said tests 3 times to reduce the effect of outlier train/valid/test splits on the reliability of our results (see Sect. 3.3 for more details). We then tested combinations of promising augmentations, making sure that they worked together well. Initially, we adopted commonly used augmentations like shear and rotate. However, they caused overfitting, as bounding boxes would include non-window parts which the model may identify as windows. Our final combination of augmentations was 90° rotate, flip, crop, greyscale, 3px Gaussian blur and brightness. Our second finding was that a greater batch size reduced the chance of overfitting. This was contrary to prior research, which suggested that a lower batch size increases the amount of noise and prevents overfitting. Our last finding was that the size of the dataset used was proportional to the number of epochs before overfitting started.

From Fig. 5, following the implementation of these measures, the validation loss fluctuated less and eventually plateaued at the graph's lowest validation loss level, indicating that the maximum accuracy was reached without any overfitting.

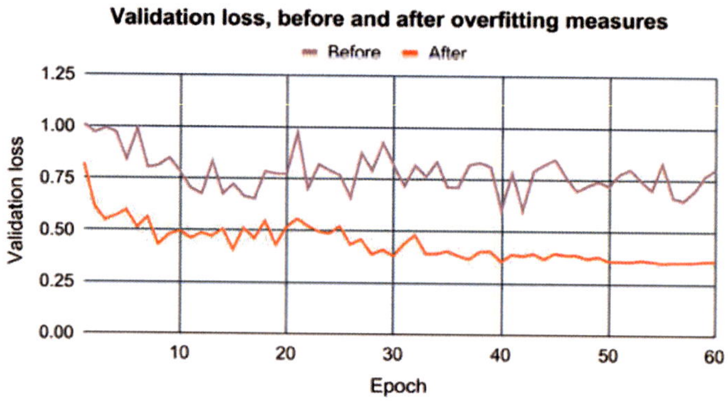

Fig. 5 Graph of validation loss as training progresses

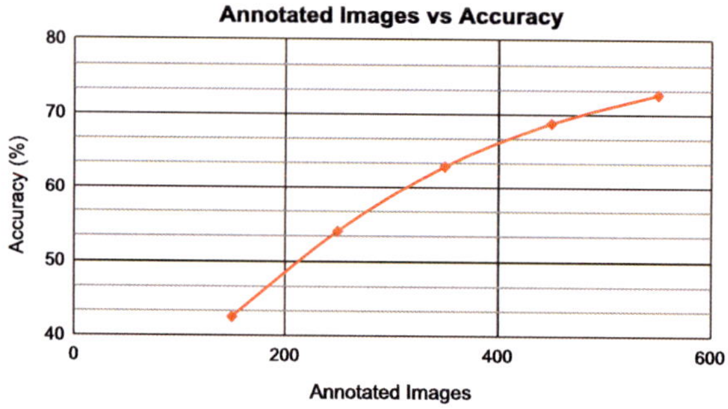

Fig. 6 Trade-off between accuracy and the number of annotated images utilised in a EfficientDetLite-D3 model

3.2 *Annotated Images Versus Accuracy*

We also decided to research how the number of annotated images affected the accuracy of the model. To ensure a fair comparison, we kept the number of images constant in the test set, while adhering to a consistent ratio between the validation and train sets. Through plotting the correlation between the image data set size and its accuracy, we can determine the minimum number of images needed for maximum accuracy, given that accuracy plateaus due to diminishing returns (Figs. 6 and 7).

Using an exponential regression and Desmos graphing software, we fit a curve to our obtained data points. This best-fit exponential curve had equation:

$$y = \left(-7.3718 * 10^{-7}\right)\left(2^{-0.00506073(x-5229.26)}\right) + 82.7362$$

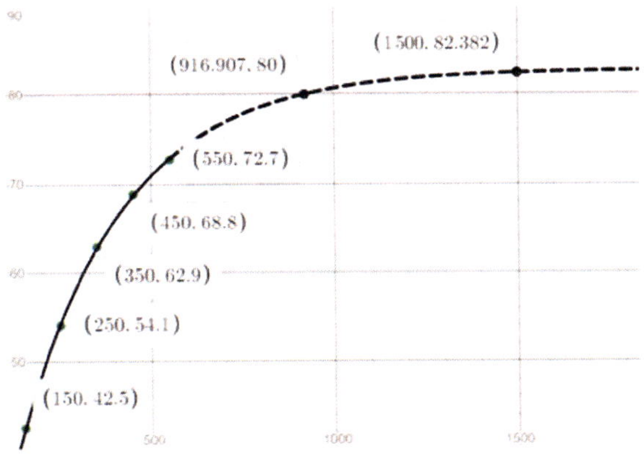

Fig. 7 Extrapolated graph illustrating the predicted correlation between accuracy and the number of annotated images

Based on this projection, it is predicted that the maximum obtainable accuracy of the model is 82.7362%, with the model reaching 80% accuracy with 917 annotated images and 82.5% accuracy at 1615 images.

3.3 *Train/Validation/Test Split*

Another key factor in the final performance is the split of the input data, as it affects the amount of training data and our ability to monitor overfitting. The format train%/validation%/test% was utilised to describe the splits in the following section (Fig. 8).

To ensure that the splits were the only independent variable, image augmentation techniques were kept constant, and the number of epochs used for training was set such that it remained proportional to the size of the training dataset (based on findings of Sect. 3.1). We also ran each test twice, except for the splits with a 5% test split, which we tested thrice. 90-5-5 was the most appropriate training split for our dataset.

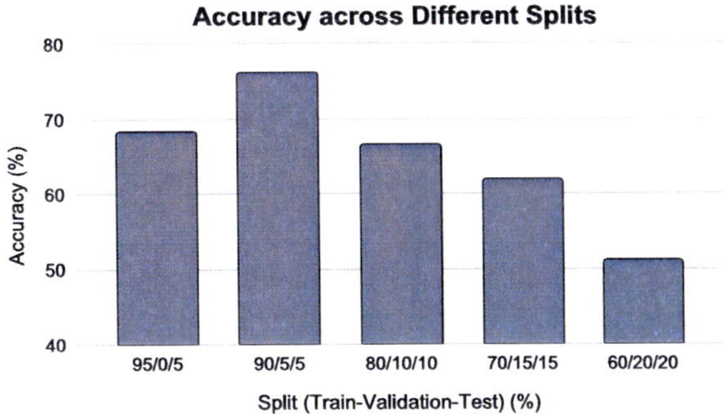

Fig. 8 Graph of split against accuracy

Table 2 Table of Quantisation has various options with different accuracy-latency trade offs

Quantisation option	Accuracy (%)	Latency (ms)
Dynamic	75.947	340
Float16	76.605	236
Int8	75.983	165

3.4 *Quantisation*

In deep learning, quantisation is a process by which a neural network's lengthy, high-latency floating point numbers are approximated into low-bit width numbers, which take up less storage space and increase the speed of inference, albeit lowering accuracy slightly. As quantisation has various approximation algorithms which have different accuracy-latency trade offs, an analysis was done to find the best option specifically for our implementation.

As the range of accuracy values was considerably lower than the range of latency values (Ref. to Table 2), we deemed it insignificant (in this analysis), and thus chose full-integer quantisation for its low-latency value.

3.5 *Final Results/Conclusion*

To train our final model, we first augmented our 547 images with the finalised augmentations (see Sect. 3.1), creating 1295 augmented images. These images were then split into train, validation and test folders in the ratio 90/5/5, which the model trained on for 60 epochs while maintaining a batch size of 16 (maximum possible on P100). Following training, we applied full-integer quantisation to the model, and achieved a final accuracy of 75.983% on the custom test set with 167 ms of latency on average on a Samsung S20.

4 Discussion

4.1 *Difficulties*

While Visipics can remove duplicate images, it is only able to detect similar copies of images; hence, images with very different quality or additional watermarks may not have been identified by the application.

Additionally, EfficientDetLite and other leading models do not currently have support for polygons, making it detrimental to train the model on diagonal windows (Ref. to Fig. 9). As such, the camera must be held vertically and parallel to the windows for accurate rectangular box detection, which is not feasible for urban

Fig. 9 An example of a window that was difficult to annotate

missions where teams may have to move at high speeds and do not have time to properly angle their phones for accurate detection.

Furthermore, due to our small team and time constraints, we were unable to label a large number of images and cannot reach the maximum potential of the model. We were also unable to get a hold of many real-life images accurately emulating field conditions and had to rely on stock photos to mimic them in our dataset. Thus, much of our data lacked realisticity. Furthermore, our limited coding experience restricted the amount of fine-tuning we could conduct, as we were unable to optimise the demo application, or specifically design an app catered to our implementation.

Lastly, the lack of focus in the object detection field limited our options, as parameters such as learning and dropout rate could only be changed on the image classification model of EfficientDetLite. There is still room for improvement if these training parameters can be modified in the future updates or with additional custom code.

4.2 Comparison to Existing Results

The unique context of our research has led to a novel approach to object detection. Firstly, all research is directed towards use in mobile devices, and as such, we are able to provide a detailed, in-depth analysis of low-latency models in this area. We also focused solely on detecting 'windows with humans', which poses a unique set of challenges, such as an increasing likelihood of overfitting, differing levels of opacity or clarity, and the distortion of the refracted image observed. Thus, our neural network had to train on 2 key identifying features, which provided an additional level of complexity. Lastly, the wide-ranging scope of the research done meant that a small dataset had to be utilised, resulting in novel research into the relationship between accuracy and the dataset size, and the training split of the images used.

Our approach of prioritising accuracy over latency differentiates this paper from existing research that focuses more on latency. With reference to our rationale stated in Sect. 2.3, we set a threshold for latency and picked the neural network with highest accuracy based on that threshold to optimise its application on mobile devices.

4.3 *Areas for Future Work*

Potential improvements include identifying other objects, such as threats armed with weapons, allowing for comprehensive scanning of areas. Modifications to our model can also be made for close quarter urban combat, where rooms of threats are cleared instead of identifying threats through windows.

Acknowledgements We would like to thank our mentor, Mr Bryan Ng, for his constant guidance and invaluable insight during the course of our project.

References

1. Tensorflow. (2021, June 8). models/tf2_detection_zoo.md at master · tensorflow/models. GitHub. Retrieved July 31, 2022, from https://github.com/tensorflow/models/blob/master/research/object_detection/g3doc/tf2_detection_zoo.md
2. Ding, L., Wang, Y., Laganière, R., Luo, X., & Fu, S. (2019). Multi-scale predictions fusion for robust hand detection and classification. *Multimedia Tools and Applications, 78*(24), 35633–35650. https://doi.org/10.1007/s11042-019-08080-4
3. Ren, S., He, K., Girshick, R., & Sun, J. (2017). Faster R-CNN: towards real-time object detection with region proposal networks. *IEEE Transactions on Pattern Analysis and Machine Intelligence, 39*(6), 1137–1149. https://doi.org/10.1109/tpami.2016.2577031
4. He, K., Zhang, X., Ren, S., & Sun, J. (2016). Deep residual learning for image recognition. In *2016 IEEE Conference on Computer Vision and Pattern Recognition (CVPR)*. https://doi.org/10.1109/cvpr.2016.90
5. Liu, R. (2020, March 16). *Higher accuracy on vision models with EfficientNet-Lite. Tensorflow Blog*. Retrieved July 31, 2022, from https://blog.tensorflow.org/2020/03/higher-accuracy-on-vision-models-with-efficientnet-lite.html
6. Geifman, A. (2021, December 14). *Are all FLOPs created equal?—Towards Data Science. Medium*. Retrieved July 31, 2022, from https://towardsdatascience.com/are-all-flops-created-equal-a-comparison-of-flops-vs-run-time-3cebb2dcb3da
7. Tensorflow. (2021, October 14). models/tf1_detection_zoo.md at master · tensorflow/models. GitHub. Retrieved July 31, 2022, from https://github.com/tensorflow/models/blob/master/research/object_detection/g3doc/tf1_detection_zoo.md

8. Lin, T. Y., Maire, M., Belongie, S., Hays, J., Perona, P., Ramanan, D., Dollár, P., & Zitnick, C. L. (2014). *Microsoft COCO: Common objects in context. Computer Vision—ECCV 2014* (pp. 740–755). https://doi.org/10.1007/978-3-319-10602-1_48
9. Huang, J., Rathod, V., Sun, C., Zhu, M., Korattikara, A., Fathi, A., Fischer, I., Wojna, Z., Song, Y., Guadarrama, S., & Murphy, K. (2017). Speed/accuracy trade-offs for modern convolutional object detectors. In *2017 IEEE conference on computer vision and pattern recognition (CVPR)*. https://doi.org/10.1109/cvpr.2017.351
10. Google. (2021, April 18). automl/efficientdet at master · google/automl. GitHub. Retrieved July 31, 2022, from https://github.com/google/automl/tree/master/efficientdet
11. Nelson, J. (2021, October 27). *Seven tips for labeling images for computer vision.* Roboflow Blog. https://blog.roboflow.com/tips-for-how-to-label-images/

D-Pen—Personalised Practice with Assistive Pen for Young Dyslexic Learners Using Machine Learning

Janessa Valencia Guo, Patrick Anand Francis, and Guoxian Tan

Abstract Dyslexia is a learning disorder that hinders one's ability to learn language. While dyslexia cannot be cured, many symptoms can be mitigated with appropriate tools. In this paper, we propose D-pen to aid dyslexic learners in the improvement of their handwriting by providing personalised penmanship guidance. As few works target handwriting recognition for dyslexic writers, we adapted machine learning models to effectively recognise dyslexic handwriting—exploring different architectures and conversion into Bezier curves for model input. For handwriting input, we prototyped a low-cost 3D-printed pen consisting of 2 major components: a rPi zero and a camera module (Pixy 2). The RPi processes the input and runs the ML model, while a graphical user interface provides immediate feedback as well as long-term practice for users on their handwriting. Evaluation of each component demonstrated the viability of this prototype. We hope that this project will serve to contribute to the field of assistive technology for learners with dyslexia, so they can benefit from more effective learning systems.

Keywords Dyslexia · Handwriting recognition · Machine learning · Image processing · 3D printing

1 Introduction

1.1 Background on Dyslexia

Dyslexia is a learning disorder that affects one's ability to learn relationships between letters and words (i.e. decoding). Dyslexia affects approximately 15% of the world

---

J. V. Guo (✉) · P. A. Francis · G. Tan
Raffles Science Institute, Raffles Institution, Singapore, Singapore
e-mail: 22yjane889j@student.ri.edu.sg

P. A. Francis
e-mail: 22yfran655i@student.ri.edu.sg

G. Tan
e-mail: guoxian.tan@ri.edu.sg

H. Guo et al. (eds.), *IRC-SET 2022*,
https://doi.org/10.1007/978-981-19-7222-5_18

population and is the most common neuro-cognitive disorder. These individuals often suffer from symptoms that reveal themselves in speaking, reading and writing: reversing sounds in words or confusing words that sound alike, as well as repetitions, additions, transpositions, omissions, substitutions and reversals in letters, numbers and/or words [1]. In today's literacy-centric society, the difficulties in language faced by dyslexic individuals, particularly that of children, may have spillover effects in other aspects of their life such as communication with others and self-image, leaving them susceptible to discrimination [2]. Neither learning disabilities can be outgrown, prevented or completely eliminated. However, the disorders are not linked to lack of cognitive ability, and many symptoms can be mitigated with early intervention as well as appropriate tools and guidance, enabling individuals to reach their full potential. For instance, writing cursive can combat some of the handwriting issues as it reduces the likelihood of letter reversals due to the lowered use of fine motor skills.

1.2 Theoretical Underpinning

As there is limited research in the area, this paper adapts machine learning solutions for the problem of dyslexic handwriting—exploring the implementation of a Feed Forward Neural Network (FNN), 1-Dimensional/2-Dimensional Convolutional Neural Network (1D-CNN/2D-CNN) and Recurrent Neural Network (RNN).

Mathematics behind FNN, CNN and RNN. The activation functions used include Sigmoid, ReLU and Softmax. Categorical cross-entropy was used to calculate loss.

In each layer of the CNNs, each activation map *Yi* is calculated as shown in Eq. (1), where l—current layer, $B_i^{(l)}$—bias matrix, $k(l-1)$—number of kernels used in the previous layer, W—current layer kernel matrix and $Y_j^{(l-1)}$—output of the previous layer.

$$Y_i^{(l)} = B_i^{(l)} + \sum_{j=1}^{k^{(l-1)}} W_i^{(l)} Y_j^{(l-1)} \tag{1}$$

In the RNN, the hidden state S_n and outputs O_n from each RNN cell are calculated with Eqs. (2), (3), respectively, where I_n—nth input, U, V, W—matrixes and b, c—biases

$$S_n = UI_n + WS_{n-1} + b \tag{2}$$

$$O_n = VS_n + c \tag{3}$$

Bézier Curves. This paper also explores the use of cubic Bézier curves to represent handwriting data, which can be defined as in Eq. (4), where P_n—control points and t—fraction of how far $B(t)$ is from P_0 relative to P_n

$$B(t) = (1-t)^3 P_0 + 3(1-t)^2 t P_1 + 3(1-t)t^2 P_2 + t^3 P_3,\ 0 \leq t \leq 1 \quad (4)$$

NiBlack's Adaptive Thresholding. Local threshold is calculated as shown in Eq. (5) based on mean and standard deviation of local neighbourhood of size w. Another applied parameter k introduces bias of variance value in Eq. (6), where I—image, T—threshold and B—binary pixel intensity

$$T(x, y) = \text{mean}(I[w, w]) - k \quad (5)$$

$$B(x) = \begin{cases} 0, \text{ if } I(x, y) > T(x, y) \\ 1, \text{ otherwise.} \end{cases} \quad (6)$$

1.3 Literature Review

Previous studies have proposed methods of addressing the challenges of dyslexia: allowing users to customise font to be dyslexia-friendly (e.g. OpenDyslexic) on reading devices, replacing complex words with simpler synonyms, text-to-speech checkers for phonetic practice, pocket spell checkers or even a combination of the above. However, none of these assistive technologies are focused on offering guidance specific to each dyslexic individual based on their written practice.

There have also been studies investigating artificial intelligence (AI) models to effectively solve the problem of handwriting recognition. In recent years, it has become increasingly popular to use curves (e.g. Bézier curves, B-splines) not just for upsampling limited data and rendering smooth handwriting on screens, but also as a form of input encoding in machine learning [3]. However, dyslexic handwriting may differ from normal handwriting due to the difficulties these learners face in writing. Few works target the area of handwriting recognition for dyslexic writers.

1.4 Engineering Goals

This project proposes a low-cost 3D-printed tool to aid dyslexic learners in improving handwriting, by utilising machine learning to recognise the handwritten practice of dyslexic users, so that personalised guidance can be provided. To make this assistive technology a reality, this project aims to develop an optimal model architecture for dyslexic handwriting recognition so as to be able to reliably assess the input

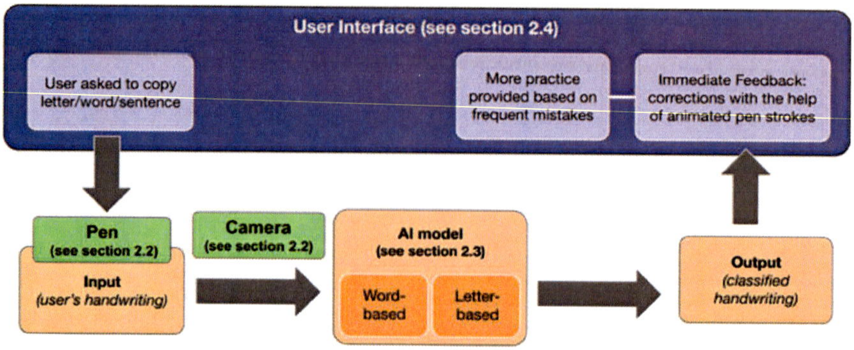

Fig. 1 Block diagram of our proposed solution

handwriting and provide suitable feedback for the user, as well as an effective data input system.

2 Methodology

2.1 *Overview of Approach*

The proposed approach comprises a pen and camera which feed input (handwriting) to the machine learning model, while users interact with this system through a python application. Figure 1 gives an overview of this approach.

2.2 *Hardware—camera System*

Figures 2 and 3 show images of the 3D-printed pen module, D-Pen (Detailed setup in Appendix IX). A Pixy2 camera was used to capture images of the user's offline handwriting, while a raspberry pi zero does the image processing and classification. The predicted output is then sent to the computer to be displayed in the user interface. To segment the handwriting from the background, Niblack's adaptive thresholding method was chosen as it accounts for the dynamic lighting condition. To reduce inconsistencies in the input such as thickness of pen strokes, down-sampling was carried out by removing data points at constant intervals until there were approximately 100–400 data points left. Another alternative considered was using an Inertial Measurement Unit (IMU) to capture the pen strokes (Detailed setup in Appendix X).

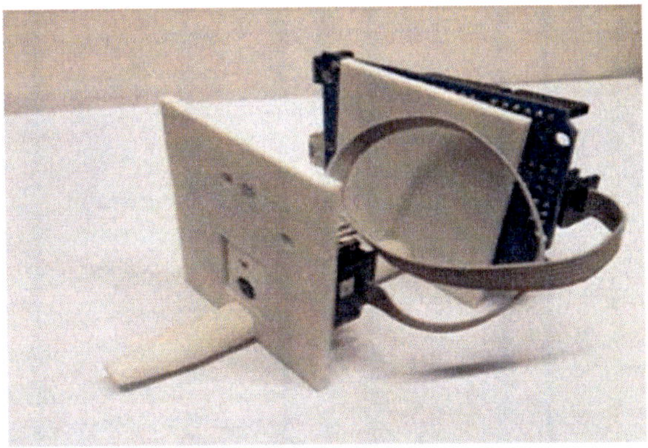

Fig. 2 3D-printed pen (side view)

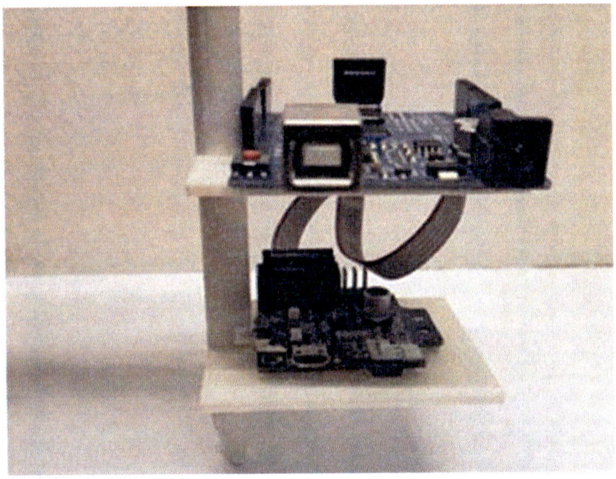

Fig. 3 3D-printed pen (front view)

2.3 *AI for Dyslexic Handwriting Recognition*

Datasets and Data Pre-processing. Online handwriting was collected from 57 dysgraphic children aged 8–15 where they copied a Slovak template in cursive handwriting [4] (Template can be found in Appendix I). We used this to create a dataset of letters [l, e, t, o, a] with 100 samples/category and dataset of letters [leto, lamoken, hrackarstvo, lete, bude, teplo] with 50 samples/category. Six normalised features (x-coordinate, y-coordinate, time, azimuth, pressure, altitude) of raw touchpoints served as input for the FNN, RNN and 1D-CNN models. The xy-coordinates were also transformed into a NumPy array of pixels which served as input for the 2D-CNN model, where pixel dimensions (height $\times$ width) were fixed at 28×28 for letters and 75×100 for words. Lastly, Bézier curves were used to represent handwriting data. Interpolating curves using every point would result in unnecessary breakpoints. Hence, the xy-coordinates were run through the Ramer–Douglas–Peucker (RDP) algorithm, where a suitable epsilon was identified to ensure that only non-essential points were decimated. Afterwards, a Bézier curve is found between each consecutive pair of points (serving as first and last control points), by solving for the 2 control points in between. Given that the Bézier coefficients may differ over a wide range, each curve in the input is a 6-dimensional vector (dx, dy, $d1$, $d2$, $\alpha 1$, $\alpha 2$), where:

- dx, dy—vector between the end points
- $d1$, $d2$—distance between the control points and the nearest end points
- $\alpha 1$, $\alpha 2$—angles between control points and end points in radians

(Detailed pre-processing process and relevant theory in Appendix II).

FNN, CNN and RNN model architecture and training. Given that the online dyslexic handwriting data was both temporal and spatial, it was of interest to explore how models such as the 2D-CNN (typically deployed for image classification) and FNN, 1D-CNN and RNN (BiLSTM) (for time-based sequences) would fare on this problem. Furthermore, the two models, especially the first, are popular choices for most handwriting recognition problems. All models were created with Tensorflow and Keras (Detailed architecture can be found in Appendix III). To train and test the models, a K-fold cross-validation approach ($i = 5$) was used, and the datasets were split into a 4:1 training to validation ratio. Each model was trained with 50 epochs and the weights corresponding to the lowest validation loss were saved based on the minimal categorical cross-entropy.

2.4 *User Interface*

A Pygame script was written to create the graphical user interface shown in Fig. 4. OpenDyslexic, a font designed to mitigate common reading errors caused by dyslexia, was selected to present the sentence the user is asked to copy. This helps our application be more accessible for the target audience of dyslexic learners.

3 Results

3.1 *Effectiveness of Camera and Image Processing Method*

Thresholding the image using Niblack's method returns a binary image with the text having a pixel intensity of 1 and the background 0. However, it is common for noises to appear after thresholding due to the presence of shadows and other lighting factors as seen in Fig. 5. Hence, the binary image has to be cleaned, leaving only the text to be extracted for coordinates. A common method used to clean images is morphological transformation. However, as seen in Fig. 6, the opening operation alone is not sufficient to remove all the noise, especially when the area of the noise is

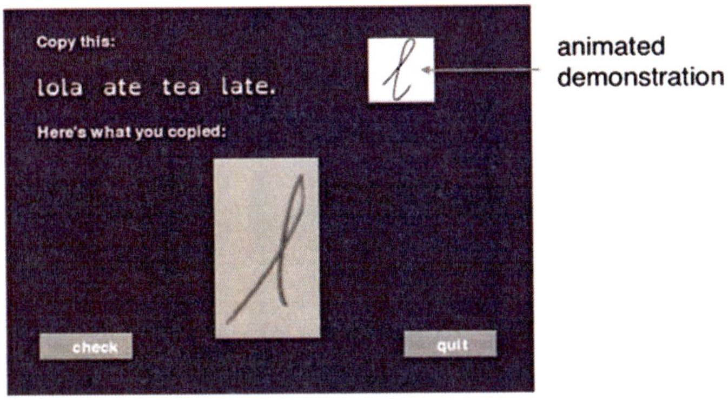

Fig. 4 User interface

significantly larger than the kernel size used. Multiple iterations of the filter should help to remove more noise. However, it also eradicates the text and hence is not suitable as seen in Fig. 7. The same effect can be observed if a larger kernel size is used, as shown in Table 1.

An alternative thresholding method is to find all contours in the final image and remove any regions with a contour area smaller than a certain threshold value. This not only helps to remove the smaller regions of noise that opening/erosion can remove, but also eradicates any larger blobs of noise. Most importantly, it is able to do this without eroding/compromising the text as shown in Fig. 8.

Effectiveness of the IMU integration method. As seen in Fig. 9, the IMU setup is not as effective as the camera setup (Detailed IMU setup can be found in Appendix X).

Fig. 5 Iteration 0 (post-threshold)

Fig. 6 Iteration 1 kernel: 5 × 5

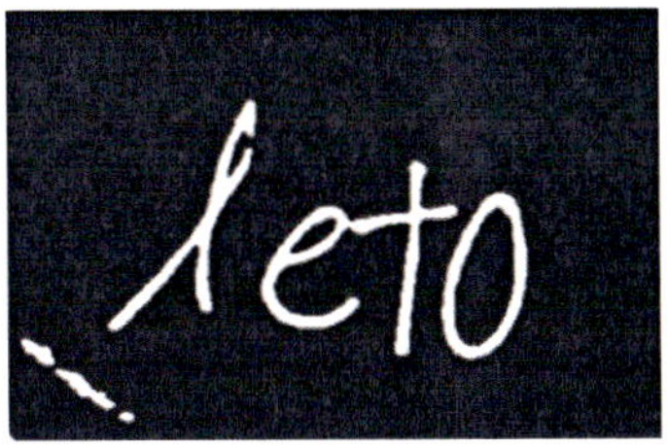

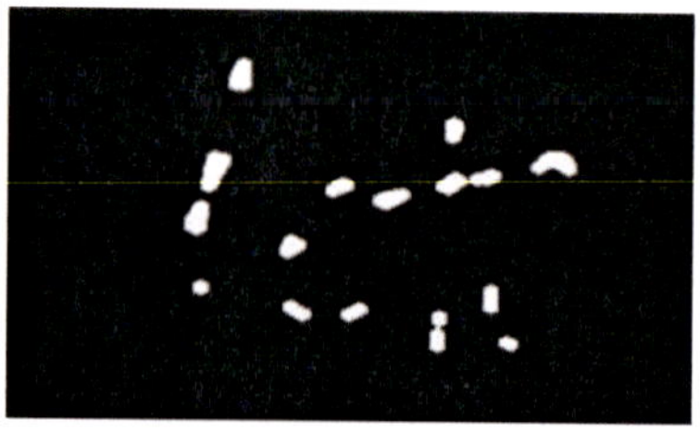

Fig. 7 Iteration 2 kernel: 5 × 5

Table 1 Effect of increasing kernel sizes on the noise and text in image

	Morphological operation (1 iteration)				Contour method
Kernel size	(3 × 3)	(5 × 5)	(7 × 7)	(9 × 9)	–
Noise removed (%)	10.32	30.01	45.79	70.06	100.00
Text eroded (%)	0.161	0.801	6.61	74.16	0.00

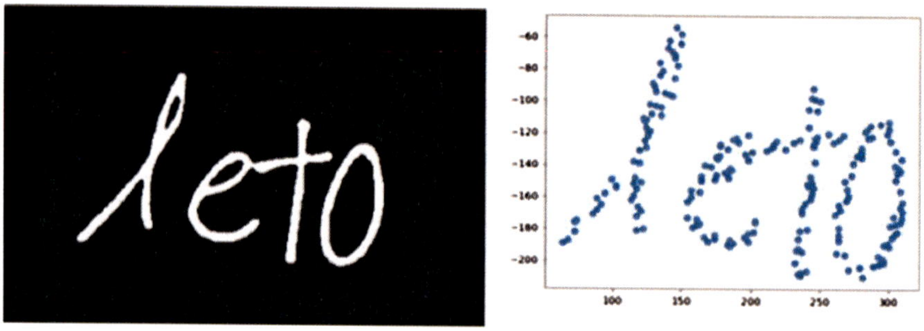

Fig. 8 Successful processing of image 'leto'

The accelerometer and gyroscope data must be fused to transform the accelerometer data from the sensor body's frame to the world frame. Hence, this means that the inaccuracies in the angular pose can largely affect the integrated values as well [5]. Given that the pen is usually held at an angle while writing, any minor drifts in angular pose largely offsets the integrated displacement and velocity estimations, producing inaccurate pen position coordinates.

3.2 *Effectiveness of Camera and Image Processing Method*

Capability of current OCR (Optical character recognition) technology. Three handwriting-to-text conversion services were given 6 samples of handwriting to process. These samples included 2 samples of dyslexic and non-dyslexic handwriting

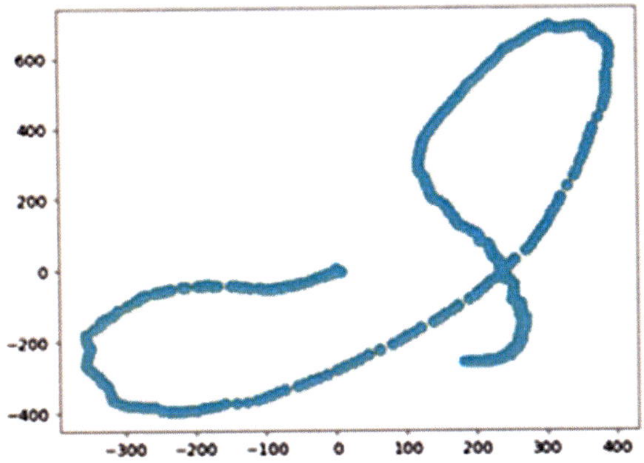

Fig. 9 Unsuccessful generation of letter 'l'

each (from Slovak schoolchildren) and 2 non-dyslexic samples from ourselves (1 cursive, 1 print). Across all the OCR services, this simple test consistently revealed that the parsed texts of dyslexic handwriting bore least resemblance to the original template, non-dyslexic cursive handwriting samples produced more similar parsed texts, and the printed handwriting sample generated text that was most error free (Detailed results are available in Appendix V). This demonstrates the limited ability of publicly available OCR technology to parse dyslexic cursive handwriting of children.

Effectiveness of RNN variants. Exploration of variants of the RNN model such as SimpleRNN, Gated Recurrent Network (GRU), Long Short Term Memory (LSTM) and Bidirectional-LSTM (BiLSTM) showed that the BiLSTM was by far the most effective. This supports theoretical expectations as Bidirectional RNNs are known to be useful when context around each datapoint (and not only that preceding it) is relevant, since the BiRNN processes sequence from start to end as well as backwards. Given that the data used to train the models in this project is particularly challenging to recognise because it was written by dyslexic individuals in cursive, the comparatively better performance of the BiLSTM model is validated by other studies investigating difficult-to-recognise handwriting. To reduce the number of independent variables, the BiLSTM model was used for the rest of the comparative experiments.

Assessing performance of FNN, CNN and RNN models. The models were evaluated based on their mean K-fold cross-validation accuracy ($K = 5$), weighted precision and recall, and F1 score across the classes, after being trained on variations of input data (letter dataset, word dataset and letter-based Bézier curve inputs dataset), as shown in Fig. 10 and Table 2 (Detailed performance report, confusion matrices, precision-recall curves and equal error rate curves can be found in Appendix IV).

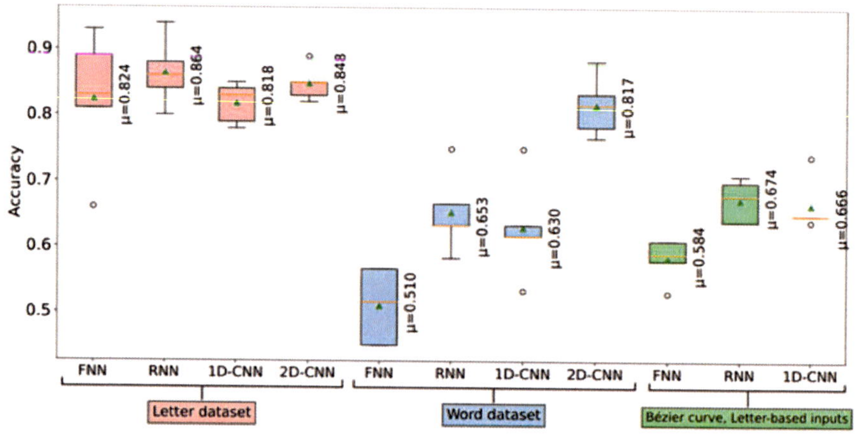

Fig. 10 Box plot comparing accuracy of models on different datasets

Table 2 Summarised performance of FNN, RNN, 1D-CNN and 2D-CNN (letter, word dataset)

	Letter dataset				Word dataset			
	Accuracy	Precision	Recall	F-1	Accuracy	Precision	Recall	F-1
FNN	82.400	0.802	0.800	0.799	51.000	0.288	0.367	0.320
RNN	86.400	0.856	0.850	0.847	65.333	0.690	0.683	0.684
1D-CNN	81.800	0.772	0.770	0.769	63.000	0.521	0.533	0.469
2D-CNN	84.800	0.894	0.890	0.891	81.667	0.843	0.833	0.827

4 Discussion

4.1 *Comparing Effectiveness of Hardware Setups: Camera Versus IMU System*

For the IMU system, as discussed in Appendix X, eliminating the sensor's noise using a low-pass filter would be effective. However, in our case the true acceleration also gets significantly damped by the filter, as seen in Fig. 11. This is due to the nature of writing. Writing is a relatively fast task involving frequencies of the motion of the tip of the pen over 2 Hz, which can overlap with the frequency of the sensor noise, resulting in inaccurate calculations. Figures 11 and 12 show the damping effect of various low-pass filters, 2, 10 and 20 Hz, on the pen acceleration data, in mm/s, when writing (true acceleration) and when stationary (noise).

Ideally when using low-cost digital IMUs, obtaining the gravity-corrected accelerometer output by using sensor fusion and noise filters is sufficient to accurately estimate the displacement over large distances as the IMU data can be fused with the GPS data (e.g. for tasks such as step count, long-distance runs etc.) as seen

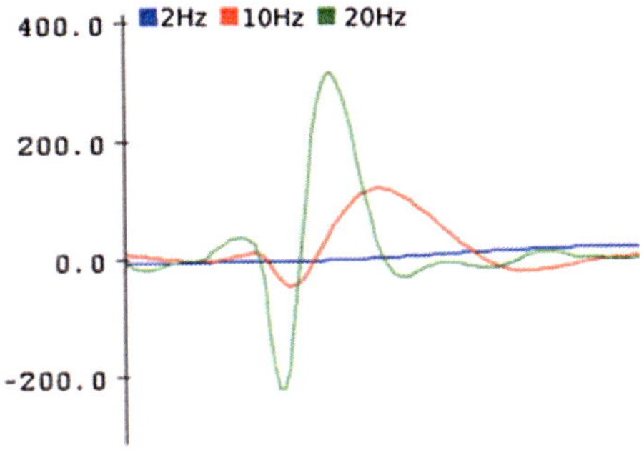

Fig. 11 True accelerations (writing)

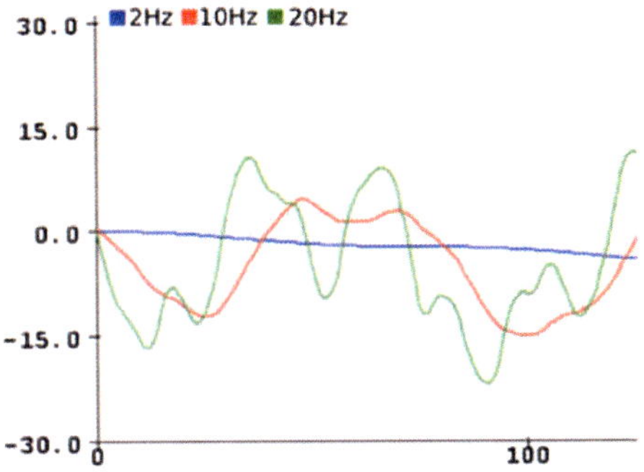

Fig. 12 Noise generated (stationary)

in smartphones [6]. However, using GPS is not suitable in our case due to the significantly smaller area typically used for writing (A4 paper spans 21.0 × 29.7 cm, whereas typical GPS precision is about 5 m). Hence, the camera system is favoured over the IMU as a handwriting input device.

4.2 Comparing Performance of FNN, 1D-CNN, 2D-CNN and RNN

Amongst the three temporal data-based architectures trained on the letter dataset (FNN, 1D-CNN, RNN), the RNN model yielded the best performance with regards to Precision, Recall and F1-Score. The poorer performance of the 1D-CNN compared to RNN in this project is corroborated by existing research indicating the tendency of RNN to outperform 1D-CNN in temporal data such as online handwriting. However, since the 1D-CNN and FNN are significantly more compact than the RNN, they may be favoured for a shorter training period and lower computational cost when more data is available for training. Meanwhile, these three models yielded poorer performance after training with the word-based dataset. There are two factors that likely contributed to this result: (1) less training data; (2) longer sequences. In particular, an

RNN can theoretically process an arbitrary long sequence; however, in practice they only take into consideration a limited number of preceding steps, and hence may be less effective in practice when longer sequences are involved.

The 2D-CNN model performed consistently well with stable performance, as seen in its K-fold cross-validation performance for both letter and word dataset. Nevertheless, the precision/recall curves also demonstrate that training on the letter dataset yielded better results than that for the word dataset. In any case, one may consider the letter-based models to be more applicable in practice because it is more feasible to train the models on the finite set of letters instead of the expansive and ever-expanding vocabulary of words.

Lastly, the performance for FNN, 1D-CNN and RNN models are diminished when trained on Bézier curve inputs instead of raw touchpoints for letters. This contradicts prior research by Carbune et al. that first proposed the notion of passing Bézier curves as input—current literature indicates that this form of input should speed up training but should not significantly affect performance of the model in handwriting recognition [3]. The divergence in findings could be explained by the smaller dataset used, as a larger dataset may have made the difference in performance negligible in other research. A closer examination of the confusion matrices and precision-recall for each class reveals that the models generally find the classification of the letters 'l', 'e' and 't' less challenging compared to that of 'o' and 'a'. This likely means that the curves characterising the former set of letters are more distinct than the latter set. Hence, the poor classification of the latter may be the main contributor to the undermined performance caused by Bézier curve inputs.

4.3 Evaluating Overall Feasibility of Assistive Tool

The results for each component demonstrate the viability of a tool that could aid dyslexic learners in their handwriting practice. The 3D-printed pen is able to effectively serve as a tool for handwriting input, while the AI model is able to reasonably perform handwriting recognition for dyslexic individuals, so that appropriate and personalised guidance can be provided. Hence, the prototype could be very useful for dyslexic users in improving their writing—in either its current form or an integration into future assistive technology.

5 Limitations and Further Work

Firstly, the error rate for all three models may be higher than some existing literature investigating the problem of handwriting recognition. This may be due to the limited data available for dyslexic handwriting in contrast to extensive datasets used in previous studies on normal handwriting, such as the MNIST handwritten digit dataset. Secondly, the assistive tool can be personalised to a greater extent such as generating more varied practises that target the users' areas of improvement based on practice over time instead of drawing from a fixed set. As they utilise the tool, their handwriting could also be used to further supplement the dataset for training the dyslexic handwriting model, thus addressing the first limitation. Lastly, it may be interesting to explore hybrid implementations of CNN/RNN as prior research has shown promising results for these hybrids, especially in spatial–temporal problems.

6 Conclusion

In conclusion, this project explored the application of artificial intelligence on the problem of dyslexic handwriting and a system to effectively input handwriting, so that they could be used to support a prototype tool for dyslexic learners for personalised handwriting practice. We hope that this project will serve to contribute to the field of assistive technology for learners with dyslexia, so that they can benefit from more effective learning systems.

Acknowledgements Our gratitude to the lab personnel of ClusterLabs for printing the 3D pen.

Appendix

(I) Dataset

For some samples in which individuals tend to write each distinct word in one pen-down stroke, the segmentation is largely effective (Fig. 13). Meanwhile, for samples where the individual writes each word in multiple pen-up/pen-down strokes, segmentation is less effective (Fig. 14).

Certain portions of the dataset may have yielded poorer recognition results, as shown in the following instances. For instance, the individuals may have either miswrote letters or wrote letters that closely resembled one another (Fig. 15).

Another example is in how the cursive letters 'l' and 'e' may resemble one another especially after data normalisation of the *x*- and *y*-coordinates to between 0 and 1. (Fig. 16).

Lastly, the letter 'o' is often misclassified as 'a', potentially because they have similar curvature compositions (Fig. 17).

Online handwriting from 57 dysgraphic children aged 8–15 was collected at the Centre for Special-Needs Education in 2018–2019 where the children copied a template of Slovak words in cursive handwriting (Fig. 18).

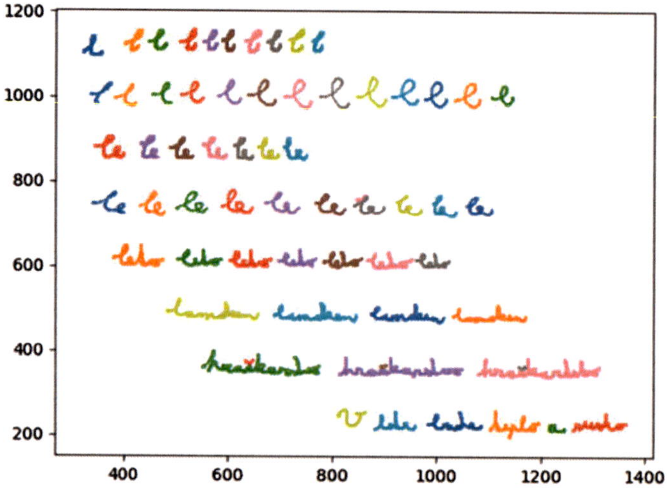

Fig. 13 Effective segmentation of a sample

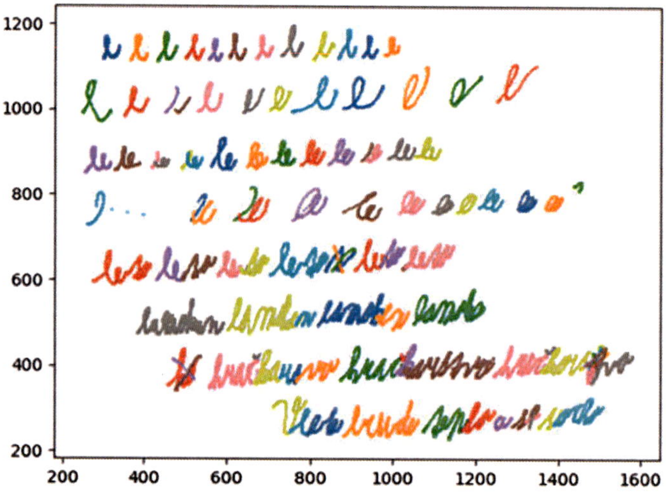

Fig. 14 Less effective segmentation of a sample

Fig. 15 't' in 'leto' versus 'd' in 'bude'

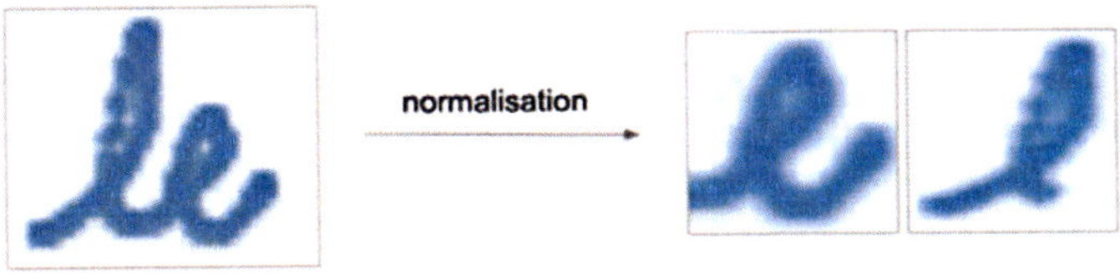

Fig. 16 'le' before and after pre-processing and normalisation to 'l' and 'e'

Fig. 17 e: Example of letter 'o' misclassified as 'a'

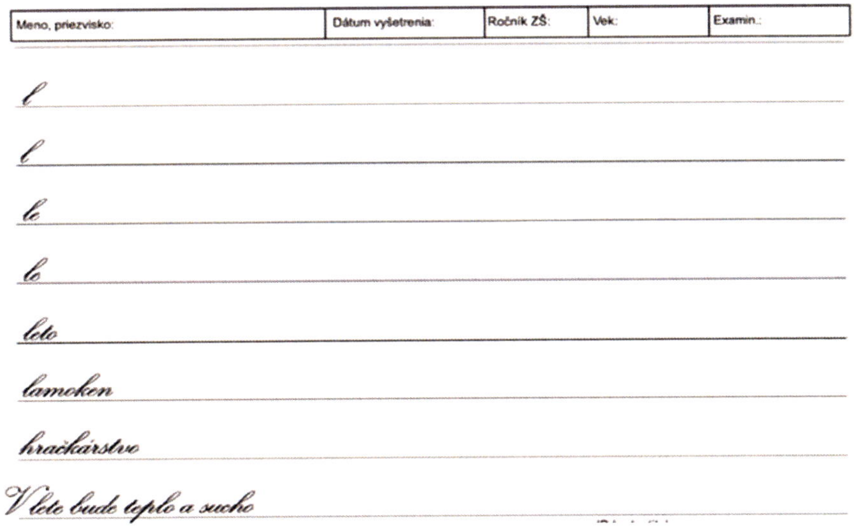

Fig. 18 Template for handwriting acquisition [4]

(II) Detailed Data-processing Methodology

The 7 features of each data point included *x*-coordinate, *y*-coordinate, time, pen-up/pen-down, azimuth, pressure and altitude. Only pen-down strokes were considered, to make the semi-automated segmentation more efficient with a python script.

The remaining 6 features were then normalised to values between 0 and 1. These raw touchpoints served as input for the FNN, RNN and 1D-CNN models. Afterwards, the *xy*-coordinates were also plotted with Matplotlib, before being converted to grayscale and transformed into a NumPy array of pixels. These 2D images served as input for the CNN 2D model, where pixel dimensions (height $\times$ width) were fixed at 28 $\times$ 28 for letters and 75 $\times$ 100 for words.

Bézier curve conversion. Bézier curves were also used to represent handwriting data. Interpolating curves using every point would result in unnecessary breakpoints. Hence, the *xy*-coordinates were run through the Ramer–Douglas–Peucker (RDP) algorithm, where a suitable epsilon was identified via observation to ensure that only non-essential points were decimated. Afterwards, a Bézier curve is found between each consecutive pair of points (serving as first and last control points), by solving for the 2 control points in between (Fig. 19). Given that the Bézier coefficients may differ over a wide range, each curve in the input is a 6-dimensional vector (dx, dy, $d1$, $d2$, $\alpha 1$, $\alpha 2$), where:

- dx, dy—vector between the endpoints
- $d1$, $d2$—distance between the control points and the nearest endpoints
- $\alpha 1$, $\alpha 2$—angles between control points and end points in radians [3].

To ensure the final curve is smooth, the first and second derivatives of consecutive curves (Eq. (7, 8)) are compared to ensure their curvatures match.

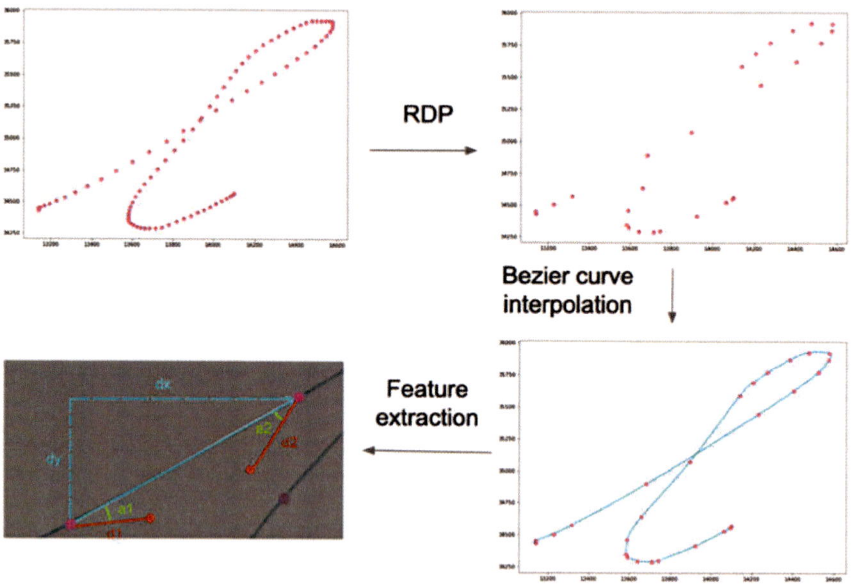

Fig. 19 Overview of procedure to convert raw data to Bézier curve features

$$B'(t) = -3(1-t)^2 \cdot P_0 + 3\left[(1-t)^2 - 2t(1-t)\right] \cdot P_1 + 3\left[2t(1-t) - t^2\right] \cdot P_2 + 3t^2 \cdot P_3$$
$$= 3\left[-(1-t)^2 \cdot P_0 + (1-3t)(1-t) \cdot P_1 + t(2-3t) \cdot P_2 + t^2 \cdot P_3 \quad (7)$$

$$B''(t) = 3\left[2(1-t)^2 \cdot P_0 + [-3(1-t) - (1-3t)] \cdot P_1 + [2-6t] \cdot P_2 + 2t \cdot P_3$$
$$= 6\left[(1-t)^2 \cdot P_0 + (3t-2) \cdot P_1 + (1-3t) \cdot P_2 + t \cdot P_3 \quad (8)$$

This is used to solve for the two control points between each two consecutive points, hence producing a series of Bezier curves from which the features dx, dy, $d1$, $d2$, $a1$ and $a2$ are extracted.

(III) **Model Architecture (letter-based)**

See Figs. 20, 21, 22 and 23

(IV) **Detailed Performance**

Performance of models trained on letter dataset.
See Figs. 24, 25, 26 and 27 and Tables 3, 4, 5 and 6.

Performance of models trained on word dataset.
See Figs. 28, 29, 30 and 31 and Tables 7, 8, 9 and 10.

Performance of models trained on Bézier curves dataset.

See Figs. 32, 33 and 34 and Tables 11, 12 and 13.

(V) **Online OCR parsing of handwriting**

Service 1: https://www.onlineocr.net/

Service 2: https://ocr.space/

Service 3: https://www.newocr.com/
See Tables 14, 15, 16 and 17

Table 3 FNN classification report (letter dataset)

	precision	recall	f1-score	support
0.0	0.864	0.950	0.905	20
1.0	0.824	0.700	0.757	20
2.0	0.850	0.850	0.850	20
3.0	0.696	0.800	0.744	20
4.0	0.778	0.700	0.737	20
accuracy			0.800	100
macro avg	0.802	0.800	0.799	100
weighted avg	0.802	0.800	0.799	100

Table 4 RNN classification report (letter dataset)

	precision	recall	f1-score	support
0.0	0.864	0.950	0.905	20
1.0	0.850	0.850	0.850	20
2.0	0.933	0.700	0.800	20
3.0	0.800	1.000	0.889	20
4.0	0.833	0.750	0.789	20
accuracy			0.850	100
macro avg	0.856	0.850	0.847	100
weighted avg	0.856	0.850	0.847	100

Table 5 1D-CNN classification report (letter dataset)

	precision	recall	f1-score	support
0.0	0.818	0.900	0.857	20
1.0	0.750	0.750	0.750	20
2.0	0.833	0.750	0.789	20
3.0	0.696	0.800	0.744	20
4.0	0.765	0.650	0.703	20
accuracy			0.770	100
macro avg	0.772	0.770	0.769	100
weighted avg	0.772	0.770	0.769	100

Table 6 2D-CNN classification report (letter dataset)

	precision	recall	f1-score	support
0	0.931	0.964	0.947	28
1	0.750	0.857	0.800	14
2	0.800	0.800	0.800	10
3	0.952	0.833	0.889	24
4	0.917	0.917	0.917	24
accuracy			0.890	100
macro avg	0.870	0.874	0.871	100
weighted avg	0.894	0.890	0.891	100

(VI) Analysis of thresholding methods

Segmentation

Thresholding is the first step of the image segmentation. It is used to separate the text from the background properly so that the coordinates can be extracted accurately.

```
Model: "sequential_2"
_________________________________________________________________
 Layer (type)                Output Shape              Param #
=================================================================
 masking_1 (Masking)         (None, 18600)             0

 dense_3 (Dense)             (None, 512)               9523712

 dense_4 (Dense)             (None, 512)               262656

 dense_5 (Dense)             (None, 6)                 3078

=================================================================
Total params: 9,789,446
Trainable params: 9,789,446
Non-trainable params: 0
_________________________________________________________________
```

Fig. 20 FNN model architecture (letter-based)

```
Model: "sequential_3"
_________________________________________________________________
 Layer (type)                Output Shape              Param #
=================================================================
 masking_3 (Masking)         (None, 3100, 6)           0

 conv1d_6 (Conv1D)           (None, 3098, 64)          1216

 conv1d_7 (Conv1D)           (None, 3096, 64)          12352

 max_pooling1d_3 (MaxPooling  (None, 1548, 64)         0
 1D)

 flatten_3 (Flatten)         (None, 99072)             0

 dense_6 (Dense)             (None, 100)               9907300

 dense_7 (Dense)             (None, 6)                 606

=================================================================
Total params: 9,921,474
Trainable params: 9,921,474
Non-trainable params: 0
_________________________________________________________________
```

Fig. 21 1D-CNN model architecture (letter-based)

We considered 3 different thresholding methods: simple thresholding, Otsu's method and adaptive thresholding.

1. Simple thresholding involves a manual trial and error of the threshold value (T) to obtain the optimal value to segment the image clearly. This method is very tedious and is not practical for our project since the background light intensity of each frame is likely to be different and using a single hardcoded threshold value would not segment the image well.
2. An alternative to this would be the Otsu's method. The Otsu's method automatically determines the optimal threshold value using the histogram of the

```
Layer (type)                  Output Shape              Param #
===============================================================
conv2d_1 (Conv2D)             (None, 73, 98, 32)        320

max_pooling2d_1 (MaxPooling   (None, 36, 49, 32)        0
2D)

flatten_1 (Flatten)           (None, 56448)             0

dense_2 (Dense)               (None, 100)               5644900

dense_3 (Dense)               (None, 6)                 606

===============================================================
Total params: 5,645,826
Trainable params: 5,645,826
Non-trainable params: 0
```

Fig. 22 2D-CNN model architecture (letter-based)

```
Layer (type)                  Output Shape              Param #
===============================================================
masking_1 (Masking)           (None, 60, 6)             0

bidirectional_2 (Bidirectio   (None, 60, 128)           36352
nal)

bidirectional_3 (Bidirectio   (None, 64)                41216
nal)

dense_1 (Dense)               (None, 5)                 325

activation_1 (Activation)     (None, 5)                 0

===============================================================
Total params: 77,893
Trainable params: 77,893
Non-trainable params: 0
```

Fig. 23 RNN model architecture (letter-based)

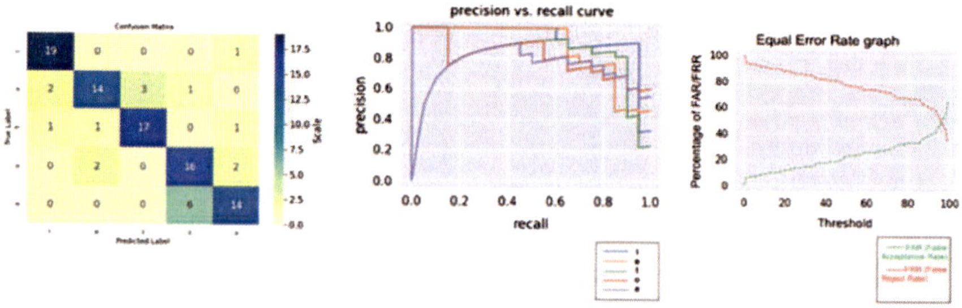

Fig. 24 FNN confusion matrix, precision-recall graph, equal error rate graph (letter dataset)

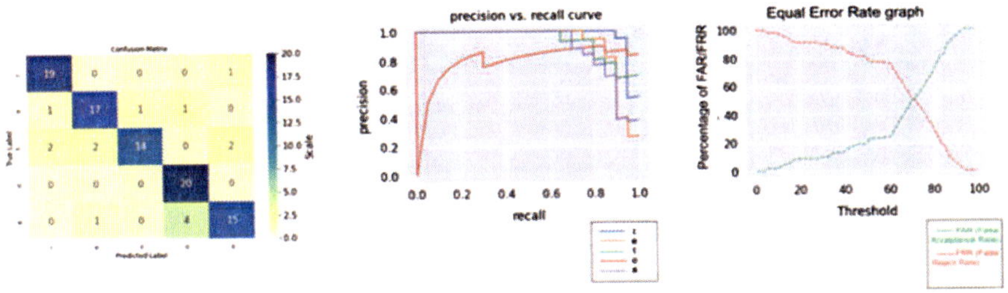

Fig. 25 RNN confusion matrix, precision-recall graph and equal error rate graph (letter dataset)

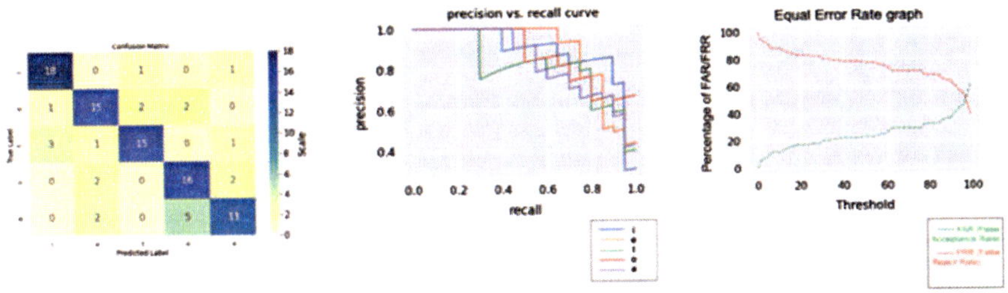

Fig. 26 1D-CNN confusion matrix, precision-recall graph and equal error rate graph (letter dataset)

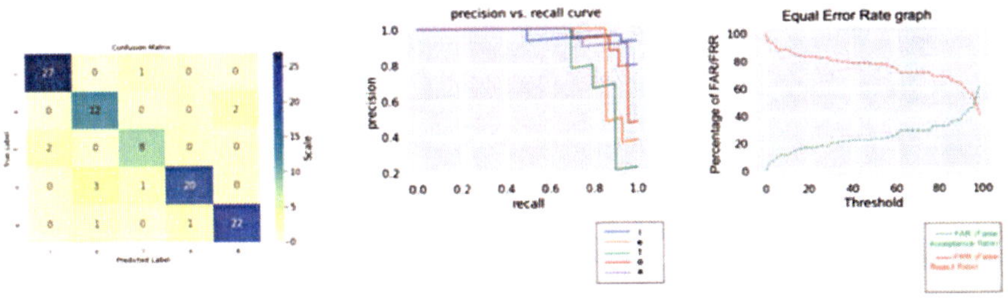

Fig. 27 2D-CNN confusion matrix, precision-recall graph, equal error rate graph (letter dataset)

Table 7 FNN classification report (word dataset)

	precision	recall	f1-score	support
0.0	0.000	0.000	0.000	10
1.0	0.600	0.600	0.600	10
2.0	0.000	0.000	0.000	10
3.0	0.538	0.700	0.609	10
4.0	0.235	0.400	0.296	10
5.0	0.357	0.500	0.417	10
accuracy			0.367	60
macro avg	0.288	0.367	0.320	60
weighted avg	0.288	0.367	0.320	60

Table 8 RNN classification report (word dataset)

	precision	recall	f1-score	support
0.0	0.667	0.800	0.727	10
1.0	0.500	0.500	0.500	10
2.0	0.545	0.600	0.571	10
3.0	0.750	0.600	0.667	10
4.0	0.778	0.700	0.737	10
5.0	0.900	0.900	0.900	10
accuracy			0.683	60
macro avg	0.690	0.683	0.684	60
weighted avg	0.690	0.683	0.684	60

Table 9 1D-CNN classification report (word dataset)

	precision	recall	f1-score	support
0.0	0.500	0.100	0.167	10
1.0	0.429	0.300	0.353	10
2.0	0.667	0.800	0.727	10
3.0	0.500	0.600	0.545	10
4.0	0.500	0.600	0.545	10
5.0	0.533	0.800	0.640	10
accuracy			0.533	60
macro avg	0.521	0.533	0.496	60
weighted avg	0.521	0.533	0.496	60

Table 10 2D-CNN classification report (word dataset)

	precision	recall	f1-score	support
0	0.833	0.500	0.625	10
1	0.786	0.917	0.846	12
2	0.700	1.000	0.824	7
3	0.778	0.778	0.778	9
4	0.923	0.857	0.889	14
5	1.000	1.000	1.000	8
accuracy			0.833	60
macro avg	0.837	0.842	0.827	60
weighted avg	0.843	0.833	0.827	60

pixel values (see Fig. 35). Hence, each frame would have its unique threshold value computed based on the contrast between the text and the background pixel intensity.

However, this model, similar to the simple thresholding method, assumes a bi-modal distribution of the greyscale pixel intensities and hence applies a global threshold. Thus, if there are multiple peaks in distribution, the threshold value may not return accurate results. As such, Otsu's threshold only works in stable lighting conditions where the image has even lighting. For this project,

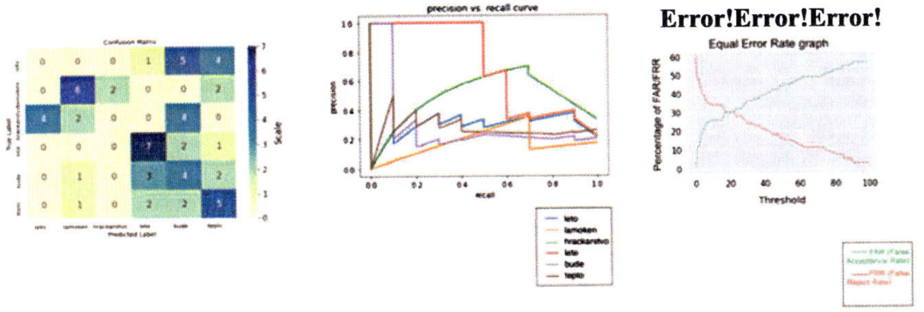

Fig. 28 FNN confusion matrix, precision-recall graph and equal error rate graph (word dataset)

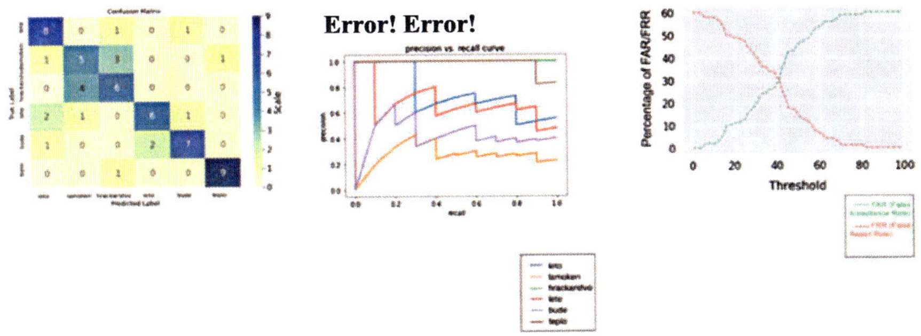

Fig. 29 RNN confusion matrix, precision-recall graph and equal error rate graph (word dataset)

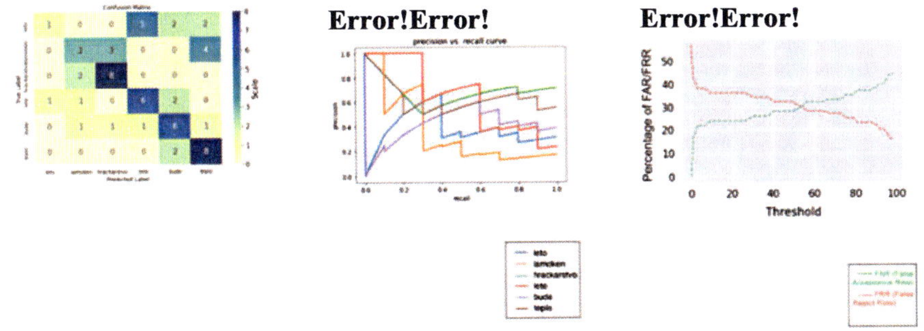

Fig. 30 1D-CNN confusion matrix, precision-recall graph and equal error rate graph (word dataset)

since the frames are taken in real time, non-uniform lighting conditions are to be expected. Different parts of the image may be illuminated more than the others due to shadows from the user's hand or other lighting factors (see Figs. 35 and 36 in Appendix VII).

3. Niblack's adaptive thresholding method is different from the global thresholding methods as it applies different thresholds on sub-regions of the image based on

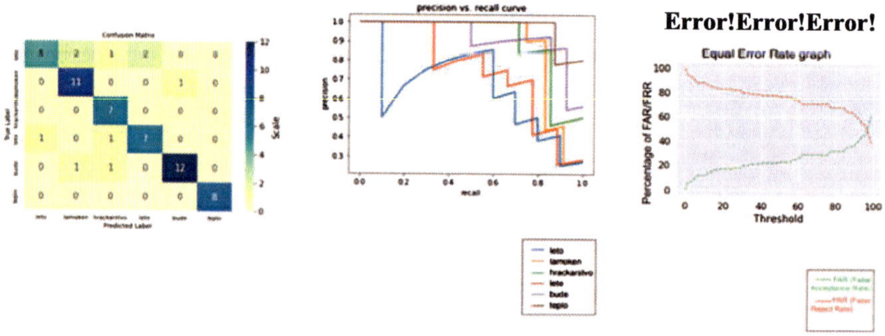

Fig. 31 2D-CNN confusion matrix, precision-recall graph and equal error rate graph (word dataset)

Table 11 FNN classification report (Bézier curves dataset)

	precision	recall	f1-score	support
0.0	0.362	0.850	0.507	20
1.0	0.765	0.650	0.703	20
2.0	1.000	0.400	0.571	20
3.0	0.444	0.200	0.276	20
4.0	0.421	0.400	0.410	20
accuracy			0.500	100
macro avg	0.598	0.500	0.494	100
weighted avg	0.598	0.500	0.494	100

Table 12 RNN classification report (Bézier curves dataset)

	precision	recall	f1-score	support
0.0	0.800	0.600	0.686	20
1.0	0.571	0.800	0.667	20
2.0	0.864	0.950	0.905	20
3.0	0.545	0.300	0.387	20
4.0	0.542	0.650	0.591	20
accuracy			0.660	100
macro avg	0.664	0.660	0.647	100
weighted avg	0.664	0.660	0.647	100

Table 13 1D-CNN classification report (Bézier curves dataset)

	precision	recall	f1-score	support
0.0	0.706	0.600	0.649	20
1.0	0.591	0.650	0.619	20
2.0	0.875	0.700	0.778	20
3.0	0.308	0.400	0.348	20
4.0	0.368	0.350	0.359	20
accuracy			0.540	100
macro avg	0.570	0.540	0.550	100
weighted avg	0.570	0.540	0.550	100

Table 14 Parsed text from scrvicc 1

Source	Output (Parsed text)
Slovakian school child—Dyslexic(1)	No recognised text!
Slovakian school child—Dyslexic(2)	No recognised text!
Slovakian school child—Non-Dyslexic(1)	1,1, LIJ i J. isi, .I,, t, t#1. ALLL LA, 11, fo, 11. ibitis ide gi, 44479/ /44444. lawkw kelas, 4n4404šyr- Jur44;str. isniaon-ntlb indt, 11,411 f,
Slovakian school child—Non-Dyslexic(2)	€.i .1 i i € 1.€ ,C .€ & ée be be .G‘e -G,e .1.e le le.ige.tt, 41-d> kto. .la″„-Ixeit ,ťeriii3O1~-€asodkot -1(4411p4ňti--vz9- ..čtŕacka~ ‘~tel<a,r-kť;v0. lĺ -(ľe,f, iouale ĎefiĽ& ot ft-éŕ-l49-
Our handwriting—Non-Dyslexic, cursive	€ .i .1 i i € 1 .€,C .€ & ée be be .G‘e -G,e .1.e le le .ige.tt, 41-d> kto. .la″„-Ixeit ,ťeriii3O1~-€asodkot -1(4411p4ňti--vz9- ..čtŕacka~ ‘~tel<a,r-kť;v0. lĺ -(ľe,f, iouale ĎefiĽ& ot ft-éŕ-l49-
Our handwriting—Non-Dyslexic, print	I I I I I le Ie le le le le Ie re Je le leto ieto 1Cto hracicarstvo hfackarstvo hfackarstvo v lete bude tepto a sucho -

the mean pixel intensity of the surrounding pixels to obtain the final output (See Eq. (5, 6) in 1.1.4 Theoretical Underpinnings). Hence, it accounts for varying light intensities across the whole image, based on the assumption that small regions of an image will have more uniform lighting.

Using an appropriate size for the local sub-region is crucial in getting an accurate threshold as it has to be large enough to include both the text and the background features, but not too large that it over-generalises the pixel intensities.

(VII) **Histogram of pixel intensities for different lighting conditions**

See Figs. 35 and 36.

(VIII) **Effectiveness of morphological operations on image segmentation**

See Fig. 37 and Table 18.

(IX) **Camera and Pen setup**

See Fig. 38

Evaluating the effectiveness of camera system. The camera system was used to capture 5 samples of offline handwriting and convert it to *x*-/*y*-coordinates. These coordinates were then plotted with Python Matplotlib. Since the plots closely resemble the offline handwriting, it is evident that the camera system is effective in capturing the required data. An example is shown in Fig. 39.

Table 15 Parsed text from service 2

Source	Output (Parsed text)
Slovakian school child—Dyslexic(1)	.L e e e e e e e e e e e A, f.. £.. .&. lt ~ 8.. .. Q: "" Q.. ~ [illegible] ..bt.. [illegible] [illegible]
Slovakian school child—Dyslexic(2)	.R.J'A Á.).l .e.e.c.£.:/...b ~.t.,A ..&~4 ..&-A..,/lv.,6;..&.,.£ .lt.lr M.,-'-4.- [illegible]
Slovakian school child—Non-Dyslexic(1)	l.l US. l j.J.l. J.~).). &l.J.. l, l.l. Ja. " 1l~ i.J., J.k. IL. (.L. J .4..Jw J.,...4..., l..J., "-"——P. ——'-,'.el.; J-c'tt,"164.r. 1flJ,, ~ 4,.L,
Slovakian school child—Non-Dyslexic(2)	..l L .l .l .. .l. .l .l ..c .l .. .l. J.. ~ .6. .6 .l. l.. A6A &A lllllO .Ma A.J. 4.nolıt. JM.... A.llLr...,r.. ~ ~ Yr .J.A .Jwl.

(continued)

Table 15 (continued)

Source	Output (Parsed text)
Our handwriting—Non-Dyslexic, cursive	l l .t t t t t t ~ .i te hhklk te.te.te.k~ leto .tel-tJJ- .,teœ lamoken lamoken lamoken hrackarstvo hrackarstvo hrackarstvo V lete bude teplo a sucho
Our handwriting—Non-Dyslexic, print	ı ı ı l ı l l ı l l Je le le Je le le le Je Je Je leto teto l e+o hro c ser s+vo hrackarstvo hrackarstvo v le te bude +epre a su cho.

(X) Alternative hardware setup—IMU

See Figs. 40 and 41.

Firstly, a low-pass filter is used to filter out any high-frequency noise from the raw accelerometer data. Next, the Madgwick filter fuses the gyroscope and accelerometer data to output a quaternion representation of the sensor's orientation, which can be used to calculate its rotation matrix, R:

$$R_{\text{sensor}} = \begin{bmatrix} 1 - 2s\left(q_j^2 + q_k^2\right) & 2\left(q_i q_j + q_k q_r\right) & 2\left(q_i q_k - q_j q_r\right) \\ 2\left(q_i q_j - q_k q_r\right) & 1 - 2s\left(q_i^2 + q_k^2\right) & 2\left(q_j q_k + q_i q_r\right) \\ 2\left(q_i q_k + q_j q_r\right) & 2\left(q_j q_k - q_i q_r\right) & 1 - 2s\left(q_i^2 + q_j^2\right) \end{bmatrix}$$

The accelerometer data can then be transformed from the sensor's body frame to the world frame using:

$$R_{\text{sensor}} = \begin{bmatrix} A_x \\ A_y \\ A_z \end{bmatrix} = \begin{bmatrix} A'_x & A'_y & A'_z \end{bmatrix}$$

Table 16 Parsed text from service 3

Source	Output (Parsed text)
Slovakian school child—Dyslexic(1)	Le 24,020 106,000 0 A La ko dV o A A Safe. ľato. ufo fannobsm—žesnobaru Lamela Arotkorefno imanie Apnastbnsatno- V tf Budu Aoyde
Slovakian school child—Dyslexic(2)	PPAAfA LBL bb ťa BĽ bá LB bálo Ah Lady Lasfo ako damotes. Lyraotemv Lanaoľsv vý tb rasa, Veteda kruhu atplru
Slovakian school child—Non-Dyslexic(1)	4Lalb, L LA AIL 44A LUK b bah ah due U o, henckv lenrkm leme TA, budu buly v
Slovakian school child—Non-Dyslexic(2)	4 LL AL LL Lád La alaéáatá Lab ba dalo hlo Abe damneéom—damolm lamobum Aractakokvo kraitakáka hnoslanidev V bá dub Ay má
Our handwriting—Non-Dyslexic, cursive	14 4 6 £ 04 4 4 4 ľe te to do le de be Le te fe betľo lebo Leto damotbon damoten Lamsten davackaitvo tuackanľvo (nacharifro V tete buse žoplo a fuche
Our handwriting—Non-Dyslexic, print	o] Ol I 10] [O 1 l Je le le le le le le le le le leto leto leto hrackarstvo hnrackarstvo hrackarsStvo v lete bude teplo a sucho

Table 17 Raw input images

Source	Input image (uploaded to OCR service)
Slovakian school child—Dyslexic(1)	
Slovakian school child—Dyslexic(2)	
Slovakian school child—Non-Dyslexic(1)	
Slovakian school child—Non-Dyslexic(2)	

(continued)

Table 17 (continued)

Source	Input image (uploaded to OCR service)
Our handwriting—Non-Dyslexic, cursive	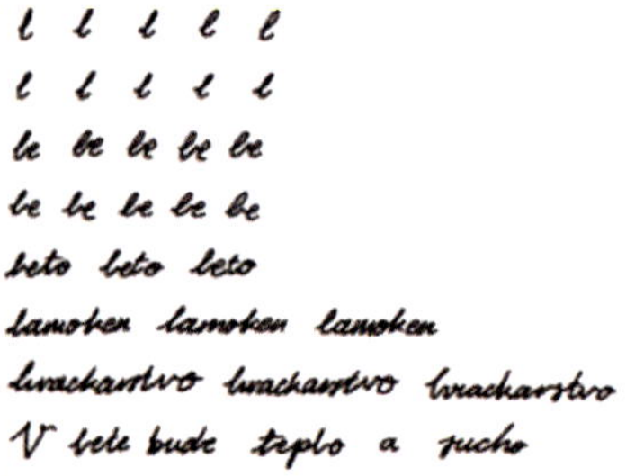
Our handwriting—Non-Dyslexic, print	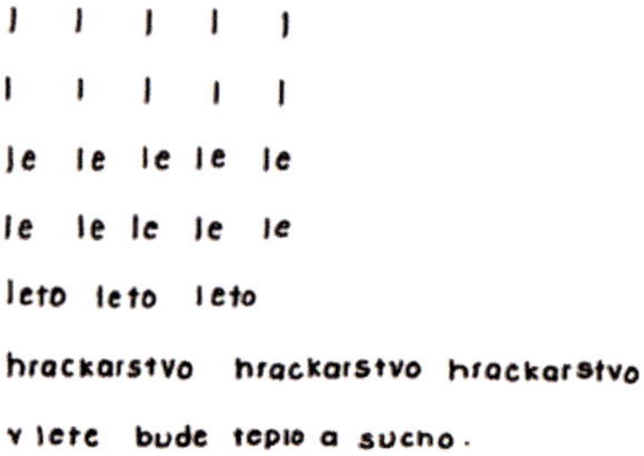

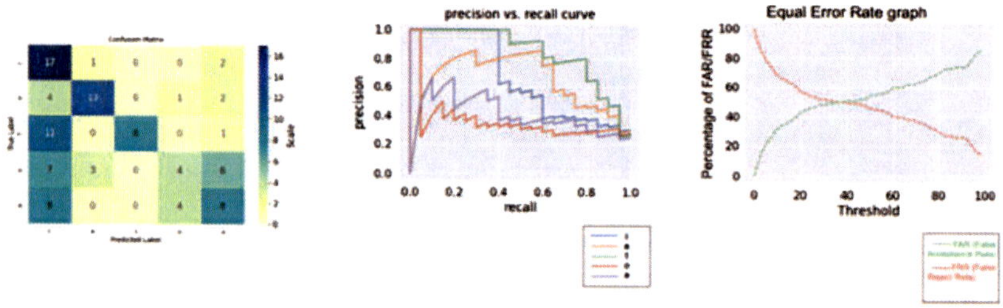

Fig. 32 FNN confusion matrix, precision-recall graph and equal error rate graph (Bézier curves dataset)

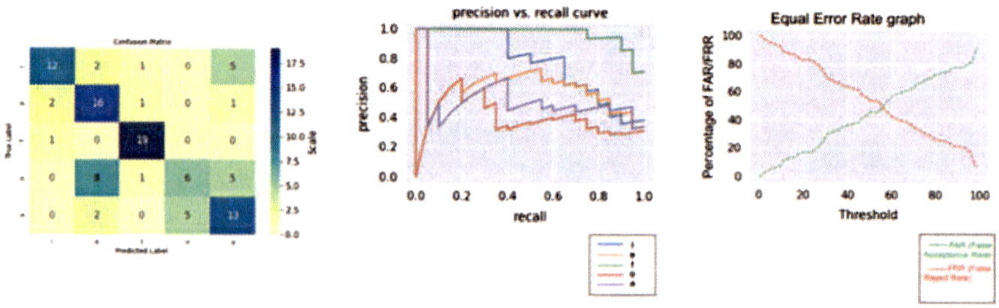

Fig. 33 RNN confusion matrix, precision-recall graph and equal error rate graph (Bézier curves dataset)

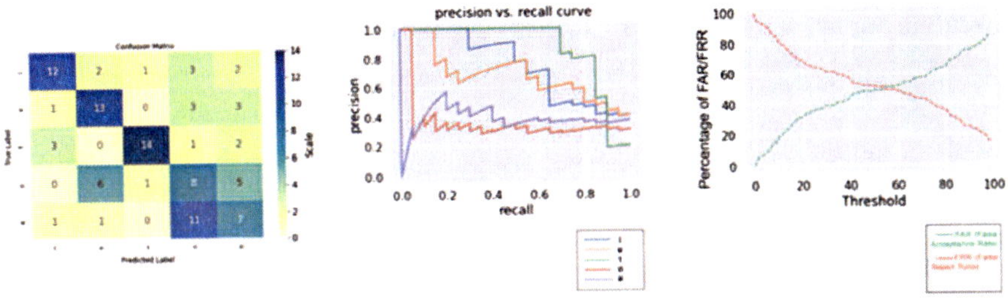

Fig. 34 1D-CNN confusion matrix, precision-recall graph and equal error rate graph (Bézier curves dataset)

Fig. 35 Uniform lighting

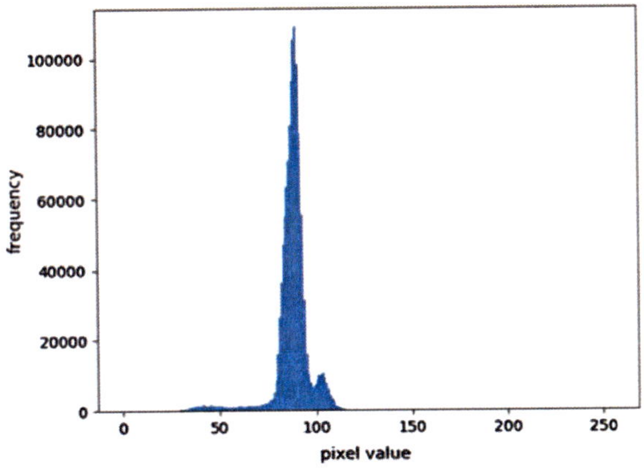

Fig. 36 Non-uniform lighting

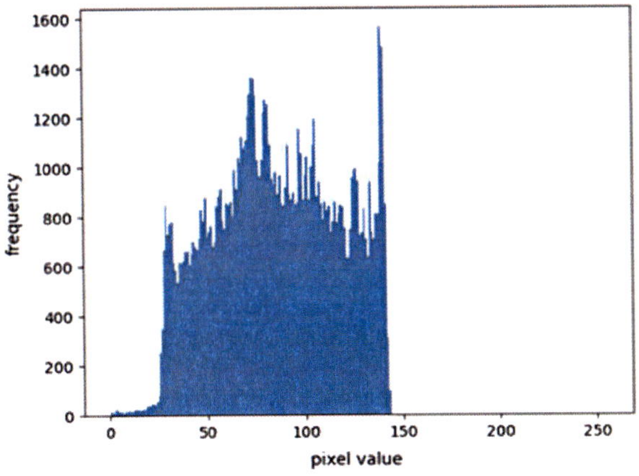

Table 18 Effect of increasing kernel sizes on the noise and text in image

Kernel size	Output		
	Text eroded	Noise left	Final output
3 × 3			
5 × 5			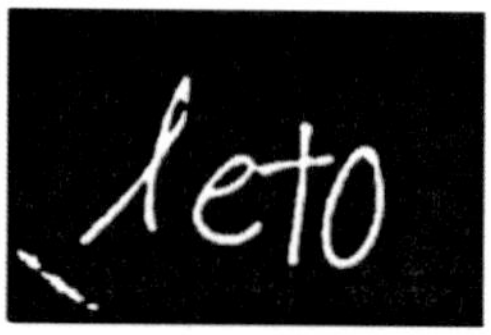
7 × 7	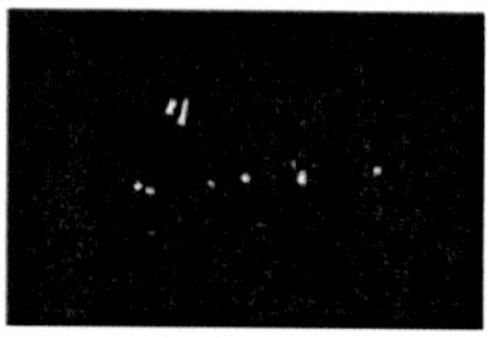		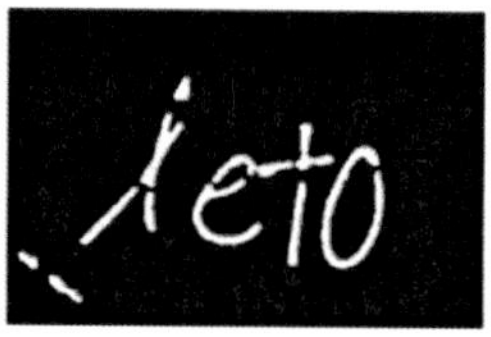
9 × 9	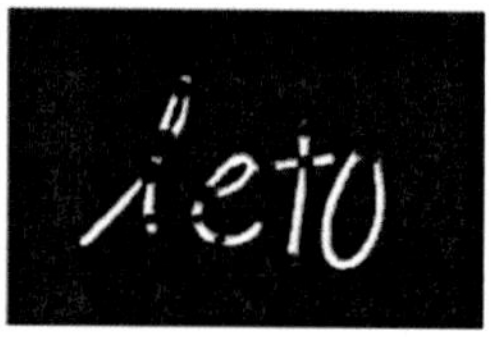		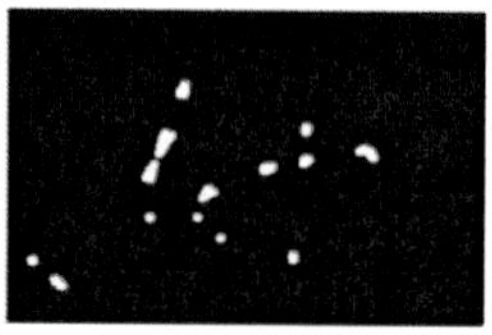
11 × 11	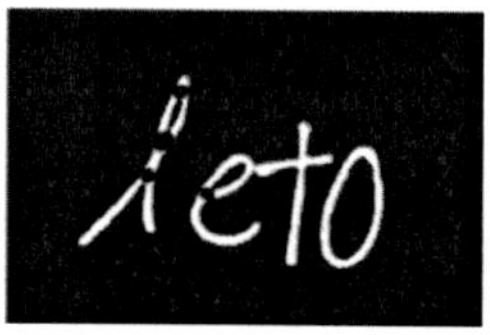		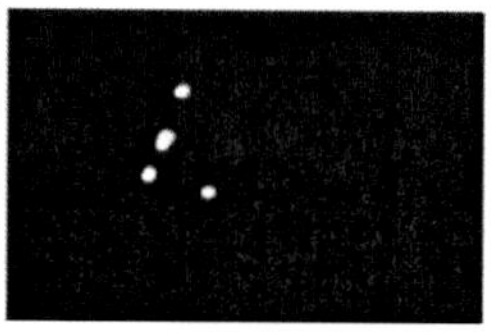

Finally the displacement is calculated by double integrating the acceleration data in each of the axes. However, this means that any small error in accelerometer data will get accumulated into a large error in the displacement data due to the double integration, as seen in Fig. 42.

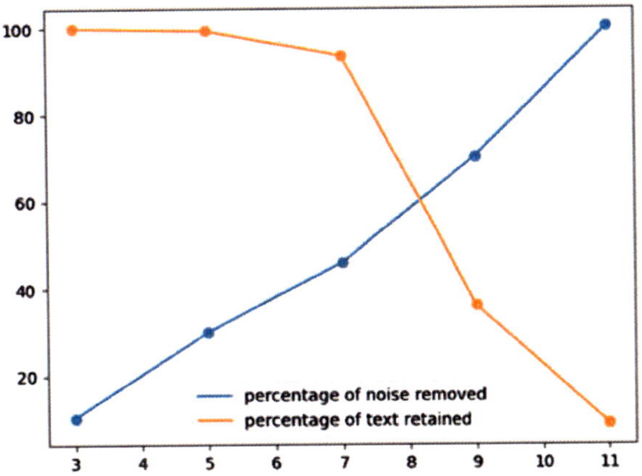

Fig. 37 Graphical representation of the effect of increasing kernel sizes on the segmentation of the image

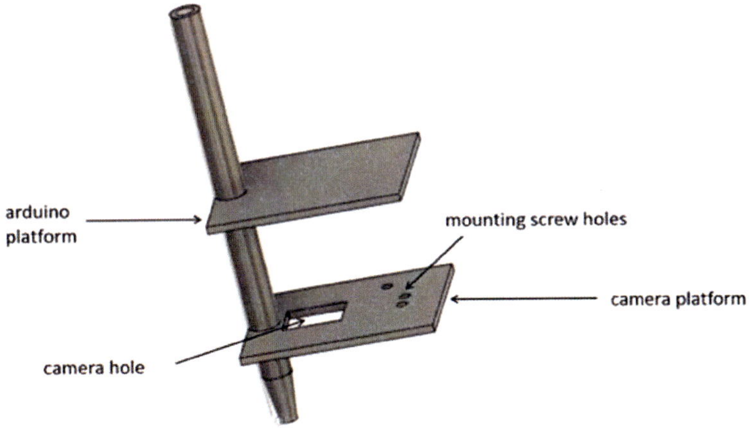

Fig. 38 Annotated diagram of the camera setup

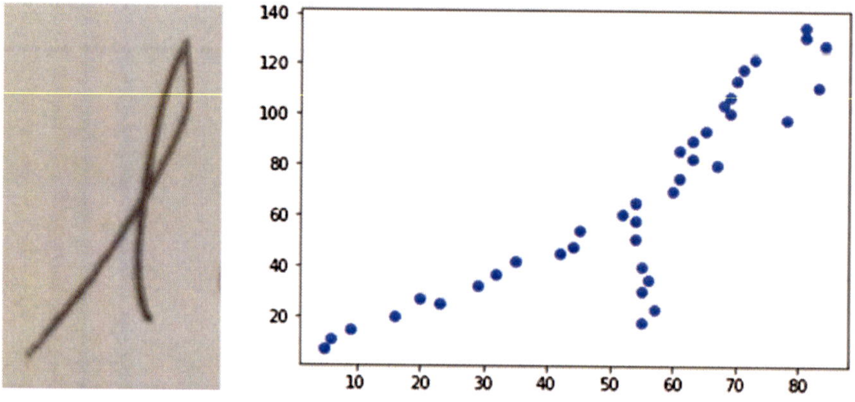

Fig. 39 Sample of coordinates extracted from offline handwriting

Fig. 40 Annotated diagram of the IMU setup

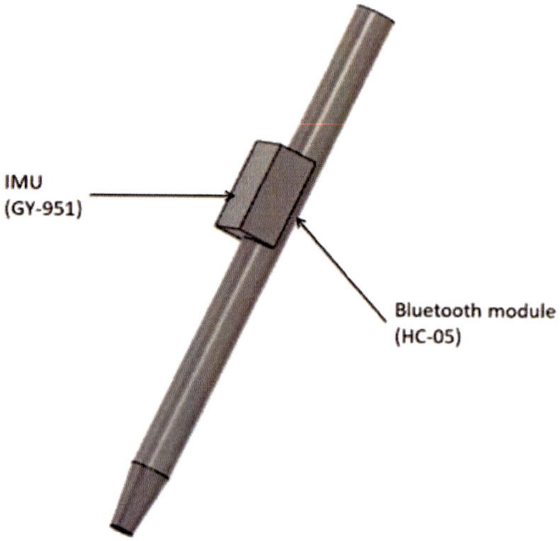

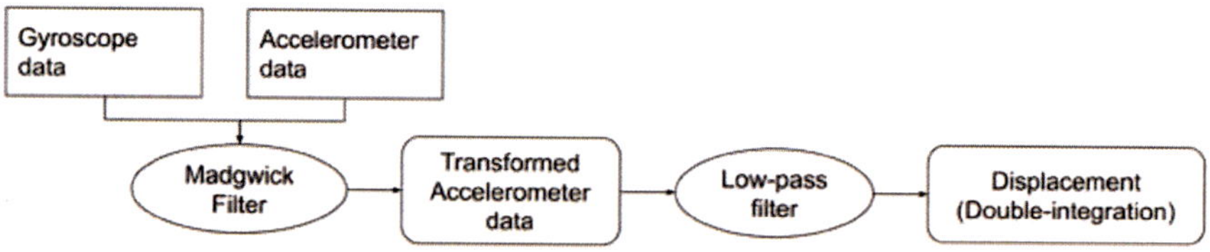

Fig. 41 Block diagram: IMU filter sequence

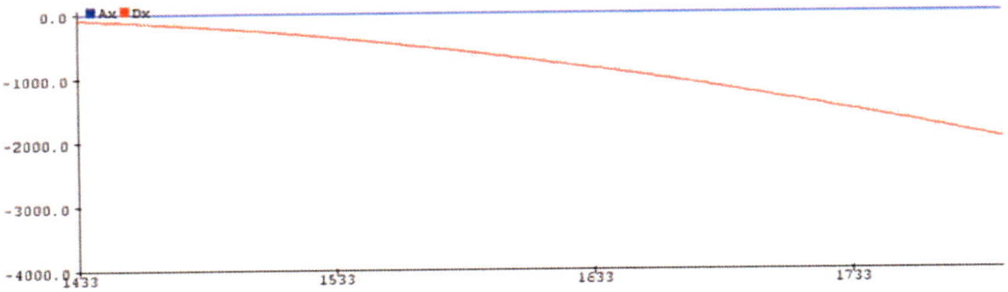

Fig. 42 Integration drift of sensor when pen is held stationary (at an angle)

References

1. Dyslexia, Mayo Clinic. Retrieved December 25, 2021, from https://www.mayoclinic.org/diseases-conditions/dyslexia/symptoms-causes/syc-20353552
2. Leitão, S., Dzidic, P., Claessen, M., Gordon, J., Howard, K., Nayton, M., & Boyes, M. E. (2017). Exploring the impact of living with dyslexia: The perspectives of children and their parents. *International Journal of Speech-Language Pathology, 19*(3), 322–334.
3. Carbune, V., Gonnet, P., Deselaers, T., Rowley, H. A., Daryin, A., Calvo, M., & Gervais, P. (2020). Fast multi-language LSTM-based online handwriting recognition. *International Journal on Document Analysis and Recognition (IJDAR), 23*(2), 89–102.
4. Drotár, P., & Dobeš, M. (2020). Dysgraphia detection through machine learning. *Scientific Reports, 10*(1), 21541.
5. Kowalczuk, Z., & Merta, T. (2015). Evaluation of position estimation based on Accelerometer Data. In *2015 10th international workshop on robot motion and control (RoMoCo)*.
6. Bai, Y.-W., Yu, C.-H., & Wu, S.-C. (2014). Using a three-axis accelerometer and GPS module in a smart phone to measure walking steps and distance. In *2014 IEEE 27th Canadian conference on electrical and computer engineering (CCECE)*.

Image Geolocation from Alternative Cues

Zoe Lee, Vellayappan Dheivanayagam Ganesan, and Teo Ching Lik

Abstract This paper presents an investigative analysis into the use of alternative cues, such as the metadata of images, for image geolocation purposes. Firstly, the metadata was retrieved, pre-processed and converted into a form that can be used for the analysis. A combination of this metadata is then used to make predictions of where the image was taken, with the help of machine learning algorithms through the use of a Support Vector Regression (SVR) model. The results of the geolocation using alternative cues were compared to those from existing image geolocation methods based on image features. The results proved that the alternative cues performed better at the higher country and continent levels, while existing geolocation methods prevail at the lower street, city and regional levels. An approach to combining the use of alternative cues and existing geolocation methods, which could help to increase the accuracy of image geolocation, is then proposed.

Keywords Image geolocation · Metadata · Support vector regression

1 Introduction

Over the years, much progress has been made to improve the accuracy of image geolocation. In this era of rapidly developing technology, image geolocation plays a critical function in enhancing a country's defence given its variety of practical applications, such as using it to fact-check images and videos released online. The primary approach to image geolocation that many notable research papers in this field [1–5] adopt is the use of image features. Such image features no doubt play an important role in determining where an image is taken since they form the basis of accurately clustering input images with their next best match. However, given

Z. Lee (✉)
Raffles Institution (Junior College), 1 Raffles Institution Ln, Singapore 575954, Singapore
e-mail: zoe.leeze@yahoo.com

V. D. Ganesan · T. C. Lik
DSO National Laboratories, 12 Science Park Drive, Singapore 118225, Singapore

H. Guo et al. (eds.), *IRC-SET 2022*,
https://doi.org/10.1007/978-981-19-7222-5_19

that many places around the world lack distinctive features (a beach in Asia and a beach in Australia could look very similar), purely relying on image features alone is insufficient.

To tackle this issue, this project aims to investigate the usefulness of *alternative* cues in image geolocation. Often, there are other forms of metadata embedded in images such as html codes, IP addresses and Exchangeable image file format (Exif) data, which can easily be retrieved. In this project, the focus is on the use of Exif data since this was readily available in our data sets. It may contain important information which gives insights relating to the country and continent the photo was taken in.

The usefulness of these alternative cues was evaluated using Machine Learning algorithms. If proven to be effective, these cues could be pivotal towards increasing the accuracy of image geolocation.

The main contributions of this work were to:

1. Determine the suitability of alternative cues to be used as a first cut to increase accuracy at the country and continent levels.
2. Eradicate largely erroneous results in the current image geolocation methods.
3. Propose a method combining the use of alternative cues and existing geolocation algorithms to increase the overall accuracy.

2 Literature Review

Some notable research papers contributing to image geolocation include IM2GPS [1], PlaNet [2] and Geolocation Estimation [3].

2.1 IM2GPS

The IM2GPS paper [1] adopts a scene matching approach. They first calculate the closeness of six scene properties (line features, gist descriptors, geometric context, tiny images, colour histograms and texton histograms) between the input image and the rest of the images in their data set. The nearest matches are then being clustered and the geolocation estimate is taken to be the largest cluster.

2.2 PlaNet

PlaNet [2] treats image geolocation as a classification problem by subdividing the Earth into geographical cells, with each cell containing the same number of images from the data set. Thereafter, they train a convolutional neural network to compute the probability of the input image being taken in each cell and the cell with the highest probability is used as the final geolocation estimate.

2.3 Geolocation Estimation

The Geolocation Estimation [3] paper uses a similar classification approach to PlaNet. However, an additional feature it includes is the use of a multi-task network that combines the complementary tasks of photo geolocalisation and scene recognition (Fig. 1).

State-of-the-art Problem

The research papers adopt similar approaches to conducting image geolocation, primarily using scene properties and convolutional neural networks. Whilst these methods draw a commendable accuracy, there are still difficulties in accurately geolocating images with few distinctive features (Fig. 2).

Additionally, a cross-comparison of the approaches used in these research papers (Table 1) show that researchers often use extremely large data sets, and their experiments require high computing power to execute.

This affects the reproducibility of the results since it is too computationally intensive and time-consuming for other researchers to replicate.

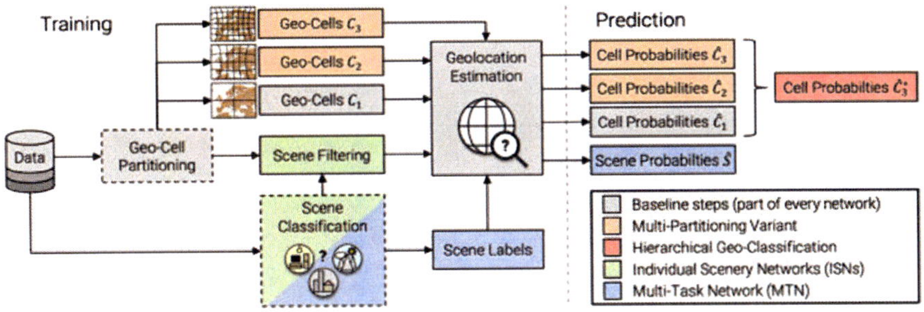

Fig. 1 Algorithm overview of geolocation estimation (*image taken from* [3])

Fig. 2 Examples of images with few distinctive features (*taken from YFCC4k data set*)

Table 1 Research approach and results comparison across papers

	IM2GPS	PlaNet	Geolocation estimation
Size of data sets			
Train	6,000,000	91,000,000	4,700,000
Test	237	2,300,000	36,500
Computing power			
CPU years	3.08	Unknown	Unknown
Number of processors	400	200	Unknown
Duration	3 days	75 days	Unknown
Cumulative accuracy[1] (on IM2GPS3k data set)			
Continent (2500 km)	51.9%	65.8%	66.0%
Country (750 km)	35.4%	45.6%	49.7%
Region (200 km)	32.1%	30.0%	36.6%
City (25 km)	21.9%	18.1%	28.0%
Street (1 km)	2.5%	6.3%	10.5%

2.4 Geolocation Using Image Metadata

Beyond the use of scene properties, other researchers have considered the use of alternative cues like metadata for image geolocation. The paper on Supervised Text-based Geolocation Using Language models on an Adaptive Grid [6] extracts word features directly from a document's text and defines an alternative grid construction to choose locations for pseudo-documents. The research on Unifying Text, Metadata and User Network Representations with a Neural Network for Geolocation Prediction [7] combines geotags from tweets, metadata and user network information trained on a neural network model to tackle geolocation prediction in Twitter. However, no concrete results have emerged from these. Researchers have acknowledged that "only a small portion of available documents being explicitly annotated in this fashion" [6], making it challenging to retrieve sufficient data for testing (Fig. 3).

If these alternative cues can be effectively combined with existing geolocation methods, this could potentially increase the accuracy of image geolocation whilst adopting a less computationally intensive approach.

[1] Refer to "Cumulative Accuracy" under Results section.

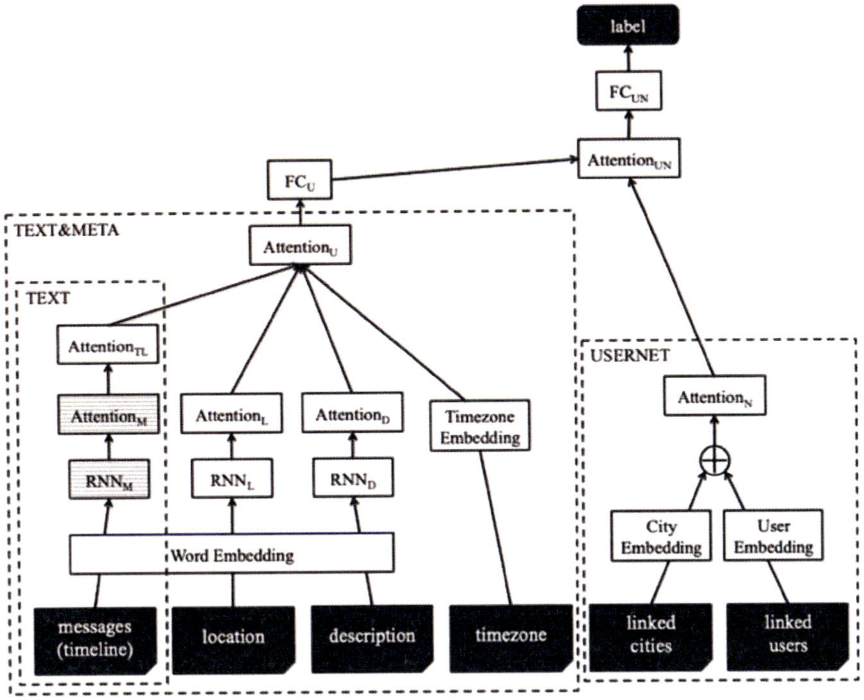

Fig. 3 Algorithm overview of unifying text, metadata and user network representations with a neural network for geolocation prediction (*image taken from* [7])

3 Methodology and Approach

We present the proposed framework which combines both alternative cues and the baseline geolocation algorithm in conducting image geolocation in this section. For the purpose of this project, the focus is on analysing the usefulness of the alternative cues through the three steps shown below, and to thereafter infer how these alternative cues best align with the framework (Fig. 4).

3.1 Baseline Geolocation Algorithm

The approach presented will make use of the baseline geolocation algorithm presented in the Geolocation Estimation paper. To ensure its reliability, we seek to reproduce the experiments on the IM2GPS3k test set as described in the research

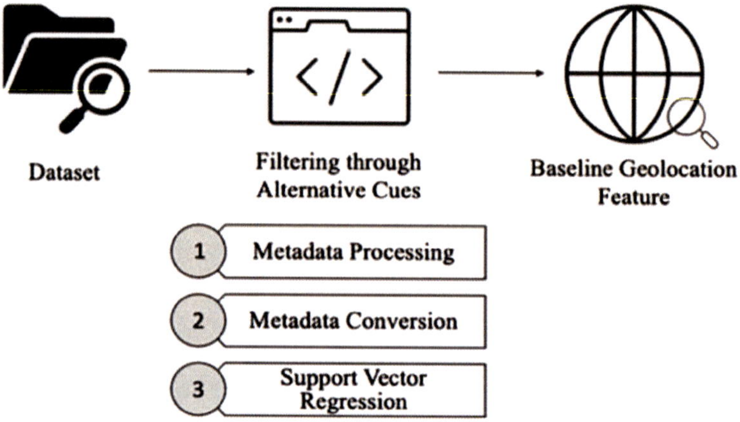

Fig. 4 Image geolocation framework

Table 2 Cumulative accuracy of baseline geolocation algorithm

	Research paper	Open source code				
Partitioning	NA (%)	Coarse (%)	Middle (%)	Fine (%)	Hierarchy (%)	Average (%)
Continent (2500 km)	66.0	66.9	66.5	66.4	66.9	66.7
Country (750 km)	49.7	51.7	51.3	51.2	51.0	51.3
Region (200 km)	36.6	36.3	35.7	36.2	36.9	36.3
City (25 km)	28.0	24.3	26.2	27.3	28.0	26.5
Street (1 km)	10.5	6.2	8.3	9.9	10.1	8.6

paper, and to confirm its accuracy. This was done by using the Geolocation Estimation open source code,[2].[3]

As seen from Table 2, the average cumulative accuracy of all four partitioning types is comparable to the results presented in the research paper. This is proof that the open source code is indeed accurate and it can therefore be reliably used as the model's baseline geolocation algorithm.

[2] https://code.google.com/archive/p/s2-geometry-library/

[3] https://github.com/TIBHannover/GeoEstimation

3.2 Metadata Processing

To narrow down the search space for geolocation predictions based on alternative cues, metadata is first being extracted from the images in the data set. We chose to focus on the use of Exif data which was available for our 2 data sets, IM2GPS3k and YFCC4k.

The IM2GPS3k data set consists of 3000 images and 10 types of Exif data, including the owner ID, file format and photo tags (descriptions of the photo which include city and country names). The YFCC4k data set comprises 4536 images and 21 unique Exif features, similar to IM2GPS, but also including the camera model used. For both data sets, the images were randomly split, with 80% of the images being used as the training data and the remaining 20% as the test set.

As part of cleaning the metadata, the GeoText function[4] was applied to the photo tags from both data sets to extract city and country names. The first version of tags (tags_v1) could contain a combination of multiple city and country names. For the second version of tags (tags_v2), each image had only one city or country name attached to.

Following the curation of the tags, the data sets were filtered to only include the tags that appeared five or more times in the data set. The above approach taken helped to decrease the number of unique tags, allowing for more precise predictions. It also ensured that there were sufficient instances of each tag during training.

3.3 Metadata Conversion

Given that the metadata consisted of various data types, there was an inconsistency in the input variables. Therefore, there is a need for standardisation which was done by converting unique categories in the data to numeric values. For this, two approaches were taken into consideration: integer encoding and one-hot vector encoding.

For the integer encoding approach, each unique category would be assigned with an integer value, e.g. "Singapore" is 1 and "Malaysia" is 2. This allows machine learning algorithms to understand and harness the natural ordered relationship between various categories. A similar integer representation is used in the one-hot vector encoding, except that it is stored in the form of a binary variable, e.g. "Singapore" is [1,0] and "Malaysia" is [0,1].

Upon testing, we found that the one-hot vector encoding approach proved to be more effective since the categorical variables we were dealing with did not carry an ordinal relationship. With the integer encoding approach, the model could have predicted a value in between the integers, leading to inaccuracies and poorer performance.

---

[4] https://pypi.org/project/geotext/

3.4 Support Vector Regression (SVR)

Following the categorisation of metadata, their usefulness was then analysed using an SVR mode. Based on the similarities of the metadata between the input test image and images from the training set, the model predicts a latitude and longitude. To do so, it uses a kernel to calculate a line or hyperplane (for higher dimensional data) to classify where a new point taken from the test data lies closest to. This model uses the Radial Basis Function (RBF) kernel, which computes the similarity between 2 points, X_1 and X_2, using the following formula:

$$K(X_1, X_2) = \exp\left(-\frac{||X_1 - X_2||^2}{2\sigma^2}\right)$$

where σ represents the variance, and $||X_1 - X_2||$ represents the Euclidean distance between the 2 points.

4 Evaluation Metrics and Results

4.1 Cumulative Accuracy

We evaluated the usefulness of the metadata features in comparison to the baseline geolocation algorithm based on their cumulative accuracy. To calculate this, we used the `vectorized_gc_distance` function in the Geolocation Estimation's open source code[1] which takes in 4 inputs (predicted latitude, predicted longitude, actual latitude, actual longitude) and calculates the great-circle distance. The percentage of images that fall within the specified radius (1, 25, 200, 750, 2500 km) of the various levels is then being calculated.

Referring to Table 2, this would mean that 66.9% of images were successfully predicted within the 2500 km radius of the groundtruth at the continent level by the baseline geolocation algorithm, of which 51% was predicted at country (750 km) or better and so on for each decreasing (finer) levels. Of note is that the *remaining 33.1%* of images that were predicted outside the continent level are lost completely: they were predicted with an error *beyond 2500 km*. The primary aim of using alternative cues here would be to recover as much of the lost 33.1% of images as possible.

4.2 Baseline Geolocation Algorithm

As shown above in Table 2, the hierarchical partitioning type performed the best on average. Hence, we will be focusing only on that for the purpose of this analysis. The baseline algorithm feature was used to predict the latitude and longitudes for both

Table 3 Cumulative accuracy of geolocation using Baseline Geolocation Algorithm

	Street (1 km) (%)	City (25 km) (%)	Region (200 km) (%)	Country (750 km) (%)	Continent (2500 km) (%)
IM2GPS3k	10.1	28.0	36.9	51.0	66.9
YFCC4k	6.6	16.5	24.2	37.5	54.9

the IM2GPS3k and YFCC4k data sets and their accuracies at the respective levels are documented below in Table 3.

4.3 Proposed Support Vector Regression (SVR) Model Using Metadata

The SVR model was used to predict the latitude and longitude based on several combinations of metadata features. The accuracies of these predictions were then calculated in a similar manner to that of the baseline geolocation algorithm. For the IM2GPS3k data set, the SVR was tested on a combination of tags and owner metadata. For the YFCC4k data set, the SVR was tested on a combination of tags, owner and license metadata.

From Table 4, it was observed that the single most useful feature amongst the metadata were the photo tags. Whilst a combination of more features helps to increase the overall cumulative accuracy as compared to using the second feature (owner/license) alone, the accuracy of this combination is still lower than using just the photo tags.

Table 4 Cumulative accuracy of geolocation using metadata features on SVR

Feature	Street (1 km) (%)	City (25 km) (%)	Region (200 km) (%)	Country (750 km) (%)	Continent (2500 km) (%)
IM2GPS3k					
tags_v1	0	0	24.0	**76.3**	97.1
tags_v2	0	**1.2**	**30.4**	75.8	**97.2**
tags & owner	0	0.3	14.6	58.8	91.8
YFCC4k					
tags_v1	0	0	24.4	67.4	**95.0**
tags_v2	0	**0.4**	**25.8**	**74.2**	91.5
owner	0	0	0.7	10.7	72.7
license	0	0	0.7	11.3	74.2
tags & owner	0	0	5.1	43.6	79.5
tags & license	0	0	9.3	50.2	88.7

Comparing Tables 3 and 4, it is observed that the metadata consistently drew more accurate predictions at the country and continent levels than the baseline geolocation algorithm across both data sets, whereas the baseline geolocation algorithm worked better at the street, city and region levels. At the continent level for the IM2GPS3k data set, the metadata was able to recover an additional 30.2% of the images in the data set, saving approximately 89% of the images that would have otherwise been lost with just the use of the baseline geolocation algorithm. At the country level, 25.3% of the images in the data set were recovered (approximately 52% of images lost). Similarly, for the YFCC4k data set, around 89% and 59% of the lost images were recovered with the use of metadata at the continent and country levels respectively.

These results show that the metadata plays an important role in increasing the cumulative accuracy of image geolocation predictions at the country and continent level, which is key to preventing the loss of images at the street, city and regional levels.

5 Discussion

Given that the metadata is more effective for geolocation at the country and continent levels, it has potential to be implemented before the baseline geolocation algorithm, serving as a first cut to classify these images on a broader country and continent scale. For example, this picture below from the IM2GPS3k data set was taken in Mueang Roi Et District, in Thailand, Asia. However, the baseline geolocation algorithm inaccurately predicts it to be taken somewhere in Ghana, Africa.

Cases like these are where the alternative cues come in handy. Should we just rely on existing methods of image geolocation through the use of image features, large proportions of images would already have been lost at the country and continent levels. In the case of the IM2GPS3k and YFCC4 data sets, as shown in Table 3, 33.1% and 45.1% of images, respectively, were predicted inaccurately at the highest continent level. This might affect the true potential of the baseline geolocation algorithm at lower levels.

Whilst the alternative cues alone might not be able to achieve the same accuracy as the baseline geolocation algorithm for the lower street, city and regional levels, it could be used as a first cut to classify images at the higher country and continent levels, thereby increasing the probability of the baseline geolocation algorithm classifying it correctly at lower levels. For instance, the photo tag retrieved from the Exif data in Fig. 5 was "Thailand", which the SVR was able to accurately predict minimally at the continent level. If this is implemented before the baseline geolocation algorithm, it would help to greatly narrow down the search space. The baseline geolocation algorithm would then pick the next best match from the same country or continent to be the final geolocation estimate, which significantly increases the chance of the predictions being right at lower levels, thereby increasing the overall accuracy too.

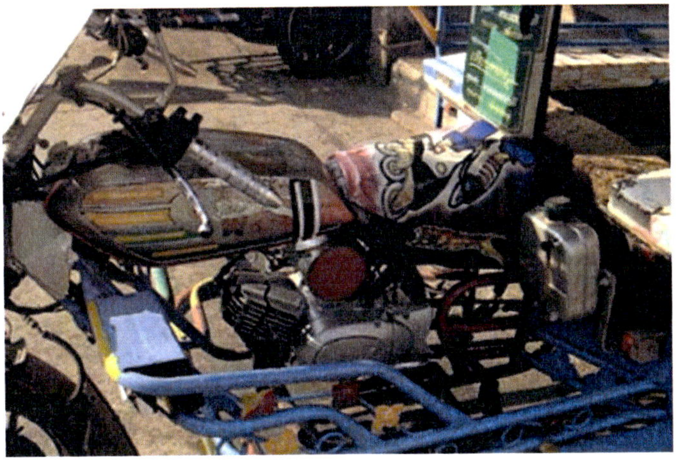

Fig. 5 Image taken in Thailand (*taken from IM2GPS3k data set*)

6 Conclusion

In conclusion, this paper presents an analysis of the usefulness of alternative cues, specifically Exif data, in image geolocation. The Exif data was pre-processed, converted into a desired form and then evaluated using an SVR model, which gave us an indication of the correlation between a specific type of data and the location where an image was taken.

From the experimentation results, we found that alternative cues performed better at the country and continent levels whilst the baseline geolocation algorithm performed better at the street, city and regional levels. Given this, an approach to combining the both methods was proposed. The alternative cues could be used as a first cut at the country and continent levels to accurately predict the country and continent, significantly narrowing down the search space. Based on this narrowed region, to predict locations at a lower level, this would require the use of image features which can be implemented using the baseline geolocation algorithm. With this proposed approach, there is a greater likelihood that the images would be predicted accurately at lower levels too, increasing the overall accuracy of image geolocation.

References

1. Hays, J., & Efros, A. A. (2008). IM2GPS: Estimating geographic information from a single image. In *Proceedings of the IEEE conference on computer vision and pattern recognition (CVPR)*.
2. Weyand, T., Kostrikov, I., & Philbin, J. (2016). PlaNet—Photo geolocation with convolutional neural networks. In *Proceedings of the European Conference on Computer Vision (ECCV)*.
3. Müller-Budack, E., Pustu-Iren, K., & Ewerth, R. (2018). Geolocation estimation of photos using a hierarchical model and scene classification. In *Proceedings of the European Conference on Computer Vision (ECCV)*.
4. Tahmasebzadeh, G., Kacupaj, E., Müller-Budack, E., Hakimov, S., Lehmann, J., & Ewerth, R. (2021). GeoWINE: Geolocation based wiki, image, news and event retrieval. In *Proceedings of the 44th international ACM SIGIR conference on research and development in information retrieval*.
5. Vo, N., Jacobs, N., & Hays, J. (2017). Revisiting IM2GPS in the deep learning era. In *Proceedings of the 2017 IEEE International Conference on Computer Vision (ICCV)*.
6. Roller, S., Speriosu, M., Rallapalli, S., Wing, B., & Baldridge, J. (2012). Supervised text-based geolocation using language models on an adaptive grid. In *Proceedings of the 2012 joint conference on empirical methods in natural language processing and computational natural language learning*.
7. Miura, Y., Taniguchi, M., Taniguchi, T., Ohkuma, T. (2017). Unifying text, metadata, and user network representations with a neural network for geolocation prediction. In *Proceedings of the 55th annual meeting of the association for computational linguistics* (vol0 1), Long Papers.

Measurement of Electromagnetic Emissions from Smart Devices Within Faraday's Box

Yu Hao Lim, Yeow Kwang Tai, and Umai Chan

Abstract Devices may pose an electromagnetic (EM) security risk where information may unintentionally leak out while in use. As such, testing on frequencies emitted by smart devices was conducted, with a DIY shielded box (Faraday's box), to mimic the classic Faraday's cage. The shielding effectiveness of said box, wrapped in bubble wrap or aluminium foil, was then checked using existing frequency modulation (FM) radio transmissions. The more effective shielded box was then used to create an environment that was free of emissions from other devices, or 'noise-free' for short. For the first part of the study, the effectiveness of shielding of EM waves using aluminium foil versus bubble wrap was evaluated. Comparing both materials, the aluminium foil was found to attenuate more FM frequencies than the bubble wrap and acted as a more effective shield. Emission measurement was then conducted on the smart devices of interest with a TV antenna as a receiving antenna, an Airspy Mini dongle as a receiver, the aluminium foil-wrapped box, and a SpectrumSpy software to analyse the emissions from the smart devices in frequency domain. The devices used in this experiment were the Samsung J2 Pro, Vivo Y21 and Samsung A32. When measuring EM emissions of the smart devices, it was found that the Samsung A32 had the most emissions of the three devices, followed by the Samsung J2 Pro and Vivo Y21.

Keywords Electromagnetic emissions · Smart devices · Faraday's box

1 Introduction

Modern information communications technology equipment can have unwanted and accidental electromagnetic emissions due to a poor design of devices and electronics, which may cause information leakage and pose as a cybersecurity threat, as attackers

Y. H. Lim (✉)
School of Math and Science, NUS High, Singapore, Singapore
e-mail: h1910085@nushigh.edu.sg

Y. K. Tai · U. Chan
DSO National Laboratories, Singapore, Singapore

H. Guo et al. (eds.), *IRC-SET 2022*,
https://doi.org/10.1007/978-981-19-7222-5_20

can utilise said emissions to use Transient Electromagnetic Pulse Surveillance Technology (TEMPEST) operations to obtain information [1]. TEMPEST cannot be detected easily and does not leave trackable traces, making it dangerous. To investigate the extent of TEMPEST leakage, the emissions from different models of mobile phones were measured and compared. TEMPEST works by detecting the electromagnetic wave emissions and reconstructing said emissions at the leakage carrier frequency. This can be done with a simple antenna and readily available software from the open commercial market. A cost-effective solution to combat it would be shielding devices containing sensitive information with a suitable lining. The classic Faraday cage is known to block external electromagnetic fields from entering as well to prevent internal electromagnetic fields from leaving [2]. We took inspiration from it, to create a DIY Faraday shield that was used to block the external electromagnetic waves to create a noise-free enclosure. This also ensures that device emission could be recorded without disturbance.

The 'noise' emitted by each device can vary due to the difference in electronics design and implementation adopted by the different manufacturers. In theory, with a higher quality of production, such as a good selection of components, proper board layout, short trace routing, proper grounding design and additional use of metallic heat sink to encapsulate the electronics, the emissions from a device could be low enough such that they cannot be detected easily. The emissions may also vary depending on what the device is doing. There are also specific screens that can be displayed to cause the device to emit an identifiable frequency [3], which is why we used white and black video screens.

2 Materials and Methods

In this study, the dimensions of the boxes recorded in centimetres were 32.5 × 25.7 × 13.5 and 36 × 35 × 35, with corrugated cardboard as the base material for the box. For reference in future, we will be referring to the former as 'Box A' and the latter as 'Box B', in terms of sizing. The antenna used was a 25dBi television antenna purchased from Shopee. To connect said antenna to the computer, we used an Airspy dongle and utilised the SDRsharp software, called SpectrumSpy, to allow us to measure and to visualise the electromagnetic emission spikes across a span of frequency range. Images of Box A and Box B wrapped in aluminium foil and bubble wrap are given in Appendix A. The devices used in this experiment are the Samsung J2 Pro and Vivo Y21 for Box A, and the Samsung A32 for Box B.

2.1 Shielded Box

In our first study, the box was wrapped with Diamond and Grillpac brand aluminium foil and secured using Steve and Leif brand aluminium tape. In our second investigation, the box was wrapped in bubble wrap and secured with Tesa brand masking tape to evaluate the possibility of shielding using the air pockets from the bubble wrap. In this experiment, it is assumed that the tapes shield and reflect the same amount of electromagnetic emissions as the material used. To evaluate the shielding effectiveness of the two choices of material used, we placed the TV antenna inside the wrapped box and used the FM radio as a 'free transmitter source' to check the effectiveness of the box shielding.

The shielding results of the different materials are illustrated below. Comparing the 'spikes' in Figs. 2 and 3, the shielded box with aluminium is much more efficient in terms of shielding as compared to the shielded box with bubble wrap, and thus we will be using the aluminium box in the next section of the experiment. The shielded box with bubble wrap is highly ineffective and did not result in much change as compared to the control setup of an unshielded box as shown in Fig. 1.

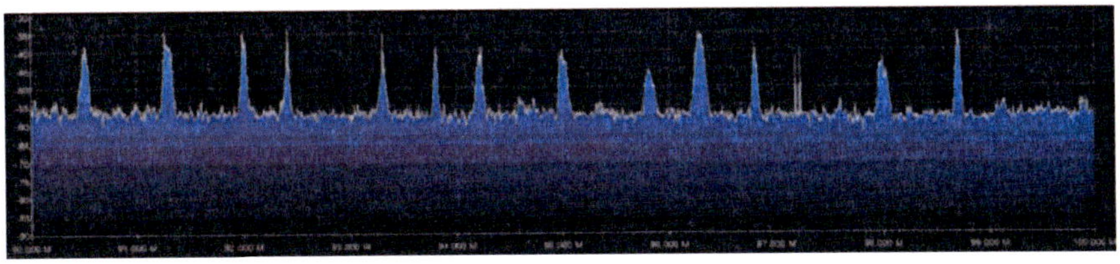

Fig. 1 FM radio range recorded within unshielded box

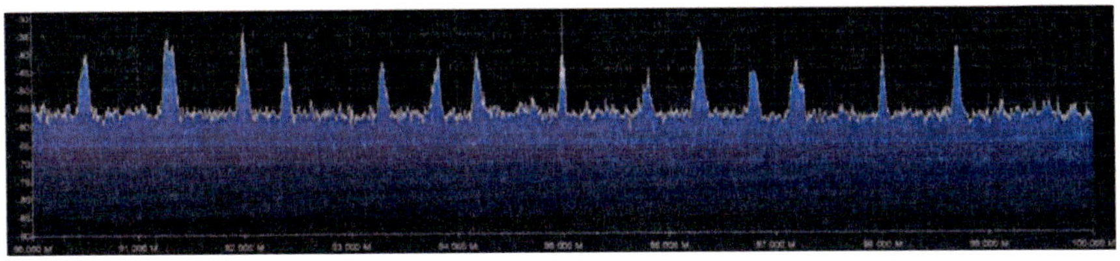

Fig. 2 FM radio range recorded within box shielded with bubble wrap

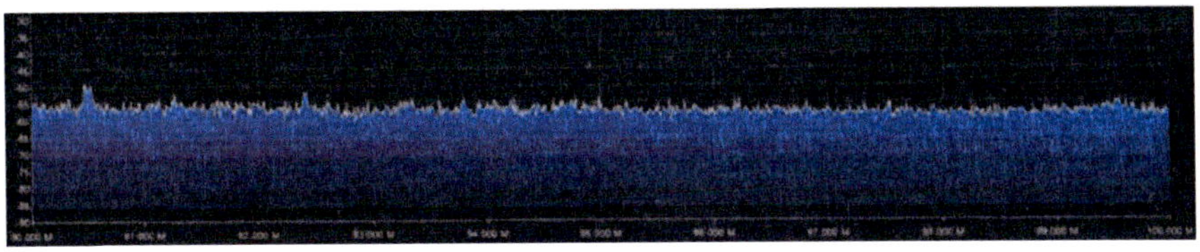

Fig. 3 FM radio range recorded within box shielded with aluminium foil

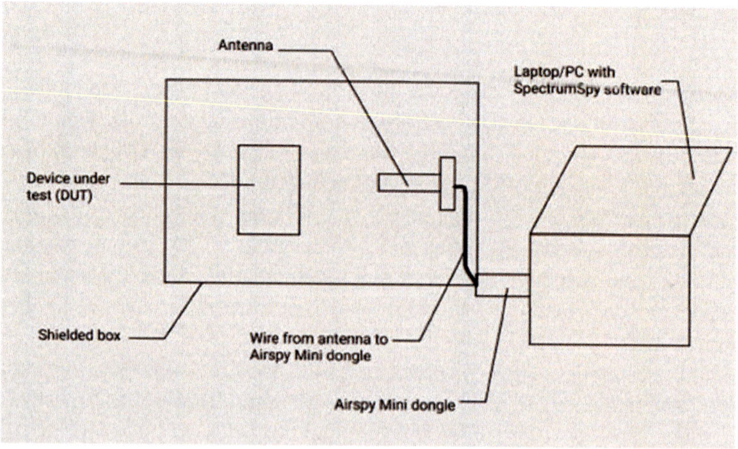

Fig. 4 Diagram of device emission experiment set-up

2.2 *Emission Recordings*

In the second part of this experiment, the shielded box was used to create a noise-free environment, in which the DUT and antenna were placed inside as seen in Fig. 4.

To measure the electromagnetic emissions of devices, we set up the DUT and antenna inside the box. Afterwards, the Airspy Mini, a software-defined radio, along with the SpectrumSpy software, was used to measure the emissions given out by the DUT. Each DUT has 3 recorded readings, one with a video with a black screen playing, a video with a white screen playing, and finally downloading applications.

3 Results

3.1 *Device Emissions—Box A*

In order to find which frequency bands are highest, a waterfall diagram was plotted as shown in Fig. 8. Similar measurements were conducted on Samsung J2 Pro as shown in Figs. 9, 10a–c and for Samsung A32. In Figs. 11 and 16, the waterfall diagrams were separated into four sections, (a) to (d). In the exact order, the letters (a) to (d) represent the graph of wavelength peaks over time for Figs. 9, 10a–c, 12, 13, 14 and 15, respectively.

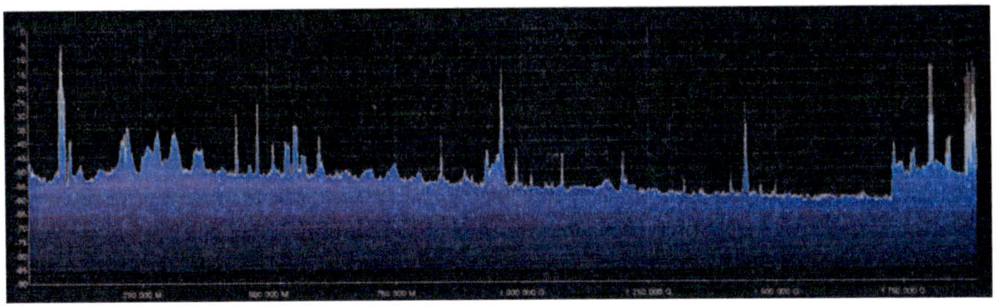

Fig. 5 Aluminium-shielded Box A with DUT, downloading applications

Vivo Y21

Vivo Y21 has little or no emissions as can be seen below in Figs. 5, 6 and 7. There is little variation in the graphs of wavelengths when the device was used for downloading applications, playing a white screen and playing a black screen, respectively. This can be seen clearly in Fig. 8, where there is little change in the waterfall diagram over time.

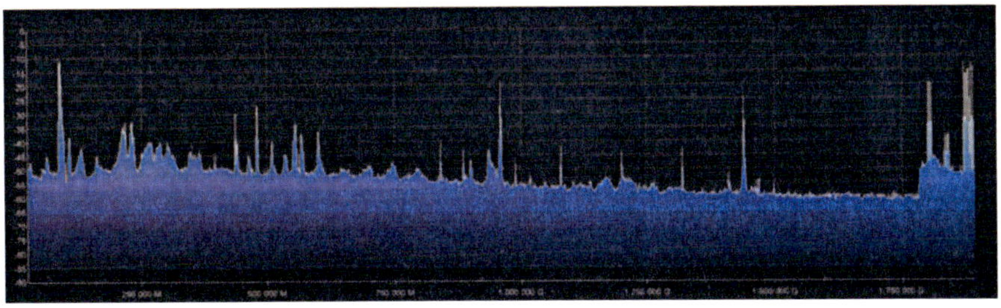

Fig. 6 Aluminium-shielded Box A with DUT, playing white screen

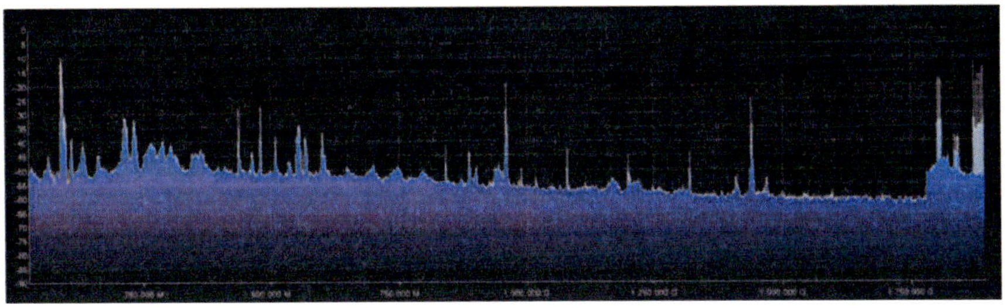

Fig. 7 Aluminium-shielded Box A with DUT, playing black screen

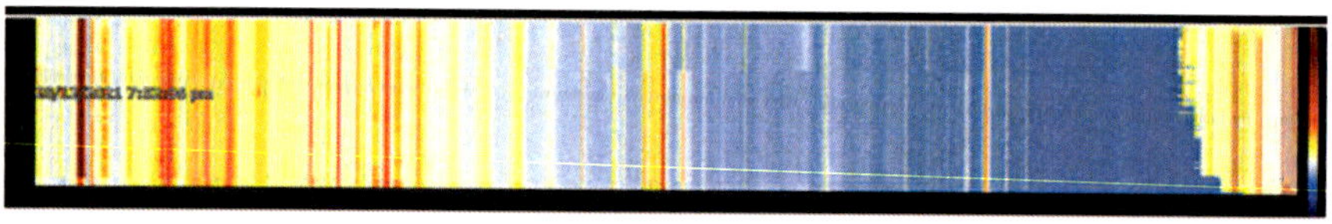

Fig. 8 Graph of wavelength peaks over time. *Note* This figure is not labelled as the borders are unclear

Samsung J2 Pro

Figure 9 is a control setup to see what wavelengths are already present due to background noise as the shielding setup used was unable to completely eliminate background noise.

Figure 10a–c show the electromagnetic emissions that were detected during testing.

Figure 11 summarises Figs. 9, 10a–c, in order, with (a) being Fig. 9 and (d) being Fig. 10c.

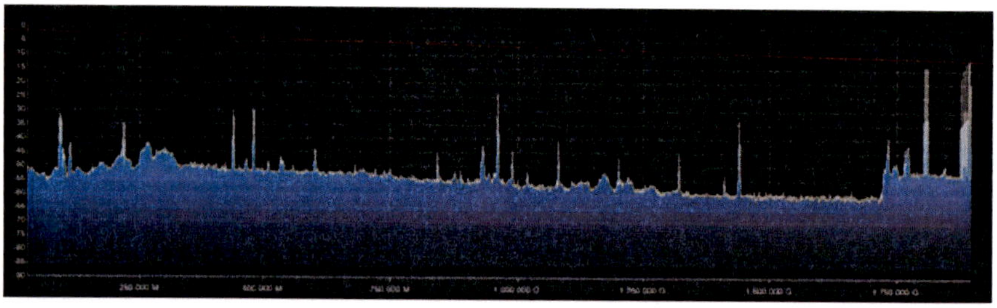

Fig. 9 Aluminium-shielded Box A with no DUT. *Note* Full graphs were cropped to only include the important parts

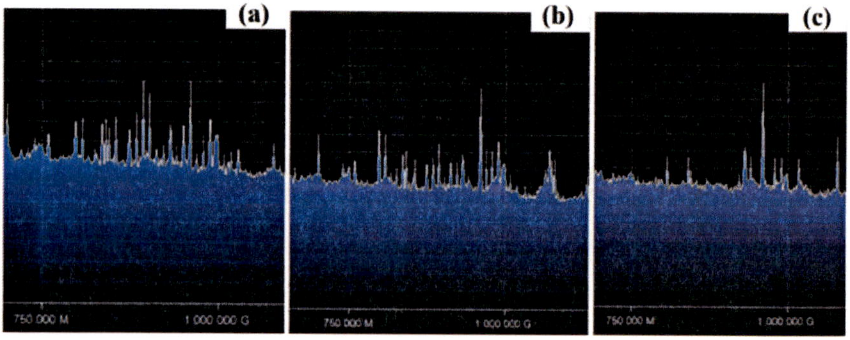

Fig. 10 **a** Aluminium-shielded Box A with DUT, playing white screen. **b** Aluminium-shielded Box A with DUT, playing black screen. **c** Aluminium-shielded Box A with DUT, downloading applications

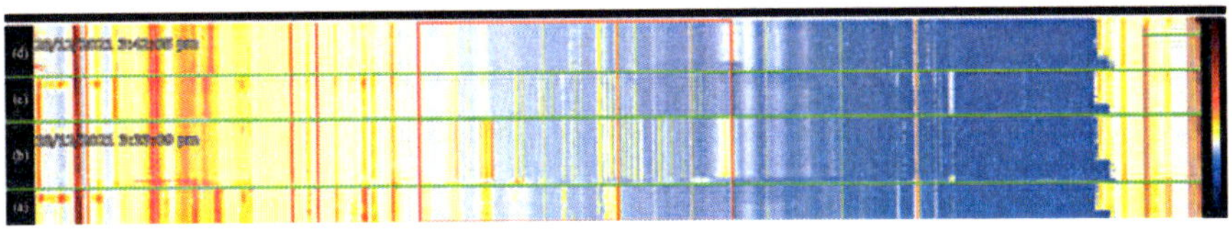

Fig. 11 Labelled graph of wavelength peaks over time. *Note* The colour gradient on the right side corresponds to the peak of emissions, with red representing a higher peak

3.2 *Device Emissions - Box B*

Samsung A32

Figure 12 is a control setup to see what wavelengths are already present due to background noise as the shielding setup used was unable to completely eliminate background noise.

Figures 13, 14, and 15 show the electromagnetic emissions that were detected during testing.

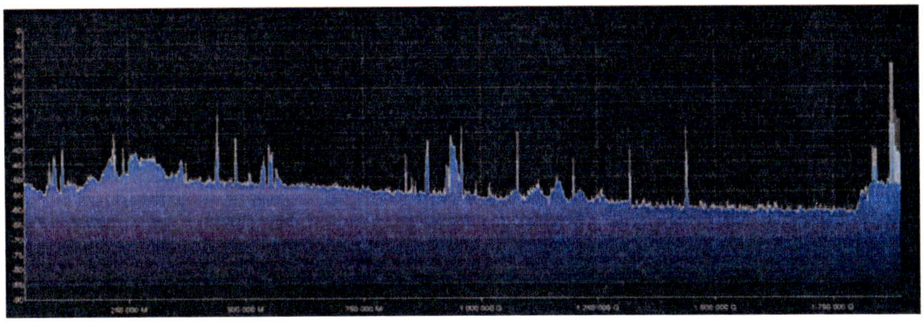

Fig. 12 Aluminium-shielded Box B with no DUT

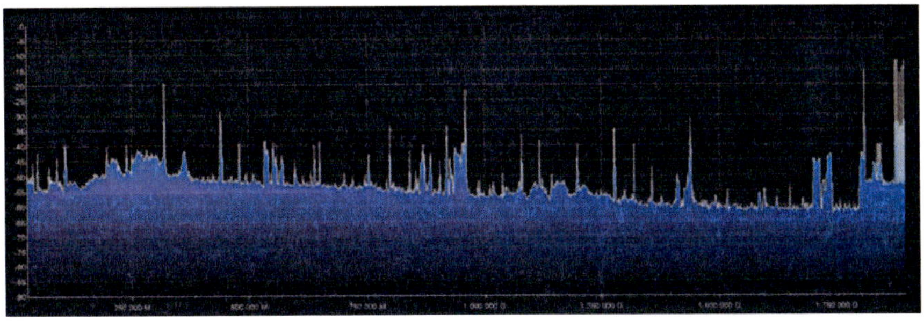

Fig. 13 Aluminium-shielded Box B with DUT, playing white screen

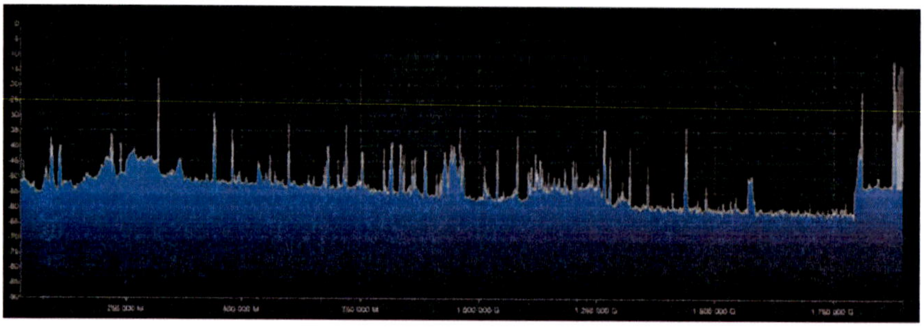

Fig. 14 Aluminium-shielded Box B with DUT, playing black screen

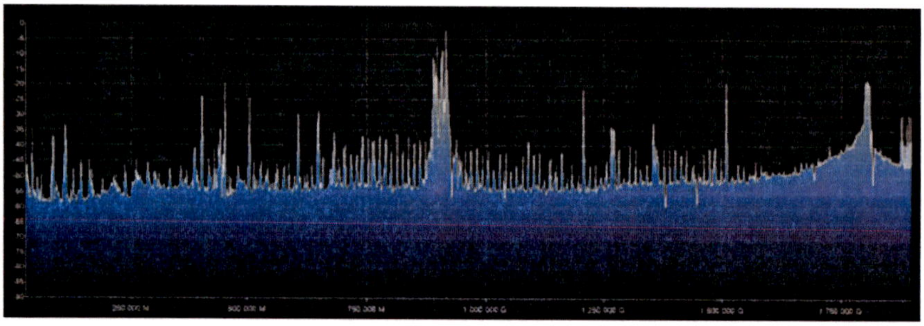

Fig. 15 Aluminium-shielded Box B with DUT, downloading applications

Figure 16 summarises Figs. 12, 13, 14 and 15, in order, with (a) being Fig. 12 and (d) being Fig. 15.

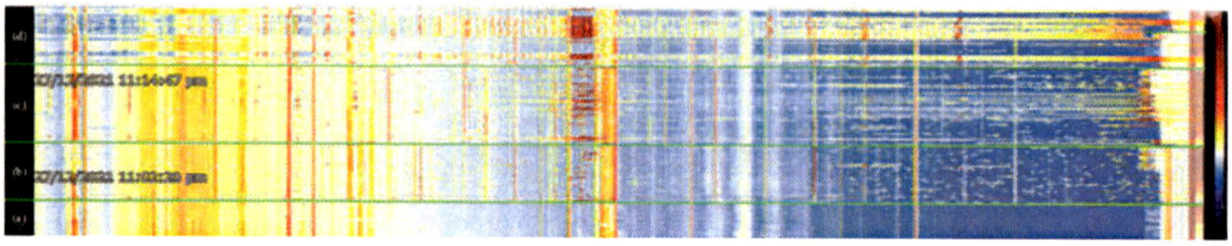

Fig. 16 Labelled graph of wavelength peaks over time. *Note* The colour gradient on the right side corresponds to the peak of emissions, with red representing a higher peak

4 Discussion

4.1 Limitations Due to COVID-19

Initially, the authors of this work were supposed to have access to better equipment and facilities such as a Spectrum Analyzer and an anechoic chamber, in which the frequency of electromagnetic emissions would be better visualised. Other devices, such as smart watches and tablets, were also planned to be used, but were not used due to home constraints. The materials used for the 'box shielding' were also constrained to materials that did not require a lot of heavy machinery and which could be handled safely at home. This arises to possible emission leakages due to poor workmanship in the wrapping of the material and the improper treatment of the opening for cable routes.

4.2 Analysis of Different Devices

Vivo Y21

Based on emissions plots from Figs. 5, 6, 7 and 8, it can be noted that the Vivo Y2 had little, or no emissions despite the different videos and downloads, as compared to the other 2 devices. A possible explanation for this could be due to the fact that the phone was released recently and may have had some better shielding implemented compared to the other devices.

Samsung J2 Pro

Figure 17 and 18 simplify Figs. 11 and 16, respectively, and focus on the electromagnetic emissions that appeared during the testing, with the red lines signifying high peaks that remained steady, orange lines signifying high peaks that remained unsteady and had a high rate of fluctuation, and the yellow lines representing low peaks that remained steady. As for low peaks that remained unsteady, they are taken as 'noise' and hence not recorded.

The white screen had higher peak readings in general as compared to the black screen, as seen by comparing sections (b) and (c). In terms of total number of peaks

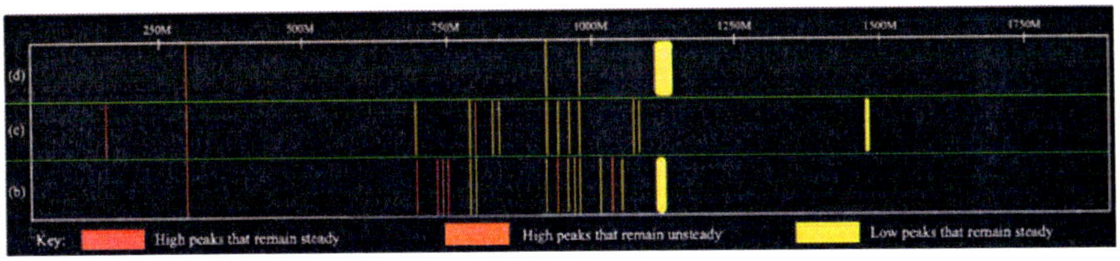

Fig. 17 Simplified recreation of Fig. 11

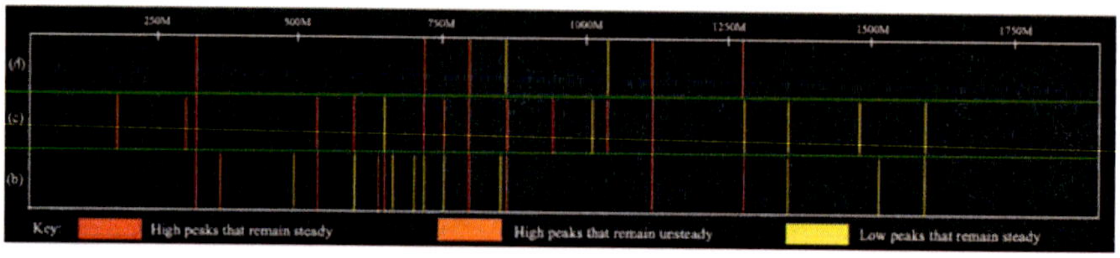

Fig. 18 Simplified recreation of Fig. 16

detected, there was also a higher number of emissions for the white screen, with a total of 16 peaks as compared to black screen having a total of 14 peaks. As for the application downloads, it had the lowest peak readings and the lowest total number of emission peaks.

Below are the results from Fig. 17 shown in a table format in Table 1.

Samsung A32

Based on the emission plot in Fig. 16, as an overview, after the DUT was placed in the box, the surrounding noise increased significantly, with the most obvious range from 1.5 to 1.75 GHz, rising up to the yellow range. However, we noticed that the instances of the noise increasing were the least while the DUT was playing a white screen, followed by the DUT playing a black screen, and the most instances of the noise increasing being the DUT downloading applications.

However, it is also worth mentioning that while the DUT is downloading applications, as seen in section (d) of Fig. 16, that the previously steady emissions between 0 and 500 M decreased from the yellow range to the blue range, which would be an anomaly and puts the accuracy of that specific section into question. Hence, when creating graphs and noting the device emissions, that section will be omitted.

The black screen had higher peak readings in general as compared to the white screen, comparing sections (b) and (c). However, there was a higher number of emissions for the white screen, with section (b) having a total of 19 peaks and section (c) having a total of 18 peaks. As for the application downloads in (d), it had the lowest peak readings, and the lowest number of emission peaks. Despite that, from Fig. 16, the noise readings for (d) were the highest among the three. This could be due to multiple reasons, such as an inaccurate shielding of Faraday's box, inadequate shielding for the Samsung A32 by the manufacturer, or that the Samsung A32 had components that released more emissions than the Samsung J2 Pro.

Below are the results from Fig. 18 shown in a table format in Table 2.

Table 1 Table of emissions of Samsung J2 Pro

Test mode	High peak (steady)	High peak (unsteady)	Low peak	Total
White screen (b)	3	4	9	16
Black screen (c)	1	2	11	14
Download (d)	0	1	3	4

Table 2 Tablc of cmissions of Samsung A32

Test mode	High peak (steady)	High peak (unsteady)	Low peak	Total
White screen (b)	6	3	10	19
Black screen (c)	5	7	6	18
Download (d)	1	4	2	7

5 Conclusion

The Samsung J2 Pro, although having less emissions in terms of number as compared to the Samsung A32, experiences large peaks that stretch over a range of frequencies, as seen by the thicker lines on Fig. 17. It is worth noting that both Samsung devices also feature a similar emission frequency at approximately 310 M, albeit at different peaks, with the Samsung A32 having a higher peak than the Samsung J2 Pro. This could be attributed to the fact that both devices' designs are made by the same company and could hence have the same shielding design flaw. The range of device emissions given off by the Samsung J2 Pro is also more concentrated at 750–1250 M, compared to the wider range of the Samsung A32, as seen by Figs. 17 and 18. Overall, due to the lower number of peaks, the Samsung J2 Pro may be less prone to information leakages as compared to the Samsung A32.

The Vivo Y21 on the other hand had little to no visible emissions and thus would be less prone to information leakage via TEMPEST. Although the release date could play a part in the device's emissions, it is likely that the date would not be a large factor, as seen by the Samsung A32 and Samsung J2 Pro. Instead, the manufacturing company would be the main factor as deduced from the three devices in this experiment, comparing the Vivo and Samsung phones. Another factor could also be the amount of processing power that the phones have. The Samsung J2 Pro has 1.5 gigabytes of RAM while the Samsung A32 has 8 gigabytes of RAM and thus may end up having a larger amount of emissions due to the higher processing power. For comparison purposes, the Vivo Y21 has 4 gigabytes of RAM. With more time, other phone models, or electronic devices such as computers or smart watches could be included in the experiment, and different metallic shielding materials other than aluminium foil.

Appendix 1: Boxes Wrapped with Different Materials

See the Figs. 19, 20 and 21.

Fig. 19 Box A wrapped in aluminium foil

Fig. 20 Box B wrapped in bubble wrap

Fig. 21 Box B wrapped in aluminium foil

Appendix 2: Resources Used

TV Antenna on Shopee. https.//shopee.sg/A..-25dBi-Antenna-TV-Digital-HD-200-Mile-Range-Skywire-TV-Indoor-1080P-Coax-i.95138046.1675233853?position = 6.

Video of Black Screen.

https.//youtu.be/AjWfY7SnMBI.

Video of White Screen.

https.//youtu.be/QggJzZdIYPI.

References

1. *What is Tempest?* APITech. Retrieved December 19, 2021, from https://www.apitech.com/brands/secure-systems-information-assurance/sst/what-is-tempest/
2. *Scientists Say. Faraday Cage.* Retrieved November 20, 2021, from https://www.sciencenewsforstudents.org/article/scientists-say-faraday-cage
3. *Soft tempest. Hidden data transmission using.* Retrieved November 11, 2021, from https://www.cl.cam.ac.uk/~mgk25/ih98-tempest.pdf

Self-supervised Learning for 3D CT Scan Segmentation

Bingquan Shen and Yew Keng Timothy Low

Abstract Deep learning's past successes in semantic segmentation of medical image data often rely on large quantities of dense and intricately labeled training data. Due to the tremendous amount of labor and expertise required for pixel-wise annotations of a single 3D medical image necessary for medical image segmentation, the accuracy of supervised segmentation models trained on the small datasets available, including the 3D COVID-19 CT scan dataset, is compromised. Self-supervised techniques take semantic information from large pools of unlabeled data through various proxy tasks to train a downstream task, feasibly resulting in a higher accuracy, label efficiency, and speed of segmentation with the same amount of labeled training data. In this paper, it was found that proxy tasks, including Contrastive Predictive Coding, context prediction, and jigsaw, greatly improve segmentation results even in small sets of data over the baseline U-Net. Maximum dice score achieved was 0.637 ± 0.188.

Keywords Self-supervised learning · Machine learning · COVID-19

1 Introduction

The 3D segmentation is a difficult yet essential problem in computer vision with widespread applications, including robotics [1], autonomous driving [2], and many other novel fields of research [3]. Semantic segmentation is a subset of segmentation which classifies each pixel of the input image to a particular label, rather than specifying a bounding box around the object classifying the image as a whole [4]. It groups and identifies pixels in an image by creating a segmentation mask, which, when overlaid on top of the original image, would classify the input image with their corresponding labels.

B. Shen
Information Division, DSO National Laboratories, Singapore, Singapore

Y. K. T. Low (✉)
Anglo-Chinese School (Independent), Singapore, Singapore
e-mail: low.tim.timothy@gmail.com

H. Guo et al. (eds.), *IRC-SET 2022*,
https://doi.org/10.1007/978-981-19-7222-5_21

Segmentation tasks can be of crucial importance for novel viruses like COVID-19 [5, p. 19]. In 2019, there was an outbreak of viral pneumonia that had emerged from Wuhan, China. The virus was named severe acute respiratory syndrome coronavirus 2 (SARS-CoV-2) by the World Health Organization, and the subsequent lung infection caused by the virus is known as Coronavirus Disease 19 (COVID-19). Recent studies have shown that infected patients demonstrate identifiable qualities within their serial chest CT examinations that could be helpful toward identification of the disease, monitoring the course of the disease, and evaluation of treatment [6, 7]. Along with MRI and X-rays, computed tomography scans (CT scans) are computer-generated images of cross-sectional slices of a body, which is a tool that is often used by radiologists to visualize all parts of the body and provide detailed information of a patients' internal organ structures [7–9].

Chest CT imaging features of COVID-19 include, but are not limited to, the presence of ground glass opacities (GGOs), consolidations, reticular patterns, "crazy paving patterns", and more [10, 11]. Although these features, also known as regions of interest (ROIs), are common throughout different forms of pneumonia, specific patterns that correspond to a COVID-19 infection can act as rapid preliminary tests for the disease, and confirmation of diagnosis to speed up the process of the infection to go into treatment. Segmentation of these CT scans through deep learning can be crucial information that allows doctors to more rapidly and unambiguously identify distribution and severity of the infection from pneumonia, helping them diagnose COVID-19 and corresponding treatment options more efficiently [12]. Furthermore, clinical studies have shown that CT scans have a high specificity of 0.97, but low sensitivity of 0.25 and overall accuracy of 0.68 in diagnosing COVID-19 [13].

Deep learning's past successes in semantic segmentation of medical image data often rely on a large amount of dense and intricately labeled training data. However, the considerable amount of medical image data is hard to come by, as the labor and expertise required for pixel-wise annotations of a single 3D medical image necessary for medical image segmentation are tremendous and frequently are prohibitively expensive [14]. Lack of data, especially for the novel virus COVID-19, could result in overfitting in classical data-hungry models [15, 16].

This paper proposes using self-supervised techniques to bridge this information need by making use of large pools of unlabeled data as a proxy task to train a downstream task, feasibly resulting in a higher accuracy and speed of segmentation with the same amount of labeled training data. Popular pretext tasks include jigsaw puzzle solving [17–20], rotation prediction [21–23], Contrastive Predictive Coding (CPC) [24, 25], and context prediction [26, 27]. This paper will generalize jigsaw, CPC, and context prediction to be used in 3D segmentation and evaluate these segmented COVID-19 CT scans for clinical use.

2 Related Work

Past work that has been done on COVID-19 CT scan segmentation has seen some success in applying classical models to the limited data available. In [28], they implemented classical 3D-Unet with various spatial, color, and noise augmentations to generate unique and random patches of data, achieving dice similarity coefficients of 0.661. Saood and Hatem [12] used 2D-Unet and SegNet, achieving high mean accuracy in both models for single and multi-class segmentation of 0.95 and 0.91, respectively. Deng-Ping Fan et al. also developed a network specifically for COVID-19, known as COVID-19 Lung Infection Segmentation Network (InfNet), a two-stage system that would ultimately perform a single or multi-class segmentation task of identifying general lung infection, or specifically classes of GGO and consolidation lesions [10]. In the same paper, semi-supervised pretext tasks on unlabeled CT scans were compared to their baseline InfNet, achieving dice similarity coefficients of 0.597 and 0.579, respectively. Fung et al. also created their own pretext task on InfNet known as self-supervised InfNet (SSInfNet) sporting a best dice score of 0.63 compared to the baseline InfNet and multi-class InfNet models of 0.55 and 0.49, respectively [29]. As such, self-supervised learning is in contention as a promising method to gain higher accuracy and dice scores for deep learning on a small COVID-19 dataset.

3 Supervised Methods

For medical image semantic segmentation, convolutional neural networks have become the modern standard for medical image segmentation. U-Net [30], V-Net [31], and other fully convolutional neural networks [32] (FCNs) are the most commonly used architectures.

The baseline for the segmentation task is chosen to be 3D-UNet [33]. The 3D-UNet is a form of encoder-decoder architecture that has shown success for dense volumetric segmentation. Originally created by Olaf et al., for biomedical image segmentation, the architecture has found applications in many other areas of computer vision. U-Net consists of down and up convolutions, activation functions, max-pooling, batch normalizations, and concatenations. The segmentation performance is performed for each epoch on cropped and augmented patches on the validation dataset, allowing for the evaluation of overfitting on our data. The loss is evaluated with a weighted cross-entropy loss function, and the overlap coefficient will be evaluated with accuracy, $\frac{TP+TN}{TP+TN+FP+FN}$, mean dice similarity coefficient, (F1 score) $\frac{2TP}{2TP+FP+FN}$, and Jaccard score (intersection over union (IOU)) $\frac{TP+TN}{TP+TN+FP+FN}$.

3.1 Self-supervised Methods

Self-supervised methods essentially aim to maximize semantic similarity between the equivalent parts of data, say, an unlabeled CT scan, and minimize similarity between those that are far apart in the trained embedding space. These embedding that can solve the pretext supervisory task could be useful for other downstream tasks, as in for segmentation of COVID-19 infection in a CT scan.

We choose three different proxy tasks based on [34]'s implementations: 3D Contrastive Predictive Coding (CPC), 3D context prediction (CP), and 3D jigsaw (Jig) and implement them for segmentation tasks on CT scans. Each task generates an encoder model g_{enc} which would later be fine-tuned in downstream tasks. Each encoder model would define a unique proxy loss which, in theory, encourages the network to derive broad "concepts" from semantic representations, allowing the model to acquire supervisory signals from the data.

Contrastive Predictive Coding

Contrastive Predictive Coding is an autoregressive model, which means it predicts the embedding space for the "future" samples, or, in the case of a 3D image, the adjacent overlapping samples [35]. As in Fig. 1, the proxy task crops equally sized and overlapping 3D patches from all the input CT scans and maps the input patch, $x_{i,j,k}$ (with positional coordinates i, j, k) to its embedded representation $z_{i,j,k}$. Another network would summarize the latent vectors of the set of adjacent (or "previous") embedding $z_{u,v,w\leq i,j,k}$ into its context vector $c_{i,j,k}$. This higher-level context predicts the embeddings of the next n patches $z_{i+n,j,k}$. The autoregressive network uses gated recurrent units (GRUs) in accordance with [36]. Thus, it is a classification problem utilizing noise contrastive estimation loss (NCE loss, or categorical cross-entropy loss) [37] to maximize the mutual embedding vectors of $c_{i,j,k}$ and $z_{i+n,j,k}$ and move the embedding away from the negative representations. Note that the context is chosen as in the inverted pyramid neighborhood to reduce excessively high computation time and memory use [38].

3D-Jigsaw Task

Jigsaw puzzle solving is a pretext task of splitting a volume to a few non-overlapping patches and shuffled arbitrarily [20]. This architecture is known by the creator as a context-free network, as each patch is separated, up till the context handling in the final fully connected layer [19]. As a 3D contrastive task, the puzzle is generated by sampling a grid of possible patches with n in width, and with all the permutations within the set of $n^3!$ possible permutations, P permutations were chosen with maximized average Hamming distance between every pair of permutations. Hamming distance refers to the number of different jigsaw tile locations between the two permutations and thus represents the similarity between these two puzzles [39]. The pretext task is thus a classification task of each of the P jigsaw puzzles into the corresponding indexes $f : y_p^k \leftrightarrow \{1, 2, ..., P\}$. In other words, the proxy task is to minimize the cross-entropy loss of the true index y_p^k and the predicted index $\hat{y}_p^k$

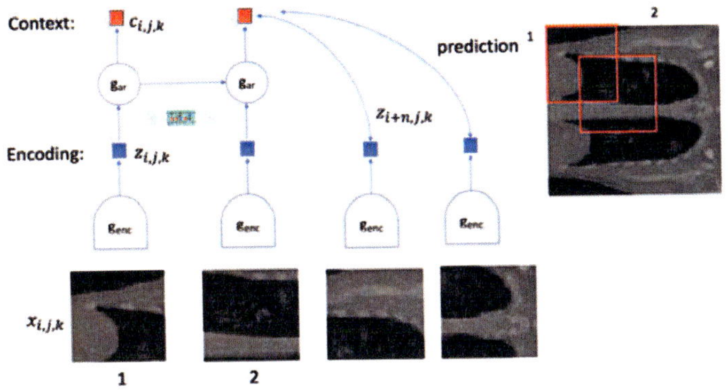

Fig. 1 Diagram of 2D contrastive predictive coding (author's own)

where k is an arbitrary puzzle from the set of P jigsaw puzzles. In the diagram above, there is a simplified model based on a 3×3 puzzle generated by a single CT scan. Each of the 9 patches, or tiles, that are sampled from the image is from a randomly reordered permutation from the permutation set of P jigsaw puzzles. (Note that the max-pooling and nonlinear layers are not indicated). Note that patch jitter of a few pixels (3) is required to ensure that the network does not learn "shortcut" solutions to the jigsaw problem, which may result in overfitting and/or 1:1 feature mapping which would not benefit downstream tasks (Fig. 2).

Context Prediction

Context prediction is another unsupervised pretext task, like jigsaw puzzle solving, being a context-free network. Initially proposed by Doersch et al. [40], this method is successful in many 2D and 3D applications, including X-ray classification [41], and many other medical non-medical applications [26]. For a 2D image, by sampling pairs of image patches randomly around the image in one of 8 spatial configurations

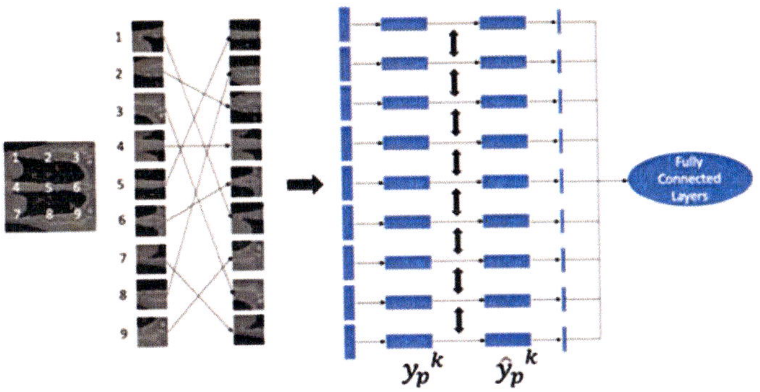

Fig. 2 3D-jigsaw pretext task architecture

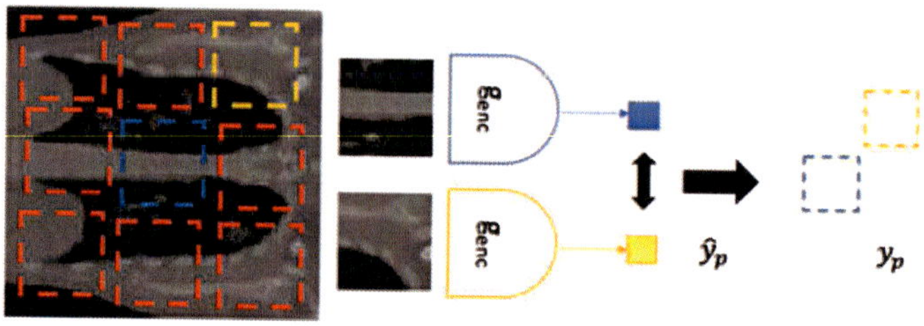

Fig. 3 2D version of context prediction

(as in the 8 cardinal directions), it is fed into the machine learner to guess the spatial relationship between these two cropped images. Just as it is easier to guess the relative location of patches once a human recognizes the object the patches are sampled from, a machine can learn such context clues from the image as a pretext task to be transferred into a downstream classification or segmentation task. Generalizing to an input 3D image, a grid of N overlapping patches is sampled around the CT scan, and pairs of cropped images are randomly selected around the central patch. The algorithm then classifies each randomly selected patch as one of the N − 1 positional patches. In other words, the algorithm aims to minimize cross-entropy loss which is taken between the true index of each positional patch y_p and the guessed index $\hat{y}_p$, while the locations of the remaining patches as negative samples in the set $\{y_{N-1}\}$. Note that similarly to jigsaw, patch jitter of a few pixels is required to not learn "shortcut" solutions (Fig. 3).

4 Experiments

See Table 1.

Table 1 Table of hyperparameters for the self-supervised pretext tasks

Proxy task	CPC	CP	Jigsaw	Proxy task	CPC	CP	Jigsaw
Dimension of latent space	1024	1024	1024	Batch size	16	4	4
Filters on encoder model	4	16	16	Patch jitter	–	3	3
Patches per side	4 × 4 × 4	4 × 4 × 4	3 × 3 × 3	Top architecture	U-Net	FCN	FCN
Patch dimension	64	–	–	Learning rate	0.0001	0.0001	0.0001
Epochs	1000	1000	1000				

4.1 Downstream Hyperparameters

For the downstream task, we train the network with 400 epochs with 25 warm-up epochs and a learning rate of 0.0001 with Adam as the optimizer. The batch size is 2. The prediction architecture is a 3D-UNet-based decoder, with 5 layers of deconvolutional blocks, max-pooling, softmax activation function, and corresponding skip connections for each layer [33].

For the baseline task, we train the U-Net encoder-decoder network with 1000 epochs and a learning rate of 0.0001 with Adam as the optimizer, batch size 2. The architecture is equivalent to the downstream task including 5 layers of residual convolutional blocks of and corresponding deconvolutional blocks with skip connections. The first 400 epochs were used for comparison in labeling efficiency and in epochs, and because after 400 epochs, the baseline network architecture did not show much improvement.

4.2 Dataset

The infected COVID-19 CT scans were taken from a Kaggle competition by Larxel [10]. There are CT scans from 20 patients, each slice being of size 512 $\times$ 512, of varying axial resolution. The CT scans were obtained from Coronacases.org [42] by Radiology & Artificial Intelligence One-Stop Shop (RAIOSS) and Radiopaedia [43] and were expertly annotated by many radiologists [44].

The CT scans used for the pre-trained model were taken from Medical Decathlon's lung segmentation challenge [45]. From the training and test sets, 80 unlabeled CT images were randomly chosen to train the contrastive model through each of the pretext tasks.

In preparing data for self-supervision and downstream tasks, we resize the data from the 512 $\times$ 512 $\times$ 302 in the Medical Decathlon 3D lung segmentation CT scan dataset or the varying sizes of the COVID-19 CT scans dataset, to a unified cubic 128 $\times$ 128 $\times$ 128 with 1 greyscale color channel as a NumPy array. This augmentation may cause shearing in the CT scan, uneven across training, validation, and testing images, which is a limitation in this experiment. However, the resulting cubic masks in training and fine-tuning can be easily resized to the appearance of a regular CT scan. Each input intensity value in the CT scans was normalized by transforming them to within a range of [0,1]. The label's intensity values were vectorized to 0 and 1.

We evaluate our experimental results, by assessing the learned representations by fine-tuning them on the downstream task of COVID-19 CT scan infection segmentation. We investigate the possible improvements in performance, speed of convergence, and data efficiency of pre-training on unlabeled data. The accuracy and weighted dice coefficient will be the metrics used to evaluate the validity of segmentation of COVID-19 infection. Weighted dice coefficient loss is the primary protocol

Table 2 Breakdown of data between training, validation, and testing

% Data on baseline/self-supervised models	100%	50%	25%
Train	12	6	3
Validation	4	4	4
Test	4	4	4

because of the class-imbalanced nature of the data; that is, most of the voxels in the CT scan have negative infection labels. The significant disparity between the number of positive and negatively labeled pixels could skew the accuracy score of the segmentation. Training was performed on a GeForce RTX 3060 Ti with 8 GB of VRAM (Table 2).

5 Data and Discussion

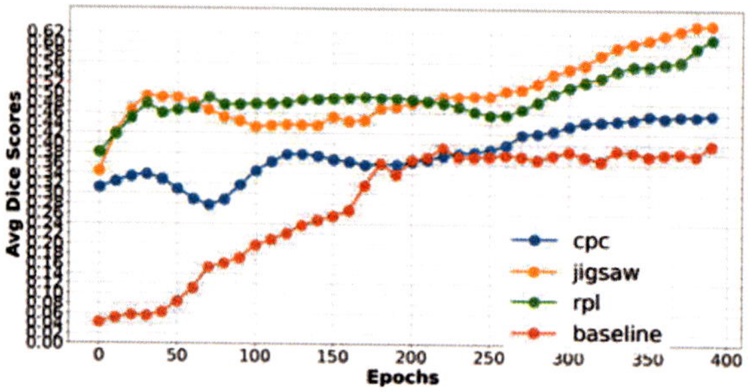

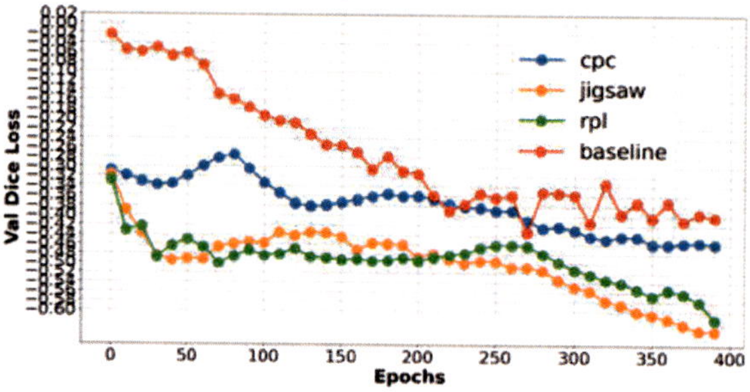

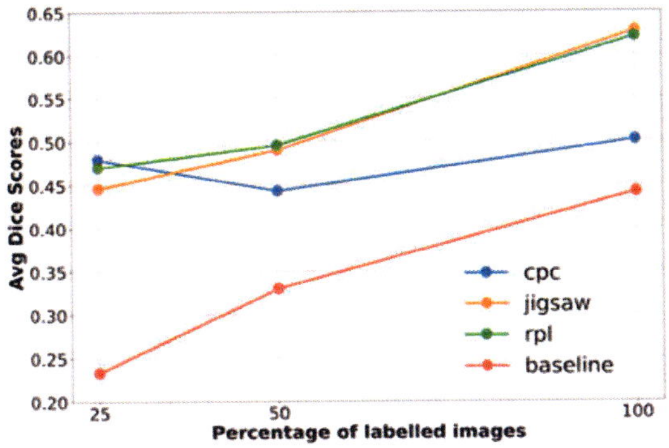

The results obtained by the 3D methods outperform the baseline supervised model across the percentages of (25%, 50%, 100%) of labeled images, corresponding to [3, 6, 12] labeled images, respectively. This is in accordance with existing literature, with data efficiency in general increasing with smaller datasets like this one. With a higher number of labeled images, the baseline model may feature average dice scores and accuracy like the supervised tasks. In addition, the self-supervised model did better than the baseline model in terms of speed of convergence, achieving higher performances for the first 180 epochs and above. However, the sporadic nature of the improvement of the dice score may indicate that more training epochs need to be done, or the model is trapped at a local minimum, or the model is overfitting. Nevertheless, the overall consistently greater dice score demonstrates the reliability and potential in each self-supervised methods to see an improvement over the baseline model (Figs. 4, 5, 6, 7 and Table 3).

As one can see from the segmentations, there is some error in the segmentation even in the most accurate of dice segmentations like in the jigsaw supervised model. The model segmented both GGO and consolidation infection with similar accuracies. Nevertheless, the localization of the pneumonia infection is relatively accurate throughout the models and often covers the general area of the infection. Some of the resolution of segmentation is lost to errors in the model, as in the small discontinuous sporadic spots where infection occurs, or the detailed edges of the infection. Still, this segmentation can be functional to doctors, whose diagnosis only needs the generalized area of infection to be identified. This base segmentation result can be

Fig. 4 Contrastive predictive coding

Fig. 5 Jigsaw

Fig. 6 Context prediction

Fig. 7 Baseline U-Net

Table 3 Tabulated accuracy and dice scores

Model	Dice score	Accuracy
CPC	0.501 ± 0.223	0.912
CP	0.620 ± 0.164	0.921
Jigsaw	0.637 ± 0.188	0.955
Baseline U-Net	0.441 ± 0.215	0.894

edited by radiologists to produce a higher number of masks for deep learning, adding to the database and allowing more accurate models to be trained.

Detection of COVID-19 infection is a challenging task even for expert radiologists. The many differing ROIs, including variations in GGO and consolidation patterns, may result in the deep learning model to be unable to successfully generalize segmentation of COVID-19 infection. Small regions of infection on top of a large background of no infection would cause any small errors in the model's predicted pixels to have a large impact on the resulting weighted dice score. In addition, any model that is trained on such small datasets is always susceptible to overfitting. It may be more effective to slice 3D CT scan to 2D images and perform segmentation tasks with the resulting large amount of image data. Regardless, the segmentation dice score of 0.637 and 0.620 for jigsaw and context prediction is high relative to existing models trained on a similar dataset of a similar size.

6 Conclusion

In conclusion, self-supervised methods in pre-training a deep learning model for a segmentation task of CT scans have proven successful in speed of convergence, label efficiency, and overall accuracy. For the small 20-element COVID-19 dataset, every self-supervised training method, including Contrastive Predictive Coding, context prediction, and jigsaw, greatly improves segmentation results even in small sets of data over the baseline U-Net.

Acknowledgements The completion of this project would not have been possible without my mentor, Dr Shen Bingquan, for introducing me to the world of machine learning and guiding me sufficiently throughout the course of the project. For their endless support and encouragement, I would like to express my endless gratefulness to my parents for being with me even throughout late nights working on this project.

References

1. Uckermann, A., Haschke, R., & Ritter, H. (2013). Realtime 3D segmentation for human-robot interaction. In *2013 IEEE/RSJ International conference on intelligent robots and systems*, Tokyo, pp. 2136–2143. https://doi.org/10.1109/IROS.2013.6696655
2. Zhou, D., Fang, J., Song, X., Liu, L., Yin, J., Dai, Y., & Yang, R. (2020). *Joint 3D instance segmentation and object detection for autonomous driving*.
3. He, Y., Yu, H., Liu, X., Yang, Z., Sun, W., Wang, Y., Fu, Q., Zou, Y., & Mian, A. (2021). *Deep learning based 3D segmentation: A survey*. arXiv:2103.05423 [cs]. [Online]. Available: http://arxiv.org/abs/2103.05423. Accessed 2022, January 06.
4. A 2021 guide to Semantic Segmentation. (2021). https://nanonets.com/blog/semantic-image-segmentation-2020/. Accessed 2022, January 06.
5. Yuki, K., Fujiogi, M., & Koutsogiannaki, S. (2020). COVID-19 pathophysiology: A review. *Clinical Immunology, 215*, 108427. https://doi.org/10.1016/j.clim.2020.108427
6. Hall, L. O., Paul, R., Goldgof, D. B., & Goldgof, G. M. (2022). *Finding Covid-19 from chest X-rays using deep learning on a small dataset*. arXiv:2004.02060 [cs, eess]. [Online]. Available: http://arxiv.org/abs/2004.02060. Accessed 2022, January 06.

7. Rubin, G. D., et al. (2020). The role of chest imaging in patient management during the COVID-19 pandemic: A multinational consensus statement from the fleischner society. *Radiology, 296*(1), 172–180. https://doi.org/10.1148/radiol.2020201365
8. Power, S. P., Moloney, F., Twomey, M., James, K., O'Connor, O. J., & Maher, M. M. (2016). Computed tomography and patient risk: Facts, perceptions and uncertainties. *WJR, 8*(12), 902. https://doi.org/10.4329/wjr.v8.i12.902
9. Vasilev, Y. A., et al. (2021). Chest MRI of patients with COVID-19. *Magnetic Resonance Imaging, 79*, 13–19. https://doi.org/10.1016/j.mri.2021.03.005
10. Carotti, M., et al. (2020). Chest CT features of coronavirus disease 2019 (COVID-19) pneumonia: Key points for radiologists. *La Radiologia Medica, 125*(7), 636–646. https://doi.org/10.1007/s11547-020-01237-4
11. Abdel-Basst, M., Mohamed, R., & Elhoseny, M. (2020). A model for the effective COVID-19 identification in uncertainty environment using primary symptoms and CT scans. *Health Informatics Journal, 26*(4), 3088–3105. https://doi.org/10.1177/1460458220952918
12. Saood, A., & Hatem, I. (2021). COVID-19 lung CT image segmentation using deep learning methods: U-Net versus SegNet. *BMC Medical Imaging, 21*(1), 19. https://doi.org/10.1186/s12880-020-00529-5
13. Ai, T., et al. (2020). Correlation of chest CT and RT-PCR testing for coronavirus disease 2019 (COVID-19) in China: A report of 1014 cases. *Radiology, 296*(2), E32–E40. https://doi.org/10.1148/radiol.2020200642
14. Xie, Y., Zhang, J., Liao, Z., Xia, Y., & Shen, C. (2022). *PGL: Prior-guided local self-supervised learning for 3D medical image segmentation.* arXiv:2011.12640 [cs]. [Online]. Available: http://arxiv.org/abs/2011.12640. Accessed 2022, January 06.
15. Shi, F., et al. (2021). Review of artificial intelligence techniques in imaging data acquisition, segmentation, and diagnosis for COVID-19. *IEEE Reviews in Biomedical Engineering, 14*, 4–15. https://doi.org/10.1109/RBME.2020.2987975
16. Litjens, G., et al. (2017). A survey on deep learning in medical image analysis. *Medical Image Analysis, 42*, 60–88. https://doi.org/10.1016/j.media.2017.07.005
17. Ahsan, U., Madhok, R., & Essa, I. (2019). Video jigsaw: Unsupervised learning of spatiotemporal context for video action recognition. *IEEE Winter Conference on Applications of Computer Vision (WACV), 2019*, 179–189.
18. Carlucci, F. M., D'Innocente, A., Bucci, S., Caputo, B., & Tommasi, T. (2019). Domain generalization by solving jigsaw puzzles. In *Proceedings of the IEEE/CVF conference on computer vision and pattern recognition*, pp. 2229–2238.
19. Noroozi, M., & Favaro, P. (2017). *Unsupervised learning of visual representations by solving jigsaw puzzles.* arXiv:1603.09246 [cs]. Accessed: Jan. 06, 2022. [Online]. Available: http://arxiv.org/abs/1603.09246. Accessed 2022, January 06.
20. Paumard, M.-M., Picard, D., & Tabia, H. (2018). Jigsaw puzzle solving using local feature co-occurrences in deep neural networks. In *2018 25th IEEE international conference on image processing (ICIP)*, pp. 1018–1022.
21. Chen, T., Kornblith, S., Norouzi, M., & Hinton, G. (2020). A simple framework for contrastive learning of visual representations. In *International conference on machine learning*, pp. 1597–1607.
22. Gidaris, S., Singh, P., & Komodakis, N. (2018). *Unsupervised representation learning by predicting image rotations*. arXiv preprint, arXiv:1803.07728.
23. Chen, T., Zhai, X., Ritter, M., Lucic, M., & Houlsby, N. (2019). Self-supervised gans via auxiliary rotation loss. In *Proceedings of the IEEE/CVF conference on computer vision and pattern recognition*, pp. 12154–12163.
24. van den Oord, A., Li, Y., & Vinyals, O. (2018). *Representation learning with contrastive predictive coding*. arXiv preprint, arXiv:1807.03748.
25. Henaff, O. (2020). Data-efficient image recognition with contrastive predictive coding. In *International conference on machine learning*, pp. 4182–4192.
26. Chen, L., Bentley, P., Mori, K., Misawa, K., Fujiwara, M., & Rueckert, D. (2019). Self-supervised learning for medical image analysis using image context restoration. *Medical Image Analysis, 58*, 101539.

27. Denton, E., Gross, S., Fergus, R. (2016). *Semi-supervised learning with context-conditional generative adversarial networks*. arXiv preprint, arXiv:1611.06430.
28. Müller, D., Soto-Rey, I., & Kramer, F. (2021). Robust chest CT image segmentation of COVID-19 lung infection based on limited data. *Informatics in Medicine Unlocked, 25*, 100681. https://doi.org/10.1016/j.imu.2021.100681
29. Fung, D. L. X., Liu, Q., Zammit, J., Leung, C.K.-S., & Hu, P. (2021). Self-supervised deep learning model for COVID-19 lung CT image segmentation highlighting putative causal relationship among age, underlying disease and COVID-19. *Journal of Translational Medicine, 19*(1), 318. https://doi.org/10.1186/s12967-021-02992-2
30. Ronneberger, O., Fischer, P., & Brox, T. (2015). U-Net: Convolutional networks for biomedical image segmentation. In Navab, N., Hornegger, J., Wells, W. M., & Frangi, A. F. (Eds.), *Medical image computing and computer-assisted intervention—MICCAI 2015*, (vol. 9351, pp. 234–241). Springer International Publishing. https://doi.org/10.1007/978-3-319-24574-4_28
31. Milletari, F., Navab, N., Ahmadi, S.-A. (2016). V-net: Fully convolutional neural networks for volumetric medical image segmentation. In *2016 Fourth international conference on 3D vision (3DV)*, 2016, pp. 565–571. https://doi.org/10.1109/3DV.2016.79
32. Long, J., Shelhamer, E., & Darrell, T. (2015) *Fully convolutional networks for semantic segmentation*. arXiv:1411.4038 [cs]. [Online]. Available: http://arxiv.org/abs/1411.4038. Accessed 2022, January 06.
33. Çiçek, Ö., Abdulkadir, A., Lienkamp, S. S., Brox, T., & Ronneberger, O. (2016). *3D U-Net: Learning dense volumetric segmentation from sparse annotation*. arXiv:1606.06650 [cs]. [Online]. Available: http://arxiv.org/abs/1606.06650. Accessed 2022, January 06.
34. Taleb, A., Loetzsch, W., Danz, N., Severin, J., Gaertner, T., Bergner, B., & Lippert, C. (2020). *3D self-supervised methods for medical imaging*. arXiv:2006.03829 [cs, eess]. [Online]. Available: http://arxiv.org/abs/2006.03829. Accessed 2022, January 06.
35. Guilly, Y. L. (2019). Representation learning with contrastive predictive coding. *Stochasticity and Chaos*. https://yann-leguilly.gitlab.io/post/2019-09-29-representation-learning-with-contrastive-predictive-coding/. Accessed 2022, January 10.
36. Cho, K., Van Merriënboer, B., Gulcehre, C., Bahdanau, D., Bougares, F., Schwenk, H., & Bengio, Y. (2014). Learning phrase representations using RNN encoder–decoder for statistical machine translation. In *Proceedings of the 2014 conference on empirical methods in natural language processing (EMNLP)*, Doha, Qatar, pp. 1724–1734. https://doi.org/10.3115/v1/D14-1179
37. Ma, Z., & Collins, M. (2018). *Noise contrastive estimation and negative sampling for conditional models: Consistency and statistical efficiency*.
38. Stollenga, M. F., Byeon, W., Liwicki, M., & Schmidhuber, J. (2015) *Parallel multi-dimensional LSTM, with application to fast biomedical volumetric image segmentation*. arXiv:1506.07452 [cs]. Available: http://arxiv.org/abs/1506.07452. Accessed 2022, January 7.
39. Bookstein, A., Kulyukin, V. A., & Raita, T. (2002). Generalized hamming distance. *Information Retrieval, 5*(4), 353–375. https://doi.org/10.1023/A:1020499411651
40. Doersch, C., Gupta, A., & Efros, A. A. (2015). Unsupervised visual representation learning by context prediction. *IEEE International Conference on Computer Vision (ICCV), 2015*, 1422–1430. https://doi.org/10.1109/ICCV.2015.167
41. Hao, H., Didari, S., Woo, J. O., Moon, H., & Bangert, P. (2021). *Highly efficient representation and active learning framework for imbalanced data and its application to COVID-19 X-ray classification*. arXiv:2103.05109 [cs, eess]. [Online]. Available: http://arxiv.org/abs/2103.05109. Accessed 2022, January 7.
42. Paiva, O. (2020). CORONACASES.ORG—by RAIOSS.com. https://coronacases.org/. Accessed 2022, January 7.

43. Radiopaedia.org, the wiki-based collaborative radiology resource. *Radiopaedia.* https://radiopaedia.org/. Accessed 2022, January 7.
44. Ma, J., et al. (2021). Toward data-efficient learning: A benchmark for COVID-19 CT lung and infection segmentation. *Medical Physics, 48*(3), 1197–1210. https://doi.org/10.1002/mp.14676
45. Antonelli, M., Reinke, A., Bakas, S., Farahani, K., Kopp-Schneider, A., Landman, B.A., Litjens, G., Menze, B., Ronneberger, O., Summers, R.M., & van Ginneken, B., (2021). The medical segmentation decathlon.

Investigation of Properties of Patterned Dielectric Superlattice Using Computational Methods

Chunyi Deng and Jingxiang Yang

Abstract Moiré Superlattices, such as twisted bilayer graphene (tBLG) and graphene on hBN, are extensively studied as promising platforms for superlattice engineering. However, they are either extremely challenging to fabricate or scarce in tunability. To overcome these drawbacks while retaining novel properties, we propose Patterned Dielectric Superlattices (PDSL) as alternative potent candidates for superlattice engineering. PDSL has been reported to demonstrate band modification and fully developed minigaps of the Hofstadter spectrum. Moreover, they are predicted to show exotic properties, such as ferromagnetism in a strong coupling regime. In this project, we apply reasonable models for different superlattice potentials and Hamiltonians of the system. We probe the band structures of graphene and transition metal dichalcogenides PDSL using a continuum model. Our band structure plots suggest that graphene PDSL may exhibit exotic insulating and superconducting properties just like tBLG. We thoroughly analyse the reasons for the absence of those exotic transport properties in previous experiments and propose improvements for the experiments. We also discuss the characteristics of TMD PDSL band structure, such as conduction-valence bands asymmetry and minigap formation. Finally, we vary the parameters of PDSL, such as biased voltage, superlattice length scale, and types of superlattice periodicity, so that we can discuss the tunability of PDSL's band structure and properties. By doing this, we aim to optimize the condition for PDSL to exhibit the most robust properties, laying the ground for future applications.

Keywords Condensed matter physics · 2D Material · Patterned dielectric superlattices

Authors 'Chunyi Deng and Jingxiang Yang' are contributed equally to this work.

C. Deng (✉)
NUS High School of Math and Science, Singapore, Singapore
e-mail: h1930002@nushigh.edu.sg

J. Yang
River Valley High School, Singapore, Singapore

H. Guo et al. (eds.), *IRC-SET 2022*,
https://doi.org/10.1007/978-981-19-7222-5_22

1 Introduction

In a periodic lattice, Bloch's theorem dictates the characteristics of band structure, which are intrinsically determined by the spacing, symmetry, and magnitude of lattice site potentials. In the last century, GaAs/AlGaAs heterostructure was shown to confine electrons in two dimensions. Externally applied periodic potentials were able to effectively alter the original lattice site potentials and modify the intrinsic band structure of the system [1–3]. Nowadays, two-dimension (2D) materials such as graphene and transitional metal dichalcogenides (TMD) have emerged as better candidates to realize synthetic band structure by superlattice pairing due to their much smaller superlattice wavelength and more robust properties [1–5]. Superlattice engineering was extensively studied. Previous efforts include electrostatic schemes [6], chemical doping [7], etching [8], and strain engineering [9, 10]. Despite these means, it's challenging to control the superlattice length scale within an optimal range. If the superlattice length scale is too large, the mini-Brillouin zone edge will approach the neutrality point and thus become indistinct. If it is too small, the zone edge energy will become too high and thus inaccessible. Moiré superlattices mitigate such challenges yet leave other experimental difficulties and shortcomings. They are formed by stacking and twisting two identical van der Waal (vdW) materials, such as twisted bilayer graphene (tBLG) or stacking two vdW materials with slightly mismatched lattice constants, such as graphene on hexagonal boron nitride (hBN). They demonstrate emergent electronic bandgaps [11], modified electron dynamics [12], Hofstadter spectra [11], to name a few. When two layers of the superlattice are strongly coupled at the magic angle, they demonstrate nearly flat electronic bands [13, 14]. However, for tBLG, it's remarkably difficult to control the twist angle within a small range near the magic angle while considering the complicating effect of lattice reconstruction [13, 14]. Besides, graphene on hBN lacks tunability in superlattice symmetry and magnitude. These limit the application of Moiré superlattices in engineering electronic structures.

Recently, an inverted electrostatic scheme has opened new opportunities for the synthesis of high-quality tuneable superlattices with less challenging experimental procedures. The patterned dielectric superlattice (PDSL) scheme introduced patterned SiO_2 pillars on top of a uniform electric gate to vary the dielectric interface in the local vicinity of the 2D electron gas, therefore creating an effective periodic superlattice potential. The graphene PDSL heterostructure showed relatively low periodicities of 35 nm and ultrahigh mobilities. The device also demonstrated the superlattice band modification effect and fully developed fractal mini-gaps of the Hofstadter spectrum [15, 16]. Another group predicted that TMD PDSL can be pushed to a strong coupling regime, where flat miniband formation will lead to exotic properties such as ferromagnetism, resembling tBLG but requiring no twisting [17]. These studies show that PDSL is a promising platform for superlattice engineering and band modification. However, more studies are needed to study the tunability of PDSL, to optimize the conditions for future application.

Our study establishes superlattice potential and Hamiltonian of graphene PDSL and TMD PDSL theoretically (See Appendix for the illustration of devices). Then we utilize the continuum method to investigate the tunability of their band structures, with varying biased voltage, superlattice length scale, and types of superlattice periodicities. Based on this information, our study aims to discuss novel properties that PDSL demonstrates and the condition for them to be most robust. In the end, we will discuss their future applications.

2 Methodology

2.1 Periodic Superlattice Potentials

Due to periodicity, we consider only position vector r in the Wigner primitive cell of the central hole. The superlattice potential profile $V(r)$ can be written in the form of Fourier expansion.

$$V(r) = \sum_{G} V_G e^{iG\cdot r}, \text{ where } G = n_1G_1 + n_2G_2 \cdot n_1, n_2 \in Z. \tag{1}$$

G_1 and G_2 are primitive reciprocal vectors of the superlattice.

For triangular superlattice potential, we only consider terms of 6 neighbouring holes in Eq. (1) as their contribution (Fourier coefficient) is most significant. Hence, we write

$$V_{\text{TSL}}(r) = V_T \sum_{i=1}^{3} \cos(G_i \cdot r), \tag{2}$$

where $G_1 = \frac{2\pi}{a}\left(-\frac{\sqrt{3}}{3}, 1\right)$, $G_2 = \frac{2\pi}{a}\left(\frac{2\sqrt{3}}{3}, 0\right)$, $G_3 = \frac{2\pi}{a}\left(-\frac{\sqrt{3}}{3}, -1\right)$, and $V_T =$ 22.1 meV. When $V_T = 22.1$ meV, the difference between the maximum and minimum potential ($\Delta\phi$) is 100 meV. Modelled triangular superlattice potential (Fig. 1a) shows asymmetric maximum and minimum potential, which agrees with previous experiments [15].

If we only consider the neighbouring four holes for square superlattice potential, it will yield equal magnitude for maximum and minimum potential, which is against previous observation [15]. As such we introduce a second-order term, and rewrite Eq. (2) as

$$V_{\text{SSL}}(r) = V_s \sum_{j=1}^{2} \cos\left(G_j \cdot r\right) + V_s \cos(G_1 \cdot r) \cos(G_2 \cdot r), \tag{3}$$

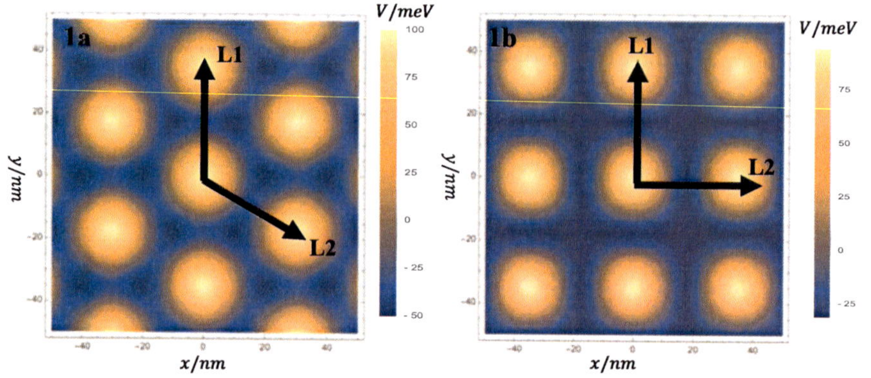

Fig. 1 Modelled triangular and square superlattice potentials, both with 5 V biased gate voltage and 10 nm radius of air holes. Regions with higher potential are air holes (orange), and regions with lower potentials are SiO_2 substrate (blue). **a** Triangular superlattice potential. Primitive lattice vectors $L_1 = L(0, 1)$, $L_2 = L\left(\frac{\sqrt{3}}{2}, -\frac{1}{2}\right)$, where $L = 35$ nm. L is the distance between adjacent two air holes, representing the superlattice length scale. **b** Square superlattice potential. Primitive lattice vector $L_1 = L(0, 1)$, $L_2 = L(1, 0)$, where $L = 35$ nm. L is the distance between adjacent two air holes, representing the superlattice length scale

where $G_1 = \frac{2\pi}{a}(1, 0)$, $G_2 = \frac{2\pi}{a}(0, 1)$ and $V_s = 25.0$ meV. The value of V_s corresponds to $\Delta\phi$ of 100 meV. Modelled square superlattice potential (Fig. 1b) also shows asymmetric maximum and minimum potential, which agrees with previous experiments [15].

Here we discuss effects of biased gate voltage and radius of air holes on V_T and V_s, the amplitude of potential profile, which will then affect the band structure. Numerical calculation shows that V_T and V_s have a linearly increasing relationship with biased gate voltage. V_T and V_s first increase then decrease with the increase of radius of air holes.

2.2 Hamiltonian Formation

Graphene has a hexagonal lattice structure with side length a. Using a Bloch-like basis, we can write the graphene Hamiltonian as a 2×2 matrix. The matrix has the form:

$$H(k) = \begin{pmatrix} 0 & -t_1 e^{ika_1} - t_2 e^{ika_2} - t_3 \\ -t_1 e^{-ika_1} - t_2 e^{-ika_2} - t_3 & 0 \end{pmatrix}, \quad (4)$$

where t_1, t_2, t_3 are the number of hopping along the respective directions to the neighbouring lattice sites. Using nearest neighbour approximation and evaluate the

Hamiltonian near high symmetry points of the first Brillouin Zone. Substituting in the coordinate of K in reciprocal lattice basis, we get the Hamiltonian in Dirac form [18]:

$$H(K+\epsilon) \propto \begin{pmatrix} 0 & k_x - ik_y \\ k_x + ik_y & 0 \end{pmatrix} = k_x\delta_x + k_y\delta_y, \tag{5}$$

where δ_x and δ_y are Pauli matrices and ϵ is a small displacement around K point (a Dirac node). The Hamiltonian is proportional to v, the fermi velocity of graphene. And we get:

$$H(k) = v\hbar\big(k_x\delta_x + \xi k_y\delta_y\big), \tag{6}$$

where $\xi = 1$ for K point and -1 for K' point. The superlattice cell dimension is around 35 nm, which is larger than the dimension of graphene cells (0.142 nm), so it is justifiable to use continuum model [19].

Usually, graphene has zero band gap, due to the inversion and time-reversal symmetry of the hexagonal lattice structure. However, the band gap of TMD was both theoretically predicted [20] and experimentally observed [21]. To split the eigenvalues of the original Hamiltonian, we can add terms on its diagonal entries:

$$H(k) = v\hbar\begin{pmatrix} \Delta/v\hbar & k_x - i\xi k_y \\ k_x + i\xi k_y & -\Delta/v\hbar \end{pmatrix} = v\hbar\big(k_x\delta_x + \xi k_y\delta_y\big) + \Delta\delta_z, \tag{7}$$

We have defined the models for superlattice potential previously. Here we take triangular SL as an example. The total Hamiltonian is the sum of the unperturbed TMD Hamiltonian and the periodic potential:

$$H(k) = v\hbar\big(k_x\delta_x + \xi k_y\delta_y\big) + \Delta\delta_z + V_T\sum_{i=1}^{3}\cos(G_i.r), \tag{8}$$

This is the Hamiltonian we used for our band structure computation.

2.3 Computational Methodology

We use Mathematica® to successfully plot the band structure of PDSL. The key to band structure plotting is to convert our Hamiltonian to a finite matrix so that we can compute its eigenvalues and eigenfunctions. To achieve that, we can define a cut-off circle for the wave vectors. $|k| \leq 4G_1$ gives a good balance between accuracy and computational cost. Equation (7) becomes a 2×2 component of a larger matrix, which includes the band structure information of all cells within the range. For a k within the first Brillouin Zone, we can define: $k_{ab} = k + aG_1 + bG_2$ to be the corresponding point in higher order zones.

The (*ab*, *cd*) component of the larger matrix is:

$$[H_T(k)]_{ab,cd} = \varphi^*_{ab}(r) H_T(k) \varphi_{cd}(r), \tag{9}$$

By Bloch's theorem,

$$\varphi_{cd}(r) = e^{ik_{cd} \cdot r} u(r), \tag{10}$$

where $u(r)$ has the same periodicity as the lattice. We can therefore write the Hamiltonian in the Bloch form, using Dirac Bra-Ket notation, we get:

$$\begin{aligned}[H_T(k)]_{ab,cd} &= < k_{ab} \left| v\hbar\left(k_x \delta_x + \xi k_y \delta_y\right) + \Delta \delta_z \right| k_{cd} > \\ &+ < k_{ab} \left| v_T \sum_{i=1}^{3} \cos(G_i . r) \right| k_{cd} > . \end{aligned} \tag{11}$$

Due to the normalization and orthogonality of wave functions:

$$\begin{aligned} &< k_{ab} | k_{cd} > = \delta_{a,c} \delta_{b,d} \\ &< k_{ab} \left| e^{i(mG_1 + nG_2) \cdot r} \right| k_{cd} > = < k_{ab} | k_{c+m,d+n} > = \delta_{a,c+m} \delta_{b,d+n} \end{aligned} \tag{12}$$

We can express Eq. (11) as a 2 × 2 matrix:

$$\begin{aligned}[H_T(k)]_{ab,cd} = &\left[v\hbar\left(k_x \delta_x + \xi k_y \delta_y\right) + \Delta \delta_z\right] k_{cd} \delta_{a,c} \delta_{b,d} \\ &+ V_T \left(\delta_{a,c+1} \delta_{b,d} + \delta_{a,c} \delta_{b,d+1} + \delta_{a,c-1} \delta_{b,d} \right. \\ &\left. + \delta_{a,c} \delta_{b,d-1} + \delta_{a,c+1} \delta_{b,d-1}\right) \end{aligned} \tag{13}$$

The next important thing is to iterate the 2 × 2 components of the matrix. *a*, *b*, *c*, *d* can all be iterated from −4 to 4, so the dimension of $H_T(k)$ will be 162 × 162. As such, we can compute the energy eigenvalue at any point *k* inside the first Brillouin Zone.

Now we will introduce the procedure to plot the high symmetry path diagram. For the triangular dielectric substrate, the Brillouin Zones are hexagonal. We can take the *Γ*–*M*–*K*–*K*′–*Γ* path. We evenly select 100 points on each path, and with all the plot points, we can get our band structure diagram.

3 Results/Discussion

3.1 Graphene PDSL

For triangular graphene PDSL, we observe a Dirac cone (DC) (green arrow) of linear dispersion relationship at Γ at $V_T = 10$ meV, $L = 35$ nm (Fig. 4, A1.1), which diminishes when V_T value increases (Figs. 2a and 4, A1.1, A1.2). We observe multiple second-generation Dirac cones (SDC) (blue arrows, not all are indicated) at different positions of mostly nonlinear dispersion relationships. Notably, their positions and dispersion relationships are highly tuneable by varying V_T value or L value. The SDCs become indistinct as V_T value increases (Figs. 2a and 4, A1.1, A1.2) or L value increases (Figs. 2a and 4, A1.3, A1.4), and eventually, diminish as bands overlaps (Fig. 4, A1.2, A1.4) or flatten (Fig. 2b). Therefore, we can infer properties, which are directly associated with the formation of SDCs, are likely to be most robust when a moderate potential difference is applied between two gates, with a low superlattice constant. These properties include the following. Theoretically, when Fermi energy is adjusted to the SDC, a sharp spike in resistance will be observed due to a drastic decrease in the density of states [11, 22]. Moreover, theoretically, in a high magnetic field, we will observe magnetoconductance of the device to form a recursive structure known as the "Hofstadter butterfly" spectrum, due to the intersection of Landau levels of original bands and superlattice bands [11, 22]. Besides, due to formation of SDCs, graphene PDSL is likely to demonstrate renormalization of the Fermi velocity [23], topological currents [24], and gate-dependent pseudospin mixing [25].

Graphene on hBN superlattice has intrinsic band gaps at Dirac points and second-generation Dirac points, effectively altering the semimetal nature of graphene [6]. In contrast, our results show that graphene PDSL has no band gap either at Dirac points or second-generation Dirac points, retaining the semimetal nature (Fig. 2a). Nonetheless, when we increase the V_T value (Fig. 4, A1.2) or L value (Fig. 4, A1.4), we observe that the valence band overlaps with the conduction band. We also note flattened mini bands, which indicates a spike in the density of states at this level. This implies the conductance of the device is likely to surge when the Fermi energy is tuned

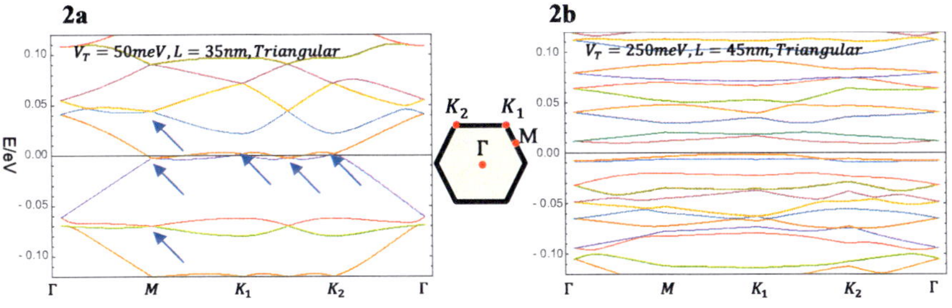

Fig. 2 Band structure of triangular graphene PDSL in the first Brillouin zone with different V_T and L. Green arrows represent Dirac cones, blue arrows represent second-generation Dirac cones

to the level [15, 26], rendering its metallic nature. When we further increase the V_T value and L value (Fig. 2b), the mini bands flattened more, but a small gap between the conduction band and the valence band opens, which implies a semiconducting nature. Given that the breakdown electric field strength for hBN is about 0.5 V/nm, we can theoretically increase V_T up to 270 meV without compromising the integrity of the device. Since graphene PDSL's conductance can be significantly altered by varying V_T and L, graphene PDSL's transport properties can be regarded as remarkably tuneable.

The graphene PDSL band structure we obtained shows remarkably flat bands near Fermi Level when V_S and L are in the optimal ranges (Figs. 2b and 4, A1.2). This resembles the band structure of tBLG at the "magic angle". In tBLG, an additional pair of correlated half-filling insulating states is observed when $n = 0.5 \cdot n_s$ (n is carrier density equals and n_s is the density of superlattice states) at Kelvin-level temperature [27]. Given the similar band structure of graphene PDSL and our control over temperature and charge carrier density, an educated conjecture tells us that graphene PDSL may also exhibit half-filling insulating states. However, only a pair of conventional superlattice insulator bands when $n = \pm 4 \cdot n_s$ was observed previously for graphene PDSL [15]. When tBLG Fermi Level is moved to be near half-filling of valence flat miniband, unconventional superconductivity was observed [28]. We can control the charge carrier density by adjusting the gate voltage. The Fermi Level is related to the charge carrier density (n) by Joyce-Dixon Approximation. Even with a wider range of n investigated, no obvious sign of superconductivity was observed previously for graphene PDSL [15]. Here we propose that the absence of exotic transport properties in previous experiments on graphene PDSL [15] may be due to the relatively weak superlattice potential. The highest superlattice potential applied was around 50 meV in the previous experiments, which may not be strong enough to push the system into a strong coupling regime. There may also be deeper reasons for insulating and superconducting phase formation, which have not been discussed in the existing literature. For example, despite the similar band structures, the reasons for flat miniband formation are different. In tBLG, this is due to the hybridization between two adjacent Dirac cones [27]. In graphene PDSL, this is due to the direct manipulation of an external electric field.

In general, the band structures of square and triangular graphene PDSL are different. For example, at $V_s = 50$ meV, $L = 35$ nm (Figs. 2a and 5, A2.1), we observe different numbers of SDCs with different dispersion relationships. At $V_s = 250$ meV, $L = 45$ nm (Figs. 2b and 5, A2.2), we note that square graphene PDSL has a larger miniband width (less flattened) than its triangular counterpart. Besides, the conduction band and valence band, as well as many mini bands, overlap in the square graphene PDSL but not its triangular counterpart. Such distinct band structures and different properties could provide graphene PDSL with higher tunability.

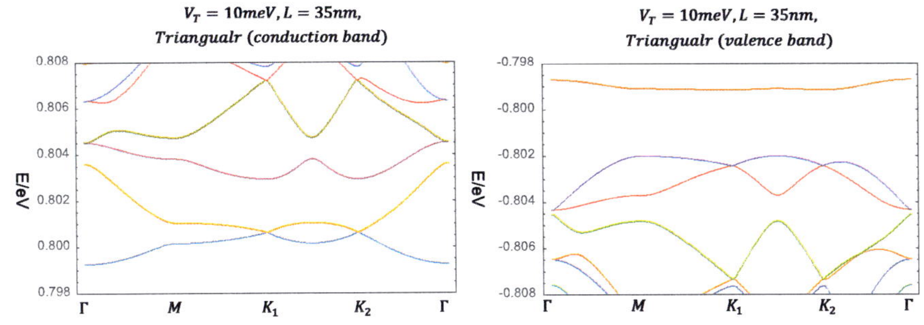

Fig. 3 Band structure of triangular TMD PDSL in the first Brillouin zone

3.2 TMD PDSL

The most profound difference between graphene and TMD PDSL is their band gap. WS_e2, the TMD material we investigate, has a band gap of around 1600 meV and Fermi velocity of around $\frac{3.94 eVA}{\hbar}$, according to previous experiments [29]. Meanwhile, unperturbed graphene is gapless.

From our band structure plots, we observe that there is an asymmetry between the conduction and valence bands of TMD PDSL (Fig. 3). When a positive gate potential is applied, a separated non-degenerate miniband is generated at the top of the valence band. Meanwhile, at the bottom of the conduction band, a pair of Dirac-like flat bands are observed. This is due to the particle-hole asymmetry caused by the positive–negative asymmetry of the external electric field doping (Fig. 1a). When we have a negative gate potential of the same magnitude, we note the band structure reflects about 0 meV and the separated non-degenerate miniband appears at the bottom of the conduction band, while a pair of Dirac-like flat bands appear at the top of the valence band (Fig. 6, A3.1). As such, we can easily manipulate the type of flat band in the TMD PDSL valence and conduction band. This extra tunability may give TMD PDSL more potential for application.

When we increase the positive superlattice potential (Figs. 3 and 6, A3.2), minibands in both valence bands are further flattened and form flat band bundles. There are minigaps which increase with the superlattice potential between the bundles. Meanwhile, minigaps are not observed for graphene PDSL, where a π berry phase prevents backscattering and gap opening [17].

We also observe that with the same V_S and L, the minibands of TMD PDSL are even flatter than those of graphene PDSL. (Figs. 2a and 6, A3.1) Due to the constraint of the large band gap of TMD materials and miniband opening, the topmost valence band is confined within a small energy range, which reduces the kinetic energy of electrons in the band. Therefore, other energy, such as electron–electron interaction energy, may dominate the electronic system. The band structure will be altered and interesting phenomena such as flat band ferromagnetism can be predicted [17].

4 Limitations/Future Studies

One limitation of our model is the independent-electron approximation. Electron–electron interaction will contribute to the total Hamiltonian and influence the band structure. Previous research [17, 29, 30] suggested models for the average interaction energy of electrons confined locally by the superlattice potential. In future, we would like to apply Hartree–Fock Theory to consider electron–electron interaction.

The continuum model we use also limits the accuracy of our band structure plotting. Being in the form of a first-order approximation of Taylor's expansion, its accuracy significantly decreases as the mini-Brillouin Zone gets larger (i.e. the superlattice period becomes smaller). In future, we would like the apply Density Functional Theory to investigate the band structure of PDSL more accurately and completely.

5 Conclusion

Our study focuses on the PDSL as a potent alternative to Moiré Heterostructure in superlattice engineering. Our study uses the continuum model to probe the band structure of graphene and TMD PDSL. For graphene PDSL, we conclude that "Hofstadter butterfly" spectrum, renormalization of the Fermi velocity, topological currents, and gate-dependent pseudospin mixing are likely to be observed due to the formation of SDCs. The phenomena are likely to be most robust with moderate superlattice potential (~50 meV) and low superlattice constant (~30 nm). Besides, we find that graphene PDSL demonstrates remarkably tuneable conductance, which may be pushed to superconductivity with high superlattice potential (> 100 meV) and high superlattice constant (~45 nm), and suitable temperature and carrier density. For TMD PDSL, we attribute asymmetric valence band and conduction band to asymmetric superlattice potential and explore the tunability brought out by this characteristic. We predict robust ferromagnetism in TMD PDSL with moderate superlattice potential (~50 meV). Furthermore, our study shows the effect of the radius of air holes on superlattice potential, so that superlattice potential can be pushed to higher values without compromising the integrity of the device. PDSLs may be applied in Quantum computers and other state-of-the-art electronic devices due to their fascinating properties and ease of manufacturing. Our study lay the ground for future applications, by evaluating the tunability of such devices and discussing the optimal condition to improve the performance of such devices.

Acknowledgements We would like to express our utmost gratitude towards Assoc Prof Gao Weibo, who amid his busy schedule, patiently guided us and helped us whenever we are in doubt throughout the entire project. Prof Gao would always be willing to have insightful discussions with us whenever we consulted him for the project. We would also le to showcase our appreciation towards Dr Wu Weikang. Dr Wu patiently helped us with formation of our theoretical framework and guided us in coding to compute band structure with Mathematica. He always gave insightful responses to our questions. Lastly, we would like to take this chancc to thank our teacher coordinators Mr Chow Ban Hoe from River Valley High School and Mr Lim Yeow Heng from NUS High School of Mathematics and Science for helping to liaise and for providing us abundant resources and guidance on research in general.

Appendix

See the Figs. 4, 5, 6 and 7.

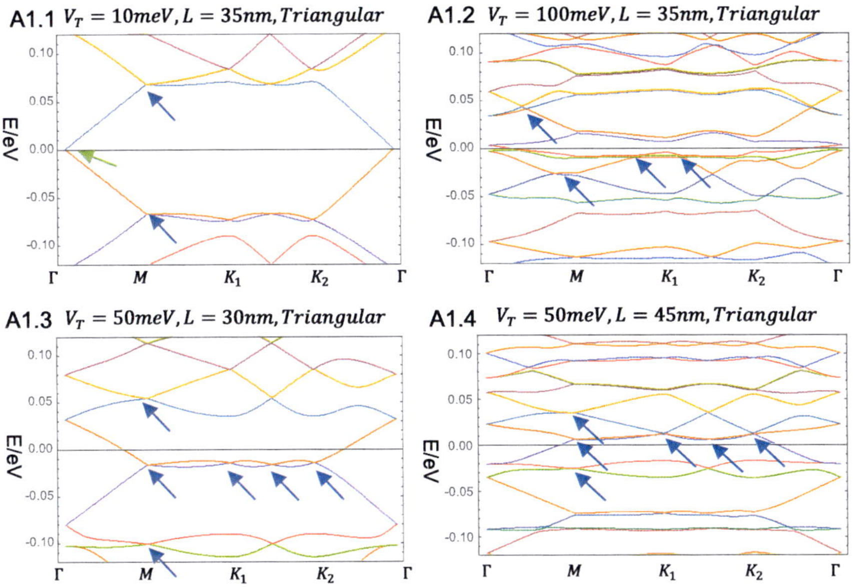

Fig. 4 Band structure of triangular graphene PDSL in the first Brillouin zone, with varying V_T (superlattice potential amplitude) and L (distance between two adjacent air holes in SiO_2)

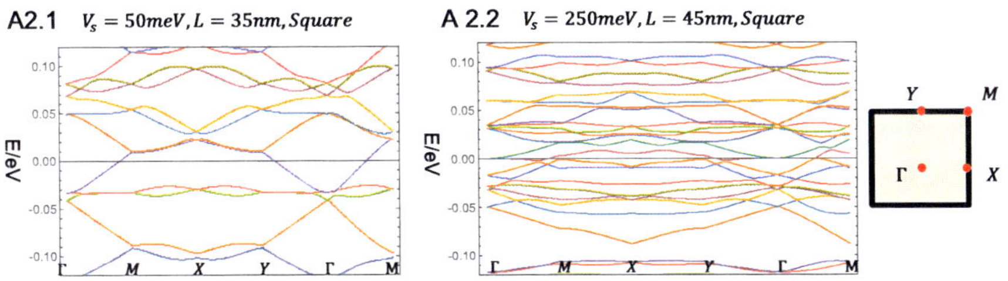

Fig. 5 Band structure of square graphene PDSL in the first Brillouin zone, with varying V_s (square superlattice potential amlitude) and L (distance between two adjacent air holes in SiO_2)

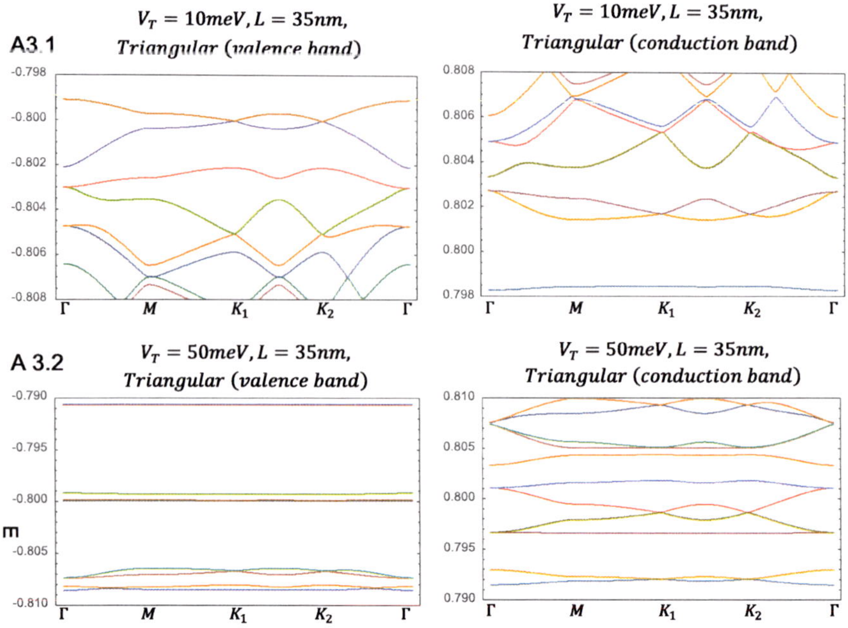

Fig. 6 Band structure of triangular TMD PDSL in the first Brillouin zone, with varying V_T (triangular superlattice potential amplitude) and L (distance between two adjacent air holes in SiO_2)

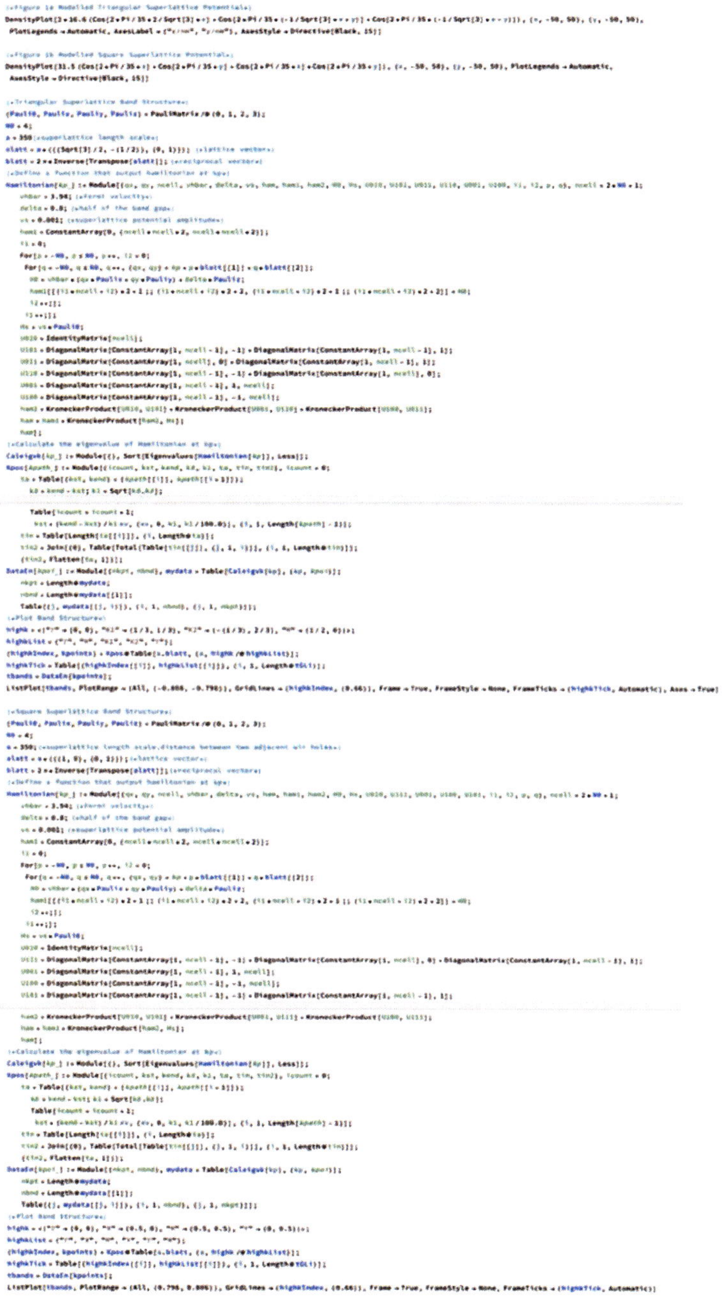

Fig. 7 Mathematica codes for all graphs in this paper

References

1. Ismail, K., Chu, W., Yen, A., Antoniadis, D. A., & Smith, H. I. (1989). Negative transconductance and negative differential resistance in a grid-gate modulation-doped field-effect transistor. *Applied physics letters, 54*(5), 460–462. https://doi.org/10.1063/1.100952
2. Weiss, D., Klitzing, K. V., Ploog, K., & Weimann, G. (1989). Magnetoresistance oscillations in a two-dimensional electron gas induced by a submicrometer periodic potential. *EPL (Europhysics Letters), 8*(2), 179–184. https://doi.org/10.1209/0295-5075/8/2/012
3. Schlösser, T., Ensslin, K., Kotthaus, J. P., & Holland, M. (1996). Landau subbands generated by a lateral electrostatic superlattice—chasing the hofstadter butterfly. *Semiconductor Science and Technology, 11*(11S), 1582–1585. https://doi.org/10.1088/0268-1242/11/11s/022
4. Nimbalkar, A., & Kim, H. (2020). Opportunities and challenges in twisted bilayer graphene: A Review. *Nano-Micro Letters, 12*(1). https://doi.org/10.1007/s40820-02000464-8
5. Ruiz-Tijerina, D. A., & Fal'ko, V. I. (2019). Interlayer hybridization and moiré superlattice minibands for electrons and excitons in heterobilayers of transition-metal dichalcogenides. *Physical Review B, 99*(12). https://doi.org/10.1103/physrevb.99.125424
6. Drienovsky, M., Sandner, A., Baumgartner, C., Liu, M. H., Taniguchi, T., Watanabe, K., Richter, K., Weiss, D. & Eroms, J. (2017). Few-layer graphene patterned bottom gates for van der Waals heterostructures. Preprint at https://arxiv.org/abs/1703.05631
7. Sun, Z., Pint, C. L., Marcano, D. C., Zhang, C., Yao, J., Ruan, G., Yan, Z., Zhu, Y., Hauge, R. H. & Tour, J.M. (2011). Towards hybrid superlattices in graphene. *Nature Communications, 2*(1). https://doi.org/10.1038/ncomms1577
8. Sandner, A., Preis, T., Schell, C., Giudici, P., Watanabe, K., Taniguchi, T., Weiss, D. & Eroms, J. (2015). Ballistic transport in graphene Antidot lattices. *Nano Letters, 15*(12), 8402–8406. https://doi.org/10.1021/acs.nanolett.5b04414
9. Wang, F., Li, S., Bissett, M. A., Kinloch, I. A., Li, Z., & Young, R. J. (2020). Strain engineering in monolayer WS2 and WS2 nanocomposites. *2D Materials, 7*(4), 045022. https://doi.org/10.1088/2053-1583/ababf1
10. Zhang, Y., Kim, Y., Gilbert, M. J., & Mason, N. (2018). Electronic transport in a two-dimensional superlattice engineered via self-assembled nanostructures. *Npj 2DMaterials and Applications, 2*(1). https://doi.org/10.1038/s41699-018-0076-0
11. Hunt, B., Sanchez-Yamagishi, J. D., Young, A. F., Yankowitz, M., LeRoy, B. J., Watanabe, K., Taniguchi, T., Moon, P., Koshino, M., Jarillo-Herrero, P. & Ashoori, R. C. (2013). Massive dirac fermions and Hofstadter butterfly in a van der waals heterostructure. *Science, 340*(6139), 1427–1430. https://doi.org/10.1126/science.1237240
12. Lee, M., Wallbank, J. R., Gallagher, P., Watanabe, K., Taniguchi, T., Fal'ko, V. I., & Goldhaber-Gordon, D. (2016). Ballistic miniband conduction in a graphene superlattice. *Science, 353*(6307), 1526–1529. https://doi.org/10.1126/science.aaf1095
13. Bistritzer, R., & MacDonald, A. H. (2011). Moire bands in twisted double-layer graphene. *Proceedings of the National Academy of Sciences, 108*(30), 12233–12237. https://doi.org/10.1073/pnas.1108174108
14. Wu, F., Lovorn, T., Tutuc, E., Martin, I., & MacDonald, A. H. (2019). Topological insulators in twisted transition metal dichalcogenide homobilayers. *Physical Review Letters, 122*(8). https://doi.org/10.1103/physrevlett.122.086402
15. Forsythe, C., Zhou, X., Watanabe, K., Taniguchi, T., Pasupathy, A., Moon, P., Koshino, M., Kim, P., & Dean, C. R. (2018). Band structure engineering of 2D materials using patterned dielectric superlattices. *Nature Nanotechnology, 13*(7), 566571. https://doi.org/10.1038/s41565-018-0138-7
16. Yang, J., & Gan, C. J. (2021). (rep.). *NRP Junior Research Log SPMSjr03.* Nanyang Technological University. Retrieved January 9, 2022, from https://www.ntu.edu.sg/docs/default-source/talent-outreach/nrp/spmsjr03_rl834398bf-60a2-4033-ab62-6ab6cd351dd2.pdf?sfvrsn=d14da6af_3
17. Shi, L.-kun, Ma, J., & Song, J. C. (2019). Gate-tunable flat bands in van der waals patterned dielectric superlattices. *2D Materials, 7*(1), 015028. https://doi.org/10.1088/2053-1583/ab59a8

18. Bernevig, B. A., & Hughes, T. L. (2013). Topological insulators and topological superconductors. https://doi.org/10.1515/9781400846733
19. Bistritzer, R., & MacDonald, A. H. (2011). Moiré bands in twisted double-layer graphene. *Proceedings of the National academy of Sciences of the United States of America, 108*, 12233–12237.
20. Seifert, G., Terrones, H., Terrones, M., Jungnickel, G., & Frauenheim, T. (2000). Structure and electronic properties of MoS2 nanotubes. *Physical Review Letters, 85*, 146–149.
21. Mak, K. F., Lee, C., Hone, J., Shan, J., & Heinz, T. F. (2010). Atomically thin MoS_2: A new direct-gap semiconductor. *Physical review letters, 105*, 136805.
22. Ponomarenko, L. A., Gorbachev, R. V., Yu, G. L., Elias, D. C., Jalil, R., Patel, A. A., Mishchenko, A., Mayorov, A. S., Woods, C. R., Wallbank, J. R., & Mucha-Kruczynski, M. (2013). Cloning of dirac fermions in graphene superlattices. *Nature, 497*(7451), 594–597. https://doi.org/10.1038/nature1217
23. Woods, C. R., Britnell, L., Eckmann, A., Ma, R. S., Lu, J. C., Guo, H. M., Lin, X., Yu, G. L., Cao, Y., Gorbachev, R. V., & Kretinin, A. V. (2014). Commensurate–incommensurate transition in graphene on hexagonal boron nitride. *Nature Physics, 10*(6), 451–456. https://doi.org/10.1038/nphys2954
24. Gorbachev, R. V., Song, J. C. W., Yu, G. L., Kretinin, A. V., Withers, F., Cao, Y., Mishchenko, A., Grigorieva, I. V., Novoselov, K. S., Levitov, L. S., & Geim, A. K. (2014). Detecting topological currents in graphene superlattices. *Science, 346*(6208), 448–451. https://doi.org/10.1126/science.125966.
25. Gorbachev, R. V., Song, J. C. W., Yu, G. L., Kretinin, A. V., Withers, F., Cao, Y., Mishchenko, A., Grigorieva, I. V., Novoselov, K. S., Levitov, L. S., & Geim, A. K. (2014). Gate-dependent pseudospin mixing in graphene/boron nitride moiré superlattices. *Nature Physics, 10*(10), 743–747. https://doi.org/10.1038/nphys3075
26. Yankowitz, M., Xue, J., & LeRoy, B. J. (2014). Graphene on hexagonal boron nitride. *Journal of Physics: Condensed Matter, 26*(30), 303201. https://doi.org/10.1088/0953-8984/26/30/303201
27. Cao, Y., Fatemi, V., Demir, A., Fang, S., Tomarken, S. L., Luo, J. Y., Sanchez-Yamagishi, J. D., Watanabe, K., Taniguchi, T., Kaxiras, E., & Ashoori, R. C. (2018). Correlated insulator behaviour at half-filling in magic-angle graphene superlattices. *Nature, 556*, 80–84.
28. Cao, Y., Fatemi, V., Fang, S., Watanabe, K., Taniguchi, T., Kaxiras, E., & Jarillo-Herrero, P. (2018). Unconventional superconductivity in magic-angle graphene superlattices. *Nature, 556*, 43–50. https://doi.org/10.1038/nature26160
29. MacDonald, A. H., Oji, H. C., & Liu, K. L. (1986). Thermodynamic properties of an interacting two-dimensional electron gas in a strong magnetic field. *Physical Review B, 34*(4), 2681–2689. https://doi.org/10.1103/physrevb.34.2681
30. Nomura, K., & MacDonald, A. H. (2006). Quantum Hall ferromagnetism in graphene. *Physical Review Letters, 96*(25). https://doi.org/10.1103/physrevlett.96.256602

Engineering Ticks and Mosquitos' Salivary Peptides into Potent and Uncleavable Thrombin Inhibitor

Daven Wong, Rishii Parthasarathy, and Sujan Palepu

Abstract Anticoagulants are administered during intravenous treatments to reduce the propensity for blood to coagulate on external surfaces. Anticoagulants mitigate thrombosis in blood-contacting medical devices, including sample tubes, dialysis equipment, blood transfusion bags, and heart–lung machines. Thrombin inhibitors are a subset of anticoagulants naturally found in hematophagous animals. The thrombin inhibitors anophelin and variegin, found in the salivary extracts of *Anopheles* and *Amblyomma variegatum,* respectively, have unique anticoagulant properties. Anophelin has a highly extensive half-life as it is resistant to thrombin-cleavage due to its uniquely reversed thrombin-binding orientation. A variegin derivative, ultravariegin, is highly potent ($K_i = 4.4$ pM) because of its tight thrombin-binding. We joined synthetic fragments of anophelin and ultravariegin by disulphide linkage to generate hybrid anophelin-ultravariegin (*HAU*) thrombin inhibitor, which was purified by liquid chromatography and validated using mass spectrometry. Through incubation with thrombin and chromogenic assay, we found that *HAU* had a superior half-life (62 h) and potency ($IC_{50} = 0.37$ nM, $K_i = 1.7$ pM) than commercial anticoagulants bivalirudin and heparin. *HAU* demonstrates strong potential as an exceptionally beneficial and cost-effective anticoagulant with faster, stronger, and lengthier antithrombic effects than commercial anticoagulants. *HAU* is a strong candidate for immobilised anticoagulant coatings on medical apparatus, so that patients undergoing intravenous treatments could be less reliant on consuming anticoagulants, mitigating anticoagulant side effects.

Keywords Anticoagulant · Thrombin inhibitor · Anophelin · Ultravariegin · Hybrid

D. Wong (✉) · R. Parthasarathy · S. Palepu
NUS High School of Mathematics and Science, Singapore, Singapore
e-mail: daven.wong@hotmail.com

H. Guo et al. (eds.), *IRC-SET 2022*,
https://doi.org/10.1007/978-981-19-7222-5_23

1 Introduction

Anticoagulants are a class of medications which are specifically administered to prevent blood coagulation inside and outside the human body. Anticoagulants directly or indirectly inhibit thrombin, a naturally occurring enzyme integral in clot formation. In vivo, thrombin is formed from prothrombin through both the intrinsic and extrinsic pathways of the coagulation cascade. Thrombin cleaves fibrinogen circulating in the blood into fibrin threads, enabling the formation of stable blood clots [1].

Anticoagulants are prescribed as preventive and treatment medications for chronic diseases such as venous thromboembolism, acute coronary syndrome, and stroke [2, 3]. Moreover, anticoagulants are applied in several hospital and laboratory equipment, including blood sample tubes, haemodialysis units, blood transfusion bags, and heart–lung machines [4–7].

Among the most widely used commercial anticoagulants are heparin and bivalirudin. Heparin, a traditional anticoagulant discovered in 1916, is extracted from small intestinal mucosa of swine and bovine heart [8, 9]. Heparin inhibits thrombin in the presence of cofactor antithrombin III, has a 90 min half-life, and a K_i value of 1.5 μM [9, 10]. Bivalirudin, the most common direct thrombin inhibitor in medical use, is a synthetic 20 amino acid congener of the naturally occurring anticoagulant hirudin, extracted from *Hirudo medicinalis* saliva [11]. Bivalirudin has a 25 min half-life, an IC_{50} value of 140 nM, and a K_i value of 2.3 nM [10, 11].

Hematophagous animals have the unique ability of inhibiting thrombosis, facilitating the acquisition of blood for feeding [12]. The arthropod vectors *Anopheles* and *Amblyomma variegatum* produce salivary peptides anophelin and variegin, respectively, as direct thrombin inhibitors [12, 13].

When a direct thrombin inhibitor competitively binds to the thrombin active site, thrombin is inhibited from cleaving fibrinogen into fibrin threads. However, thrombin has two exosite regions which facilitate the cleavage and unbinding of thrombin inhibitors [14]. From an evolutionary standpoint, the thrombin-cleavage mechanism prevents thrombin from remaining permanently inhibited, thereby reducing the time and energy expenditure in replenishing the body's active thrombin molecules. The thrombin-cleavage mechanism limits the half-life of most thrombin inhibitors to a few hours or less [12].

Anophelin, a 61 amino acid thrombin inhibitor ($K_i = 103$ pM), has a natural resistance to thrombin-cleavage unlike most other thrombin inhibitors [15]. Anophelin binds to thrombin in an atypical orientation, reversed from other direct thrombin inhibitors (see Fig. 1). Anophelin residues D32-F45 block thrombin exosite I. The atypically sparse interactions between the anophelin N-terminus and thrombin exosite I hinder the recognition of the thrombin-bound anophelin for cleavage, lowering the effective rate of anophelin thrombin-cleavage [14]. Therefore, anophelin is thought to confer a highly extensive half-life.

Variegin, a 32 amino acid thrombin inhibitor, is one of the smallest naturally occurring anticoagulants [13]. A 30 amino acid derivative of variegin, ultravariegin, is one of the most potent peptide derivatives of natural thrombin inhibitors, having

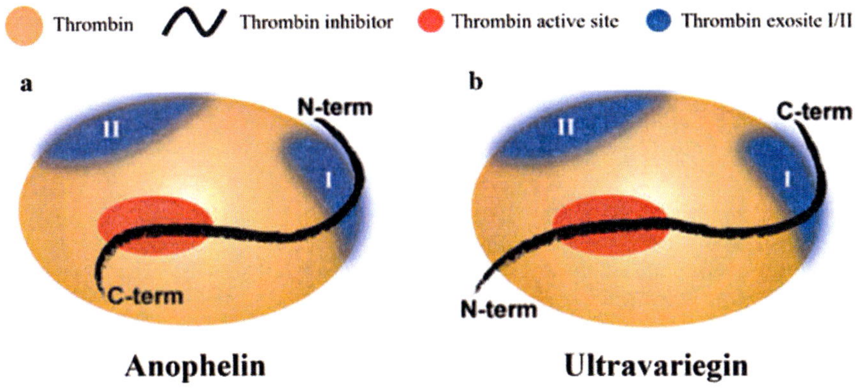

Fig. 1 Thrombin-binding mechanisms of direct thrombin inhibitors anophelin and ultravariegin. **a** Anophelin binds to thrombin in a uniquely reversed orientation, with its N-terminus bound to thrombin exosite I and its C-terminus bound to the thrombin active site, differentiating anophelin from other direct thrombin inhibitors. **b** Ultravariegin binds to thrombin in the regular orientation, with its N-terminus bound to the thrombin active site and its C-terminus bound to thrombin exosite I [13, 14]

a K_i value of 4.4 pM [16]. The superior potency of ultravariegin is thought to be explained by the tight competitive binding of the ultravariegin N-terminus to the thrombin active site (see Fig. 1) [13].

Here, we report the generation of hybrid anophelin-ultravariegin (*HAU*) thrombin inhibitors. We investigated *HAU* half-life and inhibition kinetics and found that *HAU* has an extensive half-life and superior potency, just as anophelin and ultravariegin, respectively.

2 Methodology

2.1 HAU Generation via Hybridisation of Anophelin and Ultravariegin Fragments

Short fragments of anophelin and ultravariegin were designed rationally. The crystal structure of the thrombin·anophelin complex showed a 16–17 amino acid region of the anophelin N-terminus that was responsible for the weak interactions between the anophelin N-terminus and thrombin exosite I [14]. Incorporation of this anophelin region into *HAU* would allow *HAU* to confer thrombin-cleavage resistance and have an extensive half-life.

Two fragments of the anophelin N-terminus of amino acid lengths 16 and 17 were designed. Incorporation of an additional cysteine residue at their N-termini led to the design of anophelin fragments with total amino acid length 17 and 18 that were termed *Ano17* and *Ano18*, respectively.

Similarly, the crystal structure of the thrombin·ultravariegin complex showed a 13–14 amino acid region of the ultravariegin N-terminus that was responsible for the tight binding of the ultravariegin N-terminus to the thrombin active site [17]. Incorporation of this ultravariegin region into *HAU* would allow *HAU* to bind tightly to thrombin and have a superior potency.

Two fragments of the ultravariegin N-terminus of amino acid lengths 12 and 13 were designed. Incorporation of an additional cysteine residue at their N-termini led to the design of ultravariegin fragments with total amino acid length 13 and 14 that were termed *Uvg13* and *Uvg14*, respectively.

Fragments *Ano17*, *Ano18*, *Uvg13*, and *Uvg14* were custom synthesised by commercial vendors through Fmoc-based solid phase peptide synthesis. Patent application of the amino acid sequences of *Ano17*, *Ano18*, *Uvg13*, and *Uvg14* is ongoing and will not be reported here.

Disulphide Linkage Between Anophelin and Ultravariegin Fragments. Folding buffer was prepared as a mixture of 10% acetonitrile, 10% dimethyl sulfoxide, 1 M guanidine hydrochloride, and 100 mM sodium hydrogen phosphate maintained at pH 7.4. Equal masses of *Ano17* and *Uvg13* were mixed and dissolved in the folding buffer to a concentration of 0.5 mg/ml.

The methods described in the above paragraph were repeated to prepare three other mixture solutions, *Ano17* and *Uvg14*, *Ano18* and *Uvg13*, as well as *Ano18* and *Uvg14*. All four mixture solutions were sealed with holes to allow for air exchange. The mixture solutions were stirred for 16–24 h to allow for disulphide bridge formation between the cysteine residues on the fragments' N-termini through air oxidation.

HAU thrombin inhibitors were generated from the mixture solutions of *Ano17* and *Uvg13*, *Ano17* and *Uvg14*, *Ano18* and *Uvg13*, as well as *Ano18* and *Uvg14*, and *HAU* thrombin inhibitors were termed $HAU^{Ano17-Uvg13}$, $HAU^{Ano17-Uvg14}$, $HAU^{Ano18-Uvg13}$, and $HAU^{Ano18-Uvg14}$, respectively.

At the same time, undesired products were generated from the air oxidation process, including *Ano17* dimer, *Ano18* dimer, *Uvg13* dimer, and *Uvg14* dimer. The necessary follow-up step was to purify $HAU^{Ano17-Uvg13}$, $HAU^{Ano17-Uvg14}$, $HAU^{Ano18-Uvg13}$, and $HAU^{Ano18-Uvg14}$ from the mixture solutions.

***HAU* purification and validation of *HAU* purity.** $HAU^{Ano17-Uvg13}$, $HAU^{Ano17-Uvg14}$, $HAU^{Ano18-Uvg13}$, and $HAU^{Ano18-Uvg14}$ were eluted from the mixture solutions during high performance liquid chromatography (HPLC), using Jupiter 4 μm Proteo 90 Å (250 × 4.6 mm) reversed-phase column (Phenomenex, CA, USA) and a linear gradient of aqueous 0.1% trifluoroacetic acid, 70% acetonitrile. Concentrations of peptide solutions were estimated using UV absorbance at 215 nm.

The molecular masses of the obtained *HAU* samples were calculated by electrospray ionisation mass spectrometry (ESI–MS), using an LCQ Fleet Ion Trap MS and Thermo Xcalibur 2.2 software (Thermo Fisher Scientific, MA, USA). Mass spectra were deconvoluted using ProMass for Xcalibur 3.0 software.

The calculated molecular masses of the samples were then cross-checked with theoretical molecular masses to validate the purity of the obtained *HAU* samples.

Table 1 Theoretical molecular masses of anophelin and ultravariegin fragments, hybrids, and dimers

Anophelin and ultravariegin fragment/hybrid/dimer	Constituent fragments	Theoretical molecular mass (g mol^{-1})
Ano17	–	1685.9
Ano18	–	1742.9
Uvg13	–	1524.6
Uvg14	–	1581.6
$HAU^{Ano17-Uvg13}$	*Ano17, Uvg13*	3208.5
$HAU^{Ano17-Uvg14}$	*Ano17, Uvg14*	3265.5
$HAU^{Ano18-Uvg13}$	*Ano18, Uvg13*	3265.5
$HAU^{Ano18-Uvg14}$	*Ano18, Uvg14*	3322.5
Ano17 dimer	*Ano17, Ano17*	3369.8
Ano18 dimer	*Ano18, Ano18*	3483.8
Uvg13 dimer	*Uvg13, Uvg13*	3047.2
Uvg14 dimer	*Uvg14, Uvg14*	3161.2

The theoretical molecular masses of fragments *Ano17*, *Ano18*, *Uvg13*, and *Uvg14* were known from their amino acid sequences (see Table 1).

Disulphide linkage is an oxidation reaction between the sulfhydryl (SH) groups of two cysteine residues (*R* and *R'*). The S^- anion of one sulfhydryl group acts as the nucleophile and attacks the sulfhydryl group on the second cysteine residue. As the disulphide bridge is formed, two hydrogen ions (H^+) are released (see Eq. 1) [18].

$$R - SH + HS - R' \rightarrow R - S - S - R' + 2H^+ + 2e^- \quad (1)$$

The theoretical molecular masses of $HAU^{Ano17-Uvg13}$, $HAU^{Ano17-Uvg14}$, $HAU^{Ano18-Uvg13}$, $HAU^{Ano18-Uvg14}$, *Ano17* dimer, *Ano18* dimer, *Uvg13* dimer, and *Uvg14* dimer were calculated as the sum of their constituent fragments subtracted by the molar mass of two hydrogen ions, 2.016 g mol^{-1} (see Table 1).

Samples from ESI–MS with calculated molecular masses that deviate from theoretical molecular masses by an acceptable range of $\pm$ 2 g mol^{-1} were identified as the corresponding hybrids and dimers by molecular weight. The purified *HAU* samples were lyophilised for subsequent use in the chromogenic assay.

2.2 HAU Half-Life Estimation by Incubation with Thrombin

Comparing relative $HAU^{Ano17-Uvg13}$ intensities in incubation mixtures with thrombin. Two incubation mixtures were prepared, each containing equivalent volumes of $HAU^{Ano17-Uvg13}$ and 2.5 nM of thrombin. HPLC was performed on the control and experimental incubation mixtures at 0 h and 24 h after the start of incubation, respectively. The relative $HAU^{Ano17-Uvg13}$ intensities after 0 h and 24 h of incubation with thrombin were calculated by ESI–MS.

Relative intensities as point estimates for $HAU^{Ano17-Uvg13}$ half-life estimation. Assuming that the cleavage of $HAU^{Ano17-Uvg13}$ thrombin inhibitor by thrombin follows a first-order reaction, $HAU^{Ano17-Uvg13}$ half-life ($t_{1/2}$) was approximated from final and initial relative $HAU^{Ano17-Uvg13}$ intensities, $N(t)$ and N_0, , respectively (see Eq. 2) [19].

$$N(t) = N_0 \left(\frac{1}{2}\right)^{\frac{t}{t_{1/2}}} \tag{2}$$

2.3 HAU *Potency Determination by Chromogenic Assay*

Measuring thrombin inhibition in varied *HAU* concentrations by spectrophotometry. A chromogenic substrate of thrombin, S-2238, is hydrolysed by thrombin to produce a bright yellow product, *p*-nitroaniline [13]. The rate of hydrolysis of S-2238 by thrombin in this chromogenic assay is used to represent the rate of cleavage of fibrinogen by thrombin in the human body.

In the chromogenic assay for each *HAU* thrombin inhibitor, reaction mixtures of *HAU*, thrombin, and S-2238 were prepared by dilution with Tris-NaCl-BSA buffer (50 mM trisaminomethane, 100 mM sodium chloride, and 1 mg/ml of bovine serum albumin maintained at pH 7.4). Ten *HAU* concentrations and two control mixtures were used for later calculations of per cent thrombin inhibition (see Table 2).

The rate of formation of *p*-nitroaniline was followed at 405 nm for 5 min by spectrophotometry [13]. Linear standard curves of absorbance at 405 nm versus time were plotted using GraphPad Prism software (GraphPad Software, CA, USA) to obtain sample slopes and control slopes.

***HAU* IC_{50} and K_i determination as indicators of *HAU* potency**. The IC_{50} value is a measurement of the *HAU* concentration necessary to inhibit thrombin inhibition by 50%. Smaller *HAU* IC50 values signify greater *HAU* functional strength as a thrombin inhibitor.

Nonlinear standard curves of per cent thrombin inhibition against log (*HAU* concentration) were plotted using GraphPad Prism software (GraphPad Software, CA, USA) (see Eq. 3). *HAU* IC_{50} was found by reading log (*HAU* concentration) off the graph at 50% thrombin inhibition.

$$\%\ \text{Thrombin inhibition} = \frac{\text{Average control slope} - \text{Sample slope}}{\text{Average control slope}} \times 100\% \tag{3}$$

Table 2 Preparation of reaction mixtures of varied *HAU* concentrations for spectrophotometry

Reaction mixture	*HAU* concentration (nM), volume (μl)	Thrombin concentration (nM), volume (μl)	S-2238 concentration (μM), volume (μl)
A	300, 50	2.5, 50	300, 50
B	90, 50	2.5, 50	300, 50
C	30, 50	2.5, 50	300, 50
D	9, 50	2.5, 50	300, 50
E	3, 50	2.5, 50	300, 50
F	0.9, 50	2.5, 50	300, 50
G	0.3, 50	2.5, 50	300, 50
H	0.09, 50	2.5, 50	300, 50
I	0.03, 50	2.5, 50	300, 50
J	0.009, 50	2.5, 50	300, 50
K	0, 50	2.5, 50	300, 50
L	0, 50	2.5, 50	300, 50

The inhibitor constant, K_i, describes the binding affinity between *HAU* and thrombin. Smaller *HAU* K_i values signify tighter binding of *HAU* to thrombin, and therefore a greater level of thrombin inhibition.

The apparent inhibitor constant, $K_i{}'$, was calculated from the steady-state velocity in the presence of *HAU*, v_s, the velocity observed in the absence of *HAU*, v_0, the total thrombin concentration, E_t, and the total *HAU* concentration, I_t, given that $E_t =$ 0.83 nM (see Eq. 4) [13].

$$v_s = v_0 \frac{\sqrt{\left(K_i' + I_t - E_t\right)^2 + 4K_i'E_t} - \left[K_i' + I_t - E_t\right]}{2E_t} \tag{4}$$

HAU K_i was calculated from the S-2238 concentration, *S*, and the Michaelis–Menten constant for S-2238, K_m, given that $S = 100{,}000$ nM and $K_m = 3250$ nM (see Eq. 5) [13].

$$K_i' = K_i\left(1 + \frac{S}{K_m}\right) \tag{5}$$

3 Results

$HAU^{Ano17-Uvg13}$ has an estimated half-life of 24 h. The four *HAU* thrombin inhibitors demonstrate superior potency, with IC_{50} values within nanomolar range, and K_i values within picomolar range. $HAU^{Ano17-Uvg13}$ is the most potent thrombin inhibitor

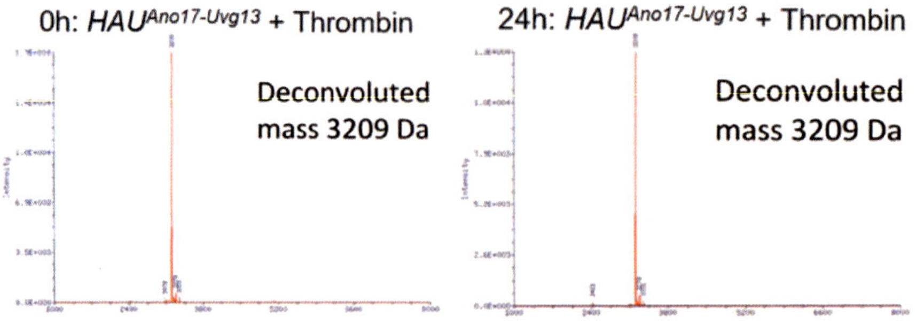

Fig. 2 Deconvoluted mass spectra of incubation mixtures of $HAU^{Ano17-Uvg13}$ with thrombin after 0 h and 24 h. $N(t) = 1.7 \times 10^4$ and $N_0 = 1.3 \times 10^4$. $t_{1/2} = 62$ h

of the four, with an IC_{50} value of 0.37 nM and a K_i value of 1.7 pM (Figs. 2, 3 and Tables 3, 4).

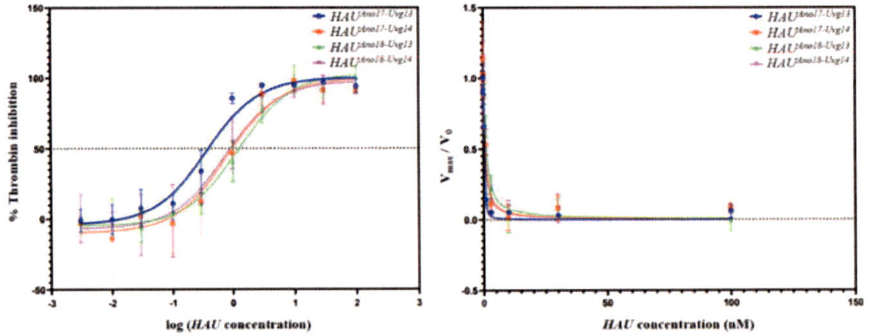

Fig. 3 IC_{50} and K_i curves of $HAU^{Ano17-Uvg13}$, $HAU^{Ano17-Uvg14}$, $HAU^{Ano18-Uvg13}$, and $HAU^{Ano18-Uvg14}$

Table 3 Molecular masses and relative abundance of purified *HAU* samples used for chromogenic assay

HAU sample	Theoretical molecular mass (g mol^{-1})	Actual molecular mass (g mol^{-1})	Relative abundance (%)
$HAU^{Ano17-Uvg13}$	3208.5	3208.1	100
$HAU^{Ano17-Uvg14}$	3265.5	3265.3	100
$HAU^{Ano18-Uvg13}$	3265.5	3265.4	100
$HAU^{Ano18-Uvg14}$	3322.5	3322.2	100

Table 4 *HAU* IC_{50} and K_i values

HAU sample	IC_{50} (nM)	IC_{50} 95% CI (nM)	K_i (pM)	K_i 95% CI (pM)
$HAU^{Ano17-Uvg13}$	0.37	(0.26, 0.54)	1.7	(0.48, 3.7)
$HAU^{Ano17-Uvg14}$	0.83	(0.49, 1.4)	14	(7.2, 26)
$HAU^{Ano18-Uvg13}$	1.2	(0.84, 1.8)	23	(15, 34)
$HAU^{Ano18-Uvg14}$	0.77	(0.45, 1.3)	14	(6.5, 26)

4 Discussion

Summary. *HAU* demonstrates superior half-life and potency in comparison with commercial anticoagulants such as heparin and bivalirudin. *HAU* serves as a cost-effective anticoagulant due to its lower effective concentration and its mass reproducibility through gene cloning. The application of *HAU* as antithrombotic coating on medical and laboratory apparatus reduced the need for anticoagulant consumption and potentially mitigates anticoagulant side effects.

Limitations and future work. *HAU* half-life estimation was performed once and only for $HAU^{Avg17-Uvg13}$. Reproducing the half-life estimation procedure for the four *HAU* thrombin inhibitors would have ensured the estimated half-life values are reliable.

The thrombin inhibition kinetics of *HAU* observed in the chromogenic assay may differ from that of in the body. Performing a blood coagulation assay would be useful for obtaining a more representative sample of *HAU* inhibition kinetics in the human body.

Applications. The terminal amino acid residues of *HAU* may be modified for use in anti-thrombotic coating. Bioconjugation is a method of promoting the formation of bonds between biomolecules. Extensive research demonstrates that oxime chemistry is a feasible method to bind inhibitors to biomembranes [20].

The biomembrane surface can be biotinylated with streptavidin micropatterns fabricated by oxime bond formation using an aldehyde-patterned polymer. Biotinylated *HAU* could be immobilised by binding to these micropatterns and forming strong adhesive forces [21].

Antithrombotic coating with the use of *HAU* benefits from its uncleavability, and therefore the reluctancy of *HAU* to leak into the bloodstream. This would mitigate undesirable anticoagulant side effects such as excessive bleeding.

To ensure that the body does not face a thrombin shortage when using an apparatus coated with *HAU*, patients can be prescribed vitamin D supplements or receive thrombin precursors during their treatment. Therefore, anticoagulant coating is a feasible real-world application of our research on *HAU*.

5 Conclusion

By hybridising anophelin and ultravariegin fragments, we successfully engineered hybrid anophelin-ultravariegin (*HAU*) thrombin inhibitors with superior half-life and potency. This was confirmed by an incubation experiment and a chromogenic assay. By utilising bioconjugation through methods such as oxime chemistry, *HAU* could be modified for use in anticoagulant coating, a technology with several applications in the medical and laboratory settings.

References

1. Palta, S., Saroa, R., & Palta, A. (2014). Overview of the coagulation system. *Indian Journal of Anaesthesia, 58*(5), 515.
2. Tiryaki, F., Nutescu, E. A., Hennenfent, J. A., Karageanes, A. M., Koesterer, L. J., Lambert, B. L., & Schumock, G. T. (2011). Anticoagulation therapy for hospitalized patients: Patterns of use, compliance with national guidelines, and performance on quality measures. *American Journal of Health System Pharmacy, 68*(13), 1239–1244.
3. Hylek, E. M., Go, A. S., Chang, Y., Jensvold, N. G., Henault, L. E., Selby, J. V., & Singer, D. E. (2003). Effect of intensity of oral anticoagulation on stroke severity and mortality in atrial fibrillation. *New England Journal of Medicine, 349*(11), 1019–1026.
4. Meinkoth, J. H., & Allison, R. W. (2007). Sample collection and handling: Getting accurate results. *Veterinary Clinics of North America: Small Animal Practice, 37*(2), 203–219.
5. Fischer, K. G. (2007). Essentials of anticoagulation in hemodialysis. *Hemodialysis International, 11*(2), 178–189.
6. Hunt, E., & Moore, J. S. (1990). Use of blood and blood products. *Veterinary Clinics of North America: Food Animal Practice, 6*(1), 133–147.
7. Dobrovolskaia, M. A., & McNeil, S. E. (2015) Safe anticoagulation when heart and lungs are "on vacation". *Annals of Translational Medicine, 3*(Suppl 1).
8. Lee, S. Y., Kang, H. J., Park, Y., & Hur, S. J. (2020). Development of effective heparin extraction method from pig by-products and analysis of their bioavailability. *Journal of Animal Science and Technology, 62*(6), 933.
9. Hirsh, J. (1991). Heparin. *New England Journal of Medicine, 324*(22), 1565–1574.
10. Cook, B. W. (2010). *Anticoagulation management*. Thieme Medical Publishers.
11. Warkentin, T. E., Greinacher, A., & Koster, A. (2008). Bivalirudin. *Thrombosis and Haemostasis, 99*(11), 830–839.
12. Koh, C. Y., & Kini, R. M. (2008). Anticoagulants from hematophagous animals. *Expert Review of Hematology, 1*(2), 135–139.
13. Koh, C. Y., Kazimirova, M., Trimnell, A., Takac, P., Labuda, M., Nuttall, P. A., & Kini, R. M. (2007). Variegin, a novel fast and tight binding thrombin inhibitor from the tropical bont tick. *Journal of Biological Chemistry, 282*(40), 29101–29113.
14. Figueiredo, A. C., de Sanctis, D., Gutiérrez-Gallego, R., Cereija, T. B., Macedo-Ribeiro, S., Fuentes-Prior, P., & Pereira, P. J. B. (2012). Unique thrombin inhibition mechanism by anophelin, an anticoagulant from the malaria vector. *Proceedings of the National Academy of Sciences, 109*(52), E3649–E3658.
15. Francischetti, I. M. B., Valenzuela, J. G., & Ribeiro, J. M. C. (1999). Anophelin: Kinetics and mechanism of thrombin inhibition. *Biochemistry, 38*(50), 16678–16685.
16. Koh, C. Y., Shih, N., Yip, C. Y., Li, A. W. L., Chen, W., Amran, F. S., & Chan, M. Y., et al. (2021). Efficacy and safety of next-generation tick transcriptome-derived direct thrombin inhibitors. *Nature Communications, 12*(1), 1–16.

17. Kini, R. M., & Chan, M. (2018). A novel class of parenteral anticoagulant agents for percutaneous coronary intervention (PCI) based on tick thrombin inhibitors. *Indian Heart Journal, 70*, S93.
18. Kastin, A. (2013). *Handbook of biologically active peptides* (2nd edn). Academic Press.
19. Hallare, J., & Gerriets, V. (2020). *Half life*. StatPearls Publishing.
20. Partha, R., & Conyers, J. L. (2009). Biomedical applications of functionalized fullerene-based nanomaterials. *International Journal of Nanomedicine, 4*, 261.
21. Maynard, H. D., Broyer, R. M., & Kolodziej, C. M. (2009). Protein and peptide conjugation to polymers and surfaces using oxime chemistry. *Biotechnology and Materials Science* 53–68.

Investigating Factors Affecting Upstream Contamination

Shen Xing Yang, Wang Hengyue, Mendoza Kieran Majel, and Bernard Ricardo

Abstract Upstream contamination is a phenomenon in fluid dynamics that is not well-understood to this day. In the presence of fluid flow from an upper reservoir to a lower reservoir, the surfactants on the lower reservoir have a tendency to climb upwards and contaminate the upper reservoir. The prevailing theory suggests that turbulent flow effects and the Marangoni effect are responsible. Our project consists of two key aspects. First, we develop a numerical model detailing the surfactant motion in the presence of fluid flow from a cylindrical upper reservoir to a cylindrical lower reservoir. Second, we conduct a series of experiments to determine the effect of surface tension gradients on the extent of upstream contamination and determine whether buoyancy effects are predominant in such a scenario. Our results suggest that reducing the surface tension in the lower reservoir increases the extent of upstream contamination, confirming that the Marangoni effect plays a key factor in this phenomenon. We also find that by reducing the surface tension of the upper reservoir, the upstream contamination effect is inhibited. This is especially relevant in industrial processes and should be taken into account when designing particle discharge systems.

Keywords Marangoni effect · Upstream contamination · Fluid dynamics · Surfactant motion

1 Introduction

When water from an upper pipe is discharged onto a lower chamber of water with surfactants, intuition would lead one to believe that these surfactants would not contaminate the pipe. However, Bianchini et al. [1] demonstrated in 2013 that contrary to intuition, such a phenomenon known as upstream contamination is observed.

S. X. Yang · W. Hengyue (✉) · M. K. Majel · B. Ricardo
NUS High School of Mathematics and Science, Singapore, Singapore
e-mail: hengyue.wang04@gmail.com

H. Guo et al. (eds.), *IRC-SET 2022*,
https://doi.org/10.1007/978-981-19-7222-5_24

Their experimental setups involved water flowing down from an upper reservoir down a channel and into a small waterfall, into the lower reservoir with mate particles. Their primary observation was that the mate particles traveled up the waterfall and up the channel, and into the upper reservoir, thus contaminating the vessel [2].

Upstream contamination is a phenomenon in fluid dynamics that is not well-understood to this day. The prevailing theory suggests that effects arising from turbulent flow and the Marangoni effect are responsible.

The 'tears of wine' phenomena is attributed to the Marangoni effect. Due to the surface tension gradient between the lower section of the glass (which has a higher concentration of alcohol and thus lower surface tension) and the higher section of the glass, the alcohol experiences an upwards flow up the wine glass and forming a liquid film [3]. However, when the accumulated weight exceeds that of this upwards force, droplets fall back down and therefore manifest as the 'tears' that are observed.

Another manifestation of the Marangoni effect is when soapy water is added to a larger reservoir of pure water. Due to the soap water having lower surface tension (as the soap particles disrupt the molecular structure of the water, reducing intermolecular forces), once added it gets 'pulled.' The images can be found in Fig. 4a, b in the Appendix.

Apart from the Marangoni effect, upstream contamination has also been linked to turbulent flow. Due to the downstream flow of water, vortices are induced at the edges on either side of the channel, assisting in the upwards movement of the mate particles. Modeling such behavior is extremely complex, as a laminar solver would not allow for appropriate convergence, and thus simplifications are required.

2 Methodology

Our project consists of two key aspects. First, we develop a numerical model detailing the surfactant motion in the presence of fluid flow from a cylindrical upper reservoir to a cylindrical lower reservoir. Second, we conduct a series of experiments to determine the effect of surface tension gradients on the extent of upstream contamination.

2.1 Theoretical Analysis

Some theoretical work has been done on the effect of buoyancy and fluid resistance on the upstream contamination effect [4]. Under some conditions, particles are able to make it upstream even without a surface tension gradient due to the effect of buoyancy. However, the particles can only do so under certain conditions, the most important of which is the density constraint (the particle *must* be less dense than water).

The paper cited used Newton's Second Law to set up an equation of motion for the particle moving upstream. However, this makes it quite hard to determine the Marangoni effect without making further approximations. Therefore, for the purposes of this analysis, we will assume that Marangoni acceleration occurs solely on an idealized boundary between the lower and upper reservoirs with different surface tensions. This boundary is likely to be found at the base of the downward flow. In other words, the Marangoni effect imparts an initial velocity onto the particle, allowing particles which are more dense than water to overcome the gravitational force to reach the upper reservoir.

The Marangoni flow rate is directly proportional to surface tension and inversely proportional to viscosity [3]. As such, the addition of a surfactant (either floating particles or soap) to the lower reservoir to decrease the surface tension would result in a higher initial velocity of the particle due to the Marangoni effect. This means that particles more dense than water will be able to make it further upstream the greater the difference in surface tension is.

For a simplified treatment, we will only consider four effects: the initial Marangoni flow rate, the gravitational force, buoyancy, and the viscous drag of the fluid. We will assume laminar flow, as well as approximating the column of falling fluid as a prism and the particulate as a sphere. This means that the hydrostatic effects of the fluid are identical to a non-moving column of fluid.

It is known that the initial flow rate due to Marangoni effect is approximately $\frac{\Delta\gamma}{\mu}$. The buoyant force acting on the particle is $\rho V g = \frac{4}{3}\pi\rho r^3 g$, while the gravitational force is mg. Assuming the fluid is descending at a speed of u, the drag force acting on the particle is $6\pi\mu r(v+u)$, where v is the upwards velocity of the particle. Putting it all together, we get the differential equation $ma = \frac{4}{3}\pi\rho r^3 g - mg - 6\pi\mu r(v+u)$. This is a first-order differential equation of the form $\frac{dv}{dt} = A - Bv$, where $A = \left(\frac{4}{3m}\pi\rho r^3 - 1\right)g - 6\pi\mu r u$ and $B = 6\pi\mu r$. Using separation of variables, we get:

$$v = \left(v_0 - \frac{A}{B}\right)e^{-Bt} + \frac{A}{B}, \text{ where } v_0 = \frac{\Delta\gamma}{\mu}$$

If $A > 0$, then we are done. It means that the buoyancy force is strong enough to account for all of the upstream contamination. However, if $A < 0$, then the velocity will eventually become negative. Hence, we may solve for the time at which $v = 0$ and then determine the maximum height that the object is able to reach. This occurs when $\left(v_0 - \frac{A}{B}\right)e^{-Bt} = -\frac{A}{B}$, which when solved gives us $t_f = -Bln\left(\frac{A}{A-Bv_0}\right) = Bln\left(\frac{A-Bv_0}{A}\right)$.

We now integrate the expression for v to find the expression for the maximum height. We get that $x = \frac{1}{B}\left(v_0 - \frac{A}{B}\right)(1 - e^{-Bt}) + \frac{At}{B}$. At the time t_f, therefore, we can just substitute the above expression for t_f ($v = 0$) into the expression for x to find h, the maximum height.

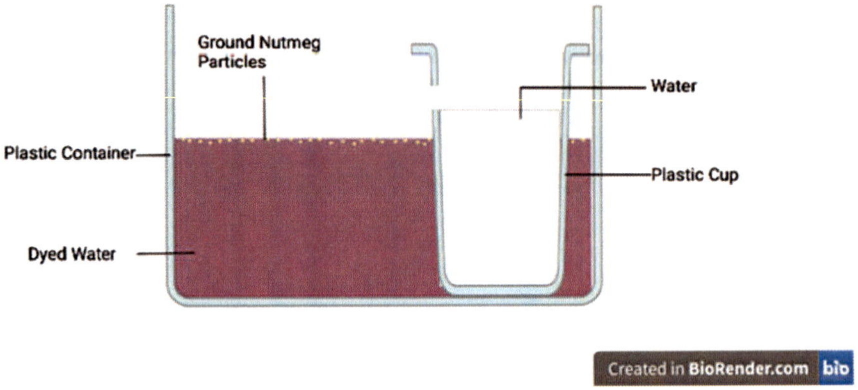

Fig. 1 Diagram of basic experimental setup

2.2 *Experimentation*

We propose a series of experiments to confirm that the Marangoni effect contributes significantly to the upstream contamination.

Consider the experimental setup given in Fig. 1, consisting of a small plastic cup nested inside of a larger container. A 0.5 cm × 0.5 cm square hole is cut into the lip of the cup, where water would flow out from. The larger container would be filled with water to a level about 2 cm below the opening of the plastic cup, and the cup would be filled with water up to the opening. Colored dye is poured into the lower reservoir as well as a sufficient amount of ground nutmeg.

We then pour water onto the plastic cup, causing the cup to overfill and water to flow to the lower reservoir. This flow is maintained until the water level of the lower reservoir is about 0.1 cm below the opening. We then check the cup for nutmeg particles and dye.

We repeat the experiment with different amounts of soap solution (50, 100 mL) added to the lower reservoir, which would reduce the surface tension of the liquid in the lower reservoir. We hypothesize that this would increase the extent of upstream contamination due to the increased Marangoni acceleration. In a separate trial, we also add 50 mL of soap solution onto the upper reservoir, and we hypothesize that the upstream contamination will be greatly inhibited, if not completely absent (Table 1).

Table 1 Summary of the experimental setup conditions

Trial No.	Amount of soap solution in lower reservoir/mL	Amount of soap solution in upper reservoir/mL
1	0	0
2	50	0
3	100	0
4	0	50

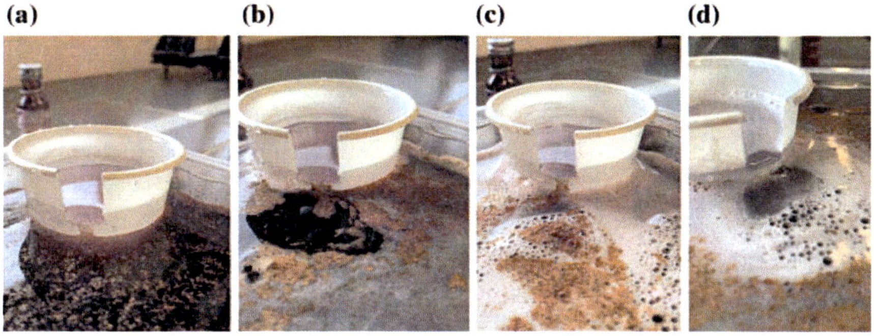

Fig. 2 **a**, **b**, **c**, **d** (left to right) End result of trials 1, 2, 3, and 4. Refer to the Appendix for the links to the videos to demonstrate the upstream contamination

3 Results

Checking the plastic cup after experimentation shows that there is no dye present in the upper reservoir. The videos of the experimentation can be found in the Appendix (Fig. 2).

Motion of the nutmeg surfactants was measured using the Tracker open-source software. Due to constraints in the auto-tracking feature, a clump of nutmeg particles was selected as the particle in question. Evidence that the extent of upstream contamination is measured by the movement of the particles in the y-direction (Fig. 3).

4 Discussion

Our experimental results confirm the significance of surface tension gradients in upstream contamination and therefore the significance of the Marangoni effect. The soap solution behaves as expected in relation to reducing the surface tension of the liquid it is poured into. Decreasing the surface tension at the lower reservoir has an apparent increase in upstream contamination, while decreasing the surface tension at the upper reservoir inhibits it.

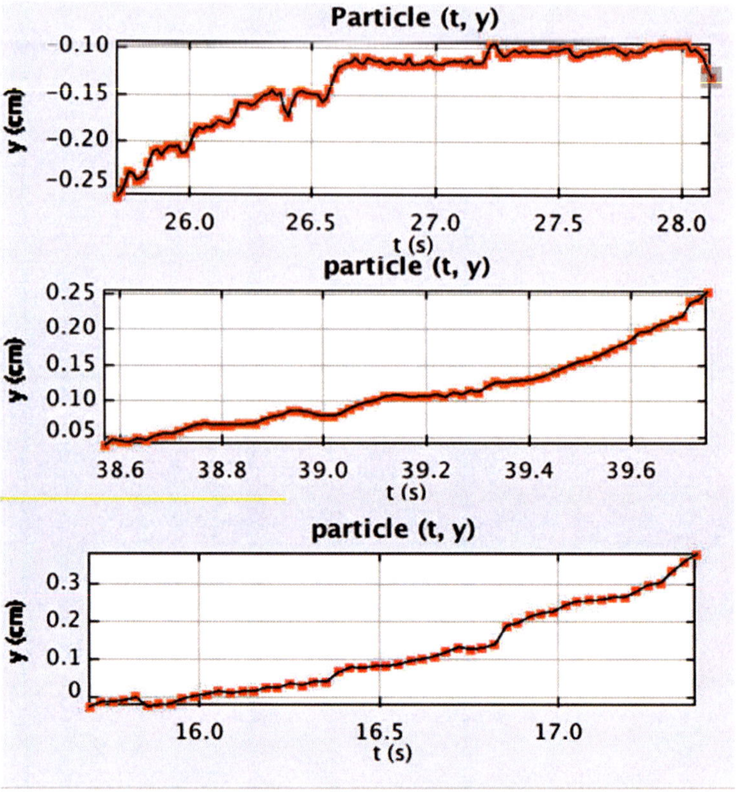

Fig. 3 **a**, **b**, **c** Tracker analysis of surfactant motion for trials 1, 2, and 3, respectively

It is worth noting that as more soap solution is added, the mixture becomes less homogenous, as seen in Fig. 2c, d especially. The apparent viscosity of the liquid increases, and it reduces the freedom of movement of the surfactant and inhibits the upstream contamination to a limited extent. Furthermore, the ground nutmeg particles are more likely to clump together due to the higher localized surface tension magnitude present. Attempts to loosen the clumps were made, though it frequently resulted in the nutmeg particles sinking due to the buoyant force being overcome. Thus future experimentation mandates a solution to the clumping and homogeneity problems present.

Our experimentation with chalk particles (most of which were only floating through the force of surface tension) showed no significant upstream contamination compared to the setups using buoyant particles, and hence, it is possible that a buoyancy effect could be significant in upstream contamination in the used setup. Further experimentation may be needed in the future to ascertain this.

5 Conclusion

We modeled the upstream contamination with regard to Marangoni flow and buoyancy effects and tested our model empirically.

We confirm that the Marangoni effect is significant in upstream contamination, but buoyancy effects are potentially a more significant factor in upstream contamination, as the trial with chalk particles shows. Further research may be required to empirically determine the relative effectiveness of both the Marangoni effect and buoyancy effects in causing or inhibiting upstream contamination of surfactants.

Appendix

Graphic Demonstration of the Marangoni Effect

See the Fig. 4.

Links to the Relevant Video Runs of Trials 1 Through 4

1. https://youtu.be/ug9Ci_bYa98
2. https://youtu.be/urG4oeyY2zk
3. https://youtu.be/bdVioVnZhWQ

(a) (b)

Fig. 4 **a, b,** A container of water with ground nutmeg particles before and after soap solution is added

4. https://youtu.be/cJy05iNgrKo

References

1. Bianchini, S., Lage-Castellanos, A., & Altshuler, E. (2011). *Upstream contamination in water pouring*. arXiv preprint: arXiv:1105.2585.
2. Bianchini, S., Lage, A., Siu, T., Shinbrot, T., & Altshuler, E. (2013). Upstream contamination by floating particles. *Proceedings of the royal society a: Mathematical, physical and engineering sciences, 469*(2157), 20130067.
3. Tadmor, R. (2009). Marangoni flow revisited. *Journal of colloid and interface science, 332*(2), 451–454.
4. Suárez-Fontanella, D., & Cabo-Montes de Oca, A. (2019). About the upstream contamination. *Revista Cubana de Física, 36*(1), 8–14.

Effectiveness of Videos and Infographics in Education of Lower Back Pain

Hoo Rae En and Kerine Sim Yee Teng

Abstract In Singapore, lower back pain is a prevalent condition affecting many. Concerns on lower back pain are addressed through public education, with educational materials—infographics and videos—playing a large role. This study aimed to investigate the effectiveness of an educational video and infographics in increasing the public's knowledge of lower back pain in Singapore. Research participants included: those who have no lower back pain experience and those who have, but yet to seek professional intervention. Fifty two participants from the general public participated in this study. A randomised survey was used to assess participants' knowledge of lower back pain and their engagement level when viewing the educational materials. It consisted of a pre-assessment (before educational intervention) and a post-assessment (after educational intervention). The grading system was measured as the percentage increase in correct answers from pre- to post-assessment, with six questions in each assessment. To calculate the significance of the percentage increase, the Student's t-test was employed to obtain the p value. Based on the results, both videos and infographics significantly increased the public's knowledge, by 25.64% and 26.28%, respectively, showing that videos and infographics are equally effective in enhancing the knowledge of the public concerning lower back pain. Additionally, both videos and infographics increased participants' confidence level of their answers. It was also found that participants' gain-in-knowledge was not affected by their educational levels.

Keywords Videos · Infographics · Educational intervention · Lower back pain · Survey

H. R. En (✉) · K. S. Y. Teng
Hwa Chong Institution, Singapore, Singapore
e-mail: hooraeen888@gmail.com

H. Guo et al. (eds.), *IRC-SET 2022*,
https://doi.org/10.1007/978-981-19-7222-5_25

1 Introduction

1.1 Background

Public education on medical-related issues has been an indispensable aspect of education in Singapore, where information is conveyed to the general public to provide them with a better understanding of various topics. Some sources of public education include websites, videos, infographics, applications (Health Buddy App [1] and SingHealth [2]) and social media platforms.

One of the frequently encountered medical conditions in Singapore is lower back pain. It is estimated that 60–90% of adults in Singapore will suffer from lower back pain at some point in their life [3]. Videos and infographics are the two more effective forms of education resources used in Singapore. A video is a system of moving graphics and audio [4] used as a medium to convey information vividly and engagingly. Infographics is a concise form of visually communicating information. Evidence suggests that the presenting information visually enhances one's understanding and ability to make decisions [5].

1.2 Videos

Many studies compared the use of videos with standard education techniques like presentations, or with control groups (absence of video), to educate both patients and the general public on medical conditions. A past study accounted for the literacy of participants as it would greatly impact the understanding and hence the gain-in-knowledge from the educational materials. Effectiveness of videos and standard education techniques was mainly compared in the aspect of gain-in-knowledge [6]. A common method seen across various studies to quantify the effectiveness of the respective educational materials was through assessments, where participants had to answer surveys before and after educational intervention [7].

From the studies reviewed, videos showed a significant increase in the gain-in-knowledge as compared to standard education techniques. One study showed a significant positive effect of 61% ($p \leq 0.002$) for the video group as compared to the control group [8], while another study also had a similar result [9].

1.3 Infographics

Infographics is useful in making information more accessible to the general public [10]. Participants are more engaged when they can process information visually. According to Garcia-Retamero et al., infographics facilitate risk communication, increase information recall, and lengthen attention span. Infographics can simplify complex concepts by providing key information in a manner that is enjoyable and straightforward [11, 12]. A study by Bradshaw [13] found that effective infographics emphasised on colour, font and consistency of shapes. Firstly, colours in infographics play a vital role in capturing the attention of the readers and also livening mood and energy levels [10, 14]. In a study by Stonbraker [14], majority of participants preferred bright colours. Another study by McCrorie et al. [10] reported that pale colours worked best as a background whilst bold colours were suited to stress important information. Next, having straightforward icons also contributes to a higher engagement level of the readers. Having visual aids can reduce comprehension differences between those with high and limited health literacy by improving their understanding [14]. Well-designed infographics should contain a breadth and depth of information presented in concise, visually attractive formats.

McCrorie et al. [10] also identified charticles as excellent alternatives to traditional articles when viewers do not have the time or patience to process large quantities of text. Charticles are a type of infographics in which the main point is purposefully positioned in the centre to attract the readers' attention while the information around it provides extra details for those interested.

1.4 Objective of Study

Studies have shown both videos and infographics to be effective in educating people on various topics. Therefore, this study aims to investigate the effectiveness of videos and infographics, related to lower back pain, on the gain-in-knowledge of the general public. Research indicated that the most effective format of videos and infographics are animated videos and charticles, respectively [8, 10]. Hence, animated videos and charticles will be compared to determine their effectiveness in gain-in-knowledge of the general public. In this study, non-patients refer to the general public who have no lower back pain experience and those who have but did not receive professional treatment.

2 Methodology

2.1 Methodological Approach

A survey was used to determine whether infographics or videos are more effective in patient education for lower back pain. It consisted of a list of questions which included multiple-choice, rating scale and short answer questions. Many studies involving educational interventions utilised surveys as the standard form of assessment of patients for the collection of relevant data [15, 16].

Based on a simulation study conducted, a recommended sample size for a randomised clinical study was between 60 and 100 participants [17]. Hence, a target to survey 65 participants from the public was set. The inclusion criterion is the general public, who are non-patients of lower back pain and at least aged 18. All participants who satisfied this inclusion criterion were included in the study [16]. The survey was disseminated through WhatsApp and Facebook. The participants were randomised into two groups, one group for infographics and one for video using a simple randomisation approach similar to Shen [18] and Robert [19].

To maximise learning, two relevant aspects were considered: cognitive load and student engagement [20]. The survey required participants to

1. Note their basic background information, which included their age group, educational background and their experience with lower back pain (yes/no), and if yes, whether they have received professional treatment.
2. Note their basic knowledge on different aspects of lower back pain via a pre-assessment.
3. Choose from two unknown choices with Path # leading to the video and Path * leading to the infographics. Since participants were unaware of which educational material the paths will lead to, selection bias was unlikely to occur and randomisation of the groups was ensured.
4. Note their gain-in-knowledge immediately after viewing the educational material via a post-assessment. The pre- and post-assessment consisted of the same six questions for participants to answer [7].
5. A section to gather participants' feedback on the level of engagement of the educational material they had viewed. The duration for the data collection was 3 days.

In view of the current COVID-19 situation, the survey was conducted online. Thus, the video, infographics and survey were assessed via an online Google form.

2.2 *Parameters Tested for Effectiveness*

Effectiveness was measured in terms of the increment of gain-in-knowledge. A study defined effectiveness to be a significant increase in the percentage of correct answers in the assessments conducted [7]. In determining the effectiveness of videos as compared to infographics, participants' age group: 18–24, 25–54, 55–64, 65 years and above, education background and their experience with lower back pain condition were considered.

2.3 Methods of Analysis

To measure the extent of knowledge gained, the results of the pre- and post-assessment for the respective educational materials were compared (i.e. comparison of pre- and post-assessment of the video group; comparison of pre- and post-assessment of the infographic group). The percentage changes in the number of correct answers from pre- to post- assessment in both groups were calculated.

Using the calculated values of the percentage change in the number of correct answers, the p value can be derived to determine its significance.

3 Results and Discussion

In both post-assessments, the % increase of correct answers significantly increased (Fig. 1).

3.1 Effectiveness of Infographics and Videos

For the video group, the percentage increase in correct answers (Δ %) was 25.64% (p value = 5.74 × 10–5, $***p < 0.001$). The gain-in-knowledge through the utilisation of videos is highly significant, which is most visible from Q1b with an increased score of 53.85% (Table 1). For the infographic group, the Δ % from pre- to post-assessment was 26.28% (p value = 2.63 × 10–6, $***p < 0.001$, Table 1). The gain-in-knowledge through the utilisation of infographics is also significant, which is seen in Q2 with an

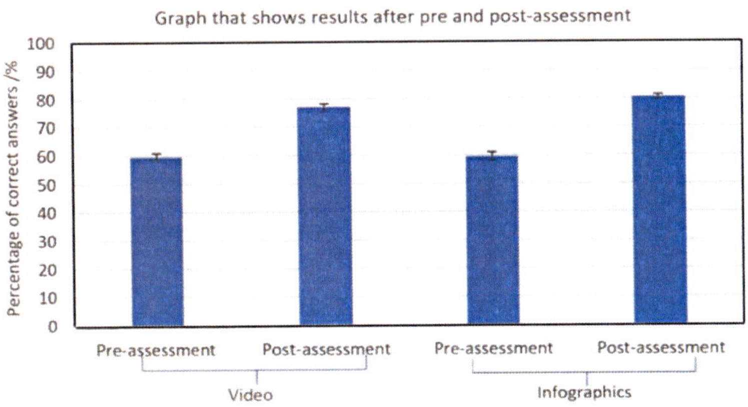

Fig. 1 Graph showing % of correct answers both pre- and post-assessment for both videos and infographics

increase of 50.00% (Table 1). Overall, both infographics and videos were found to be effective educational materials, as increments of 53.84 and 34.62% were observed for the video and infographics group, respectively, for Q1a.

Infographics accommodate participants' learning pace, providing them with sufficient time to process and understand the content provided, significantly increasing their knowledge on lower back pain. Similarly, videos can be paused and replayed infinitely, allowing participants to adjust the pace of information dissemination to one that is suitable for them. This will promote the understanding of the concepts in

Table 1 Percentage difference in correct answers for video and infographics

Questions	Videos			Infographics			*p* value
	Pre (%)	Post (%)	Δ (%)	Pre (%)	Post (%)	Δ (%)	
Q1a. When should you go to the emergency department due to lower back pain?	38.46	84.62	46.16	61.54	80.77	19.23	0.61
Q1b. When should you go to the emergency department due to lower back pain?	38.46	92.31	53.85	46.15	80.77	34.62	0.40
Q2. What kind of food should you avoid eating when experiencing lower back pain?	34.62	53.85	19.23	42.31	92.31	50.00	0.37
Q3. Should you adjust the medication dosage by yourself if the pain worsens?	92.31	96.15	3.84	73.08	92.31	19.23	0.38
Q4. Which activity should you avoid doing when you are experiencing lower back pain?	65.38	73.08	7.70	65.38	88.46	23.08	0.50
Q5. Which of the following should you avoid doing when experiencing lower back pain?	34.62	57.69	23.07	42.31	53.85	11.54	0.80
Average	50.64	76.22	25.64***	55.13	81.41	26.28***	0.95

Exposure to videos and infographics resulted in 25.64% (mean ± SD, $n = 26$) and 26.28% (mean ± SD, $n = 26$) increase in correct answers, respectively. Analysed by Student's *t*-test, asterisks indicate significant difference (*$p < 0.05$,**$p < 0.01$,***$p < 0.001$)

Δ % values represent the percentage difference in correct answers before and after educational intervention the video, resulting in a gain-in-knowledge of participants, which showed that having the autonomy to pause, ask questions and replay key segments of videos makes it more meaningful [21].

For individual questions, videos as educational materials resulted in a larger increase in Δ % in three questions (Q1a, Q1b, Q5) from pre- to post-assessment, while infographics as educational materials resulted in a larger increase in Δ % from pre- to post-assessment for the other three questions (Q2, Q3, Q4, Table 1). However, the difference in Δ % between videos and infographics for every question (Q1-5) was insignificant ($p > 0.05$, Table 1), aligning with the above deduction that both videos and infographics are equally effective in the gain-in-knowledge of participants. It is also possible to consider using infographics and videos together for education purposes. This will likely allow for optimal gain-in-knowledge.

3.2 *Engagement Level*

When designing the infographics, four aspects were incorporated: vivid colours, easily understandable icons, layout of points and more pictures than words. Most participants indicated that 'easily understandable icons'—80.77% and 'layout of points'—50.00% in an infographic were crucial to capture their attention. From feedback gathered on infographics, the two largest areas that could be improved were 'written in other languages'—69.23% and 'less wordy'—30.77%. Since minimal negative feedback was obtained, the education materials catered well to participants' learning needs, corresponding to similar studies [10] and [14] and [13].

As for the videos, four features were considered: vivid colours, animated characters and scenes, voiceovers, more animations than words. With 'animated character and scenes' at 84.62% and 'more animations than words' at 42.31%, this is in line with the study done by Dahodwala [8] which states that animated videos are the more advantageous forms of videos. However, 73.08% of participants opted for 'subtitles can be added', a significant area for future considerations when crafting an educational video. Another area of future consideration is the length of the video. Although Brame [20] reviewed that brief videos of length below 6 min were ideal for educational materials, participants indicated in the Likert scale that the duration of the 3.44-min educational video was 'long'. This suggests that perhaps educational videos may need more focus and attention to understand than typical videos.

3.3 *Confidence Level of Participants*

Participants' confidence in their answers was measured using a Likert scale with 1 being least confident and 5 being most confident, which was indicated by participants before and after the educational intervention. It was found that the confidence level rose from 3.17 arbitrary units (a.u.) to 4.12 a.u. for pre- and post-assessment, respectively. This suggests that the educational resources were effective in clearing their misconceptions, strengthening the participants' understanding of lower back pain, leading them to be more certain of their answers.

3.4 Educational Level of Participants Affects Gain-In-Knowledge

Participants' educational backgrounds ranged from primary, secondary, post-secondary and tertiary. Education levels of the participants were analysed to determine if it would affect the receptiveness of the educational resources. Although participants with tertiary education were expected to achieve higher increment of gain-in-knowledge after educational intervention, their gain-in-knowledge (19.76%) was not significantly different from those with post-secondary education (18.50%). Thus, the education level of the participants does not affect the increment of gain-in-knowledge. This might be due to the prevalence of lower back pain in Singapore [3]; hence, regardless of education level, the majority might already have a basic understanding of it.

4 Conclusion and Future Studies

Overall, significant increases in percentage of correct answers in videos (25.64%) and in infographics (26.28%) were observed. Thus, both are just as effective for the gain-in-knowledge of lower back pain for the general public, with no significant difference between them. Some limitations are present in this study. Foremost, the survey was conducted only in English, and thus is not easily understood by non-English speakers. Next, there was restricted ability to reach out to those under 18 years old due to the need for a Parental Consent Form. Although the study has not covered all age groups, it can be used as a reference for other studies with similar objectives. Thirdly, due to restrictions and recommendations in place due to COVID-19, the study was conducted via an online survey form. Thus, majority of those aged 65 and above could not participate in the study as they are technologically illiterate. For future studies, the rate of retention of educational material can be studied by adding another follow-up assessment four weeks after the post-assessment. To have a more accurate sampling of the population, the survey can also be extended to those below the age of 18 and above 65.

References

1. SingHealth. *Health Buddy—SingHealth Mobile App*. https://www.singhealth.com.sg/patient-care/patient-visitor-info/health-buddy-app. Last Accessed 28 April 2022.
2. *SingHealth Polyclinics*. Post-vaccination advice. https://polyclinic.singhealth.com.sg/vaccinations-4. Last Accessed 28 April 2022.
3. Kamil, F. M., Chng, L. S., Gan, S. L., & Lim S. L., *Overview of back pain cases in Singapore from 2013–2015*. https://www.wshi.gov.sg/-/media/wshi/posters/posterfile/overview-of-back-pain-cases-in-singapore-from-2013-2015.pdf. Last Accessed 28 April 2022.
4. Oxford Advanced Learner's Dictionary. Definition of video. https://www.oxfordlearnersdictionaries.com/definition/english/video_1. Last Accessed 28 April 2022.
5. U.S. National Library of Medicine. *Home—PMC—NCBI*. National Center for Biotechnology Information. https://www.ncbi.nlm.nih.gov/pmc/articles. Last Accessed 28 April 2022.
6. Gagliano, M. D., & Martha, E. (2000, December 18). A literature review on the efficacy of video in patient education. *Journal of Medical Education, 63*(10), 785–792.
7. Steffenino, G., Viada, E., Marengo, B., Canale, R. (2007). The nursing and the medical staff of the cardiac catheterization unit. Effectiveness of video-based patient information before percutaneous cardiac interventions. *Journal of Cardiovascular Medicine. LWW*. https://journals.lww.com/jcardiovascularmedicine/Abstract/2007/05000/Effectiveness_of_video_based_patient_information.8.aspx. Last Accessed 28 April 2022.
8. Dahodwala M., Geransar, R., Babio, J., Grood, J. D., & Sargious, P. (2018) The impact of the use of video-based educational interventions on patient outcomes in hospital settings: A scoping review. Patient Education and Counselling, https://pubmed.ncbi.nlm.nih.gov/30087021/. Last Accessed 28 April 2022.
9. Cohen, D., The effectiveness of videotape in patient education on depression. *The Journal of Biocommunication*, https://europepmc.org/article/med/6853467. Last Accessed 28 April 2022.
10. McCrorie, A., Donnelly, C., & McGlade, K. (2016). Infographiscs: Healthcare communication for the digital age. *The Ulster medical journal, 85*(2), 71–75.
11. Siricharoen, W. V., & Siricharoen, N., *How Infographic should be evaluated? ResearchGate*, http://net/publication/276595441_How_Infographic_should_be_evaluated. Last Accessed 28 April 2022.
12. Najwah, A., *Effectiveness of using infographics as an approach for teaching programming fundamentals on developing analytical thinking skills for high school students in the city of Makkah in Saudi Arabia*. https://www.macrothink.org/journal/index.php/gjes/article/view/10854/8695. Last Accessed 28 April 2022.
13. Bradshaw, M. J., & Sallie, P. (2017). Infographics a new tool for the nursing classroom. Nurse educator. *Journal, 42*(2), 57–59.
14. Stonbraker, S., Halpern, M., Bakken, S., & Schnall, R., Developing infographics to facilitate HIV-related patient-provider communication in a limited-resource setting. *Applied Clinical Informatics*. https://www.ncbi.nlm.nih.gov/pmc/articles/PMC6693997/. Last Accessed 28 April 2022.
15. Wong, D., *Effectiveness of learning through video clips and video learning improvements between business related postgraduate and undergraduate students*. https://www.researchgate.net/publication/347651896_EFFECTIVENESS_OF_LEARNING_THROUGH_VIDEO_CLIPS_AND_VIDEO_LEARNING_IMPROVEMENTS_BETWEEN_BUSINESS_RELATED_POSTGRADUATE_AND_UNDERGRADUATE_STUDENTS. Last Accessed 28 April 2022.
16. Herwati, F., Yulia, R., Arifin, B., Frasetyo, I., Setiasih, Woerdenbag, H. J., Avanti, C., & Andrajati, R. (2021). *Educational video improves knowledge about outpatients' usage of antibiotics in two public hospitals in Indonesia*, https://www.mdpi.com/2079-6382/10/5/606/pdf. Last Accessed 28 April 2022.
17. Teare, M. D., Dimairo, M., Shephard, N., Hayman, A., Whitehead, A., & Walters, S. J., Sample size requirements to estimate key design parameters from external pilot randomised controlled

trials: A simulation study. *BMC*. https://trialsjournal.biomedcentral.com/articles/https://doi.org/10.1186/1745-6215-15-264. Last Accessed 28 April 2022.
18. Shen, D., & Lu, Z., *Randomization in clinical trial studies*, https://www.lexjansen.com/pharmasug/2006/Posters/PO06.pdf. Last Accessed 28 April 2022.
19. Robert, D. (2005). Randomisation in clinical trials. *Australian Journal of Physiotherapy, 51*.
20. Brame, C. J., *Effective educational videos: Principles and guidelines for maximizing student learning from video content*. CBE life sciences education, https://www.ncbi.nlm.nih.gov/pmc/articles/PMC5132380/. Last Accessed 28 April 2022.
21. Buchholz, B., Pyles, D., Hash, P., & Hagaman, K. (2021) Teaching comprehension in the digital age. *Journal, 58*(5).

Analysis of Chemical Transport by IPSC-Derived Proximal Tubular Cell (PTC)-Like Cells and Other Emerging Human Kidney Cell Models for Drug Toxicity Screening

Treruangrachada Anantaya Kylin, Riga Tawo, and Daniele Zink

Abstract The kidney is a main target for drug-and chemical-induced toxicity, and the renal proximal tubule is frequently affected. Many widely-prescribed drugs are nephrotoxic. Reliable renal proximal tubular cell (PTC) models are needed for developing improved in vitro assays for nephrotoxicity prediction. PTC in the human kidney express various transporters for drugs and chemicals, which are often down-regulated in vitro. This renders the cells inactive to nephrotoxicants, which is a main problem in nephrotoxicity testing. Here, I addressed organic anion uptake by three different PTC models: conditionally immortalised proximal tubular epithelial cells (ciPTEC) stably transfected with OAT1 and OAT3 expression constructs (ciPTEC-OAT1 and ciPTEC-OAT3) and human-induced pluripotent stem cell (iPSC)-derived PTC-like cells. In addition, the activity of the renal organic anion transporters OAT1 and OAT3 was assessed. These transporters are important for the cellular uptake of a wide range of nephrotoxicants. The results showed that the cell culture medium has a major impact on organic anion uptake. A proprietary novel medium for PTC-like cells was formulated (Renal Tox Medium) that sustained high organic anion uptake in vitro. Furthermore, the results showed that OAT1 and/or OAT3 activity in PTC-like cells was superior compared to ciPTEC-OAT1 and ciPTEC-OAT3 cells, which are currently the PTC models favoured by the pharmaceutical industry. The results also showed that ciPTEC cell lines consisted of inhomogeneous cell populations. Together, the results suggest that PTC-like cells, in combination with the here developed proprietary Renal Tox Medium, could provide a more suitable solution for industry and academia.

Keywords Induced pluripotent stem cell · In-vitro toxicology · Proximal tubular cell · Organic anion transporters

T. A. Kylin (✉)
Raffles Institution, Singapore, Singapore
e-mail: Treru.ak04@gmail.com

R. Tawo · D. Zink
Singapore Institute of Food and Biotechnology Innovation, A*STAR, Singapore, Singapore

H. Guo et al. (eds.), *IRC-SET 2022*,
https://doi.org/10.1007/978-981-19-7222-5_26

1 Introduction

The kidney plays a key role in eliminating drugs and other chemicals from the body and is a main target for chemical-induced toxicity. A major site of damage is the renal proximal tubule, as all compounds are heavily concentrated here due to water reabsorption [1], and the proximal tubular cells (PTC) are very active in the transport, reabsorption and elimination of drugs and other chemicals from the body [2, 3]. Widely-prescribed drugs like antibiotics and immunosuppressants are often nephrotoxic [3, 4] and are the root cause of ~20% of hospital and community acquired cases of acute kidney injury [5–7]. Hence, nephrotoxicity assays with improved predictivity would be required for drug development and other applications, and reliable PTC models are needed for such in vitro assays [3, 8–11]. It is crucial for the respective PTC models to express renal transporters for drugs and other chemicals. Such transporters are often down-regulated in vitro [3, 12–14]. This impairs the uptake of drugs and chemicals and renders the cells insensitive to potential nephrotoxicants [14–16].

At the basolateral membrane of the PTC in the kidney, organic anion transporters OAT1 (solute carrier family 22 member 6; SLC22A6), OAT3 (SLC22A8) and organic cation transporter OCT2 (SLC22A2) are key transporters which mediate the uptake of anionic and cationic drugs and chemicals from the blood [17–20]. In the case of human conditionally immortalized proximal tubular epithelial cells (ciPTEC) [21], the problem of transporter downregulation in vitro was addressed by stably transfecting the cells with OAT1 and OAT3 expression constructs. The resulting cell lines are ciPTEC-OAT1 and ciPTEC-OAT3 [22]. These PTC models are currently preferred by pharmaceutical companies such as AstraZeneca and GlaxoSmithKline due to their presumably high OAT1 and 3 activity [10, 23]. Stem cell-based approaches probably also offer a solution to the problems with current renal cell models. Recently, Kandasamy et al. [12] published a simple 1-step protocol for the differentiation of human-induced pluripotent stem cells (iPSC) into PTC-like cells. These cells resemble human primary renal PTC, but larger cell numbers can be more easily obtained [12]. My mentors observed that PTC-like cells had issues with OAT1 and 3 activity (unpublished results), which was probably due to the cell culture media used.

The main goal of this study was to compare the OAT1 and OAT3 activities of three different cell types: ciPTEC-OAT1, ciPTEC-OAT3, and iPSC-derived PTC-like cells. Another goal was to identify cell culture media which sustained high OAT1 and 3 activities in vitro.

2 Materials and Methods

2.1 *Cell Culture*

Undifferentiated is070a iPSCs were obtained from Dr. Adrian Teo, IMCB, A*STAR and were differentiated into PTC-like cells. ciPTEC-OAT1 and ciPTEC-OAT3 cell lines (Accellerate, Hamburg, Germany) were cultivated according to the manufacturer's protocol in Assay Medium (AM) provided by the manufacturer. Details about cell cultivation and differentiation are outlined in Sect. 5.

2.2 *Determination of 6-carboxyfluorescein (6-CF) Uptake in Various Media*

Cellular uptake of the fluorescent organic anion 6-CF was studied in complete medium, assay medium and Renal Tox Medium. Technical details are provided in the supplementary material. A 6-CF is a high-affinity OAT1 and 3 substrate [24] that was chosen as a fluorescent tracer dye to visualize organic anion uptake by the cells. Complete medium refers to renal epithelial cell basal medium supplemented with renal epithelial growth kit (American Type Culture Collection, Manassas, VA, USA). Assay medium was provided by the ciPTEC manufacturer. Renal Tox Medium was formulated by my mentor and myself (composition not disclosed due to confidentiality issues).

2.3 *Transporter Assay/Determination of OAT1- and OAT3 Activities*

Fluorescent substrate (6-CF) and inhibitor (MK571)-based transporter assays were conducted to determine OAT1 and 3 transporter activity. These assays are explained in greater detail in Sect. 3 and technical details can be found in Sect. 5.

3 Results

3.1 *Formulation of Renal Tox Medium*

During the first half of my attachment, we worked on formulating a cell culture medium that sustained high 6-CF uptake by iPSC-derived PTC-like cells. We tested

40 different formulations. Highest 6-CF uptake was observed in what is here referred to as the Renal Tox Medium (composition not disclosed due to confidentiality issues).

3.2 Determination of 6-CF Uptake by PTC-Like Cells and CiPTEC-OAT1 and -OAT3 Cells

Next, I compared 6-CF uptake by PTC-like cells and ciPTEC-OAT1 and ciPTEC-OAT3 cell lines in different cell culture media. A 6-CF uptake was measured by flow cytometry, and the results are shown in Fig. 1.

The results showed that ciPTEC-OAT1 and -OAT3 cells have up to ~6- to ~eight-fold higher 6-CF uptake compared to PTC-like cells. For ciPTEC-OAT1 and -OAT3, highest 6-CF uptake was observed in Assay Medium that was recommended by the vendor. On the other hand, 6-CF uptake by PTC-like cells was highest in Renal Tox Medium and lowest in Assay Medium. This result suggests that medium conditions need to be carefully selected for each PTC model.

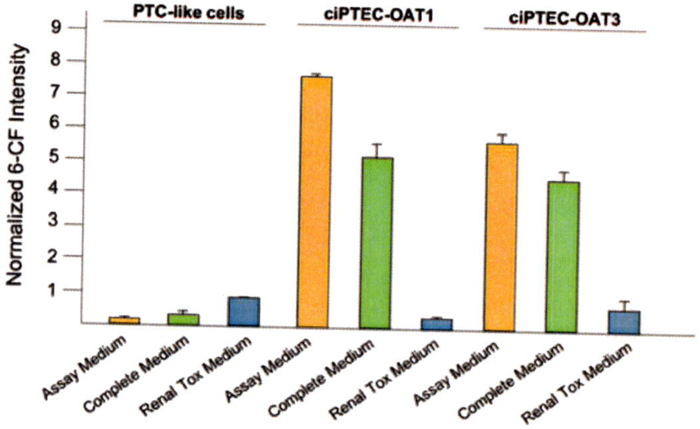

Fig. 1 For each condition (as indicated on the x-axis) the intracellular mean fluorescence intensity (MFI) of 10,000 cells was determined by flow cytometry (two technical replicates per experiment). All bars represent the mean MFIs (± standard deviation) of three independent experiments using three independently differentiated batches in the case of PTC-like cells. All data were normalized to 6-CF uptake in PTC-like cells in Renal Tox Medium, which was set to 1. The units on the *y*-axis indicate the relative normalized 6-CF-specific fluorescence intensity (1 = 6-CF intensity of PTC-like cells in Renal Tox Medium). The colour of each bar in the diagram indicates the medium in which the cells were exposed to 6-CF (labelled on *x*-axis)

3.3 *Functional Activity of OAT1 and OAT3 in Different PTC Models*

Next, the OAT1 and OAT3 transporter functionality of each cell model was evaluated using fluorescent substrate- and inhibitor-based transporter assays. These assays work by incubating cells with a transporter-specific fluorescent substrate. Cell samples are either incubated with only the fluorescent substrate, or in addition with an inhibitor for the transporter of interest. If the uptake transporter is active, cells with inhibited transporters would exhibit lower uptake of the fluorescent substrate and therefore display lower fluorescence intensity compared to the non-inhibited cells. This would be represented by a clear leftwards shift of the flow-sorting histogram (*x*-axis indicates fluorescence intensity) from the uninhibited sample (black histograms in Fig. 2) to the inhibited sample (orange histograms in Fig. 2). Here, we used the OAT1- and 3-specific substrate 6-CF and OAT1- and 3-specific inhibitor MK571.

We performed this assay with PTC-like cells, ciPTEC-OAT1 and ciPTEC-OAT3 cells. Results are shown in Fig. 2. It should be noted that due to the broadly overlapping substrate specificity of OAT1 and OAT3, this type of assay cannot distinguish between the two transporters. Both transporters could be active in PTC-like cells and so it is not clear whether both transporters or only one of them had contributed to a potentially observed activity in this cell model. However, in ciPTEC-OAT1 and -OAT3 cells, only the transporter with which the cells have been transfected (as indicated by the name of the cell line) would be potentially active, and therefore it is clear in these cases which transporter would contribute to a potentially observed activity.

Typical results as usually observed with this type of assay were obtained with PTC-like cells (Fig. 2). Single-peak, symmetrical histograms suggested a homogenous cell

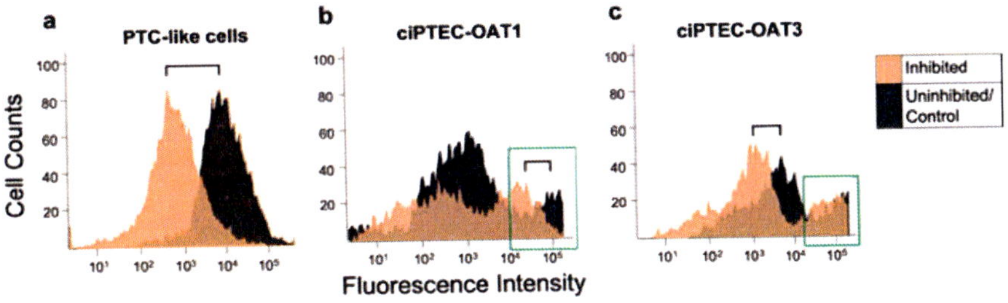

Fig. 2 Flow cytometry was performed with **a** PTC-like cells, **b** ciPTEC-OAT1 and **c** ciPTEC-OAT3 cells as indicated. Cells were incubated with the OAT1 and 3 inhibitor MK571 (orange histograms) or were left uninhibited (black histograms) before incubation with 6-CF for 40 min. 6-CF fluorescence intensity was measured by flow cytometry (scale on *x*-axis). Cell numbers with a specific level of 6-CF fluorescence intensity are indicated by the scale on the *y*-axis. The green frames highlight fractions of ciPTEC-OAT1 and ciPTEC-OAT3 cells with high 6-CF fluorescence intensity. The black brackets indicate the size of the shift in fluorescence intensity between inhibited and non-inhibited cells. For each technical replicate, 10,000 cells were analysed (two technical replicates per experiment and cell type). Three independent experiments were performed, and 3 independently differentiate batches of PTC-like cells were used (1 batch per experiment)

population. The majority of cells had a fluorescence intensity of ~102–103 (inhibited) or ~104 (uninhibited). There was a distinct shift in fluorescence intensity between the inhibited and uninhibited cells (black bracket in Fig. 2). The shift of fluorescence intensity of the entire cell population by more than an order of magnitude showed that almost all PTC-like cells had measurable and relatively high OAT1 and/or OAT3 activity (the larger the shift between the inhibited and non-inhibited samples, the higher the respective transporter activity).

Flow-sorting histograms of ciPTEC-OAT1 and ciPTEC-OAT3 cells exhibited two peaks. In both cell types, the first peak of lower fluorescence intensity was consistently higher than the second (indicating larger cell numbers in this fraction) and was slightly skewed to the right. The presence of two peaks showed that the cell population was inhomogeneous. The majority of cells had low fluorescence intensity, with only a small fraction of cells (indicated by the green frames in Fig. 2) exhibiting much higher uptake activity. The difference in fluorescence intensity between these two sub-populations was nearly 100-fold. As the majority of the ciPTEC cells had in fact low uptake activity, the high intracellular MFI of ciPTEC cells (Fig. 1; yellow bars; Assay Medium) was probably due to this small fraction of cells with very high uptake activity (green frames in Fig. 2). In the case of ciPTEC-OAT1, the main fraction of cells had ~tenfold lower 6-CF uptake than PTC-like cells (histogram of uninhibited ciPTEC-OAT1 peaked at ~103, compared to PTC-like cells where the peak was at ~104). When ciPTEC-OAT1 cells were inhibited, the histogram for the main fraction of cells (orange histogram) flattened out and shifted slightly leftwards, which suggested the presence of some OAT1 activity, albeit at relatively low levels. It is worthy to note that OAT activity here specifically refers to OAT1 activity as ciPTEC-OAT1 cells do not express OAT3 at all. For the high uptake fraction of ciPTEC-OAT1 cells (green box), there was a relatively small shift in fluorescence intensity between the inhibited and uninhibited cells (black bracket in Fig. 2). These results showed that OAT1 was mainly active in the high uptake fraction of ciPTEC-OAT1 cells (green frame in Fig. 2), but at relatively low levels (size of the shift relatively small).

The main fraction of ciPTEC-OAT3 cells showed ~ fivefold lower 6-CF uptake compared to PTC-like cells (Fig. 2, histogram of uninhibited ciPTEC-OAT3 cells showed a main peak between 10^3 and 10^4, whereas uninhibited PTC-like cells showed one peak at ~10^4). The main fraction of ciPTEC-OAT3 shifted toward lower fluorescence intensities upon inhibition (black bracket in Fig. 2), although the shift was smaller compared to PTC-like cells. This indicated the presence of OAT3 transporter functionality in the bulk of the ciPTEC-OAT3 cells (these cells do not express OAT1), although at relatively low levels. On the other hand, no clear shift was observed in case of the high uptake fraction (green box in Fig. 2). This means that 6-CF uptake by these cells was mainly OAT3-independent. Based on these flow cytometry data, we calculated the Multidrug Resistance Affinity Factor (MAF)-values of the respective cell types. MAFs are used to determine transporter activities quantitatively. Results are shown in Fig. 3.

PTC-like cells had the highest MAF-value of ~55 (Fig. 3). This is well above the threshold level of 25, at which transporters are considered as active. This matches

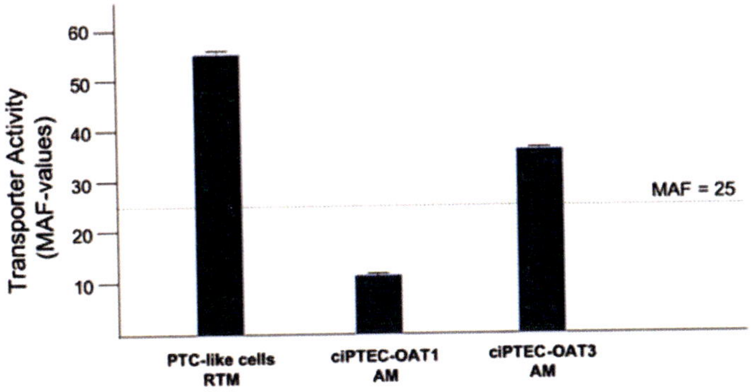

Fig. 3 Bars show the Multidrug Resistance Affinity Factor (MAF) values (scale on the *y*-axis) obtained with each cell type in the respective medium (indicated on *x*-axis; RTM—Renal Tox Medium, AM—Assay Medium). MAF-values were calculated using the formula: $MAF_{uptake} = 100 * ((MFI_{control} - MFI_{inhibitor})/MFI_{control})$, where MFI denotes the mean intracellular fluorescence intensity. MFIs were derived from the results displayed in Fig. 2. The error bars denote the standard deviation. Transporters are considered active if the MAF-value is > 25, as indicated by the dotted line

the data shown in Fig. 2, where PTC-like cells exhibited the largest distinct shift between inhibited and uninhibited cells. The results confirm the presence of high OAT1 and/or OAT3 activity in PTC-like cells in Renal Tox Medium.

The lowest MAF-value of ~10 was observed in case of ciPTEC-OAT1 cells (Fig. 3). Here, the value was well below the threshold of 25, which suggests no or only very low OAT1 activity (OAT3 is anyway known to be inactive in these cells). This is consistent with the flow-sorting histograms shown in Fig. 2, where the main fraction of cells exhibited nearly tenfold lower 6-CF uptake than PTC-like cells and no distinct shift was observed after inhibition for the main fraction of cells. Only the high fluorescence intensity fraction of cells displayed a small shift after inhibition, indicating some low level OAT1 activity in a minor fraction of cells.

ciPTEC-OAT3 cells had a MAF-value of ~35 (Fig. 3), which was above the threshold level of 25, but below the MAF value of PTC-like cells. These results are in agreement with the finding that the main fraction of ciPTEC-OAT3 cells had a ~fivefold lower 6-CF uptake compared to PTC-like cells (Fig. 2), and in case of ciPTEC-OAT3 cells the shift after inhibition was relatively small.

Together, the results showed that PTC-like cells displayed the highest OAT1 and/or OAT3 activity and that PTC-like cells represented a homogenous cell population, in contrast to ciPTEC-OAT1 and ciPTEC-OAT3 cells.

4 Conclusion

In the present study, my first mentor and I have formulated a novel cell culture medium for iPSC-derived PTC-like cells. It is planned to file a patent on this so-called Renal Tox Medium, which sustains high OAT1 and/or OAT3 activity in PTC-like cells. Comparison of organic anion uptake of ciPTEC-OAT1, ciPTEC-OAT3 and PTC-like cells in various cell culture media showed the need to select carefully the medium used with each PTC model. The activity of OAT1 and 3 was evaluated by using a fluorescent substrate- and inhibitor-based assay. Transporter activity was quantitated by calculating the MAF values, and PTC-like cells were found to exhibit highest levels of OAT1 and/or 3 transporter activity. ciPTEC-OAT3 cells showed some OAT3 activity above threshold levels (OAT1 is not active in these cells), whereas ciPTEC-OAT1 cells displayed no or only very low levels of OAT1 activity (also OAT3 is not active in these cells). The presence of active, functional drug transporters is particularly important for establishing a reliable in vitro PTC model for nephrotoxicity and drug-drug interaction studies, and OAT1 and 3 are the most important renal uptake transporters for organic anionic drugs like tenofovir or tetracycline [25, 26].

ciPTEC-OAT1 cells have been used in a study by AstraZeneca [10] to establish a novel model for nephrotoxicity prediction, and the pharmaceutical industry is currently most interested in ciPTEC-OAT1 and -OAT3 cells due to their presumably high expression of OAT1 or 3 [10, 23]. However, our results showed that ciPTEC-OAT1 and ciPTEC-OAT3 cells represent inhomogeneous cell populations and have surprisingly low OAT1 or 3 activity, especially in the case of ciPTEC-OAT1. Our flow cytometry results were in agreement with results from Nieskens et al. [22] (Fig. 2c in that publication).

The low OAT1 and 3 activity of respective ciPTEC lines was surprising in the light of the high 6-CF uptake (Fig. 1, yellow bars, Assay Medium). This can be explained by two different mechanisms, which are not mutually exclusive. Firstly, ciPTEC-OAT1 and -OAT3 are inhomogeneous cell populations, which include a minor fraction of cells with very high 6-CF uptake (green frames in Fig. 2). If the mean uptake of the entire cell population is determined, this high-uptake subpopulation would shift the mean values toward higher values. Secondly, 6-CF uptake does not necessarily reflect OAT 1 and 3 activity. PTC express various other organic anion transporters, which may have contributed to 6-CF uptake. These include OAT2 (SLC22A7), OAT4 (SLC22A11) and OATP4C1 (SLCO4C1) [27, 28]. The idea that other organic anion transporters also contributed to 6-CF uptake is supported by the finding that the flow-sorting histograms in Fig. 2 continued to display substantial fluorescence intensity in the cells inhibited with OAT1 and 3 inhibitor MK571. If 6-CF was only transported into the cells by OAT1 and 3 there would be little to no fluorescence in such cells. Also, the high uptake fraction of ciPTEC-OAT3 cells displays essentially no shift upon inhibition of OAT1 and 3 (Fig. 2, green frame). This suggests that high 6-CF uptake in this fraction of cells was mediated by other transporters. Nevertheless, although uptake of 6-CF might not only be mediated by

OAT1 and 3, the assays based on the OAT1- and 3-specific inhibitor MK571 still allowed to determine the activity of these two transporters.

Overall, the results revealed that the industry-preferred cell lines ciPTEC-OAT1 and -OAT3 showed surprisingly poor performance. Apart from the low OAT1 or OAT3 activity, the different subpopulations of cells is a problem in toxicity or drug-drug interaction studies. The behaviour of the entire cell population is typically determined in these studies, and the presence of various subpopulations comprises the interpretation of results. Findings obtained here also show that PTC-like cells, in combination with the here developed proprietary Renal Tox Medium, could provide a more suitable solution for academia and industry.

5 Supplementary Material

5.1 *Details of Cell Culture*

Undifferentiated is070a iPSCs (obtained from Dr. Adrian Teo, IMCB, A*STAR) were expanded in multi-well plates coated with growth factor-reduced Matrigel (Corning, Bedford, MA, USA). Cells were incubated at 37 °C with 5% CO_2. For differentiation into PTC-like cells, renal epithelium cell growth medium (REGM, Thermo Fisher Scientific, Singapore) supplemented with bone morphogenetic protein (BMP)2 and BMP7 (Merck/Sigma-Aldrich, Singapore) was added to the undifferentiated iPSCs as described by Kandasamy et al. [12]. This previously published protocol [12] was modified in the following way: cells were cryopreserved and stored in liquid nitrogen on day 6 after adding REGM supplemented with BMP2 and BMP7. Before usage cells were thawed and cultivated for 2 days in REGM without BMP2 and BMP7, followed by cultivation for another 2 days in REGM supplemented with BMP2 and BMP7. After this, 4-day cultivation period PTC-like cells were ready to use. Conditionally immortalized proximal tubular epithelial cells (ciPTEC)-OAT1 and ciPTEC-OAT3 cell lines (Accellerate, Hamburg, Germany) were cultivated according to the manufacturer's protocol in Assay Medium (AM) provided by the manufacturer. Cells were seeded at 62,500 cells/cm^2 and grown for 1 day in Assay Medium provided by the manufacturer at 37 °C in 5% CO_2 before the cells were applied in the assays performed here.

5.2 *Details of Determining of 6-CF Uptake*

The various PTC models tested here were incubated in cell culture medium containing 6-CF (25 μM, Merck, Singapore) for 40 min at 37 °C. Subsequently, cells were dislodged with accutase (Merck, Singapore), neutralized with 1 ml of warm phosphate-buffered saline (PBS) and transferred to flow cytometer tubes. The cell solution was centrifuged at 200 g for 1 min and the supernatant was discarded. Cells were resuspended in 500 μl ice-cold Hank's balanced buffer solution (HBSS, Gibco Life Technologies, Waltham, MA, USA) and kept on ice, until flow cytometry was conducted to analyze 6-CF uptake. In all experiments, flow cytometry was conducted using a BD FACSCelesta™ flow cytometer (BD Biosciences, Singapore).

5.3 Details of Determining of OAT1 and OAT3 Activities

Cells were first incubated with the OAT1- and 3-specific inhibitor MK571 [28] or were left untreated (non-inhibited). Cells were then incubated with the fluorescent anion 6-CF. Quantitative cellular uptake of 6-CF was determined by flow cytometry, where fluorescence-activated cell sorting would detect the 6-CF accumulated within the cells. The difference in 6-CF uptake between inhibited and non-inhibited samples shows the combined activity of OAT1 and three transporters. If the transporters were not active there would be no difference between inhibited and non-inhibited samples. Due to the broadly overlapping substrate specificity of OAT1 and OAT3 this type of assay cannot distinguish between OAT1 and OAT3 activity.

To perform the assay, cells were cultured in BD Falcon 24-well plates according to their respective protocols (see paragraph "Cell Culture"). Cells were either pre-incubated with inhibitor MK571 (50 μM, Selleck Chemicals, Houston, TX, USA) or were left untreated for 10 min at 37°C. The cell culture medium for the respective cell type containing 6-CF (25 μM, Merck, Singapore) was then added to each well, and cells were incubated for another 40 min at 37 °C. During the last 5 min of incubation, DAPI dye (4′,6-diamidino-2-phenylindole; Thermo Fisher Scientific, Singapore) was added in a 1:1250 dilution to allow for the exclusion of dead cells from the flow analysis. The reaction was stopped by washing cells with 500 μl PBS. Cells were dislodged with accutase, neutralized with 1 ml of warm PBS and transferred to flow cytometer tubes. The cell solution was centrifuged at 200 g for 1 min and the supernatant was discarded. Cells were resuspended in 500 μl ice-cold HBSS and kept on ice, until flow cytometry was conducted to analyse 6-CF uptake.

Acknowledgements I'd like to thank my two mentors, Dr. Riga Tawo (NBL-SIFBI, A*STAR) and Team Leader and Principal Research Scientist Dr. Daniele Zink (NBL-SIFBI, A*STAR) for their unwavering support, invaluable guidance and utmost dedication. I also express my deepest gratitude to A*STAR Senior Fellow Professor Jackie Y. Ying, and NBL Director and YRP Chair Ms Noreena AbuBakar for this incredible experience. They have all inspired me in immeasurable ways. This work was supported by the NanoBioLab (NBL), Biomedical Research Council, A*STAR and the Singapore Institute of Food and Biotechnology Innovation (SIFBI), Biomedical Research Council, A*STAR.

References

1. Curthoys, N. P., & Moe, O. W. (2014). Proximal tubule function and response to acidosis. *Clinical Journal of the American Society of Nephrology., 9*, 1627–1638. https://doi.org/10.2215/CJN.10391012
2. Morrissey, K. M., Stocker, S. L., Wittwer, M. B., Xu, L., & Giacomini, K. M. (2013). Renal transporters in drug development. *Annual Review of Pharmacology and Toxicology, 53*, 503–529. https://doi.org/10.1146/annurev-pharmtox-011112-140317
3. Tiong, H. Y., Huang, P., Xiong, S., Li, Y., Vathsala, A., & Zink, D. (2014). Drug-induced nephrotoxicity: Clinical impact and preclinical in vitro models. *Molecular Pharmaceutics, 11*, 1933–1948. https://doi.org/10.1021/mp400720w
4. Naughton, C. A. (2008). Drug-induced nephrotoxicity. *American Family Physician, 78*, 743–750. https://pubmed.ncbi.nlm.nih.gov/18819242/
5. Kaufman, J., Dhakal, M., Patel, B., & Hamburger, R. (1991). Community-acquired acute renal failure. *American Journal of Kidney Diseases, 17*(2), 191–198. https://doi.org/10.1016/S0272-6386(12)81128-0
6. Nash, K., Hafeez, A., & Hou, S. (2002). Hospital-acquired renal insufficiency. *American Journal of Kidney Diseases, 39*(5), 930–936. https://doi.org/10.1053/ajkd.2002.32766
7. Bellomo, R. (2006). The epidemiology of acute renal failure: 1975 versus 2005. *Current Opinion in Critical Care, 12*(6), 557–560. https://doi.org/10.1097/01.ccx.0000247443.86628.68
8. Guengerich, F. P. (2011). Mechanisms of drug toxicity and relevance to pharmaceutical development. *Drug Metabolism and Pharmacokinetics, 26*(1), 3–14. https://doi.org/10.2133/dmpk.dmpk-10-rv-062
9. Soo, J. Y., Jansen, J., Masereeuw, R., & Little, M. H. (2018). Advances in predictive in vitro models of drug-induced nephrotoxicity. *Nature Reviews Nephrology, 14*(6), 378–393. https://doi.org/10.1038/s41581-018-0003-9
10. Sjogren, A. K., Breitholtz, K., Ahlberg, E., et al. (2018). A novel multi-parametric high content screening assay in ciPTEC-OAT1 to predict drug-induced nephrotoxicity during drug discovery. *Archives of Toxicology, 92*(10), 3175–3190. https://doi.org/10.1007/s00204-018-2284-y
11. Li, S., Zhao, J., Huang, R., Steiner, T., Bourner, M., Mitchell, M., Thompson, D. C., Zhao, B., & Xia, M. (2017). Development and application of human renal proximal tubule epithelial cells for assessment of compound toxicity. *Current Chemical Genomics and Translational Medicine, 11*, 19–30. https://doi.org/10.2174/2213988501711010019
12. Kandasamy, K., Chuah, J. K., Su, R., Huang, P., Eng, K. G., Xiong, S., Li, Y., Chia, C. S., Loo, L. H., & Zink, D. (2015). Prediction of drug-induced nephrotoxicity and injury mechanisms with human induced pluripotent stem cell-derived cells and machine learning methods. *Scientific Reports, 5*, 12337. https://doi.org/10.1038/srep12337
13. Chuah, J., & Zink, D. (2017). Stem cell-derived kidney cells and organoids: Recent breakthroughs and emerging applications. *Biotechnology Advances, 35*(2), 150–167. https://doi.org/10.1016/j.biotechadv.2016.12.001
14. Zink, D., Chuah, J., & Ying, J. Y. (2020). Assessing toxicity with human cell-based in vitro methods. *Trends in Molecular Medicine, 26*(6), 570–582. https://doi.org/10.1016/j.molmed.2020.01.008
15. Jenkinson, S. E., Chung, G. W., van Loon, E., Bakar, N. S., Dalzell, A. M., & Brown, C. D. (2012). The limitations of renal epithelial cell line HK-2 as amodel of drug transporter expression and function in the proximal tubule. *Pflugers Archiv—European Journal of Physiology, 464*, 601–611. https://doi.org/10.1007/s00424-012-1163-2
16. Weiland, C., Ahr, H. J., Vohr, H. W., & Ellinger-Ziegelbauer, H. (2007). Characterization of primary rat proximal tubular cells by gene expression analysis. *Toxicology in Vitro, 21*, 466–491. https://doi.org/10.1016/j.tiv.2006.10.008
17. Wang, L., & Sweet, D. H. (2013). Renal organic anion transporters (SLC22 family): Expression, regulation, roles in toxicity, and impact on injury and disease. *The AAPS Journal, 15*(1), 53–69. https://doi.org/10.1208/s12248-012-9413-y

18. Nigam, S. K., Bush, K. T., Martovetsky, G., Ahn, S. Y., Liu, H. C., Richard, E., Bhatnagar, V., & Wu, W. (2015). The organic anion transporter (OAT) family: A systems biology perspective. *Physiological Reviews, 95*(1), 83–123. https://doi.org/10.1152/physrev.00025.2013
19. Zeng, Y., Zhang, R., Wu, J., Liu, M., Peng, W., Yu, X., & Yang, X. (2012). Organic anion transporter 1 (OAT1) involved in renal cell transport of aristolochic acid I. *Human & Experimental Toxicology, 31*(8), 759–770. https://doi.org/10.1177/0960327111424302
20. Motohashi, H., Sakurai, Y., Saito, H., Masuda, S., Urakami, Y., Goto, M., Fukatsu, A., Ogawa, O., & Inui, K. I. (2002). Gene expression levels and immunolocalization of organic ion transporters in the human kidney. *Journal of the American Society of Nephrology: JASN, 13*(4), 866–874. https://doi.org/10.1681/ASN.V134866
21. Wilmer, M. J., Saleem, M. A., Masereeuw, R., Ni, L., van der Velden, T. J., Russel, F. G., Mathieson, P. W., Monnens, L. A., van den Heuvel, L. P., & Levtchenko, E. N. (2010). Novel conditionally immortalized human proximal tubule cell line expressing functional influx and efflux transporters. *Cell and Tissue Research, 339*(2), 449–457. https://doi.org/10.1007/s00441-009-0882-y
22. Nieskens, T. T., Peters, J. G., Schreurs, M. J., Smits, N., Woestenenk, R., Jansen, K., van der Made, T. K., Röring, M., Hilgendorf, C., Wilmer, M. J., & Masereeuw, R. (2016). A human renal proximal tubule cell line with stable organic anion transporter 1 and 3 expression predictive for antiviral-induced toxicity. *The AAPS Journal, 18*(2), 465–475. https://doi.org/10.1208/s12248-016-9871-8
23. Vormann, M. K., Vriend, J., Lanz, H. L., Gijzen, L., van den Heuvel, A., Hutter, S., Joore, J., Trietsch, S. J., Stuut, C., Nieskens, T., Peters, J., Ramp, D., Caj, M., Russel, F., Jacobsen, B., Roth, A., Lu, S., Polli, J. W., Naidoo, A. A., & Vulto, P., et al. (2021). Implementation of a human renal proximal tubule on a chip for nephrotoxicity and drug interaction studies. *Journal of Pharmaceutical Sciences, 110*(4), 1601–1614.https://doi.org/10.1016/j.xphs.2021.01.028
24. Truong, D. M., Kaler, G., Khandelwal, A., Swaan, P. W., & Nigam, S. K. (2008). Multi-level analysis of organic anion transporters 1, 3, and 6 reveals major differences in structural determinants of antiviral discrimination. *The Journal of Biological Chemistry, 283*(13), 8654–8663. https://doi.org/10.1074/jbc.M708615200
25. Moss, D. M., Neary, M., & Owen, A. (2014). The role of drug transporters in the kidney: Lessons from tenofovir. *Frontiers in Pharmacology, 5*, 248. https://doi.org/10.3389/fphar.2014.00248
26. Zhu, L., Lu, L., Wang, S., Wu, J., Shi, J., Yan, T., Xie, C., Li, Q., Hu, M., & Liu, Z. (2017). Oral absorption basics: Pathways and physicochemical and biological factors affecting absorption. In Developing solid oral dosage forms (2nd edn). Academic Press. https://doi.org/10.1016/B978-0-12-802447-8.00011-X
27. Yohannes Hagos, Daniel Stein, Bernhard Ugele, Gerhard Burckhardt, Andrew Bahn, 2007. Human Renal Organic Anion Transporter 4 Operates as an Asymmetric Urate Transporter. JASN, 18 (2) 430–439; https://doi.org/10.1681/ASN.2006040415
28. Henjakovic, M., Hagos, Y., Krick, W., Burckhardt, G., & Burckhardt, B. C. (2015). Human organic anion transporter 2 is distinct from organic anion transporters 1 and 3 with respect to transport function. *American Journal of Physiology. Renal Physiology, 309*(10), F843–F851. https://doi.org/10.1152/ajprenal.00140.2015

Development of a 3D-Printable Hydrogel-Based Antimicrobial Film

Jinxia Pan, Jemimah Lee, and Reuben Cheong

Abstract Currently, most wound dressings are therapeutically inert, lacking active components to aid in the wound healing process. In addition, production is based on mass subtractive manufacturing methods and does not allow personalised application. To address these issues, we explored the formulation of a composite gellan gum (GG)-hydroxypropyl methylcellulose (HPMC) hydrogel via additive manufacturing capable as a delivery vehicle for antimicrobials. Extrusion-based 3D printing was chosen as the mode of hydrogel production. Rheological tests were conducted to optimise the hydrogel formulation starting from a GG: HPMC ratio of 1:1 derived from previous studies. The optimal formulation, found to be 1 GG: 2 HPMC, was most able to recover from stress and display consistent time-dependent mechanical behaviour. The final 3D-printing conditions comprised of pneumatic pressure of 180 kPa, printhead speed of 5 mm/s and 25% infill density resulted in a 2 cm × 2 cm × 0.8 cm film. Subsequently, this formulation was tested as a delivery system for colistin, an approved peptide antimicrobial via agar well diffusion studies using *Escherichia coli*. Our results showed that our hydrogel formulation is an effective drug delivery vehicle and can be reproducibly printed to custom specifications. Taken together, this study demonstrates the feasibility of a customisable, "on-demand" therapeutic active hydrogel wound dressing for use in clinical settings.

Keywords Hydrogel · 3D printing · Wound dressing · Drug delivery

1 Introduction

Recent research has seen hydrogels emerge as biocompatible and active drug-delivering alternatives to traditional bandage and gauze wound dressings [1]. However, most unmodified hydrogels do not act as effective drug delivery vehicles. Encapsulated antimicrobials may be burst released in high concentrations and damage adjacent somatic cells, due to incomplete antimicrobial selectivity [2]. To

J. Pan (✉) · J. Lee · R. Cheong
Hwa Chong Institution (College Section), Singapore, Singapore
e-mail: panjinxia06@gmail.com

H. Guo et al. (eds.), *IRC-SET 2022*,
https://doi.org/10.1007/978-981-19-7222-5_27

retard the release of encapsulated antimicrobials, we manipulated the physical properties of the hydrogel system via optimization of a novel polymer-matrix formulation previously reported [3]. In contrast to labour-intensive and time-consuming subtractive manufacturing methods [4], 3D printing is more time- and resource-efficient and decentralised, thus lending higher clinical translatability to custom hydrogel-based wound dressings in the future. To this end, this study will also investigate the suitability of making hydrogel-based films using extrusion-based 3D printing through rheological assessment.

1.1 Composition of Hydrogel-Based Film

Gellan Gum: Gellan gum is a hydrogel-forming exopolysaccharide produced by the bacterium *Sphingomonas elodea*. It is biodegradable [3] and does not harm living cells.

Low acyl gellan gum possesses the advantage of easy and rapid gelation under physiological conditions, in the presence of cations [5], due to a low gelation temperature (T_{gel}) of ~40 °C. In this study, given its additional anti-inflammatory property, $MgCl_2$ was used as the cross-linking agent required to form true gels below T_{gel}.

Hydroxypropyl methylcellulose (HPMC): HPMC is a FDA-approved polymer commonly used as a pharmaceutical excipient in sustained drug release matrix systems due to its ability to swell and form a gel layer controlling the rate of drug release upon contact with a dissolution medium (Fig. 1). Hence, HPMC, specifically Methocel™ K4M Controlled Release (K4MCR), designed to encapsulate common highly soluble antimicrobials [6], was incorporated into the gellan gum-based hydrogel in order to create a hydrogel-based film capable with highly tunable physicochemical properties, including controlled release [7] and mechanical robustness [8]. A 1:1 gellan gum-HPMC ratio for the composite hydrogel was used as a starting concentration in this study, as previous studies have found the 1:1 alginate-HPMC ratio to be the best in achieving constant and effective controlled drug release [2, 9].

1.2 Colistin as Encapsulated Antimicrobial

Colistin is a polymyxin antibiotic commonly used to effectively treat bacterial infections caused by Gram-negative bacteria, such as *Escherichia coli* [10], via topical application to wound sites. However, delivery systems lacking controlled release, as well as low patient compliance in the regular changing of wound dressings, often result in low colistin concentrations at the wound site. This greatly reduces its antimicrobial efficacy and allows for the selection of partially antimicrobial-resistant bacteria at colistin concentrations below the minimum selection concentration (MSC), which induces antibiotic resistance to colistin [11]. Here, the hydrogel

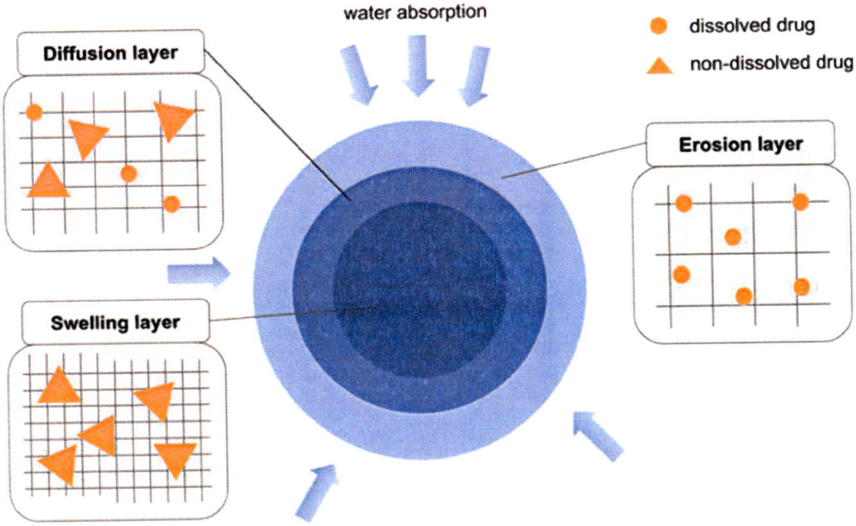

Fig. 1 Mechanism of sustained-release property of HPMC. Upon contact with a dissolution medium, HPMC forms a hydrated gel layer with three distinct zones: the innermost dry core containing the drug; a thin layer where polymer molecules are partially hydrated and swollen, slowing outward drug diffusion; and a fully hydrated outer area, allowing for the drug to fully dissolve and interact with the body [6]

may be used as a delivery vehicle with a suitable release profile to maintain colistin concentrations above the MSC.

1.3 Extrusion-Based 3D Printing

The hydrogel should have a low complex viscosity when being extruded at high shear rates, in order to avoid blocking the printhead nozzle [12]. It must also be able to quickly recover to its initial viscosity after extrusion through the printhead nozzle [13], in order to provide a stable medium for colistin and act as a foundation for further layers upon printing. Moreover, recalling the intended role of hydrogel as a controlled release excipient, it is vital that the structural integrity of its gellan gum-HPMC matrix system is not compromised after being sheared during 3D printing or due to movement as a wound dressing on the patient. Previous studies have demonstrated HPMC to have good stability and retention abilities [6], for concentrations of up to 10–12% (w/w) [12]. Thus, it is suitable as a matrix-forming excipient when mixed with gellan gum, which can be extruded and 3D printed.

1.4 Research Question

In this study, we aim to formulate a 3D-printable composite hydrogel using an optimal proportion of gellan gum and HPMC that can support the controlled release of its loaded antimicrobial.

2 Methodology

Our study adopted a three-pronged approach: a rheological study and 3D-printing trial to optimise the physical properties of the gel and an agar well diffusion study to investigate its microbiological profile/antimicrobial capability (Fig. 2).

2.1 Preparation of Gellan Gum-HPMC Hydrogel

Gellan gum (GG)-HPMC formulations with varying GG:HPMC concentration ratios (0.8%:0.4%, 0.8%:0.8%, 0.8%:1.6%, 0.8%:2.4%, 1.0%:2.0%) were tested. A standard 30 mL of GG-HPMC hydrogel was prepared by first dissolving gellan gum powder in a precursor solvent mixture of 25.5 mL of deionized water and 4.5 mL of glycerol, then heated to 80 °C for 30 min and cooled to 50 °C (Fig. 3). Next, HPMC-K4MCR powder was added via geometric dilution under continuous stirring, until fully dissolved. Lastly, 6 mg of 0.114% (w/v) $MgCl_2$ was added to crosslink the gellan gum polymers. Each hydrogel was left to set in the refrigerator overnight at 4 °C.

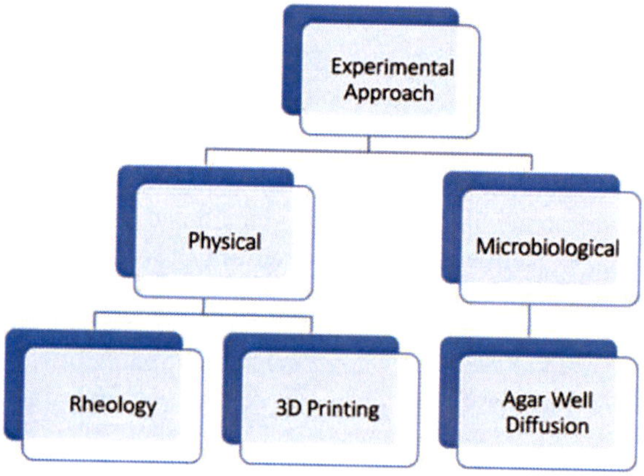

Fig. 2 Experimental approach of our study

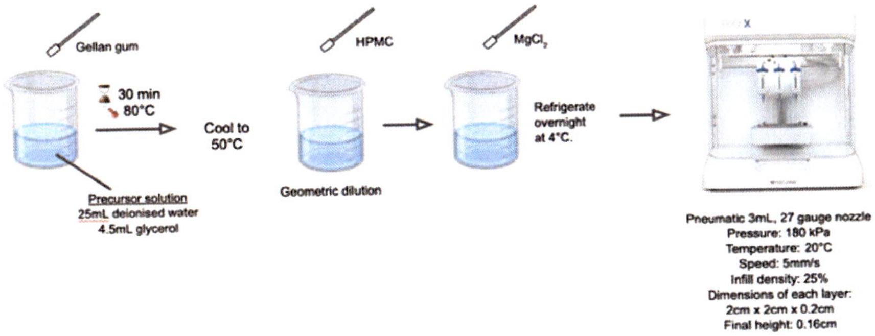

Fig. 3 Schematic drawing showing the formation and 3D-printing process of hydrogel

2.2 *Optimisation of Gel Formulation: Rheological Tests*

To establish suitability of the hydrogel for 3D printing, rheological tests were conducted on the different GG-HPMC formulations, using a modular compact rheometer (Anton Paar, MCR 302) at 25 °C. Parameters for a suitable hydrogel include physical stability, time-consistent behaviour, and a capacity for recovery.

Amplitude Sweep: To determine the linear viscoelastic region (LVR), an amplitude sweep was first conducted. Storage (G') and loss (G'') moduli were obtained in the range of 0.1% to 100% shear strain, at a constant frequency of 1 Hz. Using a straight plane, the limit of the LVR was determined, and this value was later used in frequency sweeps. A greater limit, in which G' and G'' remain constant and $G' > G''$, indicates a more stable hydrogel displaying gel-like properties.

Frequency Sweep: The hydrogel's time-dependent behaviour in its non-destructive deformation range can be determined using frequency sweeps, with an ideal hydrogel displaying time-consistent behaviour. G' and G'' moduli were measured from minimum to maximum frequency between 0.1 and 10 rad/s. Shear strain was set to a constant value of 1%, below the predetermined LVR.

Step Strain Test: Recovery of the hydrogel was tested by measuring its complex viscosity under alternating shear strains of 0.5% and 500%, respectively, with a constant shear frequency of 1 Hz. Up to five iterations of the differing shear strains, with 5 measurements each, were conducted. For optimal extrudability, the hydrogel should display sufficiently high viscosity at high shear and vice versa at low shear.

Prism software was used to organise raw rheological data into a series of prism graphs and map out storage (G') and loss (G'') modulus obtained from amplitude and frequency sweeps, as well as changes in complex viscosity under step strain for the respective hydrogel formulations.

2.3 Optimisation of Hydrogel 3D Printing

This study used the BIO X Bioprinter (Cellink Inc.), equipped with a 3 mL pneumatic syringe of size 27 gauge. The optimised GG-HPMC formulation as obtained from rheological characterisation was used.

2.4 Agar Well Diffusion Assay

Agar well diffusion method [14] was conducted using *Escherichia coli* (*E. coli*) in nutrient agar medium. 100 uL Mueller–Hinton Broth 2 was placed on a thawed *E. coli* sample and collected, with the resulting suspension made up to 2 mL with additional MHB-2, and incubated overnight. Agar plate surfaces were inoculated by spreading 50 uL of the *E. coli* suspension and put aside to dry for approximately 5 min, before 4 wells of diameter 8 mm were punched aseptically with a micropipette tip on each plate. 150 uL of 0 μM (blank gel), 20, 30 or 50 μM colistin-laden "intact" hydrogel samples were pipetted into the well. Similarly, the optimised GG-HPMC formulation as obtained from rheological characterisation was used. To investigate if colistin release is impeded by the gel, hydrogel samples were also mechanically "smashed" in situ using the sharp end of a micropipette tip, thus disrupting the hydrogel's physical structure. 3 replicates were conducted for each hydrogel type ("intact" and "smashed"). Agar plates were incubated at 37 °C, and the average diameter of the zone of inhibition (ZOI) was measured using a ruler after 24 and 48 h.

3 Results and Discussion

3.1 Optimisation of Gel Formulation: Rheological Tests

The LVR, a non-destructive deformation shear strain range, was first determined as < 1% shear strain (Fig. 4Cc). However, the absence of LVR, or a very low LVR for the 0.8: 0.4 and 0.8: 0.8 GG: HPMC gel ratios, respectively (Fig. 4b), suggests that they lose mechanical properties even at very low shear stress and break down, thus making them unsuitable for 3D printing. Next, storage (G') and loss (G'') modulus were measured over a frequency range of 0.1 to 10 rad/s (Fig. 5). All the gels displayed stable gel-like properties, as $G' > G''$ in the frequency range (Fig. 5). However, the 0.8 GG: 1.6 HPMC hydrogel displayed the best time-consistent behaviour amongst all other gels. Lastly, not only was higher shear strain able to alter the complex viscosity of the 0.8: 1.6 hydrogel to the greatest extent with a difference of 950 Pa.s, the hydrogel was also able to recover a complex viscosity close to its initial 1000 Pa.s rapidly under LVR conditions (Fig. 6). On the other hand, the 0.8–0.8 and 0.8–0.4

hydrogels had the lowest complex viscosities and significantly lower recovery ranges (Fig. 6). This shows that the two hydrogels are not suitable for 3D bioprinting, as the hydrogels would not be able to retain their shape after extrusion, under low shear.

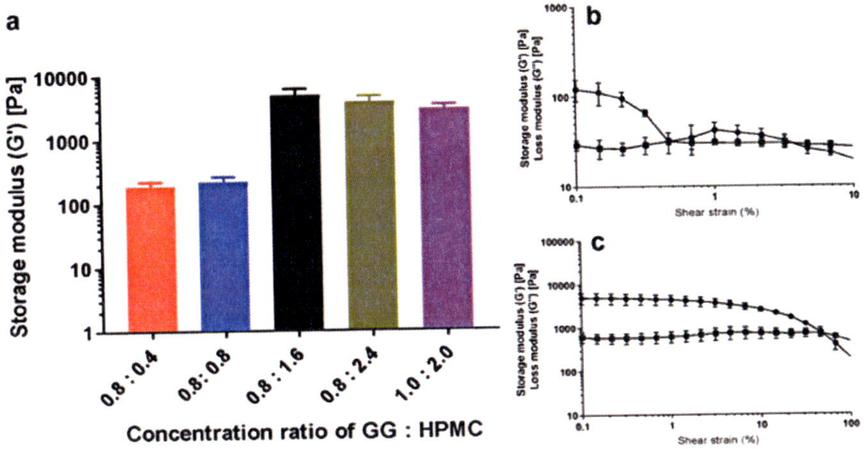

Fig. 4 Amplitude sweep results shows that the 0.8 GG: 1.6 HPMC hydrogel has the best LVR. **a** Combined results of Storage Moduli across different GG: HPMC concentration ratios. **b, c** Amplitude sweep results between 0.8 GG: 0.4 HPMC and 0.8 GG: 1.6 HPMC hydrogel formulations show significant improvement in the LVR for 0.8: 1.6 formulation. For the combined frequency and amplitude sweep graphs, loss (G'') modulus was omitted, as most of the formulations displayed gel-like behaviour within the LVR with the exception of the 0.8: 0.4 hydrogel for the amplitude sweep

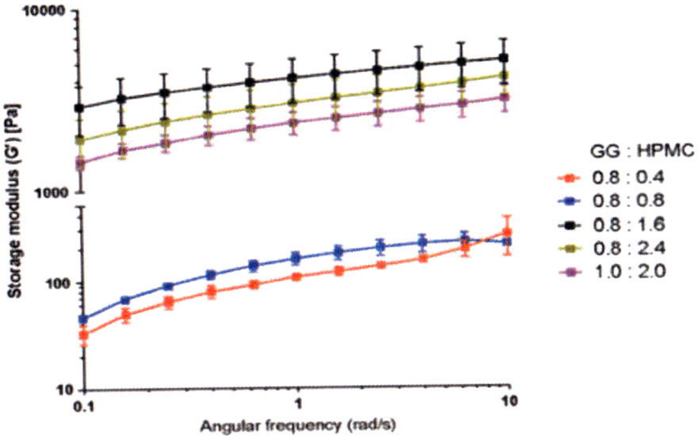

Fig. 5 Frequency sweep results show the 0.8 GG: 1.6 HPMC hydrogel exhibits the most consistent time-dependent behaviour

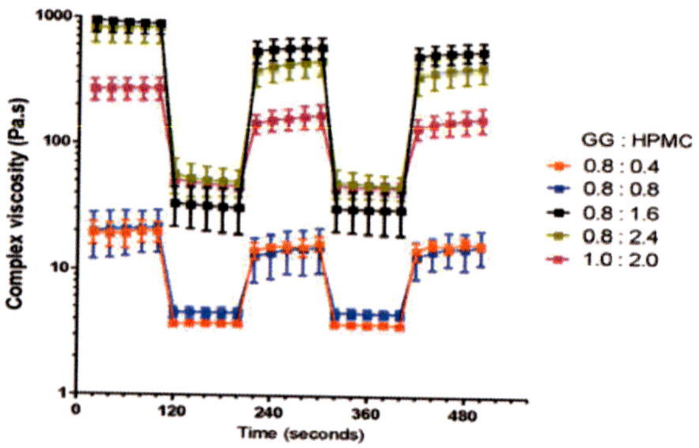

Fig. 6 Step strain results, with different shear rates at different intervals, confirm that the 0.8 GG: 1.6 HPMC hydrogel exhibits the greatest ability to recover from shearing

3.2 Optimised 3D-Printed Samples

Some samples (Fig. 8) were not suitable. For instance, the hydrogel could not be smoothly printed out at 55 kPa, 3 mm/s, 15% infill density, and had an irregular shape at a higher infill density of 35%, 100 kPa and 3 mm/s. After optimisation, the final printing parameters are as follows. To ensure optimal extrudability, air pressure was adjusted to 180 kPa. Temperature was kept constant at 20 °C to prevent changes to the physical structure of the hydrogel whilst printing. Four 2 cm × 2 cm × 0.2 cm layers were printed on top of each other, with a printhead speed of 5 mm/s and an infill density of 25% to ensure an even fill without gaps in between individual layers (Fig. 7). Figure 7 demonstrates that the hydrogel is well-suited for 3D printing, and printing parameters can be further scaled up for larger wound sizes.

Fig. 7 Hydrogel printed under optimised parameters

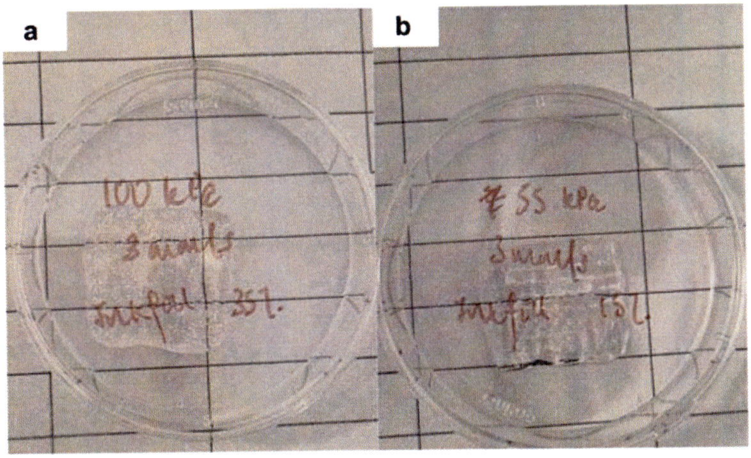

Fig. 8 **a** Hydrogel printed at 100 kPa, 3 mm/s, 35% infill density. **b** Hydrogel printed at 55 kPa, 3 mm/s, 15% infill density

3.3 *Agar Well Diffusion Assay*

Blank gels exhibited no ZOI, whilst all other wells had a clear ZOI, thus confirming that colistin is the factor that contributes to its antimicrobial activity (Fig. 9). Regardless of concentration, colistin displays effective antimicrobial activity against *E. coli*, as confirmed by Biswas et al. [10] (Fig. 10a). However, ZOIs did not seem to be affected in a concentration-dependent manner, as they remained relatively constant from 10.67 mm to 11.33 mm as concentration increases from 20 to 50 μM. Hence, results for an optimal colistin concentration for encapsulation in hydrogel are inconclusive.

From Fig. 10a, b, decreases in diameters of the ZOIs were limited to 1.2 mm. As *E. coli* has a generation time of about 20 min under aerobic, nutrient-rich conditions

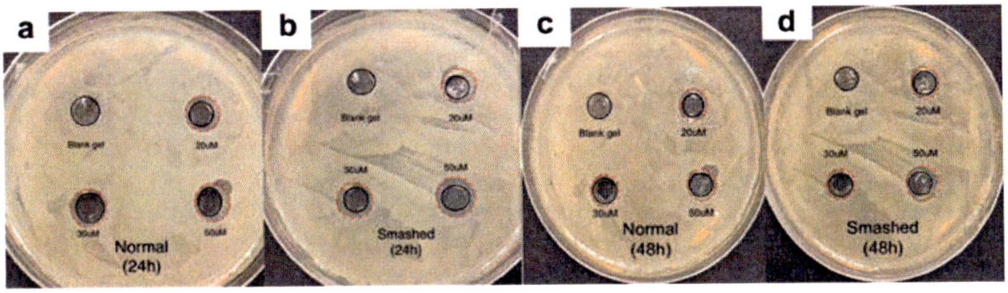

Fig. 9 Results of agar well diffusion assay. Wells are indicated with a black circle and zone of inhibitions with a dotted red circle. **a** "Intact" gel samples after 24 h. **b** *Smashed* gel samples after 24 h. **c** *Normal* gel samples after 48 h. **d** "Smashed" gel samples after 48 h

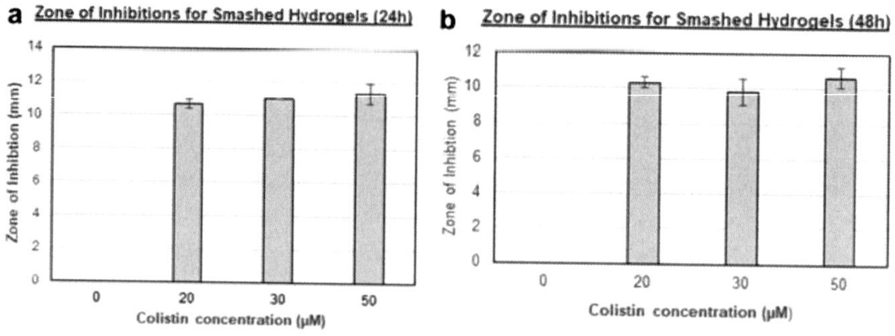

Fig. 10 Effect of different colistin concentrations on diameters of ZOI at **a** 24 h and **b** 48 h, for plates with "smashed" gels

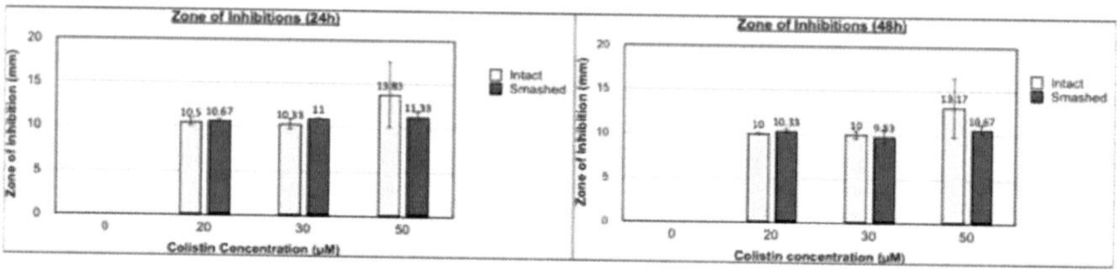

Fig. 11 Diameters of ZOIs of hydrogels for "intact" and "smashed" hydrogels, laden with different colistin concentrations. Graphs show results after **a** 24 h and **b** 48 h

[15], this likely shows that colistin was continually released in sufficient amounts to suppress antimicrobial growth across 48 h.

As seen from Fig. 11, the ZOI of intact gels and mechanically smashed gels were comparable, with differences in diameters limited to 0.7 mm. An exception is for 50 µM colistin-laden "intact" hydrogels, which exhibited a considerably larger ZOI than its "smashed" counterpart. Nevertheless, across all colistin concentrations, this suggests that release of colistin is not impeded due to the physical structure of the hydrogel matrix; hence, colistin can be encapsulated in our GG-HPMC hydrogel for release.

4 Conclusions and Future Work

0.8% GG: 1.6% HPMC hydrogel possessed the greatest physical stability, time-consistent behaviour and greatest capacity for recovery, making it most suitable to withstanding the extrusion process during 3D printing. Ideally, the gel should be printed with 180 kPa pneumatic pressure with a printhead speed of 5 mm/s and infill density of 25% to avoid collapse of the final gel dressing. Colistin can be incorporated into the hydrogel as an effective antimicrobial agent, but more experiments are needed to optimise the colistin concentration for our hydrogel formulation.

With the short timeframe being a limitation of this study, future studies have to be conducted to examine the release of colistin over a longer period (e.g. 7 days) to firmly establish a controlled release profile. Furthermore, the ZOI of smashed gel samples exhibited less variance than the ZOI of the normal gel samples, with maximum standard deviations of 0.764 and 3.82, respectively (Fig. 11). To further investigate if mechanical disruption of the gel alters its encapsulation characteristics, or if this discrepancy is a limitation of the well method, disc diffusion studies, where the drug diffuses downwards to the medium rather than through the surrounding walls of a well, may be conducted. This would also more closely simulate antimicrobial diffusion from a wound dressing placed atop a patient's wound. Lastly, in vivo antimicrobial diffusion studies could also be conducted to investigate the biocompatibility of the antimicrobial film and other aspects of its interaction with living tissue.

References

1. Tavakoli, S., & Klar, A. S. (2020). Advanced hydrogels as wound dressings. *Biomolecules, 10*(8), 1169. https://doi.org/10.3390/biom10081169
2. Nochos, A., Douroumis, D., & Bouropoulos, N. (2008). In vitro release of bovine serum albumin from alginate/HPMC hydrogel beads. *Carbohydrate Polymers, 74*(3), 451–457. ISSN 0144-8617. https://doi.org/10.1016/j.carbpol.2008.03.020
3. Ng, J. Y., Zhu, X., Mukherjee, D., Zhang, C., Hong, S., Kumar, Y., Gokhale, R., & Ee, P. L. (2021). Pristine gellan gum–collagen interpenetrating network hydrogels as mechanically enhanced anti-inflammatory biologic wound dressings for burn wound therapy. *ACS Applied Bio Materials, 4*(2), 1470–1482. https://doi.org/10.1021/acsabm.0c01363
4. Sultan, S., Siqueira, G., Zimmermann, T., & Matthew, A. P. (2017). 3D printing of nano-cellulosic biomaterials for medical applications. *Science Direct: Current Opinion in Biomedical Engineering, 2*, 29–34. https://www.sciencedirect.com/science/article/abs/pii/S2468451117300132
5. Milivojevic, M., Pajic-Lijakovic, I., Bugarski, B., Nayak, A. K., & Hasnain, M. S. (2019). Chapter 6—Gellan gum in drug delivery applications. In *Natural polysaccharides in drug delivery and biomedical applications* (pp. 146–186). Academic Press. ISBN 9780128170557. https://doi.org/10.1016/B978-0-12-817055-7.00006-6
6. Maškováa, E., Kubová, K., Raimi-Abraham, B. T., Vllasaliu, D., Vohlídalová, E., Turánek, J., & Mašek, J. (2020). Hypromellose—A traditional pharmaceutical excipient with modern applications in oral and oromucosal drug delivery. *Journal of Controlled Release, 324*, 695–727. https://doi.org/10.1016/j.jconrel.2020.05.045
7. Pan, Z., Ye, H., & Wu, D. (2021). Recent advances on polymeric hydrogels as wound dressings. *APL Bioengineering, 5*, 011504. https://doi.org/10.1063/5.0038364
8. Buwalda, S. J., Vermonden, T., & Hennink, W. E. (2017). Hydrogels for therapeutic delivery: Current developments and future directions. *Biomacromolecules, 18*(2), 316–330.
9. Conti, S., Maggi, L., Segale, L., Ochoa Machiste, E., Conte, U., Grenier, P., & Vergnault, G. (2007). Matrices containing NaCMC and HPMC. *International Journal of Pharmaceutics, 333*(1–2), 143–151. https://doi.org/10.1016/j.ijpharm.2006.11.067

10. Biswas, S., Brunel, J.-M., Dubus, J.-C., Reynaud-Gaubert, M., & Rolain, J.-M. (2012). Colistin: An update on the antibiotic of the 21st century. *Expert Review of Anti-infective Therapy, 10*(8), 917–934. https://doi.org/10.1586/eri.12.78
11. Gullberg, E., Cao, S., Berg, O. G., Ilbäck, C., Sandegren, L., Hughes, D., & Andersson, D. I. (2011). Selection of resistant bacteria at very low antibiotic concentrations. *PLoS Pathogens, 7*(7), e1002158.
12. Cheng, Y., Qin, H., Acevedo, N. C., Jiang, X., & Shi, X. (2020). 3D printing of extended-release tablets of theophylline using hydroxypropyl methylcellulose (HPMC) hydrogels. *International Journal of Pharmaceutics, 591*, 119983. https://doi.org/10.1016/j.ijpharm.2020.119983
13. Kirchmajer, D. M., Gorkin III, R., & in het Panhuis, M. (2015). An overview of the suitability of hydrogel-forming polymers for extrusion-based 3D-printing. *Journal of Materials Chemistry B, 3*(20), 4105–4117.https://doi.org/10.1039/c5tb00393h
14. Balouiri, M., Sadiki, M., & Ibnsouda, S. K. (2016). Methods for in vitro evaluating antimicrobial activity: A review. *Journal of Pharmaceutical Analysis, 6*(2), 71–79. https://doi.org/10.1016/j.jpha.2015.11.005
15. Gibson, B., Wilson, D. J., Feil, E., & Eyre-Walker, A. (2018). The distribution of bacterial doubling times in the wild. *Proceedings of the Royal Society B: Biological Sciences, 285*(1880), 20180789. https://doi.org/10.1098/rspb.2018.0789

Maritime Monitoring via Low-Earth Orbit Satellite Constellation

Zhuzhen Zhuo, Mabel You, and Xinqi Peng

Abstract With the ever-increasing volume of marine ships involved in commercial activities in an already globalised world, ensuring the efficiency and safety of these operations have become paramount. To accurately monitor their movement, satellite constellations are often used, making it crucial to optimise the performance of these constellations for cost-saving and environmental considerations. This report mainly investigates how key elements of a satellite constellation—the number of satellites, the number of orbital planes, and their inclination—affect the coverage of the system. Experimentation and configuration will be done through Freeflyer software simulations to find the most efficient configuration to maximise contact time with ground stations and minimise time needed to spot 99% of the maritime vessels. Based on the range of values taken, we found that the most optimal configuration is one with 79° of inclination, with 10 orbital planes and 11 satellites per plane.

Keywords Satellite constellation · Low earth orbit · Maritime monitoring · Maritime surveillance · Optimisation

1 Introduction

1.1 Background

Satellites orbiting on Low-Earth Orbit (LEO) trajectories provide crucial and important remote-sensing data about the sea surface status on a daily basis and at diverse resolutions. LEO is chosen as it requires the least amount of energy for satellite placement and satellites there need less powerful amplifiers for successful transmission. It is also the only range for consistent decoding and receiving of Very High Frequency type frequency, the most common type used for maritime communication [1].

Z. Zhuo (✉) · M. You · X. Peng
Hwa Chong Institution (College), Singapore, Singapore
e-mail: zhuzhen.zhuo@gmail.com

H. Guo et al. (eds.), *IRC-SET 2022*,
https://doi.org/10.1007/978-981-19-7222-5_28

Combined with the importance of ship trading and the ever-increasing goods shipped around the globe [2, 3], numerous successful maritime-monitoring satellite constellations are in LEO, such as LEMUR by Spire [4] and Iridium's NEXT [5, 6].

The information collected by these satellites plays an essential role in vessel traffic monitoring since remotely sensed data obtains and tracks information about the vessel's main features, its peculiar geometry, and even its speed. Global coverage of the satellite constellation ensures that contact with ground stations and maritime vessels will not be lost and reliable real-time monitoring of ships and the ocean can occur.

1.2 Literature Review

Existing literature on large LEO constellations is rare. Papers have been written to describe an existing satellite constellation. For instance, Sturza [7] described the technical aspects of the original Teledesic satellite system and Leopold [8] analysed Iridium's satellite constellation in-depth in his paper. Papers have also been written to compare the different existing satellite constellations. Comparetto [9] reviewed and compared the Globalstar, Iridium, and Odyssey constellations via various measures such as signal integrity and availability. Portillo et al. [10] have done an analysis of newer generations of LEO constellations, namely, comparing and analysing the proposals of OneWeb, Telesat, and SpaceX, such as on their architecture, technical comparisons, and variability in performance induced by atmospheric conditions. However, these papers do not report on the theoretically best constellation set up and do not focus only on maritime activities.

Hypothetical most efficient monitoring systems for single satellite constellations are also present, with Spangelo and Cutler [11] investigating the optimisation of single satellite in their article, though such single satellite systems do not fulfil the criteria required for monitoring maritime activities due to the limited coverage. Zong and Kohani [12] examined the most optimal LEO constellation design based on global coverage, though analysis of specifically contact with maritime vessels may yield different results due to the different concentrations of maritime vessels as compared to simply the globe.

2 Methodology

Simulations will be conduced using the FreeFlyer software. Using the different functions of the software, the following are used to determine the characteristics of the simulation and to mimic real-life requirements and conditions.

2.1 Ground Stations

To ensure continuous, real-time coverage, constant contact with ground stations was of utmost importance. To mimic global demand for such a satellite constellation, ground stations were placed in the busiest container ports of the world for a total of 5 ground stations at a diverse range of latitudes and longitudes in Singapore, Shenzhen, Valencia, Santos, and Los Angeles [13].

2.2 Contact Analysis

Maritime vessels were simulated by using point groups, which are individual points with a specific latitude and longitude and can be used to check for contact with a satellite. Due to the lack of free, publicly available data, only about 2000 maritime vessels were chosen, at random, from a database [14] in our study. Furthermore, due to the aforementioned reason, we were unable to simulate the movement of the maritime vessels.

The total simulation time is set to one day as in that time period, the majority of marine vessels would have been detected and further simulations would not render a significant difference in results. Moreover, a shorter time frame ensures that the location of the ship will likely not change drastically, making the simulation more accurate.

The total number of maritime vessels that are in contact with the sensors attached to the satellites in a day will be measured and recorded. Moreover, analysis of the time taken to observe 99% of point groups is recorded as it avoids the awkward situation where a few point groups are "unluckily" missed by the constellation. For the sake of comparison, if the number of point groups the constellation observed is less than 2000, a value of "2 days" is assigned, and if the number of point groups the constellation observed is between 2000 and 2064, a value of "1.5 days" is assigned.

To ensure minimal disruptions between satellites and ground stations, the total time whereby at least one satellite is in contact with each ground station is also recorded.

2.3 Determination of Property of Each Satellite

Given the low eccentricity of LEO orbits, we assume eccentricity as 0 and hence the argument of perigee can be ignored.

Following the design characteristics of the Iridium satellite constellation, a successful LEO satellite constellation, we will let the altitude of our satellites be 780 km, and the cone half-angle to be 17° [5, 9, 15].

To determine the right ascension of the ascending node (RAAN) and true anomaly (TA) of each satellite, Ω_k and θ_k respectively, the following equations are used (see Appendix):

$$\Omega_k = \lfloor \frac{k}{N_{\mathrm{TA}}} \rfloor \Delta\Omega \tag{1}$$

$$\theta_k = \lfloor k - \lfloor \frac{k}{N_{\mathrm{TA}}} \rfloor N_{\mathrm{TA}} \rfloor \Delta\theta \tag{2}$$

Based on the largest latitude value of a ship from our study, the range of inclination for our simulation varies from 70° to 85°. With reference to existing LEO constellations with ADS-B payloads to monitor aircraft (Spaceflight101, n.d.), the number of planes in each simulation for the monitoring of ships will range from 5 to 11, and the number of satellites per plane range from 5 to 15.

3 Results and Discussion

3.1 *Effects of Changes to Inclination*

We will use three configurations of satellite constellations for this section, with the values arbitrarily chosen. From Fig. 1, we can observe that the average contact time of the satellite constellation with the ground stations generally decreases with the increase of inclination. This is to be expected as by increasing the inclination, the constellation would spend more time orbiting at higher latitudes.

The general trend in Fig. 2a can be observed no matter the other elements. This meets our expectations as when the inclination increases, the ground track of a satellite covers more latitudes so more ships at higher latitudes can be observed. However, it is interesting that there is an anomaly, with the number of ships observed in a day decreasing for a satellite constellation with 10 orbital planes and 10 satellites

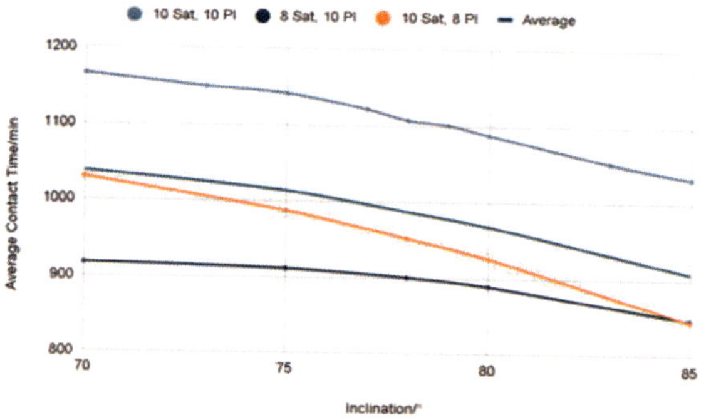

Fig. 1 Graph of average contact time versus inclination

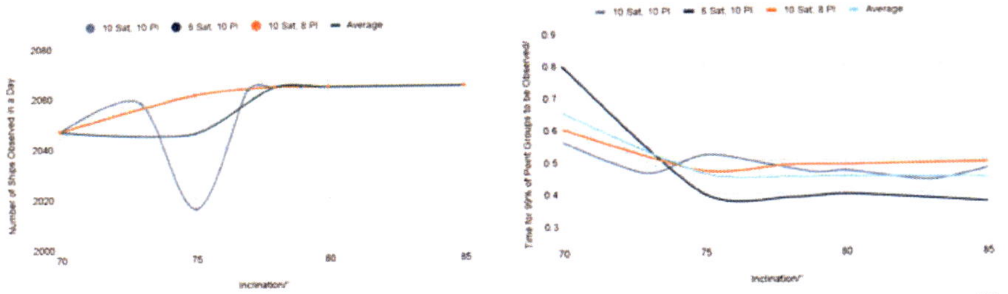

(a) Graph of the total number of ships observed against the inclination

(b) Graph of Time taken to observe 99% of point groups against the inclination

Fig. 2 Graph of the total number of ships observed against the inclination graph of time taken to observe 99% of point groups against the inclination

per plane at 75° of inclination. We hypothesise that this is due to the difference in the initial position of the satellites as the TA and RAAN are changed by the number of satellites and the number of planes, causing the constellation to unluckily "miss" some of the point groups, causing a lower number to be reported. Moreover, as the inclination increases, the time taken for the majority of the point groups to be observed decreases (Fig. 2b), with it approaching an asymptote.

Given that an inclination of 80° can observe all of the ship vessels in a day, and it is of the lowest inclination, which ensures longer contact time with ground stations and less time to observe all point groups, we will proceed with the following simulations using this value.

3.2 *Effects of Change in Number of Planes*

From Fig. 3, as the number of planes increased, the average contact time increased. However, the increase from an odd number of planes to an even number of planes is

Fig. 3 Graph of average contact time against number of planes

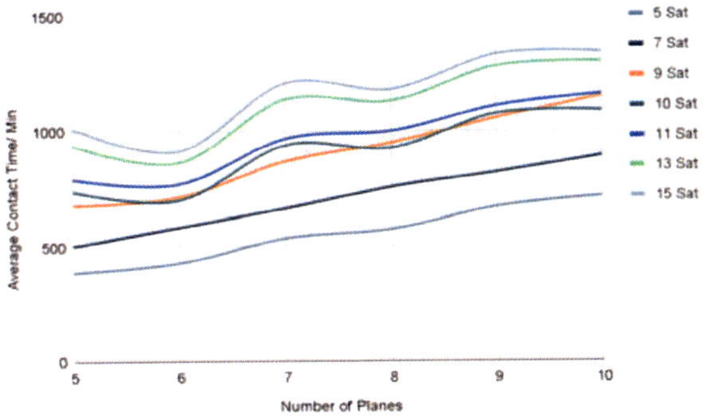

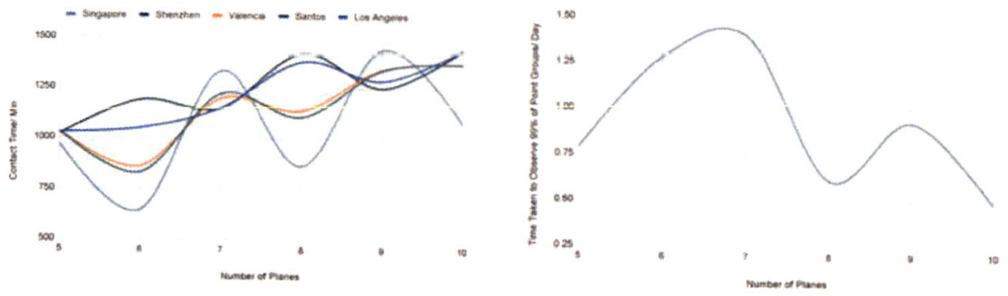

(a) Graph of the Contact Time Against Number of Planes

(b) Graph of Time Taken to Observe 99% of Point Groups Against Number of Planes

Fig. 4 Graph of time taken to observe 99% of point groups against number of planes

negligible as compared to that from an even number of planes to an odd number of planes. This trend can be seen more clearly when there are 5, 10, 11, and 13 satellites per plane when the average contact time actually decreases from an odd number of planes to an even number of planes.

Taking the number of satellites arbitrarily to be 15, it can be seen that for Singapore, Valencia, and Santos, the contact time when the number of planes increases from an odd number to an even number actually decreases (Fig. 4a), which goes against our hypothesis as we expected an increase. On the other hand, the contact time for Shenzhen and Los Angeles decreased when the number of planes increased from an even number to an odd number. Again, we hypothesise that this is due to the different initial conditions of the simulation. This reason also likely explains the behaviour in Fig. 4b, where although there seems to be a general decreasing trend with the increase in number of planes, the behaviour is rather chaotic in nature.

3.3 *Effects of Changes in Number of Satellites per Plane*

Similar to the trend observed in increasing the number of planes, increasing the number of satellites per plane increases the total number of satellites, hence increasing the contact time with the ground stations, as can be seen in Fig. 5. However, from 9 satellites per plane to 10 satellites per plane, the average contact time increased by only a negligible amount.

Interestingly, satellite constellations with an even number of planes all saw a drop in the average contact time when the number of satellites per plane increased from 9 to 10. However, no drops were observed when increasing the number of satellites per plane in satellite constellations with an odd number of orbital planes. However, for a satellite constellation with an odd number of orbital planes, some increases saw a negligible change in the average contact time. We suspect that the difference

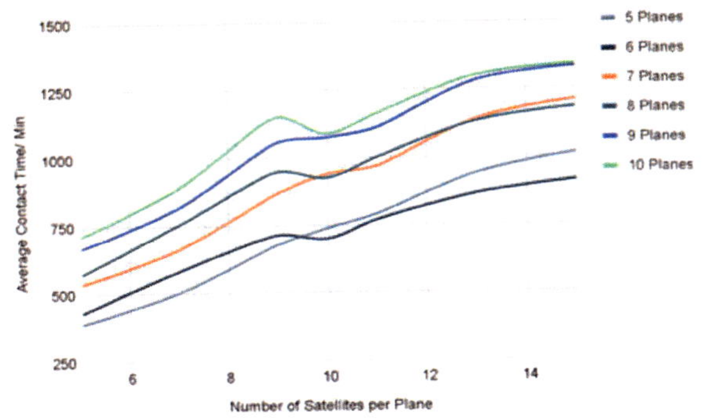

Fig. 5 Graph of average contact time against number of satellites per plane

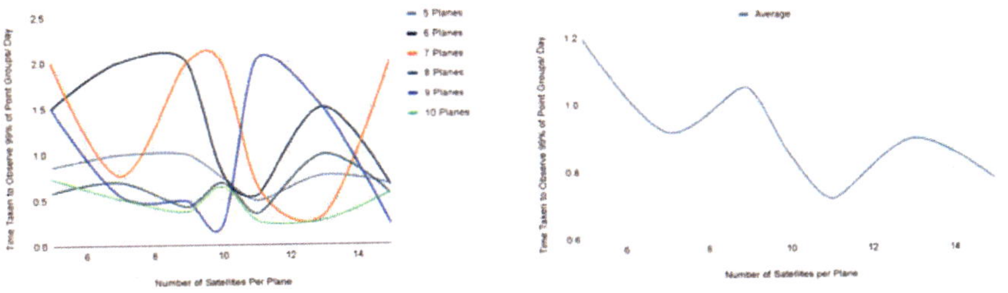

(a) Graph of Time Taken to Observe 99% of Point Groups against the Number of Planes

(b) Graph of Average Time Taken to Observe 99% of Point Groups against the Number of Planes

Fig. 6 Graph of average time taken to observe 99% of point groups against the number of planes

in initial conditions is responsible for these observations. The average contact time increased with the number of satellites per plane for the other values.

The effects of the number of planes on the time taken to observe 99% of the point groups (Fig. 6a) seem to be completely irregular. However, by taking note of the average (Fig. 6b), a pattern appears. It appears that a minimum point occurs at 7 and 11 satellites per plane, whilst a maximum point occurs at 9 and 13 satellites per plane. We are unsure of the reasoning behind this, though the likely reason is due to the different initial positions of the satellites. Additionally, an expected pattern of a general decreasing trend is observed.

Table 1 Table of average contact time/min with accordance to different number of planes and number of satellites

No. of sats.	No. of planes			
	5	7	9	10
5	–	536	670	715
7	505	–	820	892
9	679	868	–	1150
10	739	939	1075	–

Table 2 Table of min time taken to observe 99% of point groups/day with accordance to different number of planes and number of satellites

No. of sats.	No. of planes			
	5	7	9	10
5	–	1.50	1.50	0.74
7	0.99	–	0.55	0.51
9	0.99	2.00	–	0.38
10	0.74	1.50	0.22	–

3.4 *Effects of Combinations of Number of Planes and Satellites*

With reference to Table 1, with the exception of 7 planes with 5 satellites per plane, it appears that increasing the number of satellites per plane is more beneficial than increasing the number of planes to increase the average contact time between satellites and the ground stations.

When we compare the combinations that have the same total number of satellites in Table 2 (comparing it diagonally), if we only look at the top right entries of the table, we notice that for 5 and 9 satellites per plane, it appears to take more time than its corresponding entry on the bottom left of the table, whilst that of 7 satellites per plane takes shorter, most likely due to the different initial positions of the satellites.

4 Conclusion

In conclusion, most of the results prove our hypothesis correct, though many interesting anomalies were noticed.

An increase in inclination will result in a shorter time taken to observe the majority of vessels but it also shortens the average contact time of the satellite constellation with the ground stations. The increase in number of orbital planes in the satellite constellation causes the average contact time of the satellite constellation with the ground stations to generally increase, with the time taken to observe 99% of the point groups being a minimum at 10 orbital planes. When the number of satellites per plane increases, there is also an increasing trend in the contact time with the ground stations, whilst the minimum time taken to observe 99% of the point groups occurs when there are 11 satellites per plane.

Therefore, in the range of data used in our experiment, the most efficient configuration of our satellite constellation will have an inclination of 80°, with 10 orbital planes and 11 satellites per plane.

4.1 *Limitations and Future Work*

Due to the increase in inclinations ranging from 2° to 5°, depending on its location to our predicted minimum inclination for the most optimal configuration, a smaller increment can be used to increase the accuracy of the experiment. Costs were also not considered in our report, resulting in a bias towards larger values of planes and number of satellites, which may not reflect the real world due to cost restrictions and the need for cost-effectiveness.

Given the lack of sufficient data for our research, future work should be spent exploring the changes experienced by including the movement of maritime vessels across the globe. A larger number of marine vessels could also be chosen to have a more accurate and reliable representation of the effects of changes in changing the elements of the constellation as there may be inherent biases in the small sample of data.

Combined with a longer time period, the average number of marine vessels in contact with the constellation can be recorded to give a better representation of the effects of the changes in elements in the satellite constellation. A smaller step size for the simulation can also be considered to provide a more accurate simulation. Elements that were considered negligible, such as the eccentricity of the orbit, can also be considered to reflect the real world better.

Appendix: Formula for RAAN and True Anomaly

Note that k is a variable range from 0 to $(N_{\text{satellite}} - 1)$, where $N_{\text{satellite}}$ is the total number of satellites in the constellation, with each satellite being designated k, $\Delta\Omega$ is the separation between two closest orbit planes in terms of right ascension of ascending node difference, $\Delta\theta$ is the separation between two satellites on the same orbital plane in terms of true anomaly and N_{TA} is the number of satellites per orbital plane.

Derivation of Formula for RAAN. Given that each satellite is given a number k, it is important to ensure that for every N_{TA} increase in satellite number, there must be an increase of $\Delta\Omega$. Hence, it can be seen that when $0 \le k < N_{TA}, 0 \le \frac{k}{N_{TA}} < 1$, and hence it can be inferred that:

$$\lfloor \frac{k}{N_{TA}} \rfloor = 0 \Rightarrow \Omega_k = 0$$

A similar analysis can be done to $N_{TA} \le k < 2N_{TA}$, $2N_{TA} \le k < 3N_{TA}$ etc. Hence, it can be inferred from this that the formula for Ω_k is given by the equation $\Omega_k = \lfloor \frac{k}{N_{TA}} \rfloor \Delta\Omega$.

$\Delta\Omega$ is determined using the following equation, with N_Ω is the total number of planes:

$$\Delta\Omega = \frac{360}{N_\Omega} \quad (3)$$

Derivation of Formula for TA. For the TA of each satellite, it is important to ensure that for all of the satellites in a plane, the TA increases by $\Delta\theta$ for every increase in k and that the pattern repeats for every increase in N_{TA}, the number of satellites per plane.

Hence, it can be observed that when $0 \le k < N_{TA}$, $\lfloor \frac{k}{N_{TA}} \rfloor N_{TA} = 0$, and hence $\lfloor k - \lfloor \frac{k}{N_{TA}} \rfloor N_{TA} \rfloor \Delta\theta = k\Delta\theta$ which is the desired output.

Similarly, it can be observed that when $N_{TA} \le k < 2N_{TA}$, $\lfloor \frac{k}{N_{TA}} \rfloor N_{TA} = 1$, and hence:

$$\lfloor k - \lfloor \frac{k}{N_{TA}} \rfloor N_{TA} \rfloor \Delta\theta = (k - N_{TA})\Delta\theta = k\Delta\theta$$

Which is the desired output.

A similar analysis can be done to $2N_{TA} \le k < 3N_{TA}$, $3N_{TA} \le k < 4N_{TA}$ etc. to deduce the equation $\theta_k = \lfloor k - \lfloor \frac{k}{N_{TA}} \rfloor N_{TA} \rfloor \Delta\theta$.

$\Delta\theta$, the separation between two satellites on the same orbital plane, is given by:

$$\Delta\Omega = \frac{360}{N_{TA}} \quad (4)$$

The derivation of $\Delta\theta$ and $\Delta\Omega$ is trivial and hence will not be discussed.

References

1. United States Coast Guard. (n.d.). Radio information for boaters. Radio Information for Boaters. Retrieved January 19, 2022, from https://www.navcen.uscg.gov/radio-information-for-boaters
2. International Chamber of Shipping. (n.d.). Shipping and world trade: Driving prosperity. International Chamber of Shipping. Retrieved July 6, 2022, from https://www.ics-shipping.org/shipping-fact/shipping-and-world-trade-driving-prosperity/

3. Roberts. (2021, August 22). Visualization for world merchant fleet data (2016–2021). infomaritime.eu. Retrieved January 11, 2022, from http://infomaritime.eu/index.php/2021/08/22/top-15-shipowning-countries/
4. Spire. (n.d.). The Low Earth Multi-Use Receiver (LEMUR). Spire. Retrieved December 29, 2021, from https://spire.com/spirepedia/low-earth-multi-use-receiver/
5. Iridium. (2019, February 8). Iridium Next: In review. Iridium Satellite Communications. Retrieved January 18, 2022, from https://www.iridium.com/blog/2019/02/08/iridium-next-review/
6. ESA. (n.d.). Iridium NEXT. Sharing Earth Observation Resources. Retrieved July 6, 2022, from https://earth.esa.int/web/eoportal/satellite-missions/i/iridium-next
7. Sturza, M. A. (1994). The Teledesic Satellite System. In *Proceedings of IEEE National Telesystems Conference—NTC '94*. https://doi.org/10.1109/ntc.1994.316677
8. Leopold, R. J. (1991). Low-earth orbit global cellular communications network. In *ICC 91 International Conference on Communications Conference Record*. https://doi.org/10.1109/icc.1991.162525
9. Comparetto, G., & Hulkower, N. (1994). Global Mobile Satellite Communications—A review of three contenders. In *15th International Communications Satellite Systems Conference and Exhibit*. https://doi.org/10.2514/6.1994-1138
10. del Portillo, I., Cameron, B. G., & Crawley, E. F. (2019). A technical comparison of three low Earth Orbit Satellite Constellation Systems to provide Global Broadband. *Acta Astronautica, 159*, 123–135. https://doi.org/10.1016/j.actaastro.2019.03.040
11. Spangelo, S., & Cutler, J. (2012). Optimization of single-satellite operational schedules towards Enhanced Communication Capacity. In *In AIAA Guidance, Navigation, and Control Conference*. https://doi.org/10.2514/6.2012-4610
12. Zong, P., & Kohani, S. (2019). Optimal Satellite Leo Constellation Design based on global coverage in one revisit time. *International Journal of Aerospace Engineering, 2019*, 1–12. https://doi.org/10.1155/2019/4373749
13. World Shipping Council. (2019). The top 50 container ports. World Shipping Council. Retrieved January 21, 2022, from https://www.worldshipping.org/top-50-ports
14. Marine Traffic. (n.d.). Global ship tracking intelligence: AIS MARINE TRAFFIC. MarineTraffic. Retrieved July 6, 2022, from https://www.marinetraffic.com/
15. Spaceflight101. (n.d.). Iridium-next—Spacecraft & satellites. Spacecraft Satellites. Retrieved January 21, 2022, from https://spaceflight101.com/spacecraft/iridium-next/

Crowd-Clustering Detection and Dispersion Robot

Mei Xi Chia, Komati Reddy Nikhita, and Yew Keong Wong

Abstract The ongoing COVID-19 pandemic has popularised the usage of Bluetooth signals from devices to assist the process of contact tracing. The purpose of our project was to utilise Bluetooth signals for an alternate purpose: to read devices' Received Signal Strength Indicator (RSSI) values in order to create a robot that would assist in cluster detection and dispersion, hence easing the enforcement of Safe Management Measures (SMM).

Keywords Bluetooth · Smart devices · Autonomous · Obstacle avoidance

1 Introduction

Nowadays, the violation of SMMs, especially those regarding safe distancing and gathering size, is not uncommon. A Straits Times article reported in April 2021 that 57 people breached SMMs, bringing to light a serious issue [1]. Yet, given that the COVID-19 virus is primarily spread through the "respiratory droplets of infected people" [2], SMMs play a big role in reducing the spread of the virus. By limiting the number of people one could come in contact with, any spread of the virus would be more contained within one group. Despite fines and the presence of safe distancing ambassadors, there continues to be a flouting of such measures in public. Furthermore, safe distancing ambassadors have been assaulted whilst enforcing SMM. In such situations, our robot will become useful in enforcing these measures by detecting "illegal" clusters and sounding an alarm to disperse them. Thus, this would reduce the manpower needed and prevent safe distancing ambassadors from getting assaulted, easing SMM enforcement.

M. X. Chia
Nanyang Girls' High School, 2 Linden Drive, Singapore 288683, Singapore

K. R. Nikhita
Raffles Girls' School, 2 Braddell Rise, Singapore 318871, Singapore

Y. K. Wong (✉)
DSO National Laboratories, 12 Science Park Drive, Singapore 118225, Singapore
e-mail: wyewkeon@dso.org.sg

H. Guo et al. (eds.), *IRC-SET 2022*,
https://doi.org/10.1007/978-981-19-7222-5_29

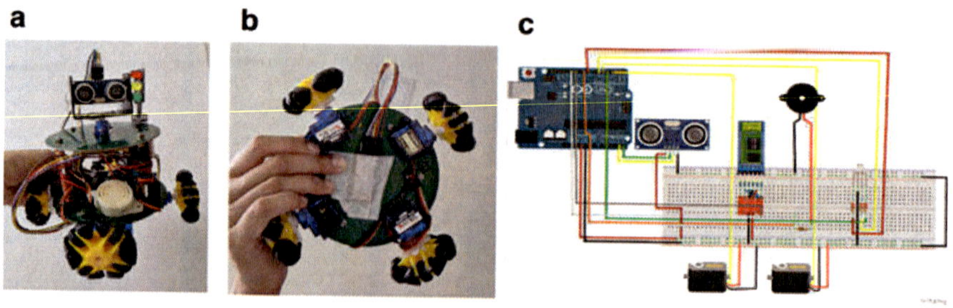

Fig. 1 **a–c** Pictures of our robot—front view, bottom view and Fritzing circuit diagram

2 Materials and Methods

2.1 *Assumptions*

- People in the region around the robot are stationary.
- The robot is deployed on flat and even terrain.
- People in the crowd are using Bluetooth-enabled devices such as mobile phones.

2.2 *Methodology*

2.2.1 Hardware Connections and Components

We utilised Fritzing to facilitate the visualisation of connections between various circuit components, followed by extensive testing on the individual circuit components before integrating the various parts onto the Printed Circuit board (PCB) of the robot. For example, calibration was done with the continuous rotation servos that drove the robot to determine the delay time needed for the robot to rotate an angle of 60°, as well as the delay time needed for the robot to move forward by a distance of 1 m. The recordings were taken with the Arduino powered by a USB-A cable, which was connected to a laptop.[1] We then integrated the wiring of the various components onto the PCB board, altering the connections mapped on Fritzing to suit the PCB (Fig. 1 and Table 1).

[1] Experimentation found that the recordings differed when different power sources were used, due to varying power inputs.

Table 1 Hardware components and purposes

Components	Purpose
Arduino UNO R3	Microcontroller to upload code and integrate hardware components together
Custom printed circuit board (PCB)	PCB card for power distribution to robot's electronics and servo motors
Voltage regulator	3–5 V boost converter mounted on PCB for power regulation to electronics and motors
HC-05 Bluetooth module	Receive RSSI signals
Voltage level shifter	Powers Bluetooth module such that it only receives 3.3 V
Continuous rotation servo motor	Servo motor connected to wheels to drive robot
HC-SR04 ultrasonic sensor	To detect obstacles in the robot's path
Traffic light module	Acts as a visual signal to what process the robot is carrying out
Buzzer	To alert people that they are in a cluster of more than the allowed number and to disperse the cluster

2.2.2 Literature Review: Calculation of Distance from RSSI Values

Gansemer et al. [3] use RSSI fingerprinting and Euclidean distance for indoor positioning and navigation specifically. This technology is especially useful in dynamically changing environments, however, has not been utilised for crowd clustering purposes.

Huang et al. [4] discuss the development of a distance estimation algorithm for Bluetooth devices using RSSI values in their paper. Their algorithm uses a median filter to filter out outliers resulting from noise present in the data collected, resulting in the following algorithm where $\text{RSSI}_{\text{refer}}$ is the RSSI value for a 1 m distance between two Bluetooth devices and $\text{RSSI}_{\text{median}}$ the median of ten most recent RSSI values collected.

$$\text{distance} = 1.37 \times \left(\frac{\text{RSSI}_{\text{median}}}{\text{RSSI}_{\text{refer}}}\right)^{6.854} - 0.372, \text{RSSI}_{\text{median}} > \text{RSSI}_{\text{refer}}$$

$$\text{distance} = 0.979 \times \left(\frac{\text{RSSI}_{\text{median}}}{\text{RSSI}_{\text{refer}}}\right)^{7.489} + 0.183, \text{RSSI}_{\text{median}} < \text{RSSI}_{\text{refer}}$$

Experimentation with the formula above resulted in inaccuracies likely due to vastly different output from different hardware. Hence, we devised our own formula for the conversion of RSSI values to distances by first using the HC-05 Bluetooth module to obtain ten RSSI values of a mobile phone when placed at distances ranging from 0.00 m to 8.00 m away from the Bluetooth module. A graph of RSSI values against distance was then plotted, and the equation of the line of best fit of the graph was determined (Fig. 2).

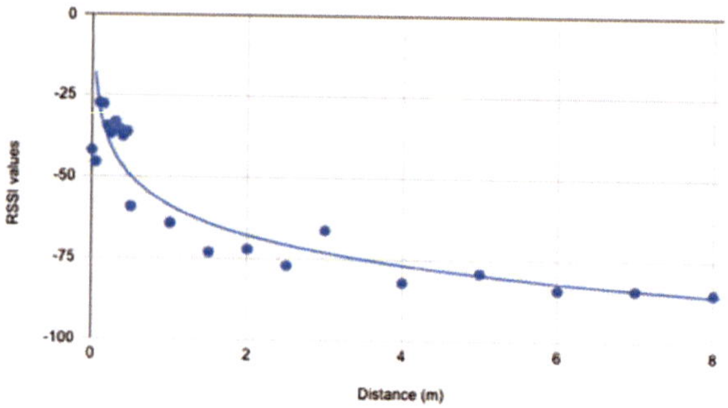

Fig. 2 Line of best fit of graph of RSSI values against distance

The formula we devised to convert RSSI values to distance is

$$\text{distance} = 0.0237e^{-0.0661 \times \text{RSSI value}}$$

2.2.3 Development of Crowd Clustering Algorithm

The crowd clustering algorithm was primarily developed on the basis of the difference in distances of people to the robot's position since as the robot moves, the distance of people with respect to the robot's position would change. Hence, the difference in distance of each person with respect to the different positions of the robot would be similar when the people belonged to the same cluster. A simulation of the robot's movement was conducted with Processing,[2] and the distances of 20 stationary "people" with reference to a roaming "robot" were collected at three different positions. This data was then plotted on a graph of distance against position, as shown in Fig. 3. As the robot moves from position to position, the people whose variation in their distances follow a similar trend will belong to the same cluster. We conducted further experimentation to determine the exact movement of the robot and collection of data (Fig. 4).

2.2.4 Development of Angle Determination Algorithm

According to the results from the experimentation, the robot will be collecting four sets of data in total, three sets of data will be used in the crowd clustering algorithm and the last set of data will be used to determine the angle of the cluster with respect to the robot after the clusters have been detected. The most optimum robot movement would be an equilateral triangle, but since four sets of data need to be collected, the robot will move in a rhombus to make the calculation of the angles easier, and to

[2] Processing is a Java-based software sketchbook that specialises in programming in the context of visual arts. We used this software to visualise the robot and the people in its surrounding environment to conduct simulations of the robot's movements and actions.

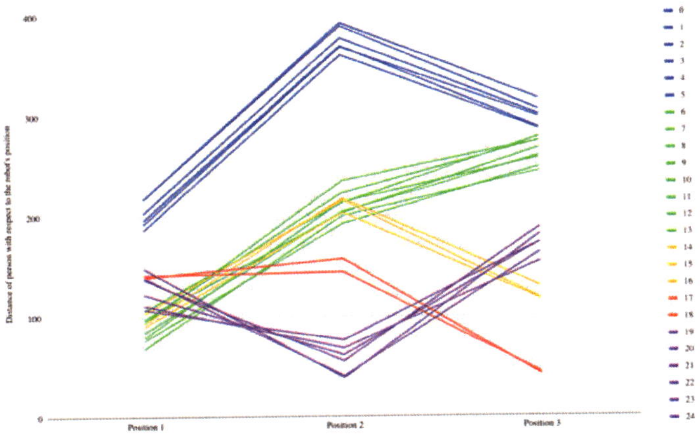

Fig. 3 Graphical representation of the distance of people with respect to the robot's position

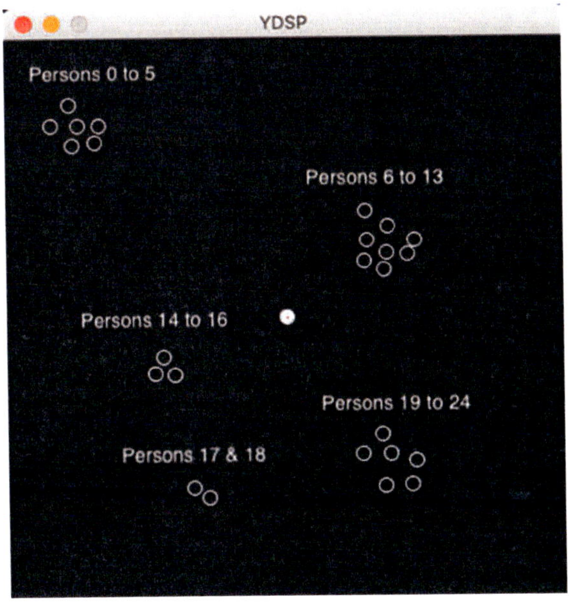

Fig. 4 Example of visualisation on Processing corresponding to the distance data in Fig. 3, where robot is represented using a closed circle and people using open circles

ensure that the robot collects three sets of data at the vertex of an equilateral triangle for the algorithm (Fig. 5).

The angles of all the people in a cluster with respect to the robot's original position (Robot Position 1) are calculated. The angle of a person with reference to the robot, θ, can be found using the cosine rule, $\theta = \cos^{-1}\left(\frac{x^2+a^2-b^2}{2xa}\right)$. The mean of the angles of all people in a cluster is taken to find the angle of the cluster with respect to the robot's original position. There are four possible scenarios of where a cluster

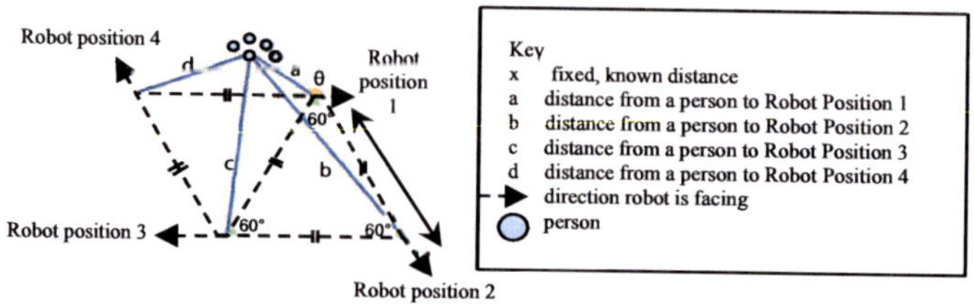

Fig. 5 Visualisation of the angle determination algorithm

could be located with respect to the robot's original position and depending on each scenario, the angle is determined. In order to be able to differentiate between each of the scenarios, a set of conditions were designed, which were based on the distances from a person in the cluster to all four positions of the robot. The conditions for each scenario are as follows[3]:

Scenario 1: $0 < \theta \leq 90$ • $\min(a, b, c, d) = a$ • $b - d < y$, where $y = 125$ • $\max(a, b, c, d) = c$ or $c - d < 20$	Scenario 3: $180 < \theta \leq 270$ • $\min(a, b, c, d) = c$ or $\min(a, b, c, d) = d$
Scenario 2: $90 < \theta \leq 180$ • $\min(a, b, c, d) = a$ or $\min(a, b, c, d) = d$	Scenario 4: $270 < \theta \leq 360$ • $\min(a, b, c, d) = a$ or $\min(a, b, c, d) = b$ • $b - d > y$, where $y = 125$

3 Results

3.1 *Experimentation on Processing (Simulation Software)*

Two experiments were conducted. Experiment 1 aimed to determine the optimum number of data sets needed to achieve a high accuracy and examine whether the density of people in an area affected the accuracy of the algorithm. For this experiment, we generated clusters of varied sizes and densities. Six cases were created where each case has a cap of 25 people:

- Case 1: Small clusters of 2 s and 3 s spaced close to each other
- Case 2: Small clusters of 2 s and 3 s spaced far apart
- Case 3: Large clusters of 5 s and 6 s spaced close to each other
- Case 4: Large clusters of 5 s and 6 s spaced far apart
- Case 5: A mixture of small and large clusters spaced close to each other
- Case 6: A mixture of small and large clusters spaced far apart.

[3] The value of y refers to that in the simulation. y is later recalibrated according to x for real-time testing.

Table 2 Accuracy values for experiment 1

Cases/accuracies for different number of sets of data used	3 data points (%)	4 data points (%)	5 data points (%)
Case 1	94.4	91.7	94.4
Case 2	86.4	90.9	89.4
Case 3	46.7	26.6	46.7
Case 4	66.7	53.3	60
Case 5	85.7	66.7	66.7
Case 6	95.2	85.7	81.0

Experiment 2 aimed to investigate the optimum positions of the robot. We compared the accuracy with which the robot detected clusters when it moved in a straight line (vertical, horizontal and diagonal), right-angled triangle, isosceles triangle and in an equilateral triangle. We also used different sizes of the shapes mentioned above in order to determine if the distance the robot moved from position to position affected the accuracy of the clustering algorithm.

A scoring metric was also developed by us, which included an assumption that detecting two clusters as one big cluster is considered as poorer performance than detecting one cluster with an additional one or two people who do not belong to that cluster.

3.1.1 Experiment 1

According to Table 2, the optimum number of sets of data needed to be collected is three. It was also found that the greater the density of clusters in the area, the poorer the performance of the algorithm except in cases that involve small clusters as it could be hard to distinguish between clusters in areas with high densities of people, leading to scenarios of low accuracy such as Case 3.

3.1.2 Experiment 2

According to our results, positioning the robot along a triangle generally resulted in a much higher accuracy than positioning it along a straight line of any orientation, as this allowed for a greater coverage of the area and hence a greater variety of values. Furthermore, to determine which specific triangle should be used, the algorithm was tested across the 6 cases generated in Experiment 1 (Table 3). Although an isosceles triangle movement led to the highest accuracies across the tested cases, we decided to place the robot on the vertices of an equilateral triangle for the crowd clustering process instead as this would lead to a better area coverage for the same distance travelled, according to Zenodorus's Theorem stating that "of all rectilinear figures

Table 3 Accuracy values for experiment 2

Type of triangle	Average accuracy across all 6 cases (%)
Right-angled triangle	69.7
Isosceles triangle	81.2
Equilateral triangle	79.2

Table 4 Crowd clustering algorithm accuracies

Cases	Accuracy of algorithm (%)
1	93.1
2	74.2
3	46.7
4	80
5	83.3
6	90.5
Average	78.0

having an equal number of sides and equal perimeter, the greatest is that which is equilateral and equiangular".

3.2 Crowd Clustering Algorithm Accuracy

The algorithm was tested in Processing for all 6 cases. By observing the accuracies in Table 4, we can say that our algorithm is generally efficient, with some exceptions in highly dense areas.

3.3 Buzzer[4]

The pitch and volume at which the buzzer is sounded can be indirectly controlled by altering its pulse-width moderation (PWM), which alters the duty cycle of the power supplied to the buzzer. To investigate this relation, the buzzer was sounded at PWM values ranging from 0 to 255, and their volumes were analysed using a sound recorder. A buzzer sound of PWM value 150 was picked by selecting a sound neither too piercing nor soft, then matching the sound with the PWM value printed on the Serial Monitor in Arduino, as shown in Video 1 located in our GitHub repository.[5] A resistor was also added to further control the volume of the buzzer.

[4] Note: The Buzzer can be attached to any PWM pin on Arduino, excluding pins 9 and 10, as the use of the Servo library on the Arduino UNO disables PWM functionality on these 2 pins.

[5] https://github.com/chiameixi/crowd-clustering-robot.

3.4 *Integrated Arduino Code and Hardware*

The robot will first check for obstacles in its path. If none are present, it will kick start its rhombus movement, collecting the RSSI values and converting them to distances. The crowd clustering and angle determination algorithms are then used to determine the presence of crowds, and if present, the angle at which the crowd(s) are located with respect to the robot's original position. The robot moves towards the cluster, and sounds a buzzer to disperse it before returning to its original position, and repeating the process with other clusters until all clusters have been alerted. The robot then moves to a new region to restart the process. A function for detecting obstacles is also put in place to ensure that the robot does not collide into any obstacles during this process. Thereafter, it would restart the algorithm to detect the presence of crowds in this new region. A flowchart of the algorithm, as well as an instance of the robot moving to and alerting a cluster has been recorded and is available in our GitHub repository. The mobile phone shown in the video represents a cluster of people who all have mobile phones producing Bluetooth signals.

4 Discussion

4.1 *Experimentation*

4.1.1 Experiment 1

A common error made by the algorithm during this experiment was to classify a cluster as two separate clusters or to merge two small clusters together, as seen in Case 2. The main reason for this could be that the clusters are too small and so far apart that even if a person is slightly closer to a small cluster, the algorithm might classify them all under the same cluster. Other cases which involve either larger clusters or a mixture of smaller and larger clusters perform better in less dense regions primarily because the algorithm is more able to distinguish between each specific cluster, yielding a higher accuracy. However, in denser regions, it becomes harder to distinguish between clusters since there are more people in the proximity of a cluster. These issues likely arose because the threshold for the difference in distance might be too high and low for the cases, respectively, causing the clustering to be either over or under specific. For such situations, increasing the threshold for the former case and lowering it for the latter one would enable the algorithm to better detect clusters in the two cases. However, altering the threshold solely for the purpose of higher accuracy in these two unique cases would lead to higher inaccuracies in detecting more typical clusters.

4.1.2 Experiment 2

For this experiment, plainly looking at the accuracy of the algorithm is insufficient. It is crucial to note the position of the robot with respect to the surrounding clusters as well. When the robot is located closer to the clusters, it is easier for it to distinguish between clusters.

4.2 *Calculation of Distance from RSSI Values*

It was found that the RSSI to distance calibration could depend on the type of mobile phones, due to different signal strengths and hardware. This inaccuracy would result in the robot occasionally moving in the incorrect direction. To overcome this issue, Bluetooth data from a greater diversity of phone types needs to be collected.

4.3 *Challenges*

One major source of error during our experimentations was that the wheels of the robot rotated at different speeds, which resulted in the path of the robot being a curve rather than a straight line, causing data collected to be inaccurate. We overcame this by programming the wheels on either side of the robot to run at different speeds. Through experimentation where the speed of the faster wheel was decreased in constant decrements, a correction factor, that is unique to each robot, was applied to the faster wheel to allow it to move at the same speed as the slower one.

Aside from that, hardware errors from the Bluetooth module occasionally caused data collected to be inaccurate: while 4-digit hexadecimal RSSI values were expected, 3-digit data was sometimes collected. When such data was used as part of the crowd clustering or angle determination algorithm, the calculations obtained were often either invalid or inaccurate, leading to errors.

Another challenge we commonly encountered was inconsistent delay times for the rotation and the movement of the servo motors, i.e. time taken for the servo motor to rotate by a certain angle was different every time we tested it. Due to this issue, the movement of the robot in a rhombus could be slightly inaccurate from time to time and since the RSSI values collected by the Bluetooth module could be inaccurate as well, this could lead to cosine law error in the angle determination algorithm where cos(angle) > 1 or < -1. Currently, there is no solution for this problem, though possibilities include either restarting the algorithm or making the movement of the robot much more consistent which we have tried to do so. Nevertheless, such challenges lead to several limitations in our robot.

4.4 *Limitations*

Arduino UNO R3 has a low RAM memory and is unable to carry out concurrent actions, using bare metal programming instead. Thus, it is unable to be constantly on the lookout for obstacles, nor able to collect greater than five devices' Bluetooth data at any one time, greatly reducing the functionality of the robot.

Additionally, Android phone's Bluetooth signal can only be discovered by other devices—such as the Bluetooth module—under a discoverable mode, which requires the installation of an additional app. The Bluetooth module's range is also only 10 m, although our experimentations showed that the RSSI of mobile phones located greater than 8.00 m away were generally too weak to be detected by the Bluetooth module which also limits the extent to which our robot can be utilised.

Another limitation of the crowd clustering algorithm is that it is less accurately able to detect clusters that are within the rhombus that the robot moves which would cause clusters within the rhombus to be considered as hidden clusters. To investigate this limitation further, we tried to place a cluster of 25 people in the middle of the rhombus, in Processing; however, the algorithm classified this cluster as three clusters with 7, 11 and 7 people each. When the angles of the cluster with respect to the robot's original position was calculated, the angles were approximately the same: 208.5°, 215.5° and 221.6°. Hence, this limitation could be overcome by checking if any clusters have similar angles with respect to the robot's original position and similar distance values.

5 Conclusion

The current COVID-19 situation continually emphasises the importance of a robot for crowd clustering purposes, and the development of which would benefit many. This project has been successful in being able to create a robot from scratch, along with original algorithms. Our robot is currently just a prototype, but it is nevertheless a successful proof of concept upon which future work can be built.

6 Future Work

To translate our robot to a fully functional one, the microcontroller could be replaced with one possessing greater RAM and processing speed, the software could be modified to allow for concurrent processing and our robot could be integrated into TraceTogether's ecosystem to allow our Bluetooth module to detect the Bluetooth broadcast of phones or TraceTogether tokens. The addition of a camera and computer vision capabilities to assist in cluster and obstacle detection would also be useful. Lastly, the use of a compass can be helpful in increasing the accuracy of the exact heading of clusters which would make the robot's direction of movement more precise.

To account for moving Bluetooth devices, multiple sets of Bluetooth signals in short periods of time can be collected. Since Bluetooth signals are unique to each device, such collection of data would allow for the precise tracking and identification of moving Bluetooth devices. Thus, this makes possible the identification of mobile clusters that can be distinguished from stationary clusters, enhancing the functionality of our robot.

The developed algorithms could also be employed for crowd monitoring and management in large-scale events where large volumes of people are expected. Such algorithms can also be useful in rescue operations where the presence of people can be determined and the location can be identified; however, this depends if each individual is carrying a functioning Bluetooth device that has not been damaged.

Acknowledgements We would like to thank our mentor, Mr. Wong Yew Keong for his invaluable guidance and support throughout the entire duration of the project. We would also like to extend our gratitude to Ms. Joyce Yao for the administrative support and to DSO National Laboratories for providing such an enriching opportunity.

References

1. Goh, T. (2021, April 30) 57 people, 15 outlets breached safe distancing measures; enforcement to be stepped up immediately. Retrieved December 29, 2021, from https://www.straitstimes.com/singapore/health/57-people-15-outlets-breached-safe-distancing-measures-recently-enforcement-to-be
2. Ministry of Health Singapore. (2021, November 14). MOH | General information about COVID-19. Ministry of Health. Retrieved December 29, 2021, from https://www.moh.gov.sg/covid-19/general
3. Gansemer, S., Großmann, U., & Hakobyan, S. (2010). RSSI-based euclidean distance algorithm for indoor positioning adapted for the use in dynamically changing WLAN environments and multi-level buildings. In *International conference on indoor positioning and indoor navigation (IPIN)*. https://www.researchgate.net/publication/224198842_RSSI-based_Euclidean_Distance_algorithm_for_indoor_positioning_adapted_for_the_use_in_dynamically_changing_WLAN_environments_and_multi-level_buildings
4. Huang, J., Chai, S., Yang, N., & Liu, L. (2017). A novel distance estimation algorithm for Bluetooth devices using RSSI. *Advances in Intelligent Systems Research, 134*, 379–381. Retrieved December 29, 2021, from https://www.atlantis-press.com/article/25881194.pdf

Synthesis of a Hyper-Porous Fluorinated Covalent Organic Framework with Customisable Sites

Solomon Lim Jun Hui

Abstract Covalent organic frameworks (COFs) have been recognised as an excellent candidate for gas-phase storage, separation and catalysis due to their high porosity and stability and has been explored as a versatile and robust molecular platform, but the use of interchangeable substituents to enhance porosity and stability while retaining a functional framework has yet to be deeply explored. Herein, we report the properties and synthesis of a fluorinated COF, showing improved porosity as well as modifiable linkers for modular functionality. This was accomplished via imine condensation between a non-fluorinated monomer (5,10,15,20-tetrakis (4-aminophenyl)porphyrin) and a fluorinated monomer (2,3,5,6-tetrafluoroterephthaldehyde) to produce the COF in good yield. This compound was then compared with related COFs in a two-way comparison to relate their properties, such as the non-fluorinated analogues of the porphyrin-based COF, as well as COFs incorporating tris(aminophenyl)benzene. Upon analysis with PXRD and UV–Vis spectroscopy, correlating UV–Vis absorbance with pore size and thus porosity, it was concluded that the incorporation of fluorine in the compound indeed resulted in higher porosity in the products, possibly owing to the formation of electron-deficient tetrafluorobenzene centres in the COF, inducing coordination with the electron-rich porphyrin and triphenylbenzene units.

Keywords Covalent organic framework · Fluorination · Site customisability

1 Introduction

Covalent organic frameworks (COFs) have been considered a subject of great interest over the past decade due to their excellent stability, crystallinity, as well as porosity, which naturally makes them suitable candidates for modern applications such as gas capture [1], photo-optical manipulation [2] and catalysis [3]. The main functionality of COFs which lend themselves to such diverse and varied applications are the

S. L. J. Hui (✉)
NUS High School of Mathematics and Science, 20 Clementi Avenue 1, Singapore 129957, Singapore
e-mail: h2010077@nushigh.edu.sg

H. Guo et al. (eds.), *IRC-SET 2022*,
https://doi.org/10.1007/978-981-19-7222-5_30

modifiable linkers of the structure, which allows chemists to customise the properties of the "walls" of the structure, and thereby customising the interactions between the COF and guest molecules. Various strategies for modifying the functionalities of the COF have been developed, including post-synthetic modification [4, 5], pore-wall modification [6], click chemistry for extended groups [7, 8], etc. However, the preservation of the porosity and crystallinity of the COF is susceptible to be degraded by the use of harsh reagents for linker modification, especially with imine-linked COFs [9], which are formed via a reversible condensation reaction.

In this project, we investigated the use of modifiable functionalities along the walls of COFs that have a dually beneficial nature, in that they are able to enhance the porosity and crystallinity of the COF, as well as able to be customised in a range of versatile modification routes. The presence of fluorine substituents on the COF linker has been shown to improve the catalytic potential of the resulting COF compared to its non-fluorinated analogue [10], as well as being able to be readily swapped for more useful functional groups, such as sulphur-based functionalisation for use in lithium–sulphur batteries [11]. The presence of aryl fluoride functional groups along the pore walls of the corresponding COFs could also be used for various substitution reactions, such as Wurtz–Fittig substitution [12] and Buchwald–Hartwig amination [13] (Fig. 1).

The presence of fluorine substituents also influences the hydrophobicity of the resultant COF, allowing for a wider range of applications of such imine-linked COFs [14]. The coordinative effect of electron-deficient tetrafluorobenzene units with electron-rich linkers has been investigated [15, 16], albeit in the context of supramolecular complexes. Hence, in this project we investigated the synthesis

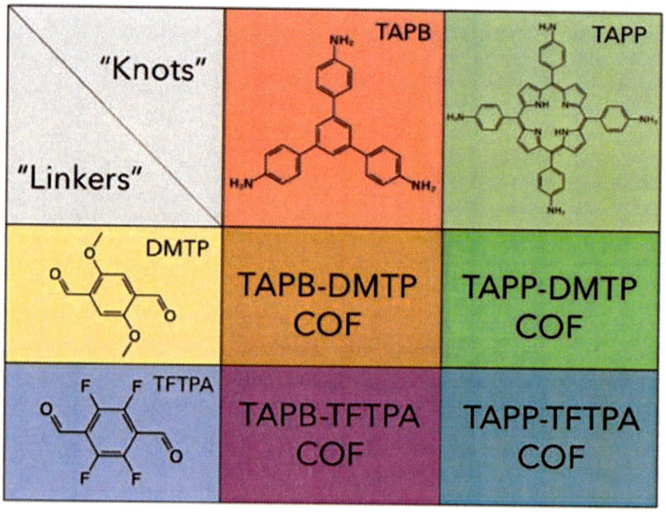

Fig. 1 Visualisation of the monomers used to prepare the compounds in a two-way comparison

of a fluorinated covalent organic framework TAPP-TFTPA COF utilising the perfluorinated linker monomer tetrafluoroterephthaldehyde (TFTPA), and compared it to related analogues, such as TAPP-DMTP COF, TAPB-DMTP COF, and TAPB-TFTPA COF to investigate the effects such functionalities have on the porosity of the COF.

2 Materials and Methods

2.1 *Synthesis of Monomers*

2,3,5,6-tetrafluoroterephthaldehyde was used as purchased from TCI, 2,5-dimethoxyterephthaldehyde was synthesised according to a literature procedure [17], and the following compounds tris(aminophenyl)benzene and tetrakis(aminophenyl)porphyrin were synthesised according to the procedures below.

Synthesis of tris(aminophenyl)benzene. 1,3,5-tribromobenzene (20 mmol, 6.300 g) and 4-(4,4,5,5-tetramethyl-1,3,2-dioxaborolan-2-ly)aniline (80 mmol, 17.35 g) were dissolved in 120 mL of DMF, before an aqueous solution of K_2CO_3 (140 mmol, 19.35 g) dissolved in 60 mL of water was added to the reaction mixture. After being flushed with argon gas three times, tetrakis(triphenylphosphine)palladium(0) (0.3 mmol, 0.347 g) was added, and the mixture is heated at 100 °C for 24 h under constant stirring, changing colour from yellowish-brown to black. The crude product is purified via recrystallisation in ethanol, affording a beige-coloured solid. Yield: 5.759 g, 81.9%. ^{1}H NMR (DMSO-*d*6, 400 MHz) δ (ppm) = 7.50 (*t*, 9H, $J = 5.0$ Hz), 6.69 (*d*, 6H, $J = 8.0$ Hz), 5.29 (*s*, 6H).

Synthesis of tetrakis(nitrophenyl)porphyrin. 4-nitrobenzaldehyde (145 mmol, 22.0 g) and acetic anhydride (254 mmol, 24 mL) were added to 600 mL of propionic acid, before refluxing the reaction mixture at 140 °C for 30 min. 10 mL of pyrrole in 45 mL of propionic acid was then added to the reaction mixture dropwise, causing a black solution to be formed. The mixture was then allowed to be cooled and stand for 24 h. A dark solid was then obtained via filtration and washing with methanol, before being refluxed at 120 °C for 1 h with 160 mL of pyridine, after which it was filtered and washed repeatedly with acetone, and dried under vacuum to yield a purple solid. Reported yield: 22%.

Synthesis of tetrakis(aminophenyl)porphyrin: Tetrakis(nitrophenyl)porphyrin (5 mmol, 4.0 g) was added to 300 mL hydrochloric acid (and the mixture de-aerated with argon gas for 30 min), before $SnCl_2$ (79.8 mmol, 18 g) in 200 mL hydrochloric acid (also de-aerated with argon gas in a similar manner) was added dropwise to the reaction mixture at 80 °C. The mixture is then cooled to room temperature for 24 h. It is then filtered and washed repeatedly with water, affording a tarry, dark-purple solid. This solid is then dissolved in 400 mL of 5% NaOH in an ice bath, and left to stir for 3 h. A purple solid is then obtained thereafter via filtering and washing with chloroform. Reported yield: 50%

2.2 Synthesis of Covalent Organic Frameworks (COFs)

Synthesis of TAPB-DMTP COF. Tris(aminophenyl)benzene and 2,5-dimethoxyterephthaldehyde (amounts in Table 1) are dissolved in a mixture of solvents (type and amounts in Table 1) in a Pyrex tube, before 6 M AcOH (amount in Table 1) is added and the tube is degassed via three freeze–pump–thaw cycles and flame-sealed. The mixture is then heated at 120 °C for 96 h, before being filtered and washed with THF to yield a yellow solid, which then undergoes Soxhlet extraction with THF for 24 h, then vacuum-dried at 120 °C overnight to obtain the product.

Synthesis of TAPP-DMTP COF. Tetrakis(aminophenyl)porphyrin and 2,5-dimethoxyterephthaldehyde (amounts in Table 1) are dissolved in a mixture of solvents (type and amounts in Table 1) in a Pyrex tube, before 6 M AcOH (amount in Table 1) is added and the tube is degassed via three freeze–pump–thaw cycles and

Table 1 Synthesis conditions for COFs

Sample No.	Monomer A/amount (mg)	Monomer B/amount (mg)	Solvent 1/amount (mL)	Solvent 2/amount (mL)	Amount of acid (mL)	Batch No
1	TAPB (12.30)	TFTPA (10.83)	n-BuOH (0.14)	N.A	0.28	1
2	TAPB (12.30)	TFTPA (10.83)	Mesitylene (0.14)	Dioxane (0.29)	0.41	
3	TAPP (17.73)	TFTPA (10.83)	EtOH (0.66)	Mesitylene (0.66)	0.13	
4	TAPP (17.73)	TFTPA (10.83)	o-DCB (0.92)	n-BuOH (0.92)	0.18	
5*	TAPB (28.10)	DMTP (23.30)	o-DCB (0.50)	n-BuOH (0.50)	0.10	
6	TAPB (28.10)	DMTP (23.30)	o-DCB (0.50)	n-BuOH (0.50)	0.10	
7*	TAPP (11.81)	TFTPA (10.83)	o-DCB (0.70)	n-BuOH (0.50)	0.18	2
8*	TAPP (11.81)	DMTP (10.19)	o-DCB (1.00)	n-BuOH (0.50)	0.20	
9*	TAPP (12.30)	TFTPA (10.83)	o-DCB (0.50)	n-BuOH (0.50)	0.10	

Samples indicated with an asterisk were used for UV Visible spectroscopy flame-sealed. The mixture is then heated at 120 °C for 96 h, before being filtered and washed with THF to yield a deep purple solid, which then undergoes Soxhlet extraction with THF for 24 h, then vacuum-dried at 120 °C overnight to obtain the product.

Synthesis of TAPB-TFTPA COF. Tris(aminophenyl)benzene and 2,3,5,6-tetrafluoroterephthaldehyde (amounts in Table 1) are dissolved in a mixture of solvents (type and amounts in Table 1) in a Pyrex tube, before 6 M AcOH (amount in Table 1) is added and the tube is degassed via three freeze–pump–thaw cycles and flame-sealed. The mixture is then heated at 120 °C for 96 h, before being filtered and washed with THF to yield an orange solid, which then undergoes Soxhlet extraction with THF for 24 h, then vacuum-dried at 120 °C overnight to obtain the product.

Synthesis of TAPP-TFTPA COF. Tetrakis(aminophenyl)porphyrin and 2,3,5,6-tetrafluoroterephthaldehyde (amounts in Table 1) are dissolved in a mixture of solvents (type and amounts in Table 1) in a Pyrex tube, before 6 M AcOH (amount in Table 1) is added and the tube is degassed via three freeze–pump–thaw cycles and flame-sealed. The mixture is then heated at 120 °C for 96 h, before being filtered and washed with THF to yield a deep purple solid, which then undergoes Soxhlet extraction with THF for 24 h, then vacuum-dried at 120 °C overnight to obtain the product.

2.3 *Characterisation of COFs*

A Bruker D2 Phaser (Cu Kα, $\lambda = 1.54$ Å) was utilised for PXRD characterisation of the first batch of COFs, and a JASCO V-750 UV Visible Spectrophotometer was used for measuring the UV–Vis spectra of the samples, which were suspended in water.

3 Results and Discussion

3.1 *COF Crystallinity and Comparison*

Out of the six initial COFs prepared in the first batch, the only samples to display significant crystallinity were TAPB-DMTP COFs, as evidenced in the PXRD graph (Fig. 2). This could have been in part due to little to no interlayer stacking due to the influence of the electron-deficient tetrafluorobenzene units on the π–π stacking between layers. Intralayer regularity could also have been affected by the presence of fluoro substituents, as the highly electrophilic nature of the linkers would disrupt the molecular arrangements during the imine-bond formation phase. Thus, the intralayer

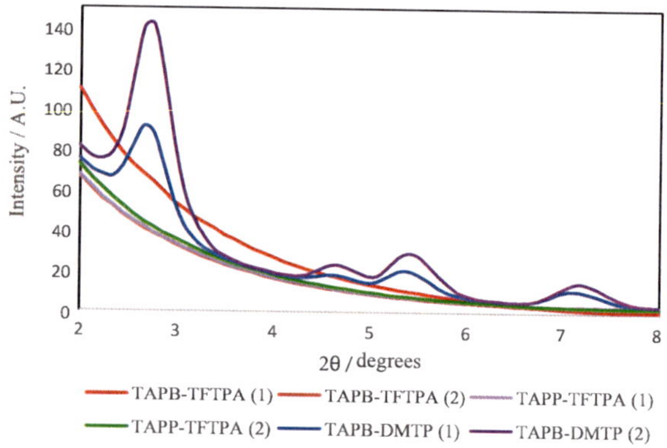

Fig. 2 PXRD graphs of each preliminary sample

and interlayer effects of the aryl fluoride substituents indicate little to no crystallinity in the fluorinated samples.

3.2 *Effect of Fluorination on COF Porosity*

As evidenced from Fig. 3, the fluorinated analogues of the compounds both had higher attenuation coefficients than the non-fluorinated analogues at the same concentration in water, showing that the fluorinated compounds possessed a smaller average pore size (correlating attenuation coefficient to pore size [18, 19]), possibly from the coordination of electron-rich tris(aminophenyl)benzene and tetrakis(aminophenyl)porphyrin units to the electron-poor tetrafluorobenzene units, forming smaller pores bounded by the polymerised monomers attracted to each other.

The fluorinated compounds also have extremely similar attenuation coefficients despite possessing different knot units, showing that the effect of fluorination produces smaller pore size, at least in amorphous samples.

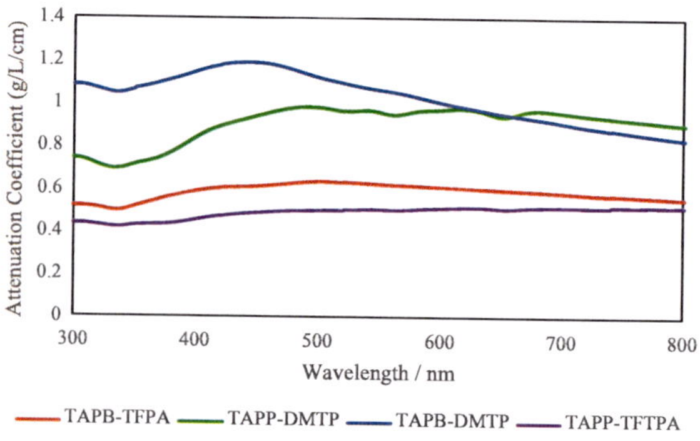

Fig. 3 UV–Vis spectra of all 4 samples. Both TAPB-based compounds are shown to possess higher absorbance than their porphyrin-based analogues

3.3 *Difference in Porosity Between Knot Units*

As observed in Fig. 3, the TAPB COFs have a higher attenuation coefficient (and thus pore size) than their porphyrin-based analogues, with an especially consistent point in attenuation coefficient magnitude seen at approximately 350 nm, which is in agreement with reported data for such compounds [20]. This could in part be due to the smaller "bonding" angle of the porphyrin unit to the linkers as compared to the TAPB unit. Hence, we can see that the porphyrin-based fluorinated compound offered the highest porosity (smallest pore size), due to a combination of the inherently more porous porphyrin framework combined with the coordinating effect of the tetrafluorobenzene units.

The non-fluorinated compounds also show greater fluctuations in attenuation coefficient (especially in the 350–450 nm range) as compared to their fluorinated analogues. This can also be seen in a macroscopic comparison between the samples, e.g. TAPB-DMTP COF is yellow in colour, whereas TAPB-TFTPA COF is orange-red. TAPP-DMTP COF is purple-blue in colour, whereas TAPP-TFTPA COF is deep purple.

Overall, we can make three inferences from the data gathered:

(1) Fluorination induces higher porosity
(2) Fluorination reduces crystallinity
(3) TAPP-based COFs usually have smaller pore size than TAPB-based analogues.

4 Future Work

As a continuation of the project, the functionalisation of the TAPP-TFTPA COF can be investigated via substitution of the fluorine substituents, and the influence of such effects on the resulting compound's porosity. Furthermore, the porphyrin units in the compound also allow for the functionalisation of the knot units with metal ions, which may serve to further enhance the practical applications of the resulting compound, especially in the fields of gas absorption/catalysis, utilizing the high surface area/porosity and customisable sites of the COF.

This would be a versatile compound to act as a baseline for future experiments which require not only high-porosity materials, but materials that can be functionalised on the molecular scale without too big a trade-off between modification and loss of porosity.

5 Conclusion

In conclusion, a fluorinated porphyrin-based COF has been synthesised, and compared to related compounds in terms of fluorination and monomer type. It has been found that fluorination results in decreased crystallinity due to the coordination effects of the electron-deficient tetrafluorobenzene units disrupting inter-layer stacking, but increases porosity/surface area, as shown through UV–Vis spectroscopy and PXRD measurements. The customisable and functionalisable properties of such a compound will indubitably be useful in future work regarding modular functions of COFs in gas-phase applications, while retaining high porosity, which is essential to such applications.

Acknowledgements Many thanks to Professor Jiang Donglin and Liu Ruoyang of the Department of Chemistry, NUS for their invaluable guidance and assistance with the project. Thank you to Dr. Maury Julien Jean Pierre and Dr. Chiam Sher-Yi from NUS High School of Mathematics and Science for their support and encouragement for the project.

References

1. Wang, C., Zhao, Y.-N., Zhu, C.-Y., Zhang, M., Geng, Y., Li, Y.-G., & Su, Z.-M. (2020, October 21). A two-dimensional conductive MO-based covalent organic framework as an efficient electrocatalyst for nitrogen fixation. *Journal of Materials Chemistry A*.
2. Yang, S., Hu, W., Zhang, X., He, P., Pattengale, B., Liu, C., Cendejas, M., Hermans, I., Zhang, X., Zhang, J., & Huang, J. (2018). 2D covalent organic frameworks as intrinsic photocatalysts for visible light-driven CO_2 reduction. *Journal of the American Chemical Society, 140*(44), 14614–14618. https://doi.org/10.1021/jacs.8b09705
3. Yusran, Y., Li, H., Guan, X., Fang, Q., & Qiu, S. (2020). Covalent organic frameworks for catalysis. *EnergyChem, 2*(3), 100035. https://doi.org/10.1016/j.enchem.2020.100035
4. Segura, J. L., Royuela, S., & Mar Ramos, M. (2019). Post-synthetic modification of covalent organic frameworks. *Chemical Society Reviews, 48*(14), 3903–3945. https://doi.org/10.1039/c8cs00978c
5. Ding, H., Mal, A., & Wang, C. (2020). Tailored covalent organic frameworks by post-synthetic modification. *Materials Chemistry Frontiers, 4*(1), 113–127. https://doi.org/10.1039/c9qm00555b
6. Grunenberg, L., Savasci, G., Terban, M. W., Duppel, V., Moudrakovski, I., Etter, M., Dinnebier, R. E., Ochsenfeld, C., & Lotsch, B. V. (2021). Amine-linked covalent organic frameworks as a platform for postsynthetic structure interconversion and pore-wall modification. *Journal of the American Chemical Society, 143*(9), 3430–3438. https://doi.org/10.1021/jacs.0c12249
7. Wu, M., Chen, G., Ma, J., Liu, P., & Jia, Q. (2016). Fabrication of cross-linked hydrazone covalent organic frameworks by click chemistry and application to solid phase microextraction. *Talanta, 161*, 350–358. https://doi.org/10.1016/j.talanta.2016.08.041
8. Li, K., Wong, N. K., Strauss, M. J., Evans, A. M., Matsumoto, M., Dichtel, W. R., & Adronov, A. (2021). Postsynthetic modification of a covalent organic framework achieved via strain-promoted cycloaddition. *Journal of the American Chemical Society, 143*(2), 649–656. https://doi.org/10.1021/jacs.0c11811
9. Haase, F., & Lotsch, B. V. (2020). Solving the COF trilemma: Towards crystalline, stable and functional covalent organic frameworks. *Chemical Society Reviews, 49*(23), 8469–8500. https://doi.org/10.1039/d0cs01027h
10. Liu, C., Xiao, Y., Yang, Q., Wang, Y., Lu, R., Chen, Y., Wang, C., & Yan, H. (2021). A highly fluorine-functionalized 2D covalent organic framework for promoting photocatalytic hydrogen evolution. *Applied Surface Science, 537*, 148082. https://doi.org/10.1016/j.apsusc.2020.148082

11. Wang, D.-G., Li, N., Hu, Y., Wan, S., Song, M., Yu, G., Jin, Y., Wei, W., Han, K., Kuang, G.-C., & Zhang, W. (2018). Highly fluoro-substituted covalent organic framework and its application in lithium–sulfur batteries. *ACS Applied Materials & Interfaces, 10*(49), 42233–42240. https://doi.org/10.1021/acsami.8b14213
12. Csákÿ, A. G., & Molina, M. T. (2011, March 15). 1,2-bis(diphenylphosphino)benzene. Wiley Online Library.
13. Luo, H., Wu, G., Xu, S., Wang, K., Wu, C., Zhang, Y., & Wang, J. (2015, July 15). Palladium-catalyzed cross-coupling of aryl fluorides with N-tosylhydrazones via C–F bond activation. *Chemical Communications*.
14. Cardoso, V., Correia, D., Ribeiro, C., Fernandes, M., & Lanceros-Méndez, S. (2018). Fluorinated polymers as smart materials for advanced biomedical applications. *Polymers, 10*(2), 161. https://doi.org/10.3390/polym10020161
15. Shu, L., Müri, M., Krupke, R., & Mayor, M. (2009). Shape-persistent macrocycles comprising perfluorinated benzene subunits: Synthesis, aggregation behaviour and unexpected µrod formation. *Organic & Biomolecular Chemistry, 7*(6), 1081. https://doi.org/10.1039/b817274a
16. Cha, M. C., Lim, Y., Choi, T. J., & Chang, J. Y. (2017). Superhydrophobic and flexible microporous polymer paper. *Macromolecular Chemistry and Physics, 218*(20), 1700219. https://doi.org/10.1002/macp.201700219
17. Palmgren, A., Thorarensen, A., & Baeckvall, J. E. (2010). Cheminform abstract: Efficient synthesis of symmetrical 2,5-disubstituted benzoquinones via Palladiumcatalyzed double Negishi coupling. *ChemInform, 29*(41). https://doi.org/10.1002/chin.199841083
18. Yang, L., Zhai, Q., Li, G., Jiang, H., Han, L., Wang, J., & Wang, E. (2013). A light transmission technique for pore size measurement in track-etched membranes. *Chemical Communications, 49*(97), 11415. https://doi.org/10.1039/c3cc45841e
19. Golovan, L. A., Timoshenko, V. Y., & Kashkarov, P. K. (2007, June 30). IOPscience. Physics-Uspekhi.
20. Xie, Y., Wang, N., Sun, X., Chu, H., Wang, Y., & Hu, X. (2021). Triple-signaling amplification strategy based electrochemical sensor design: Boosting synergistic catalysis in metal–metalloporphyrin–covalent organic frameworks for sensitive bisphenol A detection. *The Analyst, 146*(14), 4585–4594. https://doi.org/10.1039/d1an00665g

Potential for Biodiesel Production and Growth Rates of Four Microalgae Strains in Three Commonly Used Algae Growth Media

Olivier Hao Yu Lim, Aklen Wee Kai Chua, Gedeon Kusuma Ang, and Ai Fern Lye

Abstract The depletion of fossil fuel deposits remains a pressing issue in current times. Biofuel is an existing alternative to fossil fuels, but its high price continues to deter consumers and companies. Microalgae has been identified as a potential feedstock for biodiesel that may lower the cost of renewable energy in the future. To optimise biodiesel production, more research must be conducted to determine the optimum microalgal strain and culture conditions. As such, this research investigates the potential for biodiesel production and the growth rates of four microalgae strains: *Chaetoceros* sp., *Chlorella* sp., *Nannochloropsis oculata* and *Tetraselmis suecica* in *F*/2 (Guillard's) medium, Bristol medium enriched with peptone and modified Bold's basal medium. The growth rates, masses of lipids produced per 5 L of culture and qualities of biodiesel produced were measured. The best-performing growth medium for each algal strain was also noted. *Chaetoceros* sp., *N. oculata* and *T. suecica* reproduced fastest in modified Bold's basal medium and *Chlorella* sp. reproduced fastest in Bristol medium enriched with peptone. *N. oculata* in Bold's basal medium had the highest carrying capacity and *Chlorella* sp. grown in Bristol medium enriched with peptone produced the greatest mass of lipids. Overall, *Chlorella* sp. cultured in Bristol medium enriched with peptone was found to have the highest potential for biodiesel production.

Keywords Biodiesel · Algae · Biotechnology · Sustainable energy

O. H. Y. Lim (✉) · A. W. K. Chua · G. K. Ang · A. F. Lye
Raffles Institution, 1 Raffles Institution Ln, Singapore 575954, Singapore
e-mail: 25YLIMH292F@ri.edu.sg

A. W. K. Chua
e-mail: 25YANGG551E@ri.edu.sg

G. K. Ang
e-mail: 25YANGG551E@ri.edu.sg

A. F. Lye
e-mail: aifern.lye@ri.edu.sg

H. Guo et al. (eds.), *IRC-SET 2022*,
https://doi.org/10.1007/978-981-19-7222-5_31

1 Introduction

The depletion of fossil fuel deposits is a pressing issue in modern times. If fossil fuel consumption continues at current rates, it is estimated that all deposits will be completely depleted by 2060 [1]. This necessitates the search for sustainable, yet affordable energy sources. One alternative to fossil fuel is biodiesel, manufactured using oil extracted from plants and algae. The conversion from triglycerides to biodiesel takes place through a simple chemical reaction known as transesterification, resulting in fatty acid methyl esters as the final product [2]. Transesterification does not require high heat, high pressure, expensive equipment or uncommon chemicals, increasing the feasibility of commercial biodiesel production. The overall process of biodiesel production and consumption produces minimal net carbon emissions. In biodiesel production, energy is harnessed from light, making biodiesel a sustainable source of fuel. Currently, most commercial biodiesel is produced from crops such as corn and soybean [3]. However, large amounts of fertile land, manpower and water are required for their cultivation. This makes the current biodiesel produced from crops relatively expensive and unsuitable for the long term. However, biodiesel produced from microalgae does not possess many drawbacks of conventional biodiesel. In corn kernels, oil makes up only 9.0% of its dry weight [4]. When all parts of the plant are taken into account, the percentage becomes even smaller. However, this is not the case in microalgae. Each cell contains oil as stored energy, and some strains can contain up to 61% oil by dry weight [5]. Such unicellular algae reproduce quickly through asexual reproduction. Microalgae can be cultured in vertical tanks, eliminating the need for large areas of land. Microalgae are cultured in liquid media, making them easier to handle, with less manpower required and a higher potential for automation. Moreover, fertile soil is not required for cultivation. This allows microalgae to be cultivated in enclosed tanks anywhere in the world, as long as light amounts are adequate and temperature requirements are met. However, more research must be conducted to determine the best microalgal strain and culture conditions.

1.1 Literature Review

There is a multitude of research comparing different strains of algae cultured in different media. However, they mostly describe observations in biomass, lipid content and algae growth. As such, no research paper online provided a comprehensive comparison encompassing alga strain, culture medium, growth rate, lipid yield and biodiesel quality, all of which fall under the scope of our research. Many authors' findings were conclusive about several culture media that accelerated the growth of certain algae strains. For example, the *F*/2 medium was proven to enhance the growth of *Chaetoceros* sp. [6]. The growth and biomass production of *C. muelleri* and *C. calcitrans* were found to be higher using the *F*/2 medium [7], which is widely used to culture diatoms. Another research illustrates how a greater biomass was obtained from *Nannochloropsis salina* cultures at 4 g of peptone per litre of medium [8]. Peptone also supported an increased production of algal material and an enhanced rate of cell division in *Chlorella vulgaris* [9]. One research involving 13 growth media concluded that Bold's basal medium optimised the growth rate, biomass productivity and lipid yield in *C. vulgaris* [10]. Yet, none of the media studied contained peptone. Hence, we decided to include Bold's basal medium, a peptone-containing medium and *F*/2 medium in our research. We investigated the differing growth rates of four common algae strains in these growth media, then cultivated 5 L of each alga in its best-performing medium (indicated by the highest carrying capacity and shortest time taken to reach carrying capacity). Afterwards, oil was extracted and weighed, followed by its conversion to biodiesel which finally underwent a quality test.

2 Materials

2.1 Algae

Four strains of microalgae were used in this research: *Chaetoceros* sp., *Chlorella* sp., *Nannochloropsis oculata* and *Tetraselmis suecica*, due to their local availability. All four strains have been considered for commercial biodiesel production due to their high oil content. Table 1 summarises important and differing characteristics of the four strains.

The visual appearances of the algae studied are shown in Fig. 1.

Table 1 Biological characteristics of *Chaetoceros* sp., *Chlorella* sp., *N. oculata* and *T. suecica*

Alga strain	*Chaetoceros* sp.	*Chlorella* sp.	*N. oculata*	*T. suecica*
Habitat	Marine	Freshwater	Marine	Marine
Classification	Centric diatom	Chlorophyte	Ochrophyte	Chlorophyte
Main pigment	Fucoxanthin	Chlorophyll	Chlorophyll	Chlorophyll
Cell wall composition	Silicon dioxide	Cellulose	Cellulose	Cellulose
Motility	No	No	No	Yes (some life stages)
Life stages	Vegetative cell, spermatozoid, auxospore	Single-stage	Single-stage	Vegetative (immotile), Flagellated (motile), Dormant (cyst)
Reproduction method	Asexual and sexual	Asexual	Asexual	Asexual

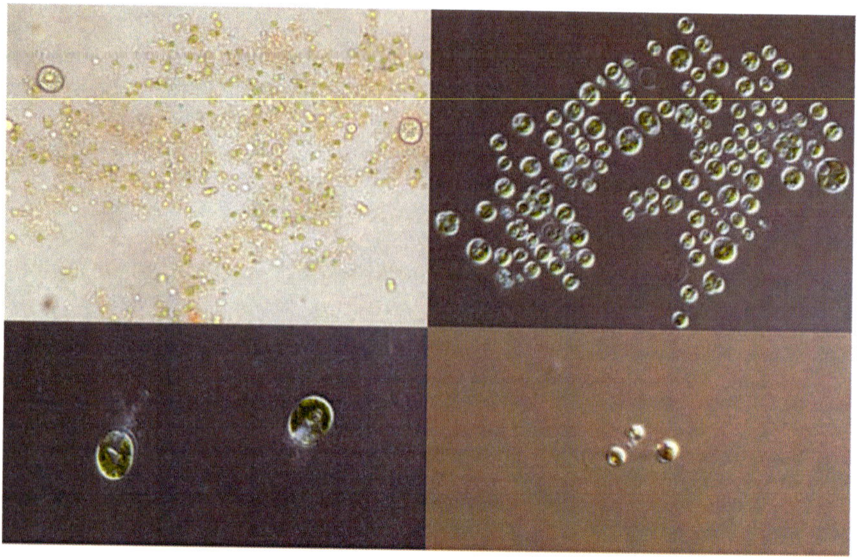

Fig. 1 Light micrographs of algae, taken by our team. Clockwise from top left: *Chaetoceros* sp. Bright field 400×; *Chlorella* sp. DIC 600×; *N. oculata* DIC 600×; *T. suecica* DIC 600×

2.2 Growth Media

The media recipes we used are described below. To prepare the growth medium for marine algae (*Chaetoceros* sp., *N. oculata* and *T. suecica*), substitute ocean water [11] was used as the solvent, instead of distilled water.

1. Standard *F*/2 Guillard's medium (*F*/2) [12]. This medium is formulated for culturing diatoms (algae with silica shell), and hence, it contains sodium metasilicate (Na_2SiO_3) as a source of bioavailable silicon. *F*/2 does not contain magnesium (Mg^{2+}) or calcium (Ca^{2+}) ions, both of which form an insoluble white precipitate in the presence of the metasilicate ion (SiO_3^{2-}). This precipitate is opaque and blocks light, reducing the rate of photosynthesis.
2. Bristol medium enriched with proteose peptone (BMEP) [13]. This medium contains peptone, a water-soluble powder containing amino acids, metal salts and vitamins. It is often used in nutrient agar to culture single-celled organisms such as bacteria.
3. Modified Bold's basal medium (BBM) [14]. It contains various trace metals which may be helpful in accelerating alga reproduction rates, carrying capacity as well as lipid yield.

3 Methods

3.1 Overview of Procedures

1. Investigation of algae growth rates
   a. Preliminary algae culturing
   b. Cell density calculation
2. Large-scale culturing
3. Harvesting and oil extraction
4. Transesterification into biodiesel
5. Testing of biodiesel for completeness of chemical reaction.

3.2 Investigation of Algae Growth Rates

Preliminary Algae Culturing

1 Combine 10 ml of algae seed culture with 190 ml of culture medium in a 250 ml conical flask.
2 Connect airline tubing to an aquarium pump. Then, insert it into the culture.
3 Leave all lights on throughout the culture duration. Keep the air-conditioning constant at 25 °C.
4 Monitor the cultures daily. If necessary, replenish evaporated water by adding distilled water until the initial culture volume of 200 ml is restored.
5 To chart the population growth rate, draw culture samples daily for cell counting. Culture until the number of cells/ml has peaked and started decreasing or fluctuating.

Cell Density Calculation

1. Swirl the conical flask until all biomass is visibly and evenly dispersed.
2. Add 100 μl of culture and 100 μl of iodine solution into an Eppendorf tube.
3. Vortex the solution using a vortex machine for at least 15 s.
4. Immediately after vortexing, draw 10 μl of the mixture and insert it into a haemocytometer.
5. Count the number of cells in the centre chamber. Multiply by 20,000 to calculate the number of cells/ml (density of cell culture).
6. Population growth graphs can now be plotted.

Fig. 2 Large-scale culture set-ups

3.3 Large-Scale Culturing

1. Combine 50 ml of algae seed culture with 950 ml of growth medium in a 1L conical flask (using the growth graphs derived, select the medium out of the three where the algal strain had the highest carrying capacity in the shortest time. Therefore, use only 1 medium for each strain). Perform this step 5 times to prepare 5 L of culture in total.
2. Connect airline tubing to an aquarium pump. Then, insert it into the culture. The large-scale culture set-ups are shown in Fig. 2.
3. Leave all lights on throughout the culture duration. Keep the air-conditioning constant at 25 °C.
4. Calculate the cell density of the large-scale culture. Using growth graphs obtained during preliminary culturing, estimate the number of days *D* that the culture takes for the algae population to reach carrying capacity. *D* is also used in considering the potential for biodiesel production.
5. Monitor the cultures daily. If necessary, replenish evaporated water by adding distilled water until the initial culture volume of 1000 ml is restored. Swirl the conical flask to evenly suspend the biomass.
6. After *D* days, the culture is ready to be harvested.

3.4 Harvesting and Oil Extraction

1. Using vacuum filtration, separate the biomass from the culture media.
2. Discard the filtrate. Scrape the residue (algal biomass) onto an evaporating dish and evaporate to dryness.
3. Pound the dry biomass with a mortar and pestle to release stored oil contents.
4. Pour the biomass into a 50 ml falcon tube and fill it completely with hexane.
5. Cap and leave for 1 week to ensure that all the oil is dissolved in the hexane.

6. Centrifuge the falcon tube and obtain the supernatant (oil–hexane mixture).
7. Heat the supernatant using a water bath until all the hexane has evaporated.
8. Obtain algal oil.

3.5 Transesterification into Biodiesel

1. Dissolve sodium hydroxide (NaOH) in methanol (CH_3OH) in the ratio 1:440 by mass. An excess of methanol is used.
2. Add oil to the mixture in the ratio 400:441 by mass.
3. Heat for 1 h at 60 °C.
4. Remove from heat. Leave the mixture to separate into two layers. Discard the bottom layer (soap, glycerol, NaOH) and retain the top layer (biodiesel).

3.6 Testing of Biodiesel for Completeness of Chemical Reaction

1. Add 10 μl of biodiesel to 10 μl of methanol or ethanol.
2. Vortex the mixture for 30 s. Examine the mixture under a microscope and check for precipitate formation. If a precipitate is formed, the biodiesel likely contains impurities in the form of triglycerides. These triglycerides did not react with methanol during transesterification, contaminating the final biodiesel product. This lowers the biodiesel quality. Therefore, the completeness of the chemical reaction is a viable indicator of biodiesel quality.

4 Results

4.1 Growth Rate

Growth graphs corresponding to the four microalgae strains are shown below (Figs. 3, 4, 5 and 6). The number of days taken to reach carrying capacity, D, is calculated by referencing the starting point of the culture, 0 days after inoculation (day 0). Hence, D varies depending on the strain and growth medium used. During days 8–13, COVID-19 measures restricted our laboratory access and we were unable to replenish the water that had evaporated from the cultures. This affected algae growth rates. The consequent abnormalities in algae cell density (Figs. 3, 4, 5 and 6) were ignored during data analysis.

From Fig. 4, *Chlorella* sp. grew best in BMEP, with 3.73 million cells/ml on day 16. In Fig. 5, *N. oculata* grew best in BBM, with approximately 5.8 million cells/ml on day 23. From Fig. 6, *T. suecica* had a similar carrying capacity in BMEP

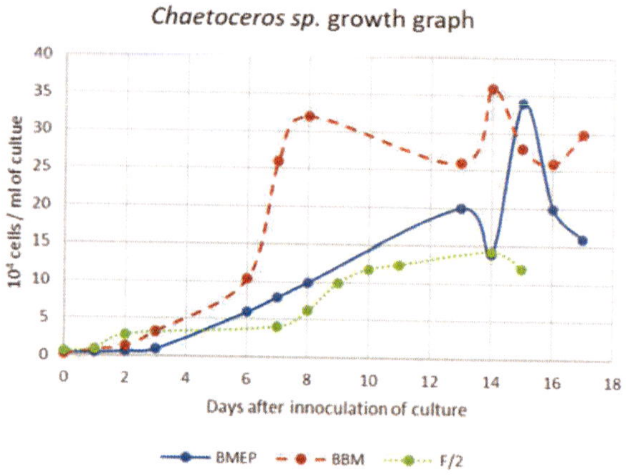

Fig. 3 Growth graph of *Chaetoceros* sp.

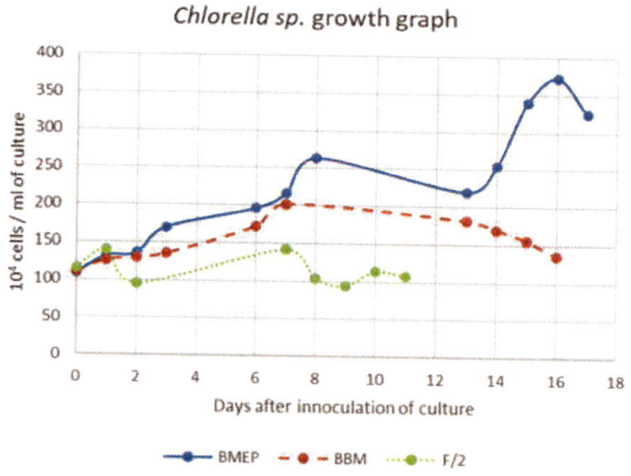

Fig. 4 Growth graph of *Chlorella* sp.

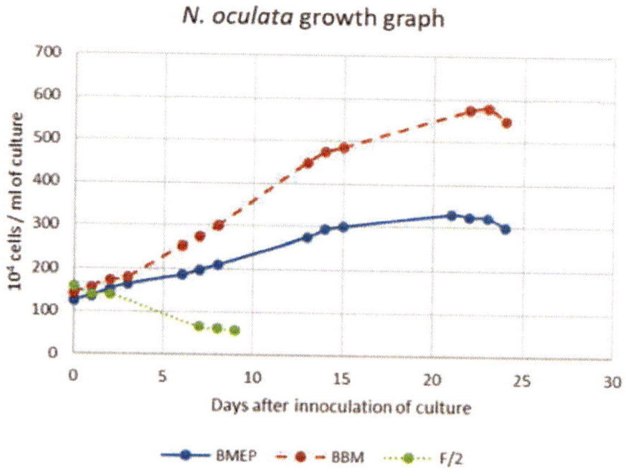

Fig. 5 Growth graph of *N. oculata*

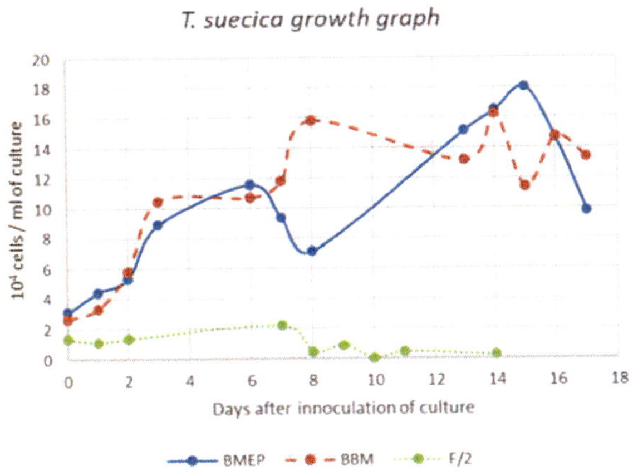

Fig. 6 Growth graph of *T. suecica*

Table 2 Varying masses of oil extracted from 5 L of each algal culture

Strain of alga	Medium used for large-scale culture	Approx. days *D*	Mass of oil/g
Chlorella sp.	BMEP	10	1.272
N. oculata	BBM	23	0.062
T. suecica	BBM	8	0.480

and BBM, but took a shorter time to reach carrying capacity in BBM. Therefore, BBM is considered the best-performing medium for *T. suecica*. *N. oculata* in BBM has the highest carrying capacity of 5.8 million cells/ml, and *T. suecica* has the lowest carrying capacity of 16,000 cells/ml. *T. suecica*, *N. oculata* and *Chlorella* sp. performed poorly when cultured in *F*/2, showing a decrease in cell density within the first 2 days of culturing. On the other hand, Fig. 3 shows that the cell density of *Chaetoceros* in *F*/2 continuously increased.

4.2 Oil Yield

Table 2 summarises the culture duration and oil yield of the microalgae strains.

4.3 Quality Test

In all three biodiesel samples, no precipitate was observed at up to 100× magnification.

5 Discussion

5.1 Growth Rate

From Figs. 4, 5 and 6, *Chlorella* sp., *N. oculata* and *T. suecica* performed poorly in *F*/2, showing a clear decrease in cell density within the first 2 days of culturing. This could be due to the absence of magnesium in the medium. The stated algae utilise chlorophyll, a magnesium-containing compound, as their main photosynthetic pigment. Without a source of magnesium, chlorophyll could not be synthesised. However, the cell density of *Chaetoceros* increased continuously, as seen in Fig. 3. In diatoms such as *Chaetoceros*, the pigment fucoxanthin is used in addition to chlorophyll. Fucoxanthin, a carotenoid, is the dominant pigment in diatoms, responsible for much of the energy transfer to chlorophyll a. Its abundance is visible from the characteristic golden-brown colour of diatoms (Fig. 1), which masks the green colour of chlorophyll. Because fucoxanthin only contains carbon, hydrogen and oxygen atoms, magnesium deficiency affects photosynthetic capability to a lesser extent in *Chaetoceros* sp.

T. suecica in BBM reached carrying capacity in the relatively short duration of 8 days. Its motile nature confers the capability of seeking out areas of higher light intensity for a higher rate of photosynthesis. Each *T. suecica* can grow up to 25 μm in length, the largest of the four strains. This could explain its carrying capacity, which is the lowest of all four strains. *N. oculata* has the smallest cell size of 2–4 μm in diameter, the smallest of the four strains, providing a plausible explanation for its high carrying capacity.

5.2 Oil Yield

From the data presented in Table 2, *Chlorella* sp. in BMEP produced the greatest mass of oil. This is followed by *T. suecica* and *N. oculata*. The variation in the masses of oil extracted can be attributed to factors including, but not limited to, the density of the cell culture, the varying oil content of each strain of alga and the volume of each cell. The mixotrophic nature of *Chlorella* likely contributed to its high oil content, as BMEP contains various biological molecules such as amino acids and vitamins.

Due to time and resource constraints, we assumed that cell density directly correlated to total oil yield. Therefore, only the medium that resulted in the highest cell density was used in large-scale culturing for each strain. However, a medium that induces quick reproduction may also cause cellular oil contents to decrease, which would explain the negligible oil yield of *N. oculata*. Therefore, carrying capacity and growth rate cannot be used as indicators of oil yield.

5.3 Quality Test

All biodiesel samples did not produce any precipitate in the quality test, showing that the chemical reaction of transesterification was performed to completion. As the rate of reaction was sufficiently high, the triglycerides had completely reacted with the methanol, resulting in the final biodiesel product being free from triglyceride impurities. It can be concluded that all three biodiesel products are similar in terms of quality.

5.4 Conclusion

Despite *N. oculata* in BBM having the highest carrying capacity, *Chlorella* sp. in BMEP yielded the greatest mass of oil. *Chlorella* sp. demonstrated reliable and consistent growth in comparison with the other strains. It also reached carrying capacity in the relatively short timeframe of 10 days. As the quality of biodiesel produced is similar between strains, it is not a factor in considering the potential for biodiesel production. *Chlorella* is the most common genus among the four strains used in this research. Additionally, BMEP uses a considerably smaller number of different nutrients as compared to BBM, adding to the ease of medium preparation. *Chlorella* is a freshwater genus, and hence, the filtering of seawater or production of substitute ocean water is not necessary. As such, it can be concluded that *Chlorella* sp. in Bristol medium enriched with peptone has the highest potential for biodiesel production. It is followed by *T. suecica* in BBM due to its high amount of oil extracted and high growth rate, and *N. oculata* in BBM for its low amount of oil extracted and low growth rate. *Chaetoceros* sp. is ranked last due to its unreliable growth (2 batches of large-scale culture failed) and difficulties faced in accurately determining cell culture density.

5.5 Limitations

There was sediment present in our media, possibly due to impurities in the chemicals used. The sediment reduced the amount of light penetrating the medium, decreasing photosynthesis rates. This likely decreased reproduction rates and microalgae oil content, rendering the results less accurate. *Chaetoceros* cells were especially difficult to count under the microscope. High amounts of sediment were present, and they absorbed the iodine, forming a colour similar to the carotenoid pigment fucoxanthin in *Chaetoceros*. Moreover, the algae are small and difficult to discern from sediment. The silica frustules of dead diatoms do not disintegrate, and these could have been counted as well. This inflated *Chaetoceros* cell densities. As such, the *Chaetoceros* growth graphs obtained are highly unreliable and inaccurate. On 2 successive batches of large-scale culturing, *Chaetoceros* failed to result in a visible increase in cell density. Therefore, we discontinued the large-scale *Chaetoceros* cultures and did not produce any results for *Chaetoceros* oil yield and biodiesel quality.

Disruptions due to the COVID-19 pandemic also affected our experimental results. During days 8–13 in Figs. 3, 4, 5 and 6, COVID-19 measures restricted our laboratory access, and we were unable to replenish water that had evaporated from the cultures. The consequent abnormalities in algae cell density are reflected in the growth graphs, and these periods were ignored during data analysis.

Acknowledgements We greatly appreciate the advice and feedback from Mdm Lye Ai Fern, our mentor for this research. We extend our gratitude to Miss Erika Chong of Blue Aqua International and Miss Punithavathi K Haridas of the Singapore Food Agency's Marine Aquaculture Centre on St. John's Island for kindly providing us with the algae seed cultures, free of charge. We also thank Mr. Goh Chun Lian, Mr. Chan Jun Hao, Mdm Musta'inah Binte Suratman and all other laboratory technicians for their invaluable help and supervision. In addition, we express our gratitude towards Mr. Haw Yuanxin of the Nikon Imaging Centre Singapore, where our DIC light micrographs were taken. Lastly, we would like to thank our families for their continuous support.

References

1. Howarth, J. (2019). When will fossil fuels run out? *Octopus Energy*. https://octopus.energy/blog/when-will-fossil-fuels-run-out/
2. Daniel, J. B. (2014). Waste cooking oil-to-biodiesel conversion for space heating applications. RIT Scholar Works.
3. Biofuels explained: Biomass-based diesel fuels. U.S. Energy Information Administration. https://www.eia.gov/energyexplained/biofuels/biodiesel.php
4. Barrera-Arellano, D., Badan-Ribeiro, A. P., & Serna-Saldivar, S. O. (2018). Corn oil: Composition, processing, and utilization. In *Corn* (3rd edn).
5. Borowitzka, M. A. (2018). Biology of microalgae. *Microalgae in Health and Disease Prevention,* 23–72.
6. Pratoomyot, J., Piyawan, S., & Noiraksar, T. (2005). Fatty acid composition of 10 microalgal species. *Songklanakarin Journal of Science and Technology, 27.*
7. Velasco, L. A., Carrera, S., & Barros, J. (2016). Isolation, culture and evaluation of *Chaetoceros muelleri* from the Caribbean as food for the native scallops, *Argopecten nucleus* and *Nodipecten nodosus. Latin American Journal of Aquatic Research, 44*(3), 557–568.
8. Marudhupandi, T., Sathishkumar, R., & Thankappan, T. A. K. (2016). Heterotrophic cultivation of Nannochloropsis salina for enhancing biomass and lipid production. *Biotechnology Reports.*
9. Griffiths, D. (1967). The effect of peptone on the growth of heterotrophic cultures of *Chlorella vulgaris* (Emerson Strain). *Planta, 75*(2), 161–163.
10. Wong, Y. K., Ho, Y. H., Ho, K. C., Leung, H. M., & Yung, K. K. L. (2017). Growth medium screening for *Chlorella vulgaris* growth and lipid production. *Journal of Aquaculture and Marine Biology, 6*(1).
11. ASTM International. (2003). ASTM D 1141: Standard practice for the preparation of substitute ocean water. https://img.21food.cn/img/biaozhun/20090815/187/11183599.pdf

12. Kang, K. H., Qian, Z. J., Ryu, B., & Kim, S. K. (2011). Characterisation of growth and protein contents from microalgae *Navicula incerta* with the investigation of antioxidant activity of enzymatic hydrolysates. *Food Science and Biotechnology, 20*, 183–191.
13. UTEX Culture Collection of Algae. (n.d.). Proteose medium. UTEX culture collection of algae. https://utex.org/products/proteose-medium?variant=30991748169818#recipe
14. Bold's basal medium (modified). Canadian Phycological Culture Centre. (2017). https://uwaterloo.ca/canadian-phycological-culture-centre/cultures/culture-media/bolds-basal-medium

Evaluation of Calibration Paradigms to Improve Wearable EEG-Based Attention Recognition

Maegan Kho Jing Ting, Nicole Yong Xue Min, and Aung Aung Phyo Wai

Abstract Attention is a complex process that is important in achieving optimal task performance. EEG-based attention recognition involves calibration tasks that subjects need to perform, as well as training of the attention model for reliable detection of attention levels. However, these calibration paradigms might not be ideal for inducing desirable attention states, as they are simple and monotonous. There is also limited evaluation of the different calibration tasks to assess their attention detection performance. To fill up this gap, we designed different calibration tasks and compared their effectiveness in inducing attention. We collected EEG data from 29 subjects using a consumer-grade EEG headband according to our experimental protocol. We used six bandpower features per channel to classify binary attention states (attention VS inattention). Based on our evaluation results, we discovered that two modified calibration tasks (Modified Flanker and Colour Flanker) achieved higher classification accuracies compared to the baseline task (Baseline Flanker). The Modified Flanker task achieved the highest mean subject-independent accuracy of $66.45 \pm 16.38\%$ across subjects. Our findings showed that the different subjects need unique calibration paradigms to achieve a high attention classification performance. In comparison with qualitative survey data analysis, we also found that the subjects' personality and learner types do not correlate highly with attention detection performance.

Keywords EEG · Attention · Calibration tasks

M. K. J. Ting (✉) · N. Y. X. Min
Nanyang Girls' High School, Singapore, Singapore
e-mail: 7114190058@nygh.edu.sg

N. Y. X. Min
e-mail: 7114190383@nygh.edu.sg

A. A. P. Wai
Nanyang Technological University, Singapore, Singapore
e-mail: apwaung@ntu.edu.sg

H. Guo et al. (eds.), *IRC-SET 2022*,
https://doi.org/10.1007/978-981-19-7222-5_32

1 Introduction

In our everyday lives, attention is crucial in decision-making, encoding things to memory, learning and so on [1–3]. There are several types of attention, including selective attention, sustained attention and divided attention [4]. Individuals who are more prone to attention deficits, such as those with Attention Deficit Hyperactivity Disorder (ADHD), might require attention training to improve their attention. Electroencephalogram (EEG)-based brain–computer interface (BCI) technology can be integrated into such training to derive one's attention and emotional state [5–7].

For data-driven attention recognition, we rely on calibration paradigms to train the attention detection model using specific machine learning algorithms for better classification performances. Existing calibration tasks include standard psychology tests, such as the Eriksen Flanker task and the Psychomotor Vigilance Task, that serve as surrogate ground-truth tasks. However, such tests lack unique features, resulting in subjects losing interest and consequently, paying less attention. These tasks result in low accuracy for attention classification, especially in subject-independent performance evaluation. Furthermore, there are limited comparative and objective assessments on these baseline calibration tasks, as well as their modifications in the attention recognition performance. Thus, our study aims to improve the EEG-based attention calibration protocol by designing new tasks and comparing their effectiveness in inducing attention.

Current research shows that personality affects social attention, with extroverts having greater gaze duration [8]. However, there is limited research on how personality affects one's attention towards other audio and visual stimuli. In terms of learner types, many educators tailor their content to suit the students' perceived learner type [9]. However, studies have shown that such teaching does not benefit the students' learning [10, 11]. Hence, we would like to investigate how personality and learner types affect one's attention during attention-training tasks.

2 Research Objectives and Experimental Protocol

We designed new calibration tasks to compare their effectiveness in modelling attentive states, so as to achieve better attention classification performances. We also collected qualitative survey data from subjects to find out the correlation between personality and learner type with attention states. Our calibration paradigms include five tasks (2 baseline, 3 modified), and each task consists of attention and inattention sub-tasks. We hypothesise that the modified calibration tasks allow for higher classification accuracies than the baseline tasks. This is because the modifications to the baseline tasks should help increase the level of interest towards the tasks, resulting in greater attention levels. Out of all the modified calibration tasks, we hypothesise that the Modified Flanker + PVT task will have the highest mean accuracy as it is the only one that demands both selective and sustained attention, so we assume it would create the highest attentive state.

We also hypothesise that personality will influence attention since personality is shown to affect social attention. Similarly, we hypothesise that the learner type is not correlated with attention since teaching materials according to one's learner type do not affect learning outcomes.

We obtained IRB approval according to our experimental protocol as shown in Fig. 1. We also obtained parental consent from our subjects before conducting the experiment according to COVID-19 guidelines [12]. Before the experiment, subjects indicated their preferred type of relaxing images (autumn, blue, greenery, mix, mountain, water bodies) and a topic which they found most boring (history, literature, politics, school, sports, weather) through an online survey. We did this to improve the subject's experience by personalising the stimuli according to their preferences.

Apart from incorporating the major attention and inattention sub-tasks, we added rest "R" tasks that allow subjects to do deep-breathing while viewing the personalised relaxing images. Studies have shown that deep-breathing, as well as different types of nature images, can improve relaxation [13–16]. The 45-s breaks in between the baseline PVT trials also require participants to look at nature images and listen to piano music, as instrumental music boosts relaxation [17].

For the Baseline PVT task, subjects do the Psychomotor Vigilance Task (PVT) [18]. For the Baseline Flanker task, subjects perform the arrowed Flanker task as the attention sub-task [19]. The inattention sub-task involves looking and listening to white noise, which disrupts attention in usually attentive students [20]. For the Modified Flanker task, subjects do a modified version of the Eriksen Flanker where the white arrow stimuli can point up, down, left or right. Flanker stimuli could also

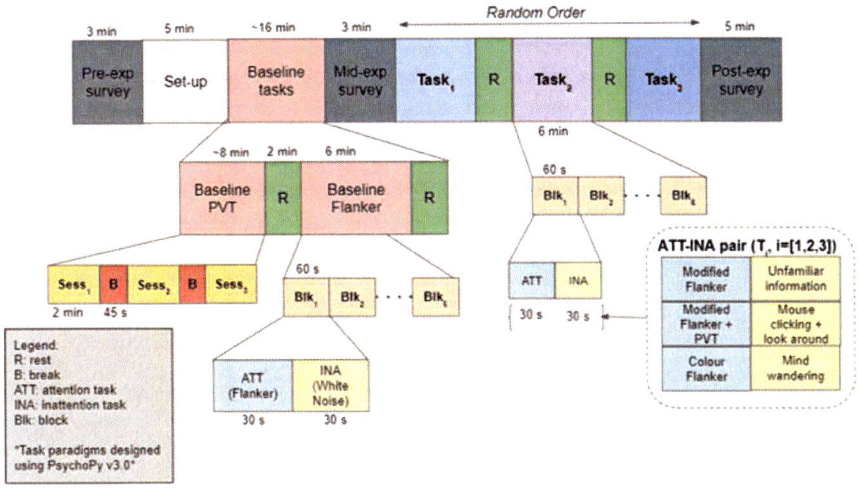

Fig. 1 Block diagram of proposed experimental protocol, including the 2 baseline and 3 modified tasks with their respective attention and inattention sub-tasks (timings listed on top of each task)

point in different directions unlike in the baseline Flanker [21]. The inattention sub-task is looking and listening to unfamiliar information as such passive activities have reported less attention among participants [22]. Auditory information is delivered using an online text-to-speech (TTS) function as mind-wandering is more likely to occur with rough and unclear speech [23]. The information was also in languages unfamiliar to the subjects (Albanian, Czech, Greek, Hindi, Java, Swahili).

For the Modified Flanker + PVT task, subjects will perform both selective (modified Flanker) and sustained attention (baseline PVT) operations simultaneously for the attention sub-task. The inattention sub-task includes random mouse clicking while looking around the computer screen's border. This mimics the use of fidget toys, which have been found distracting during learning and as a result, been banned from classrooms. Recent research has also shown that fidget toys cause poorer attention in children with ADHD [24].

For the Colour Flanker task, subjects perform a coloured version of the Flanker task to increase their interest and attention levels [25]. In the attention sub-task, Flanker stimuli were shown for a short duration before the centre stimulus, and all stimuli disappeared after a while. We assumed more attention is required to remember the colour of the centre stimulus. The inattention sub-task is brainstorming where prompts of the most boring topic, according to the subject, were shown on the screen. When a subject is bored and starts to mind wander, unrelated thoughts distract and disrupt his/her current thoughts and actions, creating an inattentive state [26].

We used the Psychopy v3.0 software package to design our experiment and used FormSG to design pre-, mid- and post-experiment surveys. These surveys allow us to collect information on the subjects' personality and learner type, their perceived emotional and mental states at different parts of the experiment as well as their perceived attention levels.

3 Data Analysis and Results

We collected EEG data using the Muse 2 headband (Interaxon Inc.) and gaze coordinates using the Tobii Eye Tracker 5 (Tobii Technology AB.) from 29 subjects, as shown in Fig. 2. We used synchronised timestamps with millisecond resolutions to segment the EEG and eye tracker data according to the task logs from Psychopy v3.0, as shown in Fig. 3. As part of pre-processing, we removed low-frequency artefacts using a series of median filters and applied a bandpass filter (0.5–45 Hz) to filter out unwanted noises. In feature extraction, we extracted six bandpower features (δ, θ, α, low-β, high-β and γ) from all channel inputs (Tp_9, Fp_1, Fp_2, Tp_{10}) [27, 28].

In the classification step, we used a support vector machine (SVM) to evaluate the binary classification accuracy of the different calibration tasks using the Leave One Block Out (L1BO) cross-validation method. The baseline PVT task was omitted from the data analysis as it had no inattention sub-task included, and it was primarily used for subjects to get familiar with the reaction time task. For each calibration task, we used four types of channel feature inputs: unipolar (Tp_9, Fp_1, Fp_2, Tp_{10}), (Tp_9,

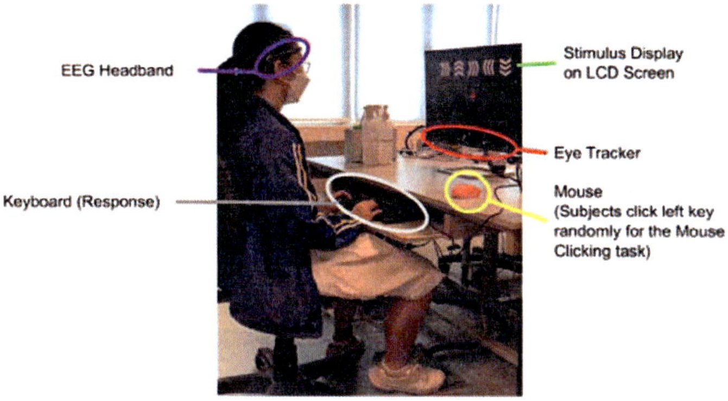

Fig. 2 Experiment set-up with a subject performing tasks according to the experimental protocol

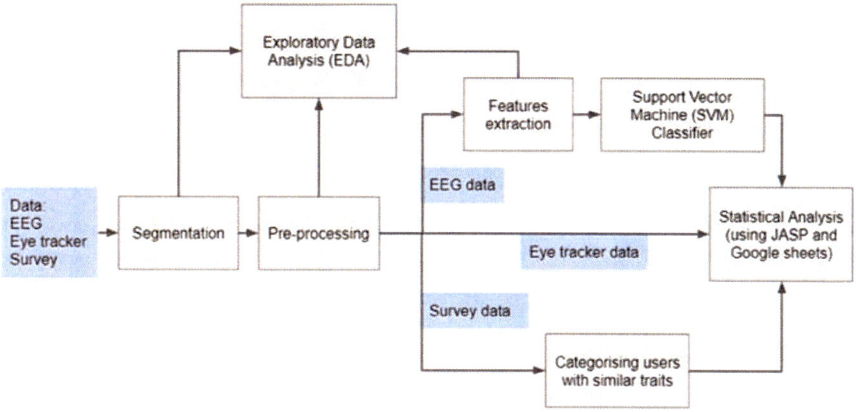

Fig. 3 Block diagram of steps involved in the data analysis procedure

Tp_{10}), (Fp_1, Fp_2) and bipolar ($Fp_1 + Fp_2$). For hyper-parameter optimisation of the SVM classifier, we used Bayesian optimisation and the optimised mean accuracies were higher for all tasks compared to the baseline non-optimised SVM results, as shown in Fig. 4. Hence, we only used the Bayesian optimised SVM mean accuracy results for the rest of the performance evaluation.

As shown in Fig. 5, the mean accuracies of the unipolar 4-channel input (Tp_9, Fp_1, Fp_2, Tp_{10}) were the highest for all of the attention tasks. Hence, we only used the unipolar 4-channel input mean accuracies using Bayesian optimised SVM classification for the rest of our analysis.

Based on the qualitative survey data, we categorised subjects based on their personality and learner types, as well as their perceived attention levels during the different tasks. We then compared the mean accuracies of these groups, as shown in Figs. 6, 7 and 8.

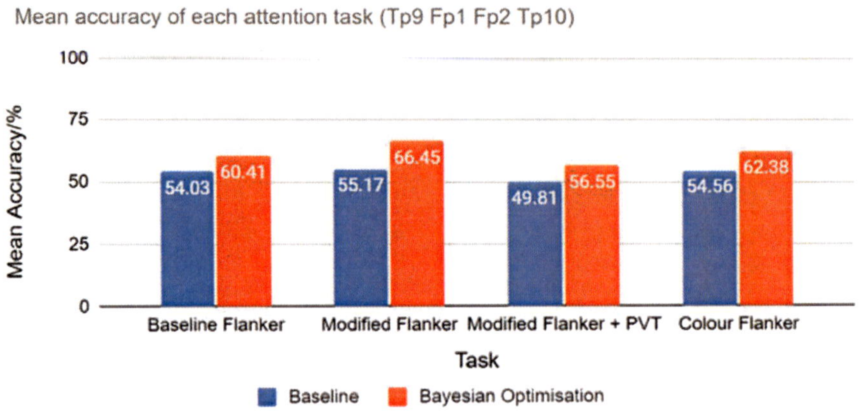

Fig. 4 Comparison of the mean L1BO accuracy of different attention tasks between baseline SVM classifier and Bayesian optimised SVM classifier

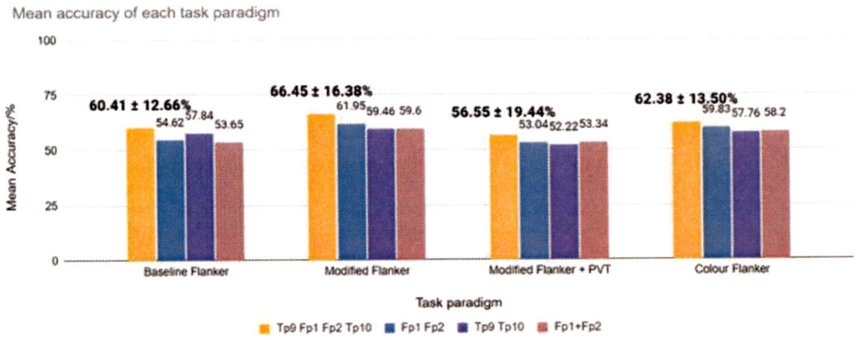

Fig. 5 Comparison of the mean L1BO accuracy of different attention tasks between the four-channel input options (using Bayesian optimised SVM classifier)

For the eye tracker gaze data, we observed the subject's gaze positions on the screen while the subject is performing the calibration tasks, as shown in Fig. 9. We also conducted Analysis of Variance (ANOVA) using the bandpower features for each calibration task, and the mean classification accuracy according to the different personality groups and learner types.

4 Discussion and Conclusion

As shown in Fig. 3, we performed exploratory data analysis and statistical analysis using quantitative sensor data and qualitative survey data. Based on our ANOVA results for each calibration task, there is a high statistical significance difference

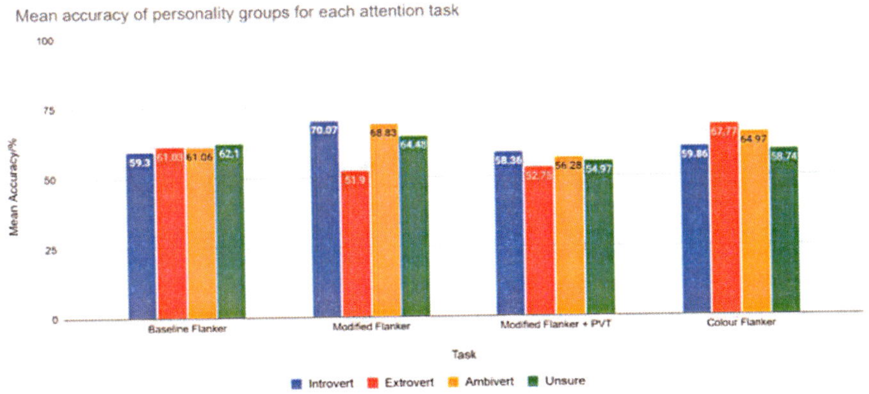

Fig. 6 Mean attention classification accuracy of subjects among the different personality groups for each attention task

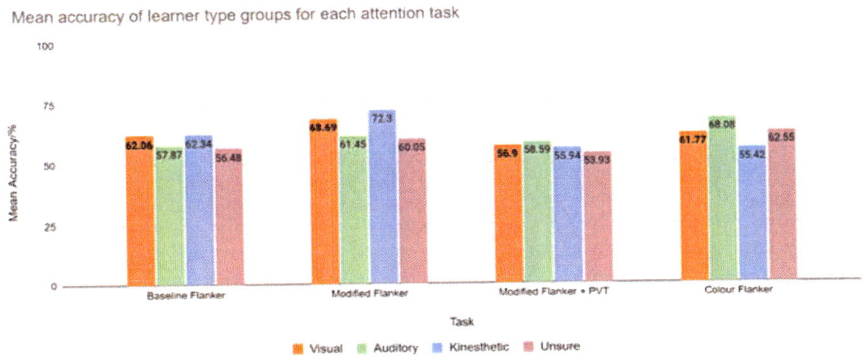

Fig. 7 Mean attention classification accuracy of subjects among the different learner types for each attention task

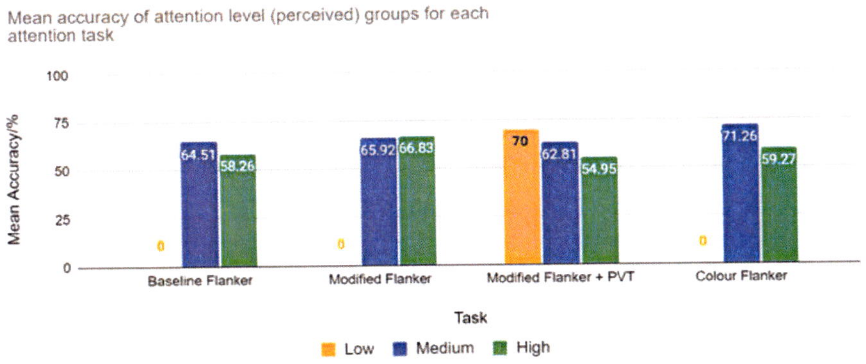

Fig. 8 Mean attention classification accuracy of subjects among the different attention level (perceived) groups for each attention task

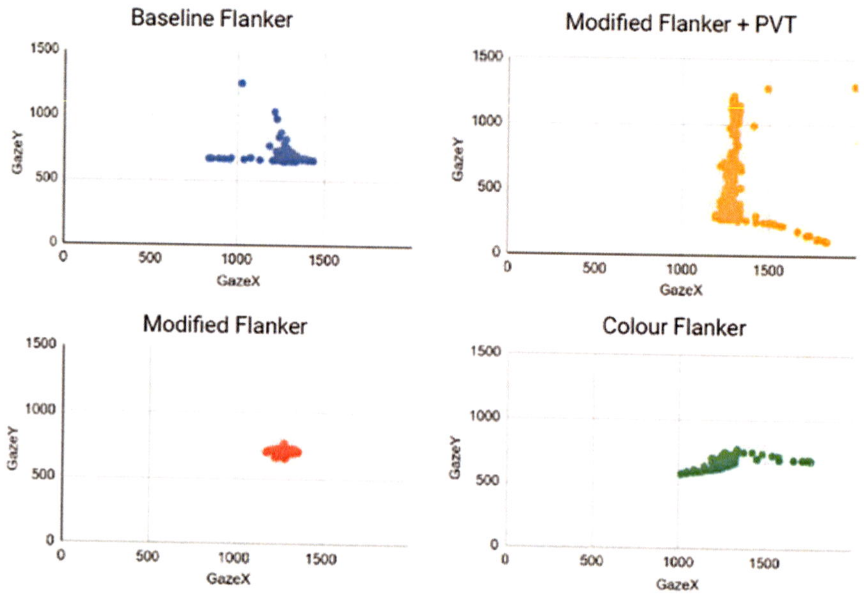

Fig. 9 Visualisation of eye tracker's gaze coordinates for the different attention tasks

($p < 0.001$) among the tasks in each bandpower feature. This concludes that the different bandpower features and channels contribute significantly towards the classification accuracy, hence the highest mean accuracy result for the unipolar 4-channel inputs. We compared the mean L1BO accuracy of each calibration task to find out which had the maximum attention classification accuracy. The Modified Flanker task achieved the highest accuracy (66.45 ± 16.38%), followed by Colour Flanker (62.38 ± 13.50%), Baseline Flanker (60.41 ± 12.66%), and Modified Flanker + PVT (56.55 ± 19.44%). This highlights that the modified tasks could induce better attentive states among our subjects and the Modified Flanker task could induce the best attentive states.

Further accuracy evaluation showed that the percentage of subjects whose highest mean accuracy falls under the unipolar four-channel input (Tp_9, Fp_1, Fp_2, Tp_{10}) was more prominent with the Modified Flanker task. This further strengthens our accuracy outcome that most subjects had high attentive states during the Modified Flanker since it is known that the unipolar four-channel input would produce the highest mean accuracies. The eye tracker gaze coordinates were also generally more condensed at the centre of the screen for the Modified Flanker tasks as shown in the visualisation of Fig. 9. This indicates that the subjects paid more attention during the Modified Flanker as they focused on looking at the stimuli.

We found no noticeable correlation between personality and attention as there are significantly different mean accuracies among different personality groups for the different tasks, as shown in Fig. 6. We also found out that the subjects' learner type does not affect attention although all of our attention tasks are using visual stimuli, as shown in Fig. 7. Furthermore, we found that the subjects' self-ratings of their emotional and alertness states do not differ across the pre-, mid- and post-experiment surveys. This potentially highlights that the accuracy differences among the attention tasks are mainly due to the subject's attentive states to stimuli.

Overall, our results prove that modified calibration tasks produce higher classification accuracies, indicating they are more effective in inducing desirable attention states. However, our hypothesis that the Modified Flanker + PVT task would achieve the highest classification accuracy, was wrong. It had the lowest mean accuracy, while the Modified Flanker task had the highest mean accuracy. We believe the Modified Flanker + PVT was too challenging for the subjects to follow the task demands, so the task was unable to induce desirable attention levels. Contrastingly, although fewer subjects perceived themselves as having high attention levels during Modified Flanker, they still paid attention to it despite it being slightly easier to do.

In terms of limitations, our attention tasks only used visual stimuli. Future research could compare the effectiveness of visual, auditory and tactile attention tasks in inducing attention. For the performance evaluation, we only used the attention–inattention sub-task pair as specified in the experimental protocol. However, cross-sub-task combinations could have rendered better classification performances. A greater sample size could have provided better representation and future research can also find out how attention varies among different age groups/demographics.

Attention training is essential for certain individuals so they can perform daily activities. The existing calibration tasks for attention model training, however, are not that effective in inducing desirable attentive states. Hence, we designed new calibration tasks and compared them with the baseline tasks for performance evaluation using wearable EEG. We collected EEG, eye tracker and survey data from 29 subjects. Based on our analysis, the modified calibration tasks produced higher classification accuracies than the baseline tasks. Despite being slightly easier in difficulty, our findings also show that the Modified Flanker task achieved the highest classification accuracy. This highlights the need to balance the difficulty of the calibration tasks with the user's interest in the task to induce desirable attentive states.

References

1. Yang, C.-T. (2017). Attention and perceptual decision making. In *Systems factorial technology* (pp. 199–217).
2. Matsukura, M., Luck, S. J., & Vecera, S. P. (2007). Attention effects during visual short-term memory maintenance: Protection or prioritization? *Perception & Psychophysics, 69*(8), 1422–1434.
3. Hidi, S. E. (1995). A reexamination of the role of attention in learning from text. *Educational Psychology Review, 7*(4), 323–350.
4. Lindsay, G. W. (2020). Attention in psychology, neuroscience, and machine learning. *Frontiers in Computational Neuroscience, 14*.
5. Guan, C., Lim, C. G., Fung, D., Zhou, H. J., Krishnan, R., & Lee, T. S. BCI facilitates the improvement of cognitive functions in children and elderly. In *2020 8th International winter conference on brain-computer interface (BCI)*.

6. Bird, J. J., Manso, L. J., Ribeiro, E. P., Ekart, A., & Faria, D. R. A study on mental state classification using EEG-based brain-machine interface. In *2018 International conference on intelligent systems (IS)*.
7. Plass-Oude Bos, D. (2006). EEG-based emotion recognition. The influence of visual and auditory stimuli.
8. Wu, D.W.-L., Bischof, W. F., Anderson, N. C., Jakobsen, T., & Kingstone, A. (2014). The influence of personality on social attention. *Personality and Individual Differences, 60*, 25–29.
9. Pashler, H., McDaniel, M., Rohrer, D., & Bjork, R. (2008). Learning styles. *Psychological Science in the Public Interest, 9*(3), 105–119.
10. Li, X., & Yang, X. (2016). Effects of learning styles and interest on concentration and achievement of students in mobile learning. *Journal of Educational Computing Research, 54*(7), 922–945.
11. Rogowsky, B. A., Calhoun, B. M., & Tallal, P. (2020). Providing instruction based on students' learning style preferences does not improve learning. *Frontiers in Psychology, 11*.
12. Simmons, A. M., & Luck, S. J. (2020). Protocol for reducing COVID-19 transmission risk in EEG research. *Research Square*, rs.3.pex-974. (2020)
13. Zaccaro, A., Piarulli, A., Laurino, M., Garbella, E., Menicucci, D., Neri, B., & Gemignani, A. How breath-control can change your life: A systematic review on psycho-physiological correlates of slow breathing. *Frontiers in Human Neuroscience, 12*.
14. AL-Ayash, A., Kane, R. T., Smith, D., & Green-Armytage, P. (2015) The influence of color on student emotion, heart rate, and performance in learning environments. *Color Research & Application, 41*(2), 196–205.
15. Brown, D. K., Barton, J. L., & Gladwell, V. F. (2013). Viewing nature scenes positively affects recovery of autonomic function following acute-mental stress. *Environmental Science & Technology, 47*(11), 5562–5569.
16. Kexiu, L., Elsadek, M., Liu, B., & Fujii, E. (2021). Foliage colors improve relaxation and emotional status of university students from different countries. *Heliyon, 7*(1).
17. Thoma, M. V., La Marca, R., Brönnimann, R., Finkel, L., Ehlert, U., & Nater, U. M. The effect of music on the human stress response. *PLoS ONE, 8*(8).
18. Dinges, D. F., & Powell, J. W. (1985). Microcomputer analyses of performance on a portable, simple visual RT task during sustained operations. *Behavior Research Methods, Instruments, & Computers, 17*(6), 652–655.
19. Servant, M., & Logan, G. D. (2019). Dynamics of attentional focusing in the Eriksen Flanker task. *Attention, Perception, & Psychophysics, 81*(8), 2710–2721.
20. Söderlund, G. B., Sikström, S., Loftesnes, J. M., & Sonuga-Barke, E. J. (2010). The effects of background white noise on memory performance in inattentive school children. *Behavioral and Brain Functions, 6*(1), 55.
21. Wickens, C. (2021). Attention: Theory, principles, models and applications. *International Journal of Human-Computer Interaction, 37*(5), 403–417.
22. Hlas, A. C., Neyers, K., & Molitor, S. (2017). Measuring student attention in the second language classroom. *Language Teaching Research, 23*(1), 107–125.
23. Gliser, I., Mills, C., Bosch, N., Smith, S., Smilek, D., & Wammes, J. D. The sound of inattention: Predicting mind wandering with automatically derived features of instructor speech. In *Lecture Notes in Computer Science* (pp. 204–215).
24. Graziano, P. A., Garcia, A. M., & Landis, T. D. (2020). To fidget or not to fidget, that is the question: A systematic classroom evaluation of fidget spinners among young children with ADHD. *Journal of Attention Disorders, 24*(1), 163–171.
25. Beaton, L. E., Azma, S., & Marinkovic, K. When the brain changes its mind: Oscillatory dynamics of conflict processing and response switching in a flanker task during alcohol challenge. *PLOS ONE, 13*(1).
26. Zheng, Y., Wang, D., Zhang, Y., & Xu, W. Detecting mind wandering: An objective method via simultaneous control of respiration and fingertip pressure. *Frontiers in Psychology, 10*.
27. Krigolson, O. E., Williams, C. C., Norton, A., Hassall, C. D., & Colino, F. L. Choosing muse: Validation of a low-cost, portable EEG system for ERP research. *Frontiers in Neuroscience, 11*.

28. Lotte, F. (2014). A tutorial on EEG signal-processing techniques for mental-state recognition in brain–computer interfaces. In *Guide to brain-computer music interfacing*, (pp. 133–161)

Jensen-Shannon Divergence of Quantum Hashing Algorithms Based on Quantum Walk

Arav Cabral

Abstract A hashing algorithm is typically a "black box" that garbles input data and makes it unreadable. Hashing algorithms are one-way programs. Consequently, the input text cannot be unscrambled and decoded easily by anyone else. Hashing protects data so even if someone gains access to your server where the hash is stored, the items stored remain unreadable. Recently quantum walks have shown promise as a new means to develop effective hashing algorithms. We studied a quantum walked based hash algorithm on a circle using Jensen-Shannon(JS) divergence. We found that the minimum JS divergence decreases with an increase in message length. This leads to a high potential of collisions when the hashing algorithm is run on noisy quantum computers. Furthermore, we found that the JS divergence only increases with length of circle till the length of the circle, $m = n + 1$, where n is the bit length, and is even and m is odd or $m = n + 2$ when n is odd.

Keywords Hashing · Jensen-Shannon divergence · Random walks · Quantum walks · Entropy

1 Introduction

A hashing algorithm is an algorithm that takes in some input data and returns a fixed length output. Hashing algorithms are widely used in modern technologies. Most modern password databases hash passwords before storing them to ensure security of the password. The passwords are passed through a hashing algorithm before being stored. At a later point, the newly typed in password can be hashed again. If the two hash values are identical, then it can be ascertained that the passwords were the same as this is a property of hashing algorithms. Hashing algorithms are also used for checksums and to detect discrepancies in databases [1].

A. Cabral (✉)
NUS High School of Math and Science, Singapore 129957, Singapore
e-mail: aravcabral31@gmail.com

H. Guo et al. (eds.), *IRC-SET 2022*,
https://doi.org/10.1007/978-981-19-7222-5_33

2 Classical Hashing Algorithms

Message digest 5(MD5) is one of the first popular hashing algorithms. The input message is converted to 128 bit output. Over time potential collisions were discovered in the MD5 output space, leading to MD5 being replaced for its use in storage of passwords.

MD5 was designed by Ron Rivest as an improvement over the MD4 algorithm [2]. MD5 input messages are padded till they are a multiple of 512 bits long so as to ensure that the hashing algorithm is more secure.

Secure Hashing Algorithm (SHA)-1 was another hashing algorithm built that was similar to MD5, created to deal with vulnerabilities from the MD4, with a 160 bit output being produced from it. Later, more versions of SHA were made, such as SHA-256, and SHA-512, having outputs of 256 bits and 512 bits, respectively.

A successful and secure hashing algorithm must fulfill 3 key principles. The same input must result in the same output. Each time the hash function is run the function and the end output must remain the same. This ensures that the hashing algorithm can act as an effective checksum. The hashing algorithm loses all use if it does not result in the same output with the same input. This means the hash cannot be randomly generated, and that the hash function cannot change each time it is used. A small change in the message that is input must lead to the output hash changing in a large and unpredictable way. If the change in output is small in nature then the hashing algorithm can be reversed by brute force searching the small change hash space. If, the change is predictable, the hashing algorithm can be reversed by pattern identification. The hashing algorithm must be irreversible in order for it to be secure, therefore, these are core properties of a secure hashing algorithm. Thirdly, hashing algorithms need to have no collisions. No two inputs can map to the same output hash value. If hashing algorithms had known collisions, collision attacks can be run on the algorithm, rendering it not secure. Hashing algorithms that have collisions or have a high probability of collisions occurring are not used for secure applications anymore.

3 Random Walk and Quantum Walk

3.1 Random Walk

The historical basis of the theory of random walk was the result of the work of the British botanist, Robert Brown, in 1827 when he looked at the random motion of individual pollen particles. This phenomenon was later described as the Brownian motion. Despite works on probability existing for a few centuries, random walks were only later described with the journal Nature's publication of a discussion between Pearson and Rayleigh [3]. Brownian motion was further studied by many scientists such as Einstein [4, 5] and Smoluchowski [6]. For instance, the mean reversion process was later applied to random walk theory [7]. Random walk theory has also been used to study many other phenomena [8, 9]. For example, it has been used for modeling of the movement of animals, micro-organisms and cells in the fields of biology, ecology and medicine.

Random walk works based on a walker walking along an infinite,discrete one-dimensional uniform lattice. The walker starts at the origin O and then moves a one point along the discrete lattice-a short distance δs to the left or right in a short time δt. We assume completely random motion with equal probabilities (1/2) of moving to the left and of moving to the right. With one time step, the walker will be a distance of δs to the left or the same distance to the right of the origin each with a probability 1/2. With a second time step, the walker will be at a distance $2\delta s$ to the left (possible only if the walker moved to the left at the first step) or to the right of the origin (possible only if the walker moved to the right at the first step) (each with probability 1/4 each) or the walker will return to the origin with a probability 1/2.

Hence, with the walker taking an even (odd) number of steps from origin, and starting at origin, the walker will end up at either even (odd) positions only. Hence, the walker will be at a distance $m\delta s$ to the right of the origin after $n\delta t$ time steps (where m and n are even) with a probability denoted by

$$\begin{aligned} p(m,n) &= \left(\frac{1}{2}\right)^n \binom{n}{(n-m)/2} \\ &= \frac{n!}{2^n\,(((n+m)/2)!\,((n-m)/2)!)} \end{aligned} \tag{1}$$

Mathematically, this is described by the binomial distribution with a mean of 0 and a variance of n. For large n, this distribution, by the law of large numbers due to mean reversion, converges to the Gaussian distribution, taking the limit $\delta s \to 0$ and $\tau \to 0$ such that $t = n\tau$ and $\frac{(\delta s)^2}{\tau} = 2D$, with D being the constant of the diffusion constant, given by

$$p(x,t) = \frac{1}{\sqrt{4\pi Dt}} \exp\left(-\frac{x^2}{4Dt}\right). \tag{2}$$

3.2 *Quantum Walk*

Following the advancements presented by random walks, quantum walks were proposed as a potential method of modeling quantum computation [10, 11]. The first literature regarding quantum walks was published by Aharonov et al. [10]. Ambainis et al. later show quantum walk spread along the line quadratically faster than classical random walks [12]. Shenvi et al. [13] found a use case of quantaum walk theory in quantum search algorithms, with quantum walks also being used in the Grover search algorithm. Additionally, quantum walks been extensively studied in other fields, with continuous quantum walks used [14] in finding hopping time distribution functions for use in excitation dynamics, for studying simple models of decoherence [15].

Similar to classical random walks, there are both continuous-time and discrete-time quantum walks. The discrete-time model of quantum walks was first used in by Feyman for his work on the discretization of Dirac equation [16]. It was later rediscovered and used [17, 18] for the work on quantum cellular automata and by Watrous [19] with his work on space-bounded computation. Subsequently, Aharonov et al. [20] introduced the formal model of discrete-time quantum walk, later researching it as a modeling and computation tool.

Discrete quantum walks along one dimension are discrete-time quantum processes on the space $\mathcal{H}_C \otimes \mathcal{H}_P$, where $\mathcal{H}_C$ and $\mathcal{H}_P$ are the Hilbert space of the coin space and the position space, respectively. The state of the particle is represented as $|c, k\rangle$, with $c \in \{L, R\}$ being the coin state, signifying whether the direction of the subsequent walk, and where k denotes the *position* of the coin, spanning the space of the integer $\mathbb{Z}$. The quantum walks consists of applying the coin operator, analogous to flipping a coin of the classical random walk, followed by the translation operator, analogous to moving on the classical random walk, on the particle state. This process is repeated n times for an n step quantum walk. This paper focuses on the Hadamard coin operator represented by H, and given by

$$H : |L\rangle \mapsto \frac{1}{\sqrt{2}}(|L\rangle + |R\rangle) \tag{3}$$

$$H : |R\rangle \mapsto \frac{1}{\sqrt{2}}(|L\rangle - |R\rangle). \tag{4}$$

The chosen translation operator, T is given by

$$T = |R\rangle\langle R| \otimes \sum_k |k+1\rangle\langle k| + |L\rangle\langle L| \otimes \sum_k |k-1\rangle\langle k|.$$

When the operator T is applied, the particle moved either to the left or to the right by one step based on the coin state associated with the particle. Finally, the unitary walk operator W is defined as $W = T(H \otimes I)$, where I is the identity matrix. This operator acts as a combination of the coin operator and the translation operator, forming one step of the quantum walk.

Let the particle start in the initial state denoting a right coin, and a position of origin, $|R, 0\rangle$. The quantum walks for t number of steps are carried out by the repeatedly applying the operator W for t times. The state of the particle with a few iterations will change as such:

$$\begin{aligned}
|R, 0\rangle &\xrightarrow{W} \frac{1}{\sqrt{2}}(|L\rangle \otimes |-1\rangle - |R\rangle \otimes |1\rangle) \\
&\xrightarrow{W} \frac{1}{2\sqrt{2}}(|L\rangle \otimes |-2\rangle - (|L\rangle - |R\rangle) \otimes |0\rangle + |R\rangle \otimes |2\rangle) \\
&\xrightarrow{W} \frac{1}{4}(|L\rangle \otimes |-3\rangle + |R\rangle \otimes |-1\rangle + |L\rangle \otimes |1\rangle - 2|R\rangle \otimes |1\rangle - |R\rangle \otimes |3\rangle).
\end{aligned}$$

Table 1 The probability distribution of the particle being observed at position n after t steps of the quantum walks on a line, with the initial state $|R, 0\rangle$

$t\ n$	−5	−4	−3	−2	−1	0	1	2	3	4	5
0						1					
1					$\frac{1}{2}$		$\frac{1}{2}$				
2				$\frac{1}{4}$		$\frac{1}{2}$		$\frac{1}{4}$			
3			$\frac{1}{8}$		$\frac{1}{8}$		$\frac{5}{8}$		$\frac{1}{8}$		
4		$\frac{1}{16}$		$\frac{1}{8}$		$\frac{1}{8}$		$\frac{5}{8}$		$\frac{1}{16}$	
5	$\frac{1}{32}$		$\frac{5}{32}$		$\frac{1}{8}$		$\frac{1}{8}$		$\frac{17}{32}$		$\frac{1}{32}$

The distribution skews from the binomial distribution and classical distribution at $t = 3$ due to interference. Furthermore, the distribution has a skew to the right and is asymmetrical

Table 2 The probability distribution of the particle at position n after t steps of the classical random walk on a line starting at 0

$t\ n$	−5	−4	−3	−2	−1	0	1	2	3	4	5
0						1					
1					$\frac{1}{2}$		$\frac{1}{2}$				
2				$\frac{1}{4}$		$\frac{1}{2}$		$\frac{1}{4}$			
3			$\frac{1}{8}$		$\frac{3}{8}$		$\frac{3}{8}$		$\frac{1}{8}$		
4		$\frac{1}{16}$		$\frac{1}{4}$		$\frac{3}{8}$		$\frac{1}{4}$		$\frac{1}{16}$	
5	$\frac{1}{32}$		$\frac{5}{32}$		$\frac{5}{16}$		$\frac{5}{16}$		$\frac{5}{32}$		$\frac{1}{32}$

With t steps, the resulting probability distribution of the particle with the quantum walks and after measurement is shown in Table 1. Measurement on the particle is only performed after the walk has evolved for t steps and no measurement is carried out before this. Due to interference based on the skew of the Hadamard function, the distribution with the quantum walks differs significantly from that of the classical random walks.

The asymmetry is caused by the Hadamard coin's difference in the $|L\rangle$ and $|R\rangle$ states; with the phase being multiplied by -1 for only the coin state of $|R\rangle$ and not for the other coin state. Furthermore, this asymmetry is dependent on the initial state. With the skew being left if the initial coin state is $|L\rangle$ and the skew being to the right if the initial coin state is $|R\rangle$ (Table 2).

4 Quantum Hashing Algorithms

It can be deduced that quantum walks may be used for a hashing scheme. Such a scheme was proposed for a two particle system using the two particle quantum walk. The proposed quantum hashing algorithm [21–24] two coin operators, with an interaction applied based on the relative position of the two particles. Two proposed interactions were used in the scheme, namely, the π phase interaction denoted by the Hadamard tensor Hadamard operator, and the I interaction denoted by the Identity tensor Identity operator. The scheme defines x as being the position of the 1st particle and y being the position of the second particle. The proposed coins used in the scheme were as follows,

$$C1 = \begin{cases} H \otimes H \text{ when } x \neq y \\ -I \otimes -I \text{ when } x = y \end{cases} \quad C2 = \begin{cases} H \otimes H \text{ when } x \neq y \\ e^{i\pi} * H \otimes H \text{ when } x = y \end{cases} \tag{5}$$

4.1 *Interacting Quantum Walks on Circles*

With quantum walks on a line, at each even step, the probability at any odd point will be zero, while the reverse is also true [21]. This is an undesirable result, as it restricts the space of the hash value, making the hashing scheme more susceptible to the birthday attack and other collision related attacks.

Instead of carrying out the quantum walk along lines, a circle can solve the aforementioned problems. For quantum walks along a circle with even circle length after any even number of steps, the probabilities will only be at even positions due to oscillating nature of the probabilities on an even circle. This too is an undesirable for a hashing scheme, therefore, odd length circles were used.

Each bit in the message results in a one step move when the hash scheme is run on the message. Thus, the bit length of the message can also be thought of as the number of steps. For our scheme on the circle, we start with an initial state of both particles being at the position 0 and coinstate $0(\downarrow)|0\rangle \downarrow; |0\rangle \downarrow$

5 Cross Entropy

Cross entropy, as well as other divergence metrics can be used to evaluate how different any two probability distributions are. Since the hash scheme used in our experiment takes a bit input and maps to a probability distribution, we can use these divergence metrics to provide a quantitative measure as to how different any two hash values are, thus providing a quantitative measure as to how safe the hash scheme is.

The first divergence metric considered was cross entropy . Suppose P and Q are two discrete probability distributions defined on the same probability space $\mathcal{X}$. The cross entropy is an asymmetric distribution, with the cross entropy of $H(P, Q)$ being different from that of $H(Q, P)$

The cross entropy of Q with respect to P is defined as

$$H(P, Q) = \sum_{x \in \mathcal{X}} -P(x) \log(Q(x)) \tag{6}$$

5.1 Kullback-Leibler Divergence

The Kullback-Leibler (KL) Divergence is another possible divergence metric. Suppose P and Q are two discrete probability distributions defined on the same probability space $\mathcal{X}$. The KL Divergence is also an asymmetric distribution, with the divergence of $P||Q$ being different from that of $Q||P$

The KL Divergence of Q with respect to P is defined as

$$D_{\mathrm{KL}}(P||Q) = \sum_{x\in\mathcal{X}} P(x)\log\left(\frac{P(x)}{Q(x)}\right) \tag{7}$$

which is equivalent to

$$D_{\mathrm{KL}}(P||Q) = -\sum_{x\in\mathcal{X}} P(x)\log\left(\frac{Q(x)}{P(x)}\right) \tag{8}$$

but different from

$$D_{\mathrm{KL}}(Q||P) = \sum_{x\in\mathcal{X}} Q(x)\log\left(\frac{Q(x)}{P(x)}\right) \neq D_{\mathrm{KL}}(P||Q) \tag{9}$$

5.2 Jensen-Shannon Divergence

In probability theory and statistics, the Jensen-Shannon(JS) divergence is a method of measuring the similarity between two probability distributions. JS divergence provides a symmetric measure of the distance between any two probability distributions. The JS divergence between P and Q is defined using the KL divergences with respect to a midway probability distribution as follows

$$M = \frac{1}{2}\Big((P+Q)\Big) \tag{10}$$

$$\mathrm{JSD}(P||Q) = \mathrm{JSD}(Q||P) = \frac{1}{2}\Big(D_{\mathrm{KL}}(P||M) + D_{\mathrm{KL}}(Q||M)\Big) \tag{11}$$

6 Results

We run the hash scheme for each odd circle length starting from 3 till a circle length of 9. The scheme was run for each possible input message for message lengths up till 8 for the circle length of 3, as well as until a message length of 7 for the circle lengths of 5, 7 and 9.

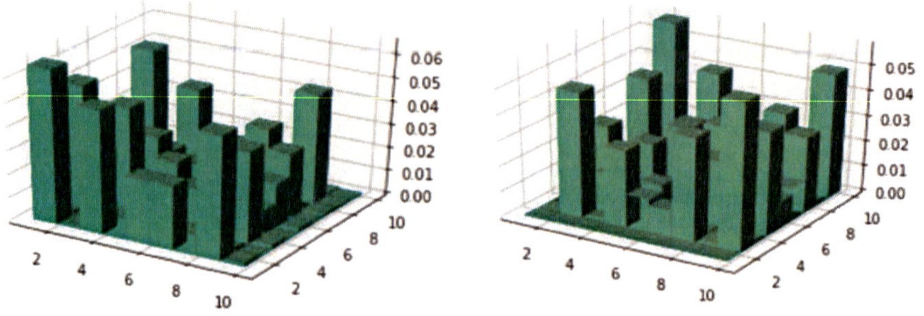

Fig. 1 Distribution with an initial input message of 10101111 (left), distribution with initial input message 1100001 (right) with a circle length of 9

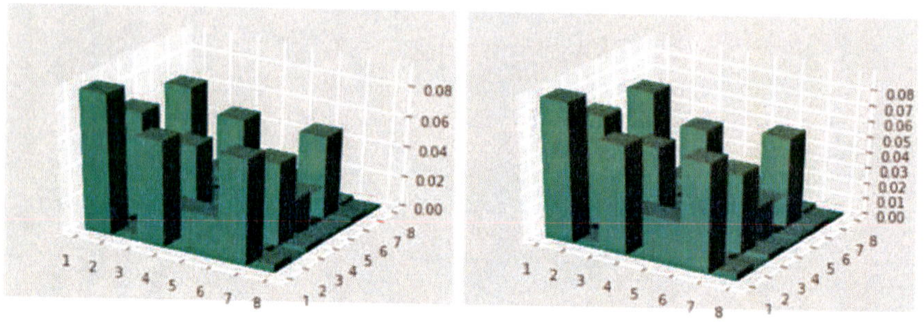

Fig. 2 Distribution with an initial input message of 111011 (left), distribution with initial input message 111101 (right)

Plotting two 8 bit distributions with a circle length of 9, we observe that two drastically different distributions result, as shown in Fig. 1.

Yet, when we input some messages like "111011" and "111101" with a circle length of 7, the landscapes looks similar to each other as can be observed in Fig. 2.

For each input message length, we also evaluate each possible JS divergence by comparing each resulting probability distribution against every other probability distribution. We then evaluated the mean, maximum and minimum divergence values for each possible set of circle lengths and message lengths. The resulting mean divergences for each message length and for each of the four circle lengths can be seen in Fig. 3.

There is a clear downtrend in the mean JS divergence as the message length increases. This makes the hashing algorithm less effective for longer messages. However, this trend seems to follow an exponential decay pattern, indicating that the mean JS divergence will not decrease by much at higher message lengths, therefore the security of the hash will not further deteriorate significantly.

The minimum and maximum JS trends for each of the circle lengths were also plotted. Through these as well as the mean trends (Fig. 3) we were able to determine: For an n bit integer, any circle of length m, where $m > n$, where m is odd, will

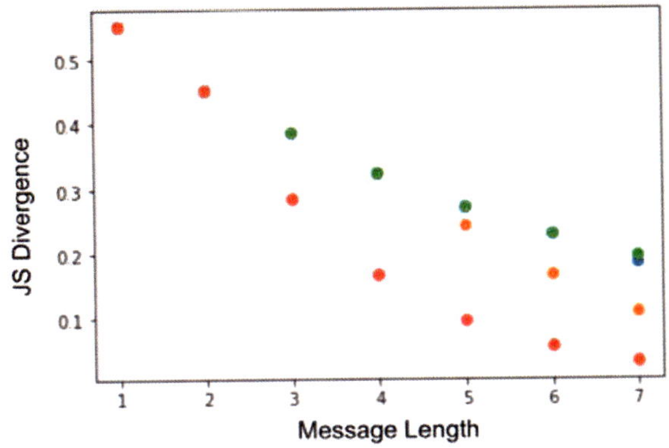

Fig. 3 Graph of mean divergence against length of message for a circle length of 3 (red), 5 (orange), 7 (blue), 9 (green)

result in the exact same JS divergence when against other probability distributions resulting from n bit integers with circle length m more than n, where m is odd.

This pattern can be seen through the alignment of the JS divergences for the 1 and 2 bit messages, and the resulting split of the 3 circle length trend from the rest at a message length of 3. Thus, for minimal memory and computing power use a length of $m = n + 1$ for even message lengths or $m = n + 2$ for odd circle lengths can be used. This would ensure that the JS divergence is the maximum it can be with the chosen length of message, with minimal memory and computing power used to generate the hash, thus making the hash maximally divergent, and thus more secure in the most efficient way possible.

7 Conclusion

In order to most efficiently use this hashing scheme in terms of memory and computing power usage while still maintaining the security of the scheme, the circle length of the scheme should be set as a variable m, based on the message length, n. If the message length is even, then $m = n + 1$ for optimal use, and if the circle length is odd, then $m = n + 2$ for optimal use. The decreasing mean JS-Divergence with message length seems to imply that the quantum hash algorithm may not be really secure since the chances of collision are high whenever the hashed texts look similar. It is advantageous that there is an increased in JS-Divergence with increasing cycle length. Unfortunately, the decreasing trend in the mean divergence implies with an increase in message length that with the noise levels in quantum computers, this algorithms is infeasible, and will effectively have many collisions. We did not manage to simulate the effects noise would have on the hash values within the time frame. We also did not manage to calculate the JS divergence values for larger message lengths as well as for quantum hashing algorithms based on other operators and we hope to do so in the future.

References

1. Pieprzyk, J., & Sadeghiyan, B. (1993). *Design of hashing algorithms*. Springer.
2. Rivest, R. (1990). *The MD5 message digest algorithm, RFC1321*, April 1992.
3. Pearson, K. (1905). The problem of the random walk. *Nature, 72*(1867), 342–342.
4. Einstein, A. (1906). Zur theorie der brownschen bewegung. *Annalen der physik, 324*(2), 371–381.
5. Einstein, A. (1956). *Investigations on the theory of the Brownian movement*. Courier Corporation.
6. Smoluchowski, M. (1924). Sur le chemin moyen parcouru par les molécules d'un gaz et sur son rapport avec la théorie de la diffusion. *Pisma Mariana Smoluchowskiego, 1*(1), 479–489.
7. Uhlenbeck, G. E., & Ornstein, L. S. (1930). On the theory of the Brownian motion. *Physical Review, 36*(5), 823, 1930.
8. Spitze, F. (2001). *Principles of random walk* (Vol. 34). Springer.
9. Lawler, G. F., Limic, V. (2010). *Random walk: A modern introduction* (Vol. 123). Cambridge University Press
10. Aharonov, Y., Davidovich, L., & Zagury, N. (1993). Quantum random walks. *Physical Review A, 48*(2), 1687.
11. Kempe, J. (2003). Quantum random walks: An introductory overview. *Contemporary Physics, 44*(4), 307–327.
12. Ambainis, A., Bach, E., Nayak, A., Vishwanath, A., & Watrous, J. (2001). One-dimensional quantum walks. In *Proceedings of the Thirty-Third Annual ACM Symposium on Theory of Computing* (pp. 37–49).
13. Shenvi, N., Kempe, J., & Whaley, B. (2003). Quantum random-walk search algorithm. *Physical Review A, 67*(5), 052307.
14. Klafter, J., & Silbey, R. (1980). Coherent and incoherent exciton motion in the framework of the continuous time random walk. *Physics Letters A, 76*(2), 143–146.
15. Kendon, V. (2007). Decoherence in quantum walks—A review. *Mathematical Structures in Computer Science, 17*(6), 1169–1220.
16. Feynman, R. P. (1985). Quantum mechanical computers. *Optics News, 11*(2), 11–20.
17. Meyer, D. A. (1996). From quantum cellular automata to quantum lattice gases. *Journal of Statistical Physics, 85*(5), 551–574.
18. Meyer, D. A. (1996). On the absence of homogeneous scalar unitary cellular automata. *Physics Letters A, 223*(5), 337–340.
19. Watrous, J. (1995). On one-dimensional quantum cellular automata. In *Proceedings of IEEE 36th Annual Foundations of Computer Science* (pp. 528–537). IEEE.
20. Aharonov, D., Ambainis, A., Kempe, J., Vazirani, H. (2001). Quantum walks on graphs. In *Proceedings of the Thirty-Third Annual ACM Symposium on Theory of Computing* (pp 50–59).
21. Li, D., Zhang, J., Guo, F.-Z., Huang, W., Wen, Q.-Y., & Chen, H. (2013). Discrete-time interacting quantum walks and quantum hash schemes. *Quantum Information Processing, 12*(3), 1501–1513.
22. Li, D., Yang, Y.-G., Bi, J.-L., Yuan, J.-B., & Juan, X. (2018). Controlled alternate quantum walks based quantum hash function. *Scientific Reports, 8*(1), 1–7.
23. Yang, Y., Zhang, Y., Gang, X., Chen, X., Zhou, Y.-H., & Shi, W. (2018). Improving the efficiency of quantum hash function by dense coding of coin operators in discrete-time quantum walk. *Science China Physics, Mechanics & Astronomy, 61*(3), 1–8.
24. Yang, Y.-G., Dong, J.-R., Yang, Y.-L., Zhou, Y.-H., & Shi, W.-M. (2021). Usefulness of decoherence in quantum-walk-based hash function. *International Journal of Theoretical Physics, 60*(3), 1025–1037.

Characterisation of Cellular Uptake of Red Blood Cell Extracellular Vesicles (RBCEVs)

Sze Chie Tan, Jing Si Loy, and Parvathy Varma

Abstract Almost all cell types, including cancer cells secrete extracellular vesicles (EVs). They play important roles in transportation of bioactive cargo and cell-to-cell communication in the body and can be used as vehicles to deliver therapeutic molecules. Previous studies have investigated the viability of using RBCEVs to kill MCF10Ca1a (Ca1a) breast cancer cells but have not studied the mechanisms in which they enter the cells to deliver their content. Here, we characterise RBCEV uptake by Ca1a cells through fluorescence-based techniques, including flow cytometry and fluorescence imaging. We incubate Ca1a cells with carboxyfluorescein succinimidyl ester-extracellular vesicles (CFSE-EVs) and measure the intensity of the wavelengths emitted. We find that RBCEVs uptake generally increases with incubation times, with an average of 690.919 RBCEVs taken up after 3.5 h. We also study the possible mechanisms through which Ca1a cells take up RBCEVs. We incubate the Ca1a cells with CFSE-EVs and various inhibitors of endocytosis and then perform flow cytometry to determine the RBCEVs uptake. Incubation with 1 and 10 μM of Cytochalasin D or 25 and 50 μM of EIPA results in significantly lower RBCEV uptake in Ca1a cells, while Filipin does not produce any significant change. The RBCEVs are likely taken up via endocytosis pathways, amongst which macropinocytosis appears to play a major role, while lipid-raft mediated endocytosis does not have any appreciable contribution. Overall, our data provide the basis for future studies to characterise RBCEV uptake by Ca1a cells in more detail, thus furthering the potential of applying RBCEVs to treat diseases.

Keywords Red blood cell extracellular vesicles · Quantification · Breast cancer · Inhibition · Flow cytometry · Endocytosis

S. C. Tan · J. S. Loy (✉) · P. Varma
NUS High School of Math and Science, Singapore, Singapore
e-mail: loyjingsi123@gmail.com

S. C. Tan
e-mail: tanszechie@gmail.com

P. Varma
e-mail: parvarma182@gmail.com

H. Guo et al. (eds.), *IRC-SET 2022*,
https://doi.org/10.1007/978-981-19-7222-5_34

1 Introduction

EVs are small spherical packages released by cells into the extracellular environment [1]. They consist of a lipid bilayer membrane encasing a small organelle-free cytosol. EVs can carry cytoplasmic and signalling proteins, small and long, coding and non-coding RNAs (miRNA, mRNA, lncRNA, etc.) and lipids [1]. It is evident through medical research that EVs are important factors involved in a range of physiological processes, including intercellular exchange of proteins and RNA, immune regulation and more [2]. There are two major classes of EVs—Exosomes and Microvesicles (MVs). Exosomes are enclosed within a single outer membrane and form via inward budding of the limiting membrane of early endosomes, which mature into multivesicular bodies (MVBs), followed by release of intraluminal vesicles. They are secreted by all cell types. MVs are formed by direct outward budding of the cell's plasma membrane. They package active cargo and deliver it to recipient cells, altering the recipient cells' functions, playing an active role in cancer metastasis [3].

RBCEVs secreted by red blood cells (RBCs) are used in this experiment. In recent years, RBCEVs are garnering attention due to advantages over other EVs. RBCEVs consist of both exosomes and MVs. The lack of DNA in RBCEVs eliminates risks of unwanted horizontal gene transfer. Moreover, RBCs are the most abundant cell type in our body, making up 84% of our cells, making RBCEVs readily available in large quantities at low cost. They can remain stable after storage, allowing for multiple freezes and thaws [1]. In addition, RBCEVs have a high uptake efficiency, allowing more proportion of the prescribed drugs to be taken up by target cancer cells [1]. Since RBCEVs have low cytotoxicity, they serve as safer modes of medical treatment.

There is both direct and indirect evidence to suggest that EVs are internalised into recipient cells. From previous research, we know endocytosis is the primary method of uptake. Identified routes include clathrin-mediated endocytosis, micropinocytosis, phagocytosis, lipid rafts and direct fusion of EV membrane with recipient cell membrane [1]. However, mechanisms of EV uptake are dependent on other factors like recipient cell type, types of glycoproteins found on EV membranes and recipient cells, present inhibitors and EV cargo [1]. Currently, there is little agreement amongst researchers as to which type of endocytic mechanisms are most important [4] (Fig. 1).

Research in EVs and EV uptake has become more popular in recent years, due to potential pharmaceutical applications of EVs in drug delivery. Considerable efforts have been made in targeting specific cell types by engineering EVs to express cell-type specific ligands, though it is inefficient [5]. Focusing on how EVs target specific cell types, and whether different EV subtypes serve different physiological roles can serve to further improve developments of EVs for drug delivery [6].

We will be working with Ca1a cells, a cell line derived from breast cancer cells, for our experiments. We seek to investigate the extent of RBCEV uptake by Ca1a cells through quantification and imaging techniques, and the pathway through which they are taken up by Ca1a cells to understand the interactions between them better.

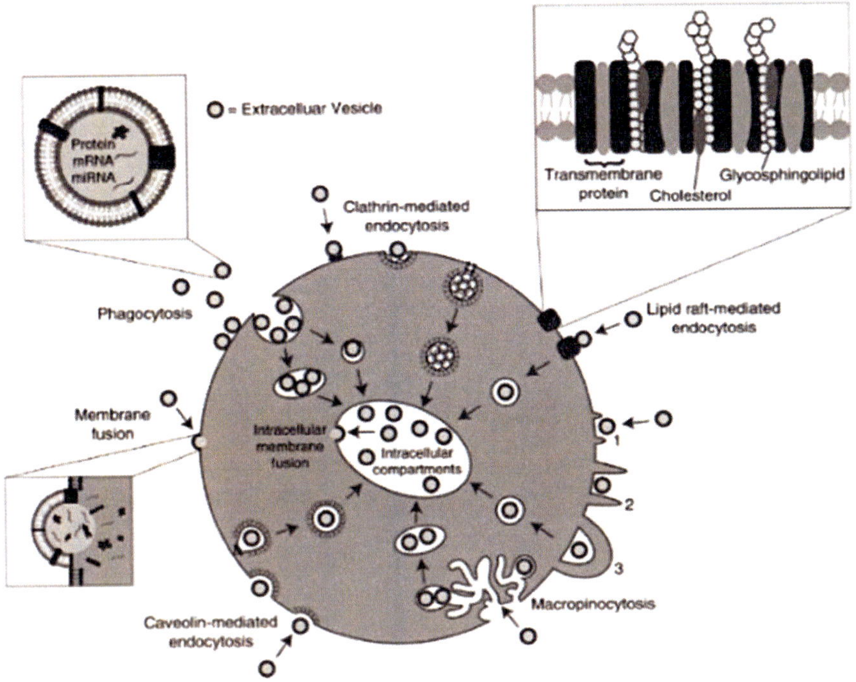

Fig. 1 Pathways shown to participate in EV uptake by target cells. EVs have been shown to be internalised by cells through phagocytosis, clathrin- and caveolin-mediated endocytosis [1]

2 Methodology and Materials

2.1 Cell Culture

Ca1a cells used for all experiments are stored in growth medium (GM) in a cell culture flask. They are "split" every few days when confluent, to ensure survival of the cells and prevent overcrowding. When splitting them, we discard "old" GM and cells are washed with Phosphate Buffered Saline (PBS) solution. They are detached and centrifuged, and the pellet is resuspended in new GM. Some of the cells are added to the flask with new GM, while the rest is discarded or used for experimentation. The cell culture flask is then stored at 37 °C for cell division.

2.2 Quantification of EV Uptake

We conducted quantification of EV uptake to obtain a numerical estimate of the EVs taken up by the cells. This information be directly compared to the extent of EV uptake by Ca1a cells with other cell types in future studies with similar conditions.

We perform quantification of EV uptake by Ca1a cells by measuring the average number of EVs taken up per cell after 2 h of incubation with 40 μg of EVs per 500 μL of GM. We used CFSE-EVs in this experiment, which are RBCEVs incubated with 20 μM CFSE for 2 h and washed after to remove free CFSE.

We seeded 75,000 cells per well into 12 wells on a 24-well plate, with 500 μL of GM (DMEM High Glucose w/L-Glutamine and Sodium Pyruvate, with 10% Foetal Bovine Serum) per well. One day after seeding, we removed the medium and added 40 μg of CFSE-EVs in 500 μL of fresh GM to 6 of the wells, and fresh GM without EVs to the other 6 wells, to act as our control. Incubate all the wells at 37 °C for 2 h.

We then count the cells. We remove the GM and wash the wells with PBS. Then, we detach the cells and wash twice, using P2 buffer (PBS with 2% FBS) and PBS. Two wells of cells from the same treatment group are pooled together as one replicate. Centrifuge and resuspend the cell pellet in 80 μL of PBS and take 10 μL for counting using C-slides and an automated cell counter. We calculate the average cell count value, then transfer an equal volume of cells from each tube to a 96-well plate.

Now, we construct a standard curve to determine the number of EVs that were taken up by the Ca1a cells. We detach the cells from the culture flask, collect them in the GM, centrifuge and resuspend in PBS. Wash with PBS before counting the cells in the sample tube with the smallest volume (identified earlier). In a 96-well plate, add the calculated number of cells to 10 empty wells. We then prepare 10 mixtures of EVs by performing 2× serial dilutions from an EV sample of known EV mass and add the same volume of these EV mixtures to the same 10 wells.

We lyse the cells to obtain a homogenous cell lysate to accurately measure any fluorescent signal from the EVs taken up. We add 10% Triton-X into each of the wells and top up with PBS to achieve a final concentration of 2% Triton-X in 200 μL of total volume. Then, measure the fluorescent signal from the samples by reading the plate using the "fluorescent intensity" settings on Spark 10M Multi-mode Microplate Reader. Set the fluorophore to Alexa Fluor 488 (H2O), the emission peak to 527 nm and the absorption peak to 482 nm. We repeat this experiment with incubation times of CFSE-EVs with Ca1a cells for 0.5 and 3.5 h.

2.3 Imaging of EV Uptake

In the next part of our experiments, we aim to visualise uptake and intracellular trafficking of RBCEVs and Ca1a cells incubated at different timepoints. The reagent setup includes GM, CFSE-EVs and 50 μg/mL Poly-D Lysine.

On day 0, we autoclave 12-mm glass coverslips. On day 1, we seed the Ca1a cells. Under sterile conditions, we use tweezers to place clean coverslips into 8 wells of a 24-well plate and add 400 μL of 50 μg/mL Poly-D Lysine into each well. Incubate at room temperature for 1 h and wash the wells with the coverslips 3 times with water and leave to dry completely in the biosafety cabinet. Then, we seed 80,000 Ca1a cells per well in 500 μL of GM. On day 2, we incubate Ca1a cells with CFSE-EVs for 0.5, 2 and 6 h, before the cells are washed and fixed with 10% formalin. We stain the cells with Hoechst dye and perform imaging using a fluorescence microscope.

2.4 *Inhibiting EV Uptake*

For the last part of our experiments, we want to observe the effect of endocytosis inhibitors: 5-(N-Ethyl-N-isopropyl) amiloride (EIPA), Filipin and Cytochalasin D, on the uptake of RBCEVs by Ca1a cells by comparing the extent of RBCEV uptake between the 3 inhibitors with DMSO control. We perform this experiment to determine uptake mechanisms of RBCEVs by Ca1a cells and understand which known mechanisms are the primary routes of uptake by Ca1a cell line.

Our reagent setup includes the GM, CFSE-EVs and P2 buffer.

On day 1, we seed 80,000 Ca1a cells per well in 500 μL of GM in 10 wells of a 24-well plate. On day 2, we use the GM to prepare solutions with appropriate concentrations of the inhibitors/DMSO control and obtain 300 μL for each replicate. We have three replicates per condition. Our conditions are: 50, 25, 10 μM of EIPA; 10, 1, 0.1 μM of Cytochalasin D; 15, 10, 5 μg/mL of Filipin; and 1:200 DMSO. We used concentrations after previous cell viability experiments.

After preparing our inhibitors/DMSO control, we replace the medium in the wells of the seeded cells with the prepared solutions and treat the seeded cells for 1 h. After that, we add 15 μg of CFSE-EVs into each well and incubate for another 1 h. We then proceed to remove the medium in the wells and wash the wells with PBS.

From here, we add 0.25% Trypsin–EDTA and incubate at 37 °C for 15 min to detach the cells. Then, we add 500 μL of P2 buffer to each well and collect the cells in centrifuge tubes. Centrifuge at 500 g for 5 min, remove the supernatant and wash again with P2 buffer. Resuspend the cells in each tube in 300 μL of P2 buffer and evaluate the extent of EV uptake in each experimental group by flow cytometry.

We analyse our raw data by comparing the DMSO control with each treatment condition using two-tailed and unpaired t-tests (assume equal SD).

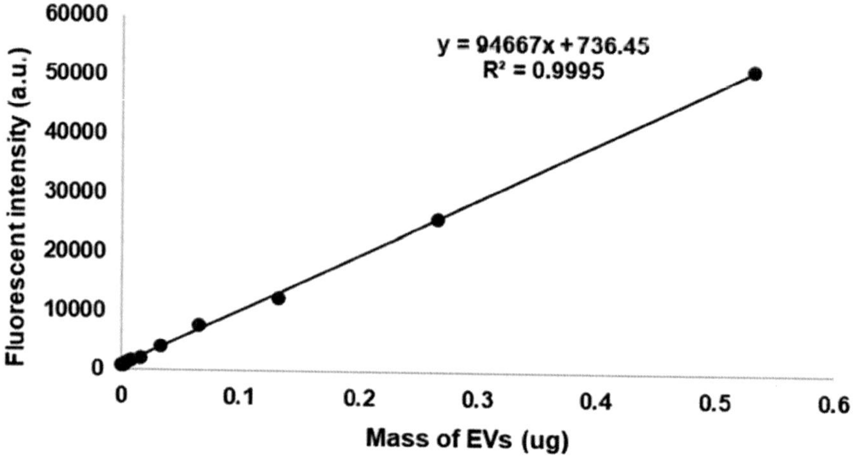

Fig. 2 Standard curve for EV uptake quantification for 2 h incubation

3 Results and Discussion

3.1 *Quantification of EV Uptake*

We plot a linear standard curve to determine the average number of EVs taken up per cell based on each well's fluorescent intensity. We find that as incubation time increases, the average number of EVs/cell increases. The value for 0.5 h is very close to 2 h. However, we note that the quantification experiment for 2 h of incubation was done on a separate occasion than for 0.5 and 3.5 h, and there was more noise for the latter experiment, which could have contributed to this closeness in value (Figs. 2 and 3).

From this we can assume that for 40 μg of EVs per 500 μL of GM, EV uptake in-creases with incubation time. A second experiment with all incubation times done in the same attempt and utilising the same standard curve is necessary, and incubation times longer than 3.5 h can be tested to investigate the maximum number of EVs a cell could take up before the cell sustains damage (Table 1).

3.2 *Imaging of EV Uptake*

The CFSE dye staining (in green) shows the fluorescence emitted by the CFSE-EVs. The Hoechst dye staining (in blue) shows the DNA and nucleus staining, so we can visualise the cells and determine whether the EVs were taken up by the cell (Fig. 4).

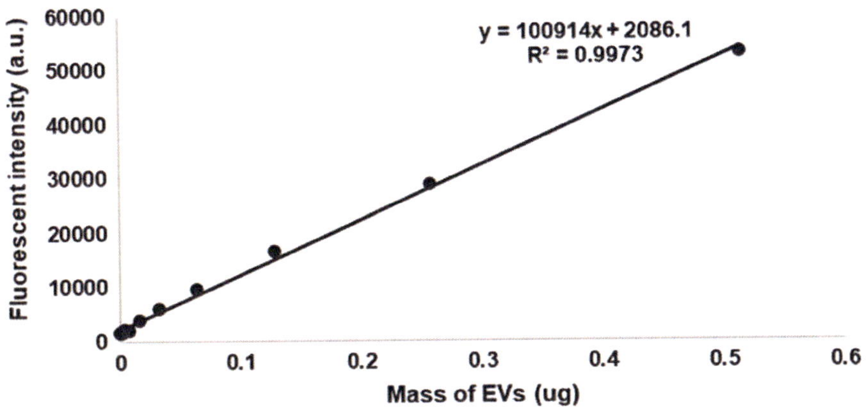

Fig. 3 Standard curve for EV uptake quantification for 0.5 and 3 h incubation

Table 1 Number of RBCEVs taken up per cell after different incubation times

Duration of incubation (hours)	Average number of EVs/cell (Sample size ($n = 3$))
0.5	132.744 ± 146.874
2	140.387 ± 19.085
3.5	690.919 ± 238.070

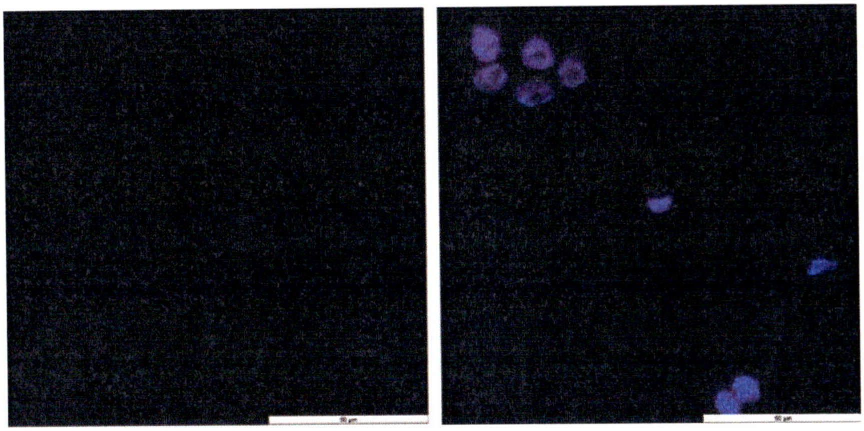

Fig. 4 Ca1a cells without RBCEV incubation

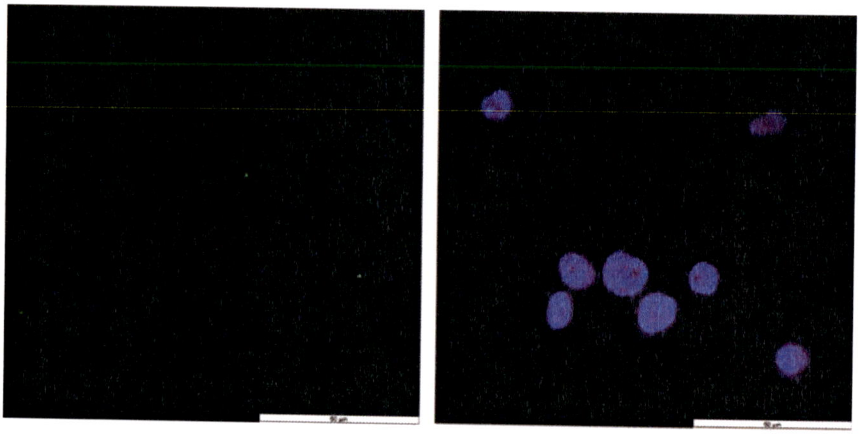

Fig. 5 Incubation of RBCEVs for 2 h

From the CFSE images, we can observe some clusters of bright green spots. These are the CFSE-EVs which have an emission spectrum that peaks at 517 nm, after receiving excitation wavelengths of around 492 nm by the microscope [7].

Based on the Hoechst images, we can observe where the cell lies in the sample, which allows us to deduce that these green spots were indeed taken up by the Ca1a cells. We can also observe several green spots found outside of the cells, showing these CFSE-EVs did not enter the Ca1a cells and are hence, left in the sample.

Based on our results, we clearly observe bright green spots inside the Ca1a cells after referencing both the green and blue images. This shows that RBCEVs were successfully taken up by the Ca1a cells. We also observe that clusters of green spots found in cells in Fig. 6 are brighter than that in Fig. 5, suggesting concentration of CFSE-EVs in the cells incubated for 6 h is higher. This complements the trend of our quantification experiment where we found that EV uptake increases with time. The CFSE-EVs are also distributed relatively evenly throughout the cells.

3.3 Inhibiting RBCEVs Uptake in Ca1a Cells

After performing the flow cytometry test with varying concentrations for each inhibitor, we get three different graphs. From here, we contrast between the different concentrations of each inhibitor and compare across inhibitors to determine the pathway in which RBCEVs enter Ca1a cells by.

Since the DMSO that the inhibitors are stored in may be a confounding variable in our experiments, we include a DMSO control, where cells are incubated with DMSO that has been diluted in the same way as the highest concentration of the inhibitors prior to incubation with EVs. Comparing our experimental results with the DMSO control is a more accurate analysis than using the GM only as the negative control.

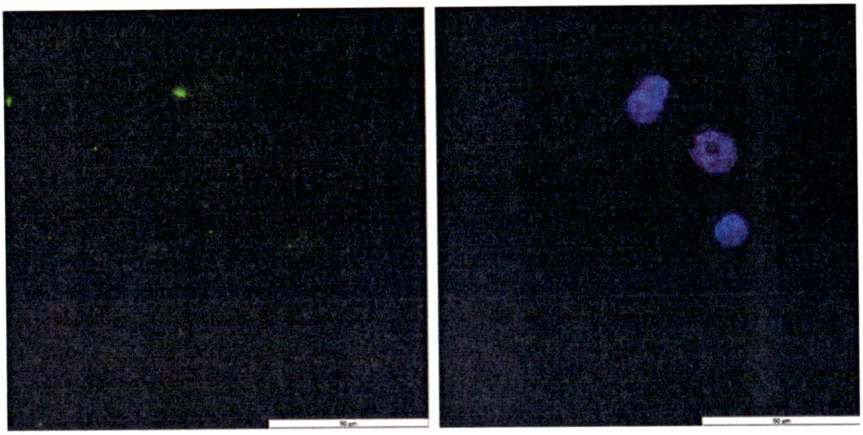

Fig. 6 Incubation of RBCEVs for 6 h

Cytochalasin D. Cytochalasin D depolymerises actin filaments and blocks general endocytosis pathways [8]. From our data, we observe that Cytochalasin D significantly decreases the RBCEVs uptake at both 10 and 1 μM when compared with the DMSO control. We can safely conclude that RBCEVs are taken up via general endocytosis mechanisms by Ca1a cells, which is similar to previous studies on ovarian cancer cells [3]. However, at 0.1 μM, the RBCEV uptake is not significantly different from the control, likely because the concentration is too low to have a significant impact on the endocytosis pathways. Hence, we can conclude that inhibition of RBCEVs uptake into Ca1a cells between 1 and 10 μM of Cytochalasin D is significant. Further testing with Cytochalasin D concentrations between 1 and 10 μM should be carried out to observe any dose-dependence in the response of the uptake extent.

EIPA. EIPA targets Na+ and H+ exchange occurring at the plasma membrane of cells, inhibiting macropinocytosis [8]. The results we obtained for EIPA inhibition are similar to that of Cytochalasin D. For higher concentrations of EIPA, e.g. 25, 50 μM, the decrease in RBCEV uptake is highly significant. We conclude that RBCEVs are likely to be taken up via macropinocytosis. However, for 10 μM of EIPA, it enhances the RBCEV uptake slightly. This disparity in our results could be because low concentrations of EIPA are ineffective in inhibiting macropinocytosis in Ca1a cells and may trigger another compensating mechanism, causing a slight overall increase in EV uptake. Hence, we conclude that RBCEVs uptake by EIPA is significant between 25 and 50 μM of EIPA, and EIPA reduces this uptake in a dose-dependent manner.

Filipin. Filipin binds to cholesterol in lipid rafts on plasma membranes, inhibiting lipid-raft mediated endocytosis [8]. In Fig. 7, we see the RBCEV uptake by Ca1a cells do not differ significantly from the DMSO control. This gives sufficient information that for 5–15 ug/ml of Filipin, it has little effect on RBCEV uptake and we conclude

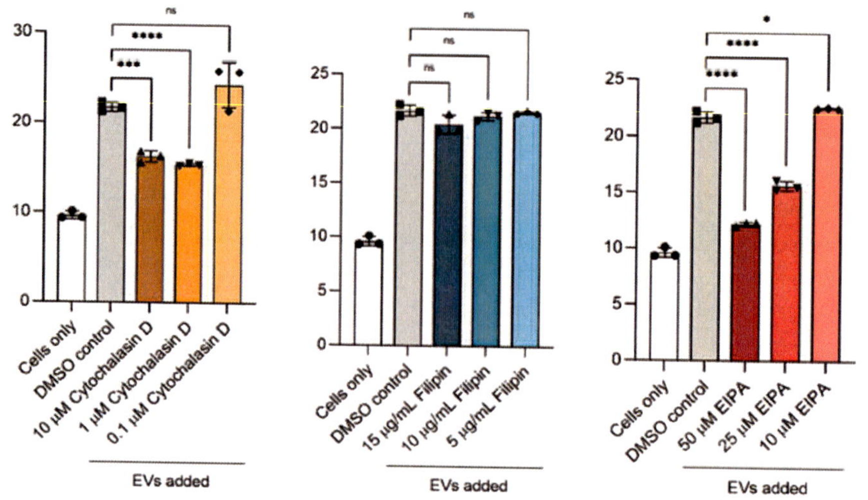

Fig. 7 Flow cytometry results (Sample size ($n = 3$), ns: $P > 0.05$, *: $P \leq 0.05$, **: $P \leq 0.01$, ***: $P \leq 0.001$, ****: $P \leq 0.0001$)

they are not likely taken up by lipid-raft mediated endocytosis in Ca1a cells. However, we should conduct more experiments to test different concentrations of Filipin.

Overall discussion. EVs are known to transport proteins, lipids and nucleic acids [9], especially miRNA, mRNA and lncRNA [10] and play essential roles in cell–cell communication, cell maintenance and transportation of cargo [9]. Hence, EVs have potential to deliver drugs to specific target cells and combat various diseases. However, due to limited agreement on the significant mechanisms of EV uptake and its many effects on different cell types, it is crucial that researchers continue to investigate interactions between different EVs and recipient cells [4].

In this project, we investigate the efficiency and mechanisms of RBCEV uptake by Ca1a cells. Using data obtained from our quantification and imaging experiments, we confirm Ca1a cells successfully take up RBCEVs when incubated together, and that RBCEV uptake increases with time. However, it is crucial to note that conclusions drawn from the imaging experiment may be inaccurate as we only compared one image from each incubation time, and there may be outliers. Further experimentation is needed to confirm our results and more images should be compared.

In addition, although RBCEVs can be taken up by Ca1a and other cancer cells, their contents may not be taken up completely. However, recent experiments have shown exosomes released by RBCEVs work effectively as carriers to facilitate effective deliveries of short and long miRNA to target cancer cells [11]. Others have showed miRNAs (miR-125b-5p, miR-451a and miR-4454) were abundant in exosome samples released from RBCs [12].

From our data, we deduce that Cytochalasin D and EIPA can significantly reduce RBCEVs uptake, but not fully prevent this process. This suggests that RBCEVs are mostly taken up by endocytosis, specifically macropinocytosis, with lipid-raft mediated endocytosis being a less likely uptake route. However, it is necessary to note that the effects of inhibitors on the cells can truly be dose-dependent, and the mechanism of uptake can differ with larger doses of the inhibitors. More experimentation is needed to determine whether other mechanisms are used in RBCEV uptake in Ca1a cells.

The potential that EVs can be future therapeutic agents means that it is imperative that the EV uptake mechanism is understood to aid prospective therapeutic design [4]. Our research brings insight into RBCEVs uptake by Ca1a breast cancer cell line, including the mechanisms of uptake and the dose-dependent nature of inhibitors. The market for EV research is continually expanding, hence greater understanding of the EV uptake pathway is certainly achievable in the foreseeable future.

4 Conclusions and Future Work

We conclude that RBCEV uptake increases with time in Ca1a cells and RBCEVs enter Ca1a cells through more than one mechanism. They are more likely taken up via endocytosis, especially macropinocytosis, and less likely taken up via lipid-raft mediated endocytosis. Along with the modifications to experiments previously mentioned, we aim to expand the inhibition experiment to other modes of EV uptake like clathrin-mediated endocytosis, phagocytosis and endosomal membrane fusion to analyse further possible routes RBCEVs enter Ca1a cells.

Acknowledgements We would like to extend our gratitude to Dr. Le Thi Nguyet Minh and Ms. Hong Anh under the Department of Pharmacology in National University of Singapore for guiding us in this research project as our research mentors. We would also like to thank Dr. Chiam Sher-Yi and Dr. Tang Hock Chun from NUS High School for helping us with our research project.

References

1. Mulcahy, L. A., Pink, R. C., & Carter, D. R. F. (2014). Routes and mechanisms of extracellular vesicle uptake. *Journal of Extracellular Vesicles, 3*(1), 24641.
2. Vader, P., Mol, E. A., Pasterkamp, G., & Schiffelers, R. M. (2016). Extracellular vesicles for drug delivery. *Advanced Drug Delivery Reviews, 106*, 148–156.
3. Doyle, L. M., & Wang, M. Z. (2019). Overview of extracellular vesicles, their origin, composition, purpose, and methods for exosome isolation and analysis. *Cells, 8*(7), 727.
4. Kumari, S., Mg, S., & Mayor, S. (2010). Endocytosis unplugged: multiple ways to enter the cell. *Cell Research, 20*, 256–275.
5. de Jong, B., Barros, E. R., Hoenderop, J. G., & Rigalli, J. P. (2020). Recent advances in extracellular vesicles as drug delivery systems and their potential in precision medicine. *Pharmaceutics, 12*(11), 1006.
6. Herrmann, I. K., Wood, M. J. A., & Fuhrmann, G. (2021). Extracellular vesicles as a next-generation drug delivery platform. *Nature Nanotechnology, 16*(7), 748–759.

7. CFSE cell division tracker kit. https://www.biolegend.com/de-de/products/cfse-cell-division-tracker-kit-9396. Last accessed June 28, 2022.
8. Escrevente, C., Keller, S., Altevogt, P., & Costa, J.: Interaction and uptake of exosomes by ovarian cancer cells. *BMC Cancer, 11*(1).
9. Xu, R., Rai, A., Chen, M., Suwakulsiri, W., Greening, D. W., & Simpson, R. J. (2018). Extracellular vesicles in cancer—Implications for future improvements in cancer care. *Nature Reviews Clinical Oncology, 15*(10), 617–638.
10. Usman, W. M., Pham, T. C., Kwok, Y. Y., Vu, L. T., Ma, V., Peng, B., Chan, Y. S., Wei, L., Chin, S. M., Azad, A., He, A. B.-L., Leung, A. Y. H., Yang, M., Shyh-Chang, N., Cho, W. C., Shi, J., & Le, M. T. N. (2018). Efficient RNA drug delivery using red blood cell extracellular vesicles. *Nature Communications, 9*(1).
11. Huang, H., Zhu, J., Fan, L., Lin, Q., Fu, D., Wei, B., & Wei, S. (2019). MicroRNA profiling of exosomes derived from red blood cell units: implications in transfusion-related immunomodulation. *BioMed Research International, 2019*, 1–10.
12. Fukumoto, I., Kinoshita, T., Hanazawa, T., Kikkawa, N., Chiyomaru, T., Enokida, H., Yamamoto, N., Goto, Y., Nishikawa, R., Nakagawa, M., Okamoto, Y., & Seki, N. (2019). Identification of tumour suppressive MicroRNA-451a in hypopharyngeal squamous cell carcinoma based on MicroRNA expression signature. *British Journal of Cancer, 111*(2), 386–394.

Vortex-Ring Interactions with Lattice-Based Wall

Chuanhua Liu, Darius Koi Yik Tham, Keshav Peri, and Daniel Tze How New

Abstract The study of the vortex-ring structure in fluid mechanics has a long history, and it is only in the recent decades that its collision with different surfaces was explored. This paper reports on the fundamental study of vortex-ring collision with planar surfaces with lattice-based cylindrical protrusions. Using the "cylindrical slug" method, vortex-rings with Reynolds number $Re = Ud/v = 2000$ were piston-generated in a water tank through a $d = 20$ mm nozzle and driven towards different lattice-based surfaces. An experimental time-resolved particle image velocimetry (TR-PIV) study had been conducted to measure the behaviour of circular vortex-rings impinging on the lattice-based surfaces. For these experiments, pins of either $D = 1$ mm or $D = 2.5$ mm diameter were used. The pins were placed at 5 mm or 10 mm intervals to form a square lattice. The study comprised of four test cases and each varied in terms of pin diameter and interval distance. Cross comparisons of test cases showed that greater pin diameter $D = 2.5$ mm or greater area density (5 mm interval) favoured earlier vortex-ring laminar to turbulent transitions. These results could be helpful towards developing bollard-like structures for the purpose of dissipating wingtip vortices and therefore improving safety and reducing the wait time of aircraft on the tarmac.

Keywords Vortex-ring · Collision · Particle image velocimetry · Reynolds number · Laminar-turbulent transition · Wing-tip vortices · Aircraft

1 Introduction

Vortex-rings are important fundamental flow structures in fluid mechanics. Their unique dynamics has attracted much attention and has been explored for the

C. Liu · D. K. Y. Tham
NUS High School of Mathematics and Science, 20 Clementi Avenue 1, Singapore 129957, Singapore

K. Peri · D. T. H. New (✉)
School of Mechanical and Aerospace Engineering, Nanyang Technological University, 50 Nanyang Avenue, Singapore 639798, Singapore
e-mail: dthnew@ntu.edu.sg

H. Guo et al. (eds.), *IRC-SET 2022*,
https://doi.org/10.1007/978-981-19-7222-5_35

past century. It was in the last four decades that the phenomenon of a vortex-ring colliding with a solid wall was investigated. Earlier works done by Walker et al. [1] and Lim et al. [2] have explored this phenomenon and are in agreement with each other on the evolution of vortical structures upon collision with a flat wall. The instability mechanics of a vortex-ring impinging onto a flat surface are far more complex than that of an undisturbed vortex-ring [3]. Numerical simulation conducted by Orlandi and Verzicco [4] confirmed variations in the stability of vortex-rings at different phases of its evolution. They revealed that the rate of the secondary vortex-ring instability (SVR) increases after being induced is greater than that of the primary vortex-ring (PVR). Hence, they concluded that the SVR is less stable than the PVR. Vortex-ring interactions with non-planart surfaces has also been investigated. Studies done by Adhikari and Lim [5] and New et al. [6] involving porous and inclined surfaces respectively showed several differences in vortical interaction in addition to the general phases of vortical evolution.

Lately, vortex-ring interactions with rough surfaces has seen a few experimentations and analysis. Li and Bruecker [7] reported on the head-on collision of a laminar vortex-ring with a rough surface formed by cylindrical pins arranged in a hexagonal lattice. They discovered that the azimuthal instability mode $N = 6$ of the vortex-ring locked in with the geometry and orientation of the hexagonal lattice, which had six preferential pathways, causing the vortex-ring to transition into turbulence quicker. In our study, which is inspired by previous works on collision with flat and rough surfaces, we investigate vortex-ring interactions with rough surfaces formed by square pin lattices, with an emphasis on the effects of varying pin diameter and area density on the vortex-ring behaviour rather than the arrangement of pins. We hypothesise that a greater pin diameter and area density will result in higher rate of PVR dissipation. We theorise that the results of this experiment could be useful in determining the effectiveness of bollard-like structures in dissipating persistent trailing vortices from an aircraft's wingtips, which can be modelled as the rotating vortex cores of a circular vortex-ring. These wingtip vortices can cause trailing aircrafts to roll out of control. By dissipating wingtip vortices, not only will the safety of following aircrafts improve, the wait time for a safe takeoff or landing may also reduce.

2 Methodology

The experimental set-up shown in Fig. 1 closely follows that in New et al. [6] and will be discussed here. The vortex-ring collision took place in a transparent water tank and was recorded from the outside of the tank. The "cylindrical slug" technique was used to generate the vortex ring, where the motion of the linear actuator drove the piston forward, pushing a volume of water through the nozzle with an internal diameter of $d = 20$ mm that was fitted into the side of the water tank. The motion of the linear actuator was achieved using a motor system consisting of a high torque micro-stepping motor and a motor driver, both of which were acquired from Applied

Motion Products. The motor system was controlled remotely by the workstation that programmed the motor to follow a trapezoidal velocity profile and hence generate vortex-rings with Reynolds number of Re = 2000. The 200 mm × 200 mm × 5 mm square plates with lattice-based protrusions, as shown in Fig. 2, were placed 6d away from the end of the nozzle and were held upright by an acrylic stand. The plates were designed in-house using Autodesk Inventor before they were manufactured by an external vendor using acrylic. They were designed such that the holes into which the cylindrical pins would be inserted follows a square lattice arrangement, with the middlemost pin located at the centre of the plates. The pins were fitted into the plates and then spray painted black for better contrast during flow visualisation. For this experiment, 4 different plates were used, each consisting of either diameter $D = 1$ mm or $D = 2.5$ mm pins arranged at 5 mm or 10 mm intervals. To capture the behaviour of the vortex collision with the lattice-based plates, an IDT high-speed CMOS camera was placed at the midpoint between the plate and the nozzle such that the field-of-view covered the entire distance. It was controlled remotely by the workstation to run at 400 frames per second to capture the flow dynamics of the vortex-ring. Polyamide Seeding Particles (PSP) with diameters of 20 μm were also homogenously mixed with the water in the tank prior to the experiment. The laser system, comprising of a laser driver, a diode-pumped solid-state (DPSS) laser, reflector mirrors, and light sheet optics have been set up such that a laser sheet is projected along the centre line of the direction of travel of the vortex-ring for streamwise flow measurements. At the time of experiment, the motor system and the camera system were activated at the same time, though the camera was programmed to continue recording even after the collision to capture the post-collision interaction with the lattice-based surfaces. The images were then exported for TR-PIV processing using PIVlab, a digital particle velocimetry tool for MATLAB. The images were first masked and calibrated before a two-pass processing with interrogation windows of 64 × 64 pixels and 32 × 32 pixels was conducted, both of which were performed with a 50% overlap ratio. To better present the velocity vectors in our context, the display parameter was set to vorticity and the colormap range was limited to + 15 and − 15/s.

3 Results and Discussion

3.1 *Baseline Case: Impingement on a Flat Wall*

To fully appreciate the additional impact that surface protrusions have on circular vortex-ring collisions with a wall shown in the four test cases, one must first understand the dynamics displayed in the experimental control. Hence, we will briefly discuss the behaviour of a circular vortex-ring impinging on a flat wall, which will function as a standard of comparison to the four test cases. For the sake of simplicity, impingement on a flat wall will be referred to as the baseline case. Figure 3 is a laser-induced fluorescence (LIF) flow visualisation of a Re = 2000 vortex-ring colliding

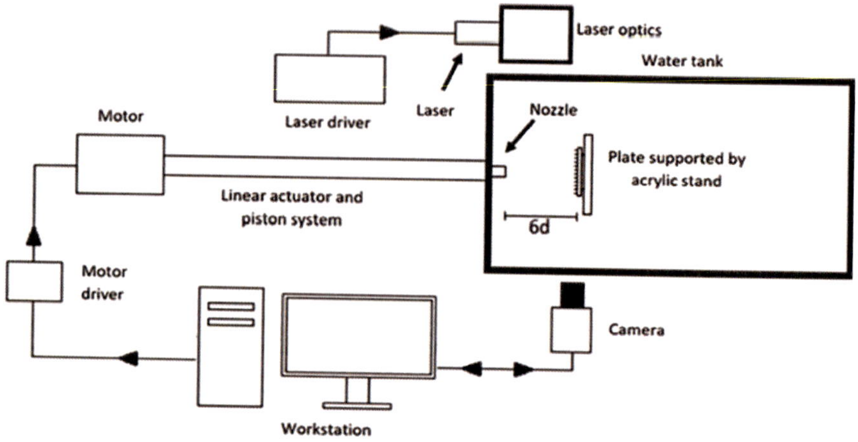

Fig. 1 Schematics of experimental set-up

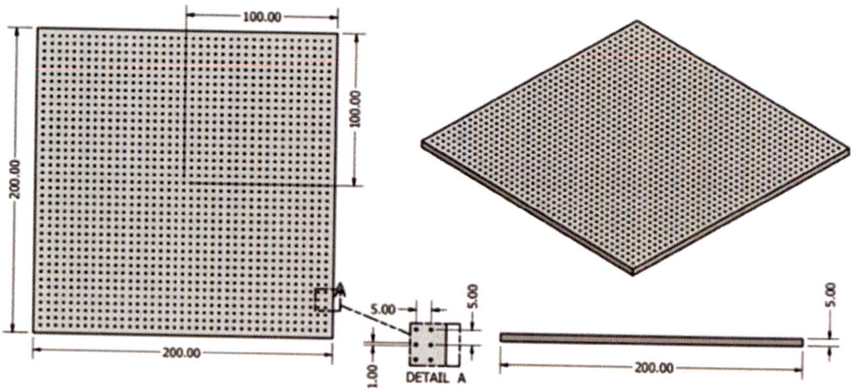

Fig. 2 2D drawing of the 200 mm × 200 mm × 5 mm lattice-based plates

with a flat wall taken from New et al. [6]. The vortex behaviours shown in the images are in agreement with those described in Walker et al. [1] and Lim et al. [2]. Those images have been selected to illustrate only the major events that takes place during a vortex-ring collision with a flat wall. The first observable change upon contact with the wall is that the vortex cores move radially outwards, resulting in a larger core-to-core distance. When the radial expansion is coupled with an adverse pressure gradient along the edge of the vortex-ring, the wall boundary layer separates from the flat surface, shown in Fig. 3c, before rolling up to form a secondary vortex-ring (SVR) of opposite vorticity, as shown in Fig. 3d. The SVR is formed adjacent to the primary vortex-ring (PVR) cores and undergoes ejection from the surface. It leapfrogs over the top of the PVR in Fig. 3e, due to entrainment of the SVR by the PVR. The SVR then proceeds to move into the interior of the PVR and towards the

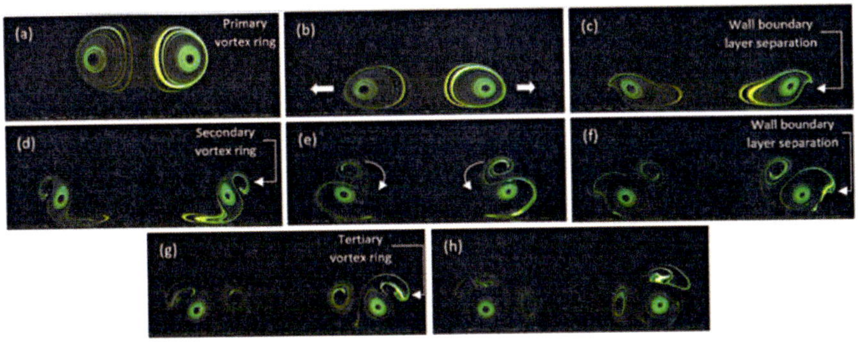

Fig. 3 Laser-induced fluorescence visualisation of Re = 2000 vortex-ring impinging on a flat surface

flat surface, during which a second separation of the wall boundary layer surrounding the exterior of the PVR takes place as depicted in Fig. 3f. The result of a second wall boundary layer separation is that a tertiary vortex-ring (TVR) forms, and like the SVR, it also experiences ejection and then entrainment but to a lower extent due to the presence of the SVR in the confines of the PVR.

3.2 *Test Case 1: Impingement on Square Lattice of $D = 1$ mm Pins Placed at 10 mm Intervals*

Figure 4 contains the vorticity maps of the TR-PIV flow measurements of vortex impingement on surface comprising a square lattice of 1 mm diameter cylindrical pins arranged with 10 mm interval spacing. Do note that in the vorticity maps in all test cases the vortex-ring is travelling downwards, and the pin lattice is along the bottom of the maps. The early stages of the vortex-ring collision seem to be identical to those observed in the baseline case. We can see that the vortex cores moving radially outwards, resulting in an increasing core-to-core diameter. This couples with an adverse pressure gradient along the peripheral of the vortex-ring to cause the separation of the wall boundary layer as shown in Fig. 4b and as observed in the baseline case. As expected, the SVR is soon induced at $t = 540$ ms and migrates over to the top of the PVRs cores. However, unlike the collision with a flat surface, the SVR only experiences partial entrainment as it can be seen to remain stagnant over a relatively long period of time, which is depicted in Fig. 4d–h. Furthermore, the coherence of the SVR reduces over time whilst it remains relatively stationary. Prior to the stagnation, the wall boundary layer separates a second time as seen in Fig. 4d, long before a weak TVR forms. There was little to no entrainment as well, unlike the TVRs observed in flat surface collision. There is little change after the induction of the weak TVR besides the loss of coherence of the PVR and the PVR finally loses most of its coherence at around $t = 2247.5$ ms.

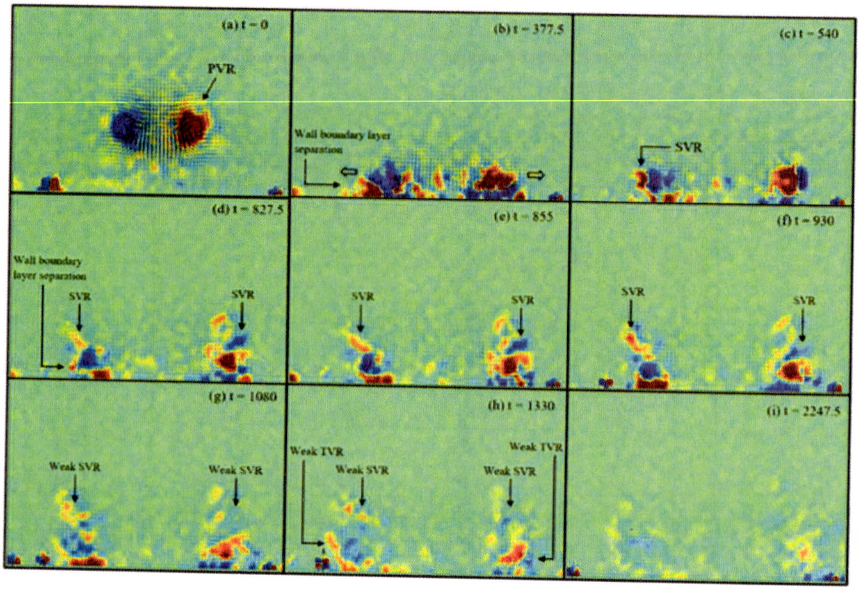

Fig. 4 TR-PIV flow measurements of Re = 2000 vortex-ring colliding with a planar surface containing square lattice of 1 mm cylindrical protrusions spaced 10 mm apart. The first frame is taken to be $t = 0$, where t is in milliseconds

3.3 *Test Case 2: Impingement on Square Lattice of $D = 1$ mm Pins Placed at 5 mm Intervals*

As show in Fig. 5, the early stages of the circular vortex-ring are in parallel with that of the baseline case. The vortex-ring cores first shift radially outwards, resulting in an increasing core-to-core diameter. However, the separation of the wall boundary layer is weak, as shown in Fig. 5c, but strong enough to induce a SVR at $t = 540$ ms, which is observable in Fig. 5e. Interestingly, there was little to no entrainment of the SVR, and it gradually lost coherence as shown in Fig. 5f–i. Unlike in the baseline case, a second wall boundary layer separation does not occur and hence there is no generation of a TVR. The PVR can be seen to be losing coherence from the very outset of the collision, until most of its coherence has been lost after 1037.5 ms, at which not much observable changes can be detected.

3.4 *Test Case 3: Impingement on Square Lattice of $D = 2.5$ mm Pins Placed at 10 mm Intervals*

As shown in Fig. 6, in the early phases of the collision, the PVR cores move radially outwards, resulting in a growing core-to-core diameter. This joins with an adverse

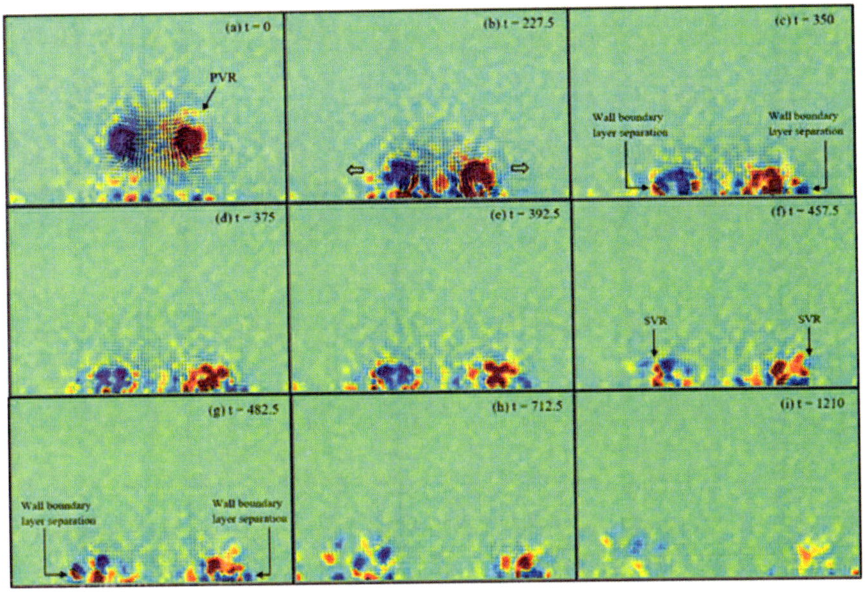

Fig. 5 TR-PIV flow measurements of Re = 2000 vortex-ring colliding with a planar surface containing square lattice of 1 mm cylindrical protrusions spaced 5 mm apart

pressure gradient along the peripheral of the PVR to cause the separation of the wall boundary layer as shown in Fig. 6c. At $t = 457.5$ ms, the SVR is induced adjacent to the PVRs cores. The SVR experiences a brief wall ejection but no entrainment, as depicted in Fig. 6f. Furthermore, the coherence of the SVR reduces over a short period of time whilst it remains relatively stationary. Prior to the stagnation, the wall boundary layer separates a second time as seen in Fig. 6g, but too weak to induce a TVR. It could be observed that there was little change after the second wall boundary layer separation. The PVR has lost most of its coherence after 1210 ms, as displayed in Fig. 6i.

3.5 *Test Case 4: Impingement on Square Lattice of $D = 2.5$ mm Pins Placed at 5 mm Intervals*

As shown in Fig. 7, in the early stages of the collision, the vortex cores move away from each other, resulting in a growing core-to-core diameter. This is coupled with an adverse pressure gradient along the peripheral of the vortex-ring to cause the separation of the wall boundary layer as shown in Fig. 7b. Surprisingly, there is extensive entrainment of the weak SVR, which forms at approximately $t = 380$ ms and finally settles within the confines of the PVR at $t = 480$ ms. The wall boundary layer separates a second time at $t = 380$ ms, although it is too weak to evolve into a

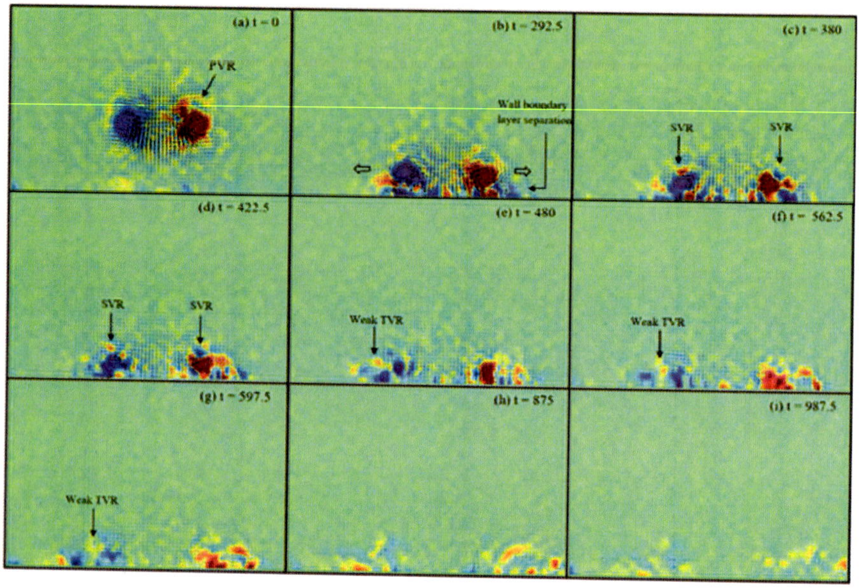

Fig. 6 TR-PIV flow measurements of Re = 2000 vortex-ring colliding with a planar surface containing square lattice of 2.5 mm cylindrical protrusions spaced 10 mm apart

TVR. After the entrainment of the SVR ceases, the PVR gradually loses coherence and has lost most of its coherence at around $t = 987.5$ ms. Most of the collision events in a flat wall collision are shown here, apart from the entrainment of the TVR.

3.6 *Effects of Varying Pins Size and Spacing*

We conducted experiments on four test cases as presented above, with each involving a different combination of diameter and area density of pins. We will first discuss the effects of increasing area density of pins on PVR dissipation. In Test Case 1, the PVR loses most of its coherence at around $t = 2247.5$ ms and the formation of an SVR is observed at around $t = 540$ ms. In Test Case 2, the PVR loses most of its coherence at around $t = 1037.5$ ms which is significantly earlier than in Test Case 1. Furthermore, only the formation of SVRs at about $t = 540$ ms, albeit weaker than those produced in Test Case 1, is observable in Test Case 2. In Test Case 3, the PVR loses most of its coherence after about 1210 ms and the formation of a weak SVR takes place at $t = 457.5$ ms. On the other hand, the PVR loses most of its coherence after around 987.5 ms in Test Case 4, which is earlier than in Test Case 3. A weak SVR forms at $t = 380$ ms, which is earlier than what we observe in Test Case 3, in Test Case 4 but not a TVR. From these, we infer that an increase in the area density promotes PVR transition to turbulence. However, it also rules out the possibility that

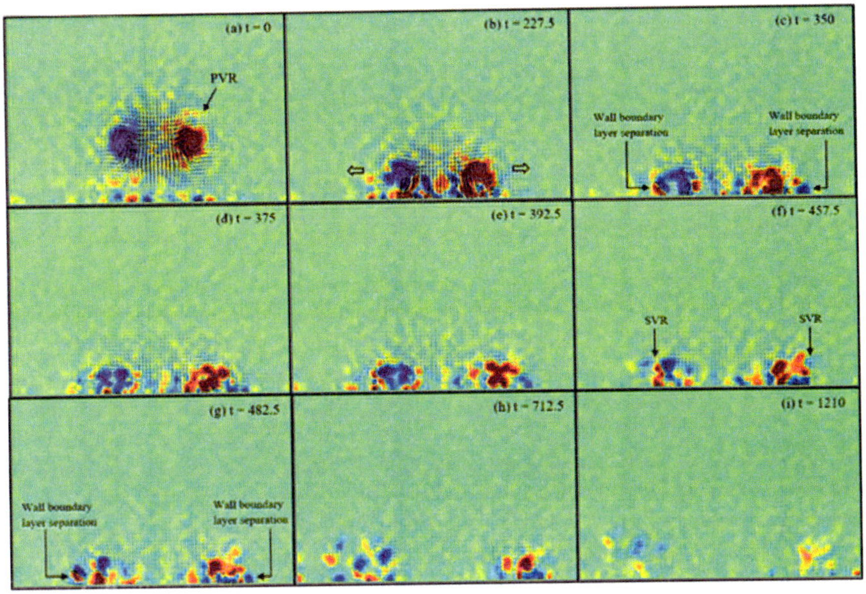

Fig. 7 TR-PIV flow measurements of Re = 2000 vortex-ring colliding with a planar surface containing square lattice of 2.5 mm cylindrical protrusions spaced 5 mm apart

the increase in the rate of laminar-turbulent transition is mostly due to the SVR being more unstable than PVRs. The earlier formation of a stronger SVR in Test Case 1 does not reduce the time taken for laminar-turbulent transition as compared to Test Case 2. Although the induction of a SVR is earlier in Test Case 4 than in Test Case 3, there is significant delay between the dissipation of SVR and PVR in both cases. This again shows that the inherently greater instability of the SVR could not have significantly caused the earlier dissipation of the PVR in Test Case 4 as compared to other potential factors associated with the protrusions. We will now proceed to compare the test cases based on the variable of increasing cylindrical pin diameter. By comparing the time at which the PVR loses most of its coherence, the data from Test Case 1 and Test Case 3 shows that a greater pin diameter will result in a greater rate of PVR dissipation. However, comparison of the formation of SVRs shows that a greater pin diameter speeds up the formation of SVRs. By doing the same for Test Case 2 and Test Case 4, the data again shows that an increase in pin diameter results in quicker dissipation of the PVR and generation of an SVR. However, the delay between the dissipation of the SVRs and that of the PVRs in all four cases is large. Hence it is unlikely that the earlier formation of the less stable SVRs is the only factor that contributed to earlier dissipations in the test cases. Our observations provided evidence that both greater diameter and area density of the pins in a square lattice contributes to an earlier loss of coherence of PVRs. However, we cannot confirm whether the rise in dissipation rates can be solely attributed to the greater rate of instability growth native to the SVR. There is likely to be other ways with a longer period of effect through which dissipation rates can be reduced. We theorise that one major contributing factor can simply be the greater physical barrier conferred by the denser and thicker pins to extensively interrupt the coherent fluid flow within the PVR throughout its entire evolution.

4 An Alternative Option for Wake Turbulence Dissipation

Whilst there have been numerous attempts to alleviate vortex generation via aircrafts modifications, attempts to dissipate via passive ground systems have been relatively rare. However, the few experiments on the effectiveness of ground-based devices have shown promising results [8, 9]. The installation of bollard-like structures as proposed previously could be another ground-based method for vortex dissipation. These bollard-like structures can be strategically placed near the apron, taxiways and runways to accelerate vortex dissipation. This structure could also offer greater flexibility in terms of the shape of the lattice that can be formed as well as its positioning. The symmetry of the lattice could provide disruption of the wake turbulence generated in different directions, which is seen in airports with intersecting runways.

5 Conclusions

We conducted TR-PIV measurements and analysis on four test cases of Re = 2000 circular vortex-ring impinging on rough surfaces comprising plates with regularly arranged cylindrical protrusions. We note that the sooner formation of the secondary vortex-ring (SVR) may not result in faster dissipation of the primary vortex-ring (PVR) although the former is inherently more unstable than the latter. We believe it is possible that the significant delay between the dissipations of the SVR and PVR facilitates other factors that would speed PVR dissipation, apart from the high instabilities of the SVR. Moving on to our main focus, which is to observe whether dimensions and arrangements of the surface protrusions can be altered to promote vortex-ring dissipation, we are able to draw some preliminary conclusions from our experiments. As we round up our analysis of the above test cases, we conclude that the dissipation of the vortex-ring takes place earlier with greater pin diameter and area density of the pins, which is in parallel with our hypothesis. Hence, we believe that designing bollard-like structures with a greater diameter and area density may reduce the hazard posed by wingtip vortices and aircraft waiting time on the tarmac.

References

1. Walker, J. D. A., Smith, C. R., Cerra, A. W., & Doligalski, T. L. (1987). The impact of a vortex-ring on a wall. *Journal of Fluid Mechanics, 181*, 99–140.
2. Lim, T. T., Nickels, T. B., & Chong, M. S. (1991). A note on the cause of rebound in the head-on collision of a vortex-ring with a wall. *Experiments in Fluids, 12*(1), 41–48.
3. Xu, Y., & Wang, J. (2013). Recent development of vortex-ring impinging onto the wall. *Science China Technological Sciences, 56*(10), 2447–2455.
4. Orlandi, P., & Verzicco, R. (1993). Vortex-rings impinging on walls: Axisymmetric and three-dimensional simulations. *Journal of Fluid Mechanics, 256*, 615–646.
5. Adhikari, D., & Lim, T. T. (2009). The impact of a vortex-ring on a porous screen. *Fluid Dynamics Research, 41*(5), 051404.
6. New, T. H., Shi, S., & Zang, B. (2016). Some observations on vortex-ring collisions upon inclined surfaces. *Experiments in Fluids, 57*(6), 1–18.
7. Li, Q., & Bruecker, C. H. (2018). Vortex interaction with a rough wall formed by a hexagonal lattice of posts. *Physics of Fluids, 30*(5), 054107.
8. Morris, S., & Williamson, C. (2020). Impingement of a counter-rotating vortex pair on a wavy wall. *Journal of Fluid Mechanics, 895*, A25.
9. Holzäpfel, F., Stephan, A., Rotshteyn, G., Körner, S., Wildmann, N., Oswald, L., Gerz, T., Borek, G., Floh, A., Kern, C., Kerschbaum, M., Nossal, R., Schwarzenbacher, J., Strobel, M., Strauss, L., Weiß, C., Kauczok, S., Schiefer, C., Czekala, H., & Smalikho, I. (2021). Mitigating wake turbulence risk during final approach via plate lines. *AIAA Journal, 59*(11), 4626–4641.

Calcium Phosphate/Collagen Ratio in Bone Grafts Influences Bone Repair in a Rabbit Femoral Condyle Defect Model

Sivakumar Avanthika and Xiaoman Luo

Abstract Bone grafts are used in orthopaedics for treatment of conditions including tumours, trauma, and infection, and critical-sized defects that do not spontaneously heal without surgical intervention. While autografts harvested from the patient's body and allografts from donors are often used as graft material, a wide variety of synthetic bone graft substitutes are also available as alternatives. In this study, we aimed to determine how calcium phosphate/collagen ratio would influence the capacity of a biphasic bone graft to support bone repair. The in vitro degradation of scaffolds with three different calcium phosphate/collagen ratios were evaluated in PBS over 7 days. It was found that higher calcium phosphate/collagen ratios resulted in less pH decrease, and scaffolds with a lower calcium phosphate/collagen ratio plateaued at higher pH. The in vivo performance of the three implants were evaluated in a femoral condyle defect model in New Zealand White Rabbits over 12 weeks, and the results obtained revealed that calcium phosphate/collagen ratios have no significant impact on degradation rates of the calcium phosphate component of the biphasic scaffold. Percentage of bone growth, on the other hand, increased from 15.4 to 25.8% from the lowest to highest calcium phosphate/collagen ratios, potentially due to calcium phosphate's osteoconductive properties. Overall, higher proportions of calcium phosphate to collagen led to better outcomes in defect healing due to greater bone formation, and matching implant resorption and bone growth rates.

Keywords Bone graft substitutes · Tricalcium phosphate · Collagen · Bone repair

S. Avanthika (✉)
Raffles Institution, Singapore, Singapore
e-mail: avanthika.s04@gmail.com

X. Luo
Institute of Molecular and Cell Biology, A*STAR, Singapore, Singapore

H. Guo et al. (eds.), *IRC-SET 2022*,
https://doi.org/10.1007/978-981-19-7222-5_36

1 Introduction

Bones possess an intrinsic ability to regenerate during the repair process in the healing of injuries such as fractures, where original tissue is completely regenerated and bone remodelling enhances the mechanical stability of newly grown bone as well [1]. However, the treatment of critical-sized bone defects, which do not heal completely without surgical intervention [2], together with other conditions such as tumours and certain types of trauma, often requires the usage of bone grafts. Autografts have been widely regarded as the 'gold standard' as they possess favourable osteoconductive, osteoinductive, and osteogenic properties, along with a low risk of immunoreactions [1]. They are, however, associated with limitations such as donor site discomfort and morbidity [3], restricted quantity, and considerable costs. Allografts from donors circumvent the aforementioned limitations of autografts but pose an increased likelihood of immunogenicity and infection transmission [1]. Bone graft substitutes have been developed as alternatives, composing a multitude of synthetic and natural biomaterials.

This study focuses on Calcium Phosphate (CaP)-collagen scaffolds as bone graft substitutes for defect healing. CaP, specifically β-Tricalcium Phosphate (β-TCP), has similar characteristics to that of bone mineral and is biodegradable. However, it cannot be used alone in bone implants due to its brittleness [4]. Similarly, collagen performs poorly as a standalone graft material, but significantly enhances graft incorporation when coupled with CaP [5]. There appears to be, however, insufficient data regarding the influence of CaP/collagen ratios in these bone grafts on their performance in defect healing, in terms of bone formation and implant resorption. The rate of scaffold degradation and resorption is critical in determining its biological performance, as where the persistence of graft material is linked to poor clinical outcomes, overly rapid resorption impairs the regenerative ability of natural bone by providing inadequate scaffold support for bone apposition [6].

Ideally, the degradation profile of the implant should match the rate of bone formation for optimum defect healing. CaP and collagen have differing degradation rates and result in the production of different end products, where CaP degradation releases calcium and phosphate ions which stimulate bone mineral formation and cell adhesion, and collagen is resorbed by macrophages and osteoblasts through the action of collagenase [7]. CaP/collagen ratio is hence significant when considering biological performance of the implants.

This study examined the in vitro degradation and in vivo performance of three implants of different CaP/collagen ratios in a critical-sized femoral condyle defect in New Zealand White Rabbits, where we aimed to identify the optimum ratio of CaP to collagen that most successfully promotes bone healing.

Since the release of calcium and phosphate ions from CaP regulates the activation of osteoclasts and osteoblasts to facilitate bone regeneration [7], a larger proportion of CaP in the bone graft will increase osteoconductivity. As such, we hypothesised that the scaffold that contains the highest ratio of CaP in relation to collagen will most successfully promote bone defect healing.

Table 1 Components of each group of samples used for in vitro degradation study

Group	No. of replicates	Scaffold used	Treatment	Total volume/ml
1	3	Low-CaP	PBS + Collagenase (16 U/ml)	5
2			PBS	
3		Mid-CaP	PBS + Collagenase (16 U/ml)	
4			PBS	
5		High-CaP	PBS + Collagenase (16 U/ml)	
6			PBS	
7		–	PBS + Collagenase (16 U/ml)	
8			PBS	

2 Materials and Methods

Cylindrical scaffolds of three different CaP/collagen ratios with diameter of 6 mm and length of 8 mm were synthesised prior to the commencement of this study. As the exact composition and ratios of CaP to collagen utilised cannot be disclosed due to confidentiality rules of our host organisation, the scaffolds will be named 'Low-CaP', 'Mid-CaP', and 'High-CaP', corresponding to the increasing CaP/collagen ratio.

2.1 In vitro *Degradation Assay*

The in vitro degradation profiles of the CaP-collagen scaffolds were assessed by measuring pH changes over time. Eight groups of scaffolds with a diameter of 6 mm and length of 4 mm were immersed in phosphate-buffered saline (PBS) solution intended to mimic in vivo conditions with or without 16 U/ml of collagenase at 37 °C with gentle shaking (Table 1). Groups 1, 3, and 5 contained collagenase, while Groups 7 and 8 were used as controls to confirm that any changes in pH in groups 1 to 6 were solely due to scaffold degradation. pH was measured at 0, 0.5, 1, 2, 4, 6, 8, 24, 48, 72, and 168 h. The assays were performed in triplicates.

2.2 In vivo *Biological Evaluation*

The in vivo biological evaluation was done prior to the commencement of this study. A cylindrical defect measuring 6 mm in diameter and 8 mm in depth was created in the femoral condyle of New Zealand White Rabbits in both the left and right knees and was filled with one CaP-collagen scaffold of the same dimensions (Ø6mm × 8 mm). After a period of 12 weeks, the animals were euthanised via a lethal injection of pentobarbital and the implanted scaffolds were retrieved. The harvested implants were fixed, dehydrated, and embedded in Poly(methyl methacrylate) (PMMA) prior to conducting this study. Each group included eight replicates and four replicates from each group were analysed in this project due to time constraints.

Micro-Computed Tomography (μ-CT) of Scaffolds. μ-CT scans were performed on Low-CaP, Mid-CaP, and High-CaP scaffolds prior to and at 12 weeks after implantation along the axial plane using the Bruker Skyscan 1176. The scans were processed using CtAn software with a standardised circular region of interest (ROI) of diameter 6 mm and threshold level of 55–255. The following parameters were determined: the total volume of the ROI (TV), total bone volume (BV), and bone volume relative to total volume (BV/TV%). Since CaP has almost identical x-ray attenuation to bone [8], it was not possible to distinguish bone from residual scaffold granules post-implantation; hence, the BV comprised both bone and scaffold material volume. Three replicates of unused scaffolds for each ratio were also scanned using μ-CT as 0-week controls and were analysed using the same parameters to find the TV, BV, and BV/TV%, where the BV/TV% would represent the starting percentage of CaP in the scaffold.

Histological and Histomorphometric Evaluation. The explants were processed through ethanol dehydration and embedded in PMMA following μ-CT analysis for histology and histomorphometric analysis. Acidic ethanol, basic fuchsin, and methylene blue dyes were prepared for histological staining using procedures detailed in the Appendix. The embedded scaffolds were sectioned with a thickness of 10–20 μm using a circular saw microtome and stained. The slides were sectioned parallel to the long axis of the inserted implant and at least 10 stained sections were obtained per sample. Scanned images of the prepared slides were imported into Adobe Photoshop and a standardised ROI was defined. Histomorphometry data was analysed using the histogram function. Material degradation rate was calculated by comparing material percentages at 0 weeks and 12 weeks using the formula:

$$\text{Material degradation rate} = \frac{\text{Initial material\%} - \text{Final material\%}}{\text{Initial material\%}} \times 100$$

2.3 *Statistical Analysis*

Statistical analysis was performed using Student's T-test, where $p < 0.05$ was considered statistically significant, with data presented in the form of mean ± standard deviation. $p < 0.05$ was represented with a single asterisk '*' on graphs.

3 Results

3.1 In vitro *Degradation Assay*

It can be observed that in the Low-CaP groups, there was a pH increase within the first 6 h, and the group lacking collagenase reached a higher peak pH of 8.75 ± 0.06 at 4 h than the group with collagenase, which reached a peak of 8.45 ± 0.03 at 6 h (Fig. 1a, $p < 0.05$). The effect of collagenase was apparent beyond the 6 h mark for Low-CaP, where without collagenase the pH plateaued at around 8.70 ± 0.02, while with collagenase the pH declined gradually to 7.89 ± 0.06 at the 168 h mark. For groups containing the Mid-CaP scaffold, there was an initial increase in pH regardless of the presence of collagenase until a peak pH of 8.25 ± 0.16 was reached at the 24 h mark (Fig. 1b). Beyond the 72 h mark, the pH of the Mid-CaP group with collagenase decreased slightly to 8.09 ± 0.11 at 168 h, while the pH for the Mid-CaP group without collagenase did not decrease ($p < 0.05$). For groups containing the High-CaP scaffolds, the pH remained constant at around 7.10 ± 0.03 throughout 7 days, with the presence of collagenase showing no effect on pH (Fig. 1c).

Across the three groups without collagenase, scaffolds with a lower CaP/collagen ratio plateaued at higher pH and at earlier time points. The Low-CaP and Mid-CaP scaffolds plateaued at pH 8.70 ± 0.02 and 8.25 ± 0.03 ($p < 0.05$), at time points 8 h and 24 h, respectively. Across the three groups with collagenase, Low-CaP and Mid-CaP scaffolds saw a peak pH of 8.45 ± 0.03 and 8.25 ± 0.16 at 4 h and 24 h, respectively. The final pH value at 168 h for groups containing the Low-CaP scaffold with collagenase was 7.89 ± 0.06, which was lower than that of the Mid-CaP

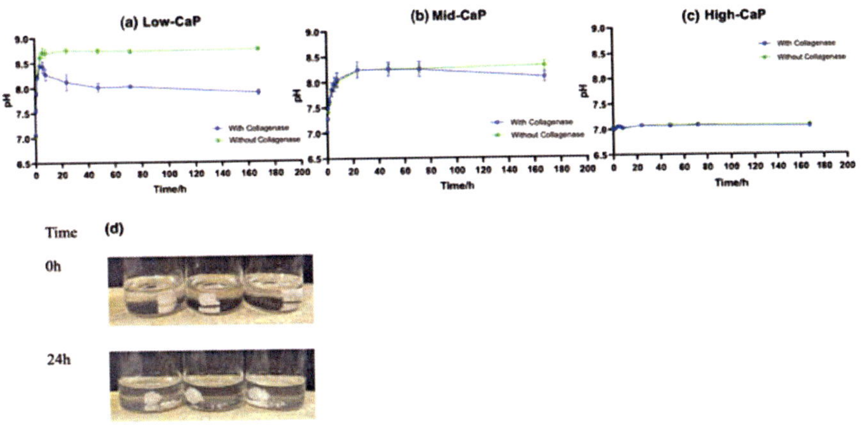

Fig. 1 **a** pH value of Low-CaP in PBS over time ± 16 U/mL collagenase at 37 °C. **b** pH value of Mid-CaP in PBS over time ± 16 U/mL collagenase at 37 °C. **c** pH value of High-CaP in PBS over time ± 16 U/mL collagenase at 37 °C. **d** Visual appearance of Low-CaP scaffold in PBS ± 16 U/mL collagenase at 0 and 24 h

scaffold with collagenase, which was 8.09 ± 0.11 ($p < 0.05$). The High-CaP scaffold maintained a constant pH of around 7.10 ± 0.03 throughout the 168 h regardless of presence of collagenase.

White CaP granules collapsed from the cylindrical scaffolds in all three replicates of Low-CaP scaffolds in the presence of collagenase at 24 h, with part of the structure no longer being maintained (Fig. 1d). Mid-CaP and High-CaP scaffolds, on the other hand, showed no visible structural changes throughout.

3.2 In vivo *Biological Evaluation*

μ-CT of Scaffolds Before and After Implantation. μ-CT scans only visualised high-density tissue, that is hard bone tissue and CaP, due to the threshold of 55–255 used for analysis. High-density bone and CaP cannot be differentiated due to their similar x-ray attenuation values [8]. Visual observation on the unused scaffolds (Fig. 2a) portrayed differences in the starting amount of CaP in the Low-CaP, Mid-CaP, and High-CaP scaffolds. The packing of CaP granules also differed amongst the scaffolds, where scaffolds with lower CaP content contained larger spaces between granules. From the scans taken at 12 weeks, it can be observed trabecular bone had formed for all three scaffolds, indicating that bone regeneration had occurred. Comparison of the high-density tissue volume percentages revealed that there were no significant differences between 0 and 12 weeks ($p > 0.05$) regardless of CaP/collagen ratio (Fig. 2b). This demonstrated that while there was no change in total high-density tissue volume, the composition of the high-density tissue had changed, such that CaP had been replaced partly by new bone.

Histomorphometric Evaluation. In the histological slides depicted in Fig. 3a, trabecular bone tissue was stained pink by basic fuchsin, whereas CaP granules appeared as black granules. The CaP granules appeared to be rounded after 12 weeks of implantation, indicating degradation of CaP in vivo. Fewer granules were present for the Low-CaP scaffold than the Mid-CaP and High-CaP scaffolds. Bone growth appeared around and incorporated CaP granules. Large empty spaces devoid of both bone and material were observed in the slides for Low-CaP and Mid-CaP scaffolds.

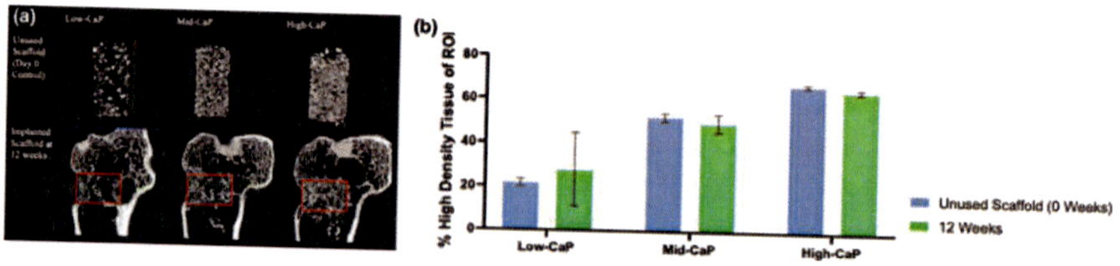

Fig. 2 **a** μ-CT scans of unused and harvested scaffolds where for unused scaffolds, dimensions are Ø6 × 8 mm, and for scaffold at 12 weeks, dimensions of ROI as defined by red frame are Ø6 × 8 mm. **b** % High-density tissue volume in the ROI before and after implantation

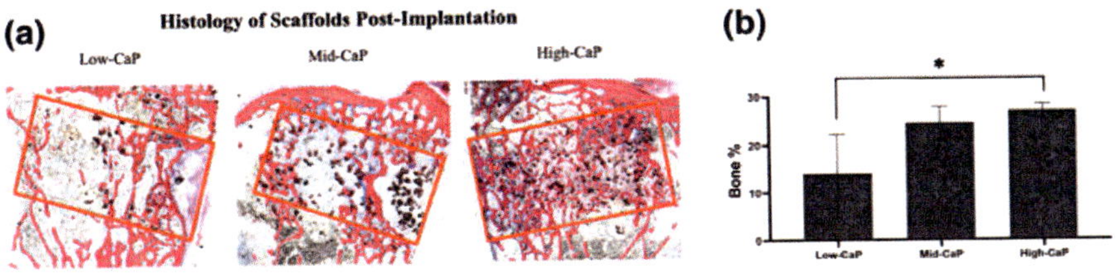

Fig. 3 **a** Representative histology of scaffolds post-implantation. **b** Bone % in ROI after 12 weeks based on histomorphometry

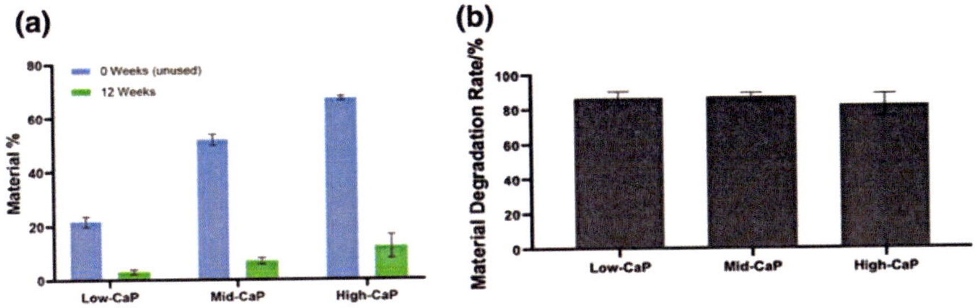

Fig. 4 Material degradation analysis based on histomorphometry. **a** Material % in ROI at 0 and 12 weeks. **b** Material degradation rate after 12 weeks

The percentage of bone present in the ROI at 12 weeks was the lowest for Low-CaP scaffolds (14.0 ± 8.5%), followed by Mid-CaP scaffolds (24.4 ± 3.39%), and was the highest for High-CaP scaffolds (26.9 ± 1.5%) (Fig. 3b). There was a significant difference between Low-CaP and High-CaP scaffolds, with nearly twice the amount of bone in High-CaP scaffolds in comparison to Low-CaP scaffolds ($p < 0.05$). Bone percentages were observed to be inconsistent for Low-CaP scaffolds, with a relatively larger standard deviation value than Mid-CaP and High-CaP scaffolds.

Material % in the ROI clearly decreased from 0 (blue bars) to 12 weeks (green bars) for all three scaffold ratios (Fig. 4a). Surprisingly, the material degradation rates were similar across the three scaffolds (Fig. 4b, $p > 0.05$), illustrating that the influence of CaP/collagen ratio on the degradation rate of CaP was not significant. Overall, CaP/collagen ratio appeared to have little impact on the CaP degradation rate but influenced the amount of bone formation and residual granules.

4 Discussion

The in vitro degradation assays demonstrated that higher CaP/collagen ratios resulted in reduced change in pH. The constant pH observed for the High-CaP scaffold-containing groups indicated that CaP is not the main contributing factor to pH changes. The pH decrease observed for groups containing Low-CaP and Mid-CaP scaffolds but not High CaP revealed that collagen was the component of the composite scaffold that caused both the initial increase in pH, and subsequent decrease in pH for groups with collagenase. The initial increase in pH may have occurred due to the presence of residual NaOH used for the neutralisation of collagen during scaffold preparation. The separation of CaP granules from the cylindrical scaffold for Low-CaP group treated with collagenase may indicate the collagen fibrils that held the granules in place have broken down. Therefore, the subsequent decrease in pH in the presence of collagenase is most likely due to the collagen breakdown products' acidic nature. However, literature shows that amino acids that form collagen such as glycine and hydroxyproline are not acidic [9]. The specific causes of pH decrease upon the addition of collagenase hence cannot be definitively explained because the in vitro assay conducted in this study was not designed to investigate so, in addition to differing information found in existing literature. The separation of CaP granules from the cylindrical scaffold could also potentially lead to inadequate mechanical support [4] in the ROI. It is worth noting that though the environment became more acidic, the overall pH continued to remain basic, possibly due to the buffer capabilities of PBS, which mimicked the pH, osmolarity, and ion concentrations of the human body [10].

μ-CT scans suggested that CaP resorption and new bone formation had occurred in the ROI. This could potentially explain the similar percentage of high-density tissue volume between 0 and 12 weeks, respectively, observed for all three scaffold ratios, where CaP granules have been partly degraded and replaced by newly formed bone. Bone percentages post-implantation demonstrate that bone formation was greater in groups containing more CaP. This can be attributed to the osteoconductive properties of CaP, where its molecular interactions with surrounding tissues can cause the formation of an interfacial, apatite layer on the CaP granule surface, leading to the incorporation of biological molecules and attachment of cells to induce bone formation [11]. Bone formation occurring directly around the CaP granules is in line with literature which states that osteoid tissue forms directly on the CaP surface, without intervening soft tissue [12]. Similar material degradation rates across scaffolds highlight that the CaP/collagen ratio did not impact CaP resorption significantly. This may be because CaP degradation appears to be influenced by its solubility and chemical stability, which have been kept constant in this experiment as the same type of CaP was used amongst all three scaffold ratios [13]. The large standard deviation of bone formation in Low-CaP scaffold suggested inconsistencies in in vivo performance, reducing its reliability as a bone graft material.

When viewed together, the in vivo μ-CT scans and histomorphometry evidence suggested that High-CaP scaffolds may be the most successful in promoting defect healing, where defect healing is evaluated based on two criteria: one being the extent to which the material degradation profile matched bone growth rates, and the other being the amount of bone formation. μ-CT analysis revealed that for all three scaffolds, total high-density tissue volume remained relatively unchanged, indicating a CaP degradation profile complementary to bone growth rates, while histomorphometry results indicated greater bone growth at higher CaP/collagen ratios. This demonstrates that scaffolds with higher CaP/collagen ratios are able to both provide mechanical support during initial stages of bone formation, after which it degrades to provide sufficient space for new bone tissue [14].

One limitation of this project was the lack of scaffolds consisting of pure collagen and pure CaP to serve as controls for the in vitro assay. Including such scaffolds will allow us to confirm that pH change was solely due to collagen and not CaP. Moreover, harvesting some implants over shorter time intervals such as 3 weeks and 6 weeks during the in vivo study could provide information on the direction and speed of bone remodelling processes.

5 Conclusion

Ultimately, the results of our study generally confirm the hypothesis that higher ratios of CaP to collagen better support defect healing in terms of scaffold resorption and bone formation. While the conditions for an 'ideal' bone graft substitute have not yet been established, identifying the impact of CaP/collagen ratio on the biological performance of the biphasic scaffold may prove useful when preparing composite implant materials for future experimentation. Future research may include investigating the effects of varying collagen and CaP types on biological performance, and the influence of additional factors such as bone morphogenetic protein (BMP) and human mesenchymal stem cells (hMSC) in providing osteoinductive properties to the scaffolds.

Acknowledgements I would like to express my gratitude towards my research mentor, Dr. Xiaoman Luo, and my teacher mentor, Dr. Tan Guoxian, for their support and patient guidance throughout this project. I would also like to thank A*STAR RAP for the invaluable research opportunity.

Appendix

Preparation of Dyes for Staining of Slides. Chemicals and dyes to be prepared: acidic ethanol, methylene blue, basic fuchsin.

96% Ethanol

1. Using a pipette, measure 96 ml 100% ethanol using a pipette and transfer into a glass bottle.
2. Using a pipette, add 4 ml of water to the glass bottle.
3. Mix well.

Acidic Ethanol

1. Using a pipette, measure 98 ml of 96% ethanol.
2. Add 2 ml of concentrated hydrochloric acid (37%).
3. Mix well and close bottle to prevent the solution from evaporating.

Methylene blue 1% in 0.1 M Borax pH 8.5

1. Using a pipette, measure 160 ml distilled water and add to a beaker.
2. Using an electronic weighing scale, weigh 7.62 g borax and add to water.
3. Stir until borax is completely dissolved.
4. Measure the pH, using a pH metre and adjust the pH to 8.5 with either hydrochloric acid (if overly basic) or sodium hydroxide (if overly acidic).
5. Using a pipette, add distilled water to make up the volume to 200 ml.
6. Using an electronic weighing scale, measure 2.0 g Methylene blue and add to the Borax solution.
7. Stir until the Methylene blue powder is dissolved.
8. Using a funnel and paper filter, filter the solution into a glass bottle.

0.3% Basic fuchsin in distilled water—prepare dye in a fume hood

1. Using a pipette, measure 200 ml distilled water and add to a beaker.
2. Using an electronic weighing scale, weigh 0.6 g basic fuchsin and add to the distilled water.
3. Place the beaker on a heating plate and heat the solution while stirring using a magnetic stirrer until boiling.
4. Let the solution cool down to room temperature.
5. Using a funnel and paper filter, filter the solution into a glass bottle.

References

1. Dimitriou, R., Jones, E., McGonagle, D., & Giannoudis, P. V. (2011). Bone regeneration: Current concepts and future directions. *BMC Medicine, 9*(1). https://doi.org/10.1186/1741-7015-9-66
2. Zhang, X., Xu, M., Liu, X., Zhang, F., Wei, Y., Meng, S., Dai, X., Duan, A., & Deng, X. (2013). Restoration of critical-sized defects in the rabbit mandible using autologous bone marrow stromal cells hybridized with nano-β-tricalcium phosphate/collagen scaffolds. *Journal of Nanomaterials, 2013*, 1–8. https://doi.org/10.1155/2013/913438
3. Younger, E. M., & Chapman, M. W. (1989). Morbidity at bone graft donor sites. *Journal of Orthopaedic Trauma, 3*(3), 192–195. https://doi.org/10.1097/00005131-198909000-00002
4. Sarikaya, B., & Aydin, H. M. (2015). Collagen/beta-tricalcium phosphate based synthetic bone grafts via dehydrothermal processing. *BioMed Research International, 2015*, 1–9. https://doi.org/10.1155/2015/576532
5. Giannoudis, P. V., Dinopoulos, H., & Tsiridis, E. (2005). Bone substitutes: An update. *Injury, 36*(3). https://doi.org/10.1016/j.injury.2005.07.029
6. Hing, K. A., Wilson, L. F., & Buckland, T. (2007). Comparative performance of three ceramic bone graft substitutes. *The Spine Journal, 7*(4), 475–490. https://doi.org/10.1016/j.spinee.2006.07.017

7. Jeong, J., Kim, J. H., Shim, J. H., Hwang, N. S., & Heo, C. Y. (2019). Bioactive calcium phosphate materials and applications in bone regeneration. *Biomaterials Research, 23*(1). https://doi.org/10.1186/s40824-018-0149-3
8. Lewin, S., Barba, A., Persson, C., Franch, J., Ginebra, M. P., & Öhman-Mägi, C. (2017). Evaluation of bone formation in calcium phosphate scaffolds with μCT-method validation using SEM. *Biomedical Materials (Bristol, England), 12*(6), 065005. https://doi.org/10.1088/1748-605X/aa801d
9. Prockop, D. J. (2004). Collagens. *Encyclopaedia of Biological Chemistry*, 482–487.https://doi.org/10.1016/b0-12-443710-9/00131-9
10. Mendoza García, M., Izadifar, M., & Chen, X. (2017). Evaluation of PBS treatment and PEI coating effects on surface morphology and cellular response of 3D-printed alginate scaffolds. *Journal of Functional Biomaterials, 8*(4), 48. https://doi.org/10.3390/jfb8040048
11. Ducheyne, P. (1999). Bioactive ceramics: The effect of surface reactivity on bone formation and bone cell function. *Biomaterials, 20*(23–24), 2287–2303. https://doi.org/10.1016/s0142-9612(99)00181-7
12. Larsson, S. (2010). Calcium phosphates: What is the evidence? *Journal of Orthopaedic Trauma, 24*(Suppl 1), S41–S45. https://doi.org/10.1097/BOT.0b013e3181cec472
13. Sheikh, Z., Abdallah, M.-N., Hanafi, A., Misbahuddin, S., Rashid, H., & Glogauer, M. (2015). Mechanisms of in vivo degradation and resorption of calcium phosphate based biomaterials. *Materials, 8*(11), 7913–7925. https://doi.org/10.3390/ma8115430
14. Hutmacher, D. W., Schantz, J. T., Lam, C. X., Tan, K. C., & Lim, T. C. (2007). State of the art and future directions of scaffold-based bone engineering from a biomaterials perspective. *Journal of Tissue Engineering and Regenerative Medicine, 1*(4), 245–260. https://doi.org/10.1002/term.24

Adversarial Attacks Against Detecting Bot Generated Text

Evan Hong Jun Lim

Abstract Contemporary text generation models have produced neural text that is increasingly indistinguishable from human text. This is a threat if used maliciously to produce misinformation or extremist content. Recent work has explored building detectors to identify neural text. In this paper, I present an approach to generate adversarial text that fools detectors whilst remaining fluent to humans. This approach decreases recall on detecting GPT-2 generated text from 99 to 0.4% with an average of 6.4 words perturbed and recall on bot generated tweets from 93 to 33.9% with an average of 4.2 words perturbed. My findings also suggest that such attacks can be performed randomly and at low cost and still achieve a significant success rate. This indicates that present detectors are not robust and many real-world systems are vulnerable to augmented neural text. Finally, I experiment with possible defences, including training with adversarial examples, and using TF-IDF and stylometric features.

Keywords Machine learning · Natural language processing · Neural text generation

1 Introduction

With the introduction of the transformer architecture by Vaswani et al. [18], contemporary Text Generation Models (TGMs) have shown incredible capabilities in generating neural text that for humans, is nearly indistinguishable from human text [8, 12, 20]. Although TGMs have many potential positive uses in writing, entertainment and software development [16], there is also a significant threat of these models being misused by malicious actors to generate fake news [17, 20], fake product reviews [1], or extremist content [10].

TGMs like GPT-2 generate text based on a given prompt, which limits the degree of control over the topic and sentiment of the neural text [12]. However, other TGMs

L. Evan Hong Jun (✉)
Raffles Institution, Singapore, Singapore
e-mail: 23YEVAN577G@student.ri.edu.sg

H. Guo et al. (eds.), *IRC-SET 2022*,
https://doi.org/10.1007/978-981-19-7222-5_37

like GROVER and CTRL allow for greater control of the content and style of generated text, which increases its potential for misuse by malicious actors [8, 20]. Additionally, many state-of-the-art pre-trained TGMs are available freely online and can be deployed by low-skilled individuals with minimal resources [16]. There is therefore an immediate and substantial need to develop methods that can detect misuse of TGMs on vulnerable platforms like social media or e-commerce websites.

Several methods have been explored in detecting neural text. The main approach is to formulate the problem as a classification task to distinguish between neural text and human text and train a classifier model (henceforth a 'detector'). Consistent through most research thus far is that fine-tuning the BERT or RoBERTa language model for the detection task achieves state-of-the-art performance [1, 4, 12, 17]. I will therefore be focussing on attacks against a fine-tuned RoBERTa model.

Although extensive research has been conducted on detecting generated text, there is a significant lack of research in adversarial attacks against such detectors [6]. However, the present research that does exist preliminarily suggests that neural text detectors are not robust, meaning that the output can change drastically even for small changes in the text input and thus that these detectors are vulnerable to adversarial attacks [19].

In this paper, I extend on [19] work on adversarial attacks on neural text detectors by proposing a series of attacks designed to counter detectors as well as an algorithm to optimally select for these attacks without compromising on the fluency of generated text. I do this with reference to a fine-tuned RoBERTa detector and on two datasets: (1) the GPT-2 WebText dataset [12] and (2) the Tweepfake dataset [4]. Additionally, I experiment with possible defences against these attacks, including (1) using count-based features, (2) stylometric features and (3) adversarial training.

2 Related Work

2.1 *Attacks in Natural Language Processing (NLP)*

Adversarial attacks have been extensively explored primarily in computer vision tasks, where extremely minor perturbations on select pixels can easily trick the model whilst remaining unnoticeable to humans [3]. However, adversarial attacks on natural language present a greater challenge due to its discrete nature, meaning all changes to the text are noticeable and can result in jarring issues in the syntax or coherence of the entire sentence.

There have been multiple attempts at generating adversarial examples for various language tasks, including sentiment analysis and textual entailment [2, 5]. These attacks generally formulate the problem as generating adversarial text that remains semantically and syntactically similar to the original text. Alzantot et al. [2] do this by randomly selecting a word in the current sentence, and considering a set of candidate synonyms that fit in the context of the word, before selecting the synonym that will maximise the target label prediction probability. This general framework remains consistent across many adversarial attacks in NLP [2, 7, 9].

2.2 *Attacks on Neural Text Detectors*

Although there is some research into adversarial attacks in NLP, there is very little research specifically about attacks against neural text detectors, which could become a significant tool for malicious actors to evade detection [6]. Wolff [19] explores two attacks on detectors: replacing characters with unicode homoglyphs (swapping English 'a's to Cyrillic 'a's) and words with commonly misspelt words (swapping 'except' to 'exept'). He randomly modifies a percentage of words in the neural text of the GPT-2 WebText dataset to attack a fine-tuned RoBERTa detector. Although the homoglyph attack achieved close to 100% success rate and would be indistinguishable from regular characters to a human, such attacks could be easily defended in the real-world setting by banning homoglyphs or flagging accounts that use them. As such, I will not be exploring homoglyph attacks in this paper. Additionally, the misspelling attack is reliant on the size of the misspelling dictionary, significantly limiting potential attacks. Instead, I extend on [19] work by (1) suggesting alternative character-level and word-level attacks, (2) proposing a score to measure sentence fluency and (3) an algorithm to optimally select words to attack.

2.3 *Context of the Attack*

Similar to work by Alzantot et al. [2], I better simulate the real-world by using the black-box setting, meaning the attacker has no prior knowledge of or access to model architecture, parameters or training data. Thus, the attacker may only query the model and get the model prediction and confidence scores. The attacker seeks to (1) fool the detector with an adversarial attack whilst (2) ensuring that the text remains recognisable and comprehensible to humans.

3 Method

3.1 *Datasets*

(1) *GPT-2 WebText Dataset.* Similar to [19], I use 5000 test samples generated by GPT-2 large with top-k 40 sampling provided by Solaiman et al. [16]. The texts in this dataset have an average length of 2091 words.

(2) *Tweepfake Dataset.* Fagni et al. [4] compiled real-world tweets taken from known human and bot accounts. It includes text generated by a variety of TGMs, which Fagni et al. [4] separate into GPT-2, RNN and others (Markov chains, LSTM). The Tweepfake dataset is marked by a few characteristics that contrast the GPT-2 WebText dataset, namely the text is (1) very short, (2) has many rare and out-of-vocab words, (3) does not always comply with standard sentence structure and grammar, and (4) is 'real' in the sense that all the tweets were actually posted to Twitter. I use the Tweepfake dataset in conjunction with the GPT-2 WebText dataset to gain more realistic and generalisable results for adversarial attacks in the real-world. It is overall balanced between human and bot text and is split to 20,712 training, 2558 test and 2302 validation samples. The texts in this dataset have an average length of 110 words.

Neural Text Detector. For the GPT-2 WebText dataset, I only use the pre-trained RoBERTa model provided by Solaiman et al. [16] due to the cost constraints of fine-tuning a model on the large training set. For the Tweepfake dataset, I fine-tune a RoBERTa model and train for 3 epochs, similar to Fagni et al. [4].

3.2 Problem Formulation

Given a pre-trained classification model $F : X \to Y$, which maps from input text X to a corresponding set of labels Y, for a particular input text $x \in X$, we have to find a valid adversarial example x_{adv} such that

$$F(x) \neq F(x_{\text{adv}}), \text{ and } S(x) - S(x_{\text{adv}}) \leq \epsilon \tag{1}$$

where $S : X \to \mathbb{R}$ is a function that measures the fluency of input text X and $\epsilon \in \mathbb{R}^+$. In other words, for a given input text X, we have to search for an adversarial text X_{adv} which fools the detector to giving a different prediction label and decreases in fluency by less than ϵ.

In conventional adversarial attacks in NLP, the attacker seeks to, as far as possible, preserve semantic similarity between the original text and the adversarial text [2, 7]. In our case, since the specific meaning of x is originally generated by the TGM on which the attacker has limited control, the attacker is not concerned with preserving the specific semantics of x. Semantics is also less of a concern if the TGM is being used to generate some posting history over time, so that the corresponding social media account looks credible. Consequently, we can treat x_{adv} as generated text on its own. Although we can assume that x_{adv} will generally share a similar overall meaning to x, we do not have to measure and preserve specific semantic similarity between x and x_{adv} in our algorithm. Instead, the primary objective of x_{adv} is to fool the detector whilst seeming fluent and 'natural' to humans. Thus, unlike previous work, I measure and preserve the change in fluency instead of semantic similarity.

Table 1 Examples for PLL fluency scoring

Text	PLL score
My friend is a citizen of Singapore	−16.3 (most fluent)
A citizen of Singapore my friend is?	−24.7
My friend is a Singapore of citizen	−49.9 (least fluent)

Fluency scoring. In this paper, I define fluency as linguistic acceptability, which generally measures the syntactic and semantic validity of the text [15]. I use pseudo-log-likelihood (PLL) scores as proposed by Salazar et al. [13] to measure fluency. In masked language modelling, a token w_t is replaced by the [MASK] token and predicted by the model using all preceding and succeeding tokens. I define the new tokenized sentence $W_{/t} = \left(w_1, \ldots, w_{t-1}, w_{t+1}, \ldots, w_{|W|}\right)$, and $P_{\mathrm{MLM}}(w_t|W_{/t})$ as the probability that the masked language model predicts w_t as the mask token in the new sentence $W_{/t}$. Iterating through all tokens, the pseudo-log-likehood score for the sentence is then given as the sum of log probabilities for each token w_t:

$$\mathrm{PLL}(W) = \sum_{t=1}^{|W|} \log P_{\mathrm{MLM}}\left(w_t|W_{/t}\right) \tag{2}$$

This serves as a good approximation for the acceptability and thus fluency of the text as determined by a masked language model. I use the BERT masked language model to compute the PLL score. In Table 1, we observe that by switching the order of words in the second and third examples to make the sentence grammatically incorrect, the PLL score drops.

3.3 Attacks

I propose four character-level attacks and two word-level attacks for neural text detection. (1) Space Insertion (Cspace): Inserts a space randomly in the target word. (2) Character swap (Cswap): Randomly swaps two adjacent characters in the target word. (3) Character deletion (Cdel): Randomly deletes a character in the target word. (4) Character substitute (Csub): Randomly swaps a character in the word with visually similar characters or characters nearby on a QWERTY keyboard. These four character-level attacks can be commonly found in real-world settings in human typos which serves both to mimic human text to detectors whilst remaining natural to other humans. The two word-level attacks are (5) Misspelling (Mis): Randomly replace the target word with a corresponding commonly misspelt word. This is the same attack performed by [19] and uses the Wikipedia list of commonly misspelt words.[1] (6) Synonym (Syn): Randomly replace the target word with a synonym. I use NLTK's

[1] https://en.wikipedia.org/wiki/Wikipedia:Lists_of_common_misspellings/For_machines.

Table 2 Examples of attacks used

Original	Cspace	Cswap	Cdel	Csub	Mis	Syn
heredity	heredity	heerdity	herediy	herrdity	heridity	genetic endowment
familiar	familiar	fmailiar	famliar	familisr	familliar	companion

interface to wordnet[2] to find a list of synonyms for a given target word. I also do not consider stopwords. I show a list of examples of each attack in Table 2.

Attack Algorithm. My attack algorithm follows the general framework proposed by previous work [7]. The critical part of the algorithm is calculating the importance score of target words. The importance score of word w_t is calculated as the change in confidence of the model after removing word w_t:

$$c_t = F\big(w_1, \ldots, w_{|W|}\big) - F\big(w_1, \ldots, w_{t-1}, w_{t+1}, \ldots, w_{|W|}\big) \tag{3}$$

where $|W|$ is the total number of tokens and $F(W)$ returns the confidence that the input text W is generated by a TGM. Tokens with higher importance scores mean that they make the detector more confident that the text is bot generated. Since we hope to fool the detector into thinking the text is human written, we should focus on tokens with higher importance first.

The algorithm works by first calculating the importance scores of every word in the given text. Then, we target the words from most important to least important. For every target word, we try different attacks on the target word and chooses the attack which decreases confidence the most. After every attack on a target word, we check if the fluency score for the perturbed text has decreased below a threshold ϵ compared to the original. We continue the attacks until the prediction for the perturbed text switches to the human label or the fluency score change drops below the threshold.

3.4 Defence

I study two approaches to defend against adversarial attacks. Previous work has shown that adding TF-IDF features [11] and stylometric features [14] can enhance detector performance. Hence, my first approach pre-trains a simple classifier based on these features, then concatenates the final hidden layer with the outputs from RoBERTa before feeding that into another classifier (Fig. 1).

I also experiment with adversarial training, where I augment the training set with pre-generated adversarial text so that the detector can learn to be robust to such noise. I first generate 6000 adversarial examples from the training set, then append these examples back to the existing training set. In order to ensure that the training set remains balanced between bot and human text, I also random sample for 6000

[2] https://www.nltk.org/_modules/nltk/corpus/reader/wordnet.html.

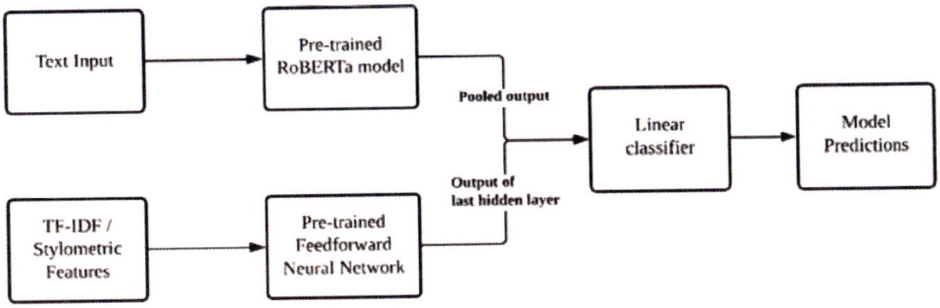

Fig. 1 Architecture of ensemble model with TF-IDF or stylometric features

human text and duplicate it back into the training set. I then use the updated training set to adversarially train a new detector.

4 Results

'All' is an optimised attack where all word and character-level attacks are performed, whereas 'Conly' only performs the character-level attacks. 'Avg. perturb' is the average number of words perturbed in each successful attack.

All attacks were conducted with a fluency score change threshold ϵ of 100, which was determined experimentally through qualitative analysis of adversarial text at different thresholds. From Table 3, the adversarial attack with all 6 attacks causes the recall to drop from 99 to 0.4% with only minimal perturbations. Additionally, I achieved a far higher drop in recall with the misspelling attack at 2.4% compared [19] reported 22.68%, likely due to the inclusion of importance scoring. Finally, character substitution has the best performance with a recall at 0.8%, suggesting that imitating human typos in text can successfully fool detectors.

From Table 4, although misspelling and synonym attacks individually decrease bot recall by a small amount relative to the character attacks (93% to 78.2% and 81.4% respectively), adding these two word-level attacks to 'character only' attack further decreases recall from 43.7 to 33.9%. However, this also increases the average number of perturbations. This indicates that word-level attacks are critical to the overall algorithm, possibly because they mimic human error better than character-level attacks. Additionally, we see that GPT-2 generated neural text drops in recall (86.1 to 15.8%) more than RNN generated neural text (98.7 to 66.5%). This is

Table 3 Results of adversarial attack on GPT-2 WebText dataset

	Original	All	Conly	Cspace	Cswap	Cdel	Csub	Mis	Syn
Recall	99.0	0.4	0.4	1.6	1.4	1.0	0.8	2.4	19.8
Avg. perturb	0	6.4	6.7	9.1	8.2	8.5	6.7	8.3	13.8

Table 4 Results of adversarial attack on Tweepfake dataset

	Original	All	Conly	Cspace	Cswap	Cdel	Csub	Mis	Syn
Overall	93.0	33.9	43.7	68.1	67.5	66.3	63.2	78.2	81.4
GPT2	86.1	15.8	20.3	37.5	38.8	39.0	37.5	54.9	59.3
RNN	98.7	66.5	75.4	92.7	93.4	91.9	91.0	95.1	97.0
Others	93.5	20.6	35.1	71.4	68.1	66.1	60.1	82.4	85.7
Avg. perturb	0	4.2	3.3	2.4	3.5	3.9	2.8	7.2	5.4

'Overall' is bot recall, i.e. percentage of bot text correctly detected as bot. 'GPT2', 'RNN', 'Others' is the percentage of text generated by each respective TGM that was detected.

sensible since GPT-2 text is more human-like than RNN text, which makes it easier to be augmented to fool the detector.

From Table 5, we clearly see that importance scoring is critical to improve the performance of the adversarial attack compared to without importance scoring. However, the performance of the attack without importance scoring at 38.2% recall also suggests that randomly selecting target words is enough to achieve considerable results with minimal perturbations.

I investigate the effects of defending against adversarial attacks in Table 6. We observe that including TF-IDF and stylometric features does not defend against adversarial attacks. In fact, the detectors with TF-IDF and stylometric features perform significantly worse compared to the standard RoBERTa detector, with recall dropping from 33.9 to 24.2% and 21.4%, respectively. However, adversarial training is shown to greatly improve the robustness of the detector, with recall increasing to 64.8%.

Table 5 Comparison of running adversarial attack algorithm with importance scoring and without importance scoring (i.e. random selection of target words) on Tweepfake dataset

	Overall	GPT2	RNN	Others	Avg. perturb
With importance scoring	33.9	15.8	66.5	20.6	4.2
Without importance scoring	38.2	21.3	69.4	25.2	4.7

Table 6 Results of running adversarial attack on standard RoBERTa model, ensemble model with TF-IDF and Stylometric features, and adversarially trained RoBERTa model on Tweepfake dataset

	Overall	GPT2	RNN	Others	Avg. perturb
RoBERTa	33.9	15.8	66.5	20.6	4.2
RoBERTa + TF-IDF	24.2	6.2	47.0	19.0	4.5
RoBERTa + Stylo	21.4	5.5	50.9	8.8	4.2
Adversarial training on RoBERTa	64.8	37.2	93.9	61.9	2.9

5 Discussion

GPT2 WebText versus Tweepfake. The adversarial attack on the GPT2 WebText dataset is more successful than the attack on the Tweepfake dataset because the GPT-2 WebText dataset is much longer than the Tweepfake dataset, giving the attacker more space to perform augmentations to the text. Additionally, the Tweepfake dataset includes other TGMs like RNNs which produce text that is less coherent and human-like than GPT-2.

Attack performance. A small number of augmentations to neural text is shown to achieve a very high success rate at fooling detectors. Four simple character-level attacks and two word-level attacks can cause recall to drop near 0% for long text (GPT-2 WebText) and 33.9% even for very short, irregular and 'real-world' text (Tweepfake). This suggests that detectors are very vulnerable to simple attacks.

Importance scoring. Although importance scoring is shown to improve the success rate of the adversarial attack, it is not essential to the attack. This suggests that a malicious actor could easily make random minor changes to the neural text and still fool a detector.

Possible defences. Although past work has shown that adding count-based and stylometric features to a classifier can enhance performance [11, 14], they are not successful and are in fact significantly detrimental in defending against adversarial attacks. This is likely because the adversarial attacks are designed to produce what would be interpreted as 'rare' words by the detector, which could disproportionately skew results towards classifying the text as human. My results also indicate that adversarial training can significantly improve the robustness of detectors against attacks. However, this approach requires the defender to know the types of attacks employed by the attacker and may be difficult to employ in practise. Future work could explore using language models that are also character-aware as detectors, as they could be more robust against character-level augmentations in the text.

6 Conclusion

In this paper, I propose adversarial attacks on neural text detectors, including an algorithm to optimally fool detectors whilst remaining fluent to other humans. My findings suggest that current neural text detectors are not robust against adversarial attacks and highlight a significant vulnerability in existing defences against bot generated text, given the ease at which malicious actors could evade detectors. This represents a significant threat to many real-world applications. I also experiment with various possible defences against the attacks. Further research should explore improving the robustness of neural text detectors.

References

1. Adelani, D., Mai, H., Fang, F., Nguyen, H., Yamagishi, J., & Echizen, I. (2020). Generating sentiment-preserving fake online reviews using neural language models and their human-and machine-based detection. In *International conference on advanced information networking and applications*, pp. 1341–1354.
2. Alzantot, M., Sharma, Y., Elgohary, A., Ho, B.-J., Srivastava, M., & Chang, K.-W. (2018). Generating natural language adversarial examples. In *Proceedings of the 2018 conference on empirical methods in natural language processing*, pp. 2890–2896.
3. Carlini, N., & Wagner, D. (2016). Towards evaluating the robustness of neural networks. *CoRR*. abs/1608.04644.
4. Fagni, T., Falchi, F., Gambini, M., Martella, A., & Tesconi, M. (2021). Tweepfake: About detecting deepfake tweets. *PLoS ONE, 16*(5).
5. Iyyer, M., Wieting, J., Gimpel, K., & Zettlemoyer, L. (2018). Adversarial example generation with syntactically controlled paraphrase networks. In *Proceedings of NAACL*.
6. Jawahar, G., Muhammad, A.-M., & Lakshmanan, L. (2020). Automatic detection of machine generated text: A critical survey. In *The 28th international conference on computational linguistics (COLING)*.
7. Jin, D., Jin, Z., Zhou, J., & Szolovits, P. (2019). Is BERT really robust? A strong baseline for natural language attack on text classification and entailment. *CoRR*. abs/1907.11932.
8. Keskar, N., McCann, B., Varshney, L., Xiong, C., & Socher, R. (2019). CTRL: A conditional transformer language model for controllable generation.
9. Li, J., Ji, S., Du, T., Li, B., & Wang, T. (2019). TEXTBUGGER: Generating adversarial text against real-world applications. In *Network and distributed systems security (NDSS) symposium 2019*.
10. McGuffie, K., & Newhouse, A. (2020). The radicalization risks of GPT-3 and advanced neural language models.
11. Prakash, A., & Madabushi, H. (2020). Incorporating count-based features into pre-trained models for improved stance detection. In *Proceedings of the 3rd NLP4IF workshop on NLP for internet freedom: Censorship, disinformation, and propaganda*, pp. 22–32.
12. Radford, A., Wu, J., Child, R., Luan, D., Amodei, D., & Sutskever, I. (2019). Language models are unsupervised multitask learners.
13. Salazar, J., Liang, D., Nguyen, T., & Kirchhoff, K. (2021). Masked language model scoring. *CoRR*. abs/1910.14659.
14. Sari, Y., Stevenson, M., & Vlachos, A. (2018). Topic or style? Exploring the most useful features for authorship attribution. In *Proceedings of the 27th international conference on computational linguistics*, pp. 343–353.
15. Schütze, C. T. (1996). *The empirical base of linguistics: Grammaticality judgments and linguistic methodology*. Language Science Press.
16. Solaiman, I., Brundage, M., Clark, J., Askell, A., Herbert-Voss, A., Wu, J., Radford, A., Krueger, G., Kim, J.W., Kreps, S., & McCain, M. (2019). Release strategies and the social impacts of language models. *OpenAI Report*.
17. Uchendu, A., Le, T., Shu, K., & Lee, D. (2020). Authorship attribution for neural text generation. In *Proceedings of the 2020 conference on empirical methods in natural language processing (EMNLP)*, pp. 8384–8395.
18. Vaswani, A., Shazeer, N., Parmar, N., Uszkoreit, J., Jones, L., Gomez, A.N., Kaiser, Ł., & Polosukhin, I. (2017). Attention is all you need. *Advances in Neural Information Processing Systems*, 5998–6008.
19. Wolff, M. (2020). Attacking neural text detectors. *CoRR*. Retrieved from abs/2002.11768.
20. Zellers, R., Holtzman, A., Rashkin, H., Bisk, Y., Farhadi, A., Roesner, F., & Choi, Y. (2019). Defending against neural fake news. *Advances in Neural Information Processing Systems*, 9054–9065.

Reducing the Diagnostic Odyssey for Patients with Neuromuscular Disorders (NMDs)

Travis Yong Han Tan, Xin Xiang Lee, Ethan Kee Kiat Leo, Stacey Kiat Hong Tay, Chun Ping Liu, Grace Li Xuan Tan, and Poh San Lai

Abstract Fast and accurate diagnosis of neuromuscular disorders (NMDs) remains challenging for clinicians. Consequently, patients with NMDs typically face a diagnostic odyssey. Recently, Next-Generation Sequencing (NGS) has become a promising technology to shorten the average diagnostic odyssey. However, the lack of a standardised diagnostic workflow for end-users without extensive bioinformatics background impedes its usage. Therefore, our project aims to design a simplified yet comprehensive pipeline to accurately identify disease-causing variants in NMD patients more quickly. The pipeline comprises filters for quality control score, population frequency and disease-phenotype associations. By filtering Whole-Exome Sequencing (WES) data through our customised pipeline, a total of 630,297 variants were prioritised for disease-causing variants in 3 unrelated, undiagnosed NMD patients. Variant curation was then conducted based on the American College of Medical Genetics and Genomics (ACMG) guidelines. In Case 1, a missense mutation in Fused in Sarcoma (*FUS*) gene was found as a strong candidate variant most likely associated with Amyotrophic Lateral Sclerosis 6. For Case 2, a missense variant in Kinesin Family Member 1B (*KIF1B)* gene was the highest-ranking causative variant most likely associated with Charcot-Marie-Tooth disease. However, no high-ranking causative variants were found in Case 3. Thus, although the proposed pipeline did not yield conclusive pathogenic genes, this WES-based pipeline narrowed down causative candidates for the cases and can reduce efforts in clinical validation by providing clinicians with a more refined dataset for a more accurate diagnosis of NMD patients. It can be successfully adapted to a clinical setting for the diagnosis of other rare NMDs, thereby shortening the diagnostic odyssey of many patients.

Keywords Neuromuscular disorders (NMDs) · Next-generation sequencing (NGS) · Diagnostic odyssey

---

T. Y. H. Tan · X. X. Lee · E. K. K. Leo (✉) · S. K. H. Tay · C. P. Liu
Raffles Institution, 1 Raffles Institution Ln, Singapore 575954, Singapore
e-mail: ethanleo734@gmail.com

G. L. X. Tan · P. S. Lai
Department of Paediatrics, Yong Loo Lin School of Medicine, National University of Singapore, Singapore 119228, Singapore

H. Guo et al. (eds.), *IRC-SET 2022*,
https://doi.org/10.1007/978-981-19-7222-5_38

1 Introduction

Neuromuscular disorders (NMDs) refer to a group of more than 200 rare genetic diseases that affect the nervous and muscular systems [1]. Currently, their low incidence rates combined with the presence of unidentified NMDs make their diagnosis extremely challenging using conventional methods such as Sanger sequencing. As a result, a diagnostic odyssey often occurs for many NMD patients, its length averaging 5–7 years in the United States, the United Kingdom and Singapore [2]. Thus, in order to minimise delayed treatments, reduce mortality rates and reduce the financial burden on patients' families, generating novel diagnostic methods to better identify and diagnose NMDs is critical to shortening the average diagnostic odyssey. Over the past decade, Next-Generation Sequencing (NGS) has emerged as a promising alternative to identify causative variants of NMDs. As NMDs are genotypically and phenotypically extremely heterozygous, high-throughput NGS would allow one to identify causative variants more quickly, resulting in more time-effective, cost-saving diagnosis [3]. However, although NGS is able to achieve high rates of accuracy, there is currently no standard workflow for simplified diagnosis with every laboratory using its own customised protocols. Therefore, navigating the various accessible in silico tools for bioinformatics analysis remains challenging for the majority of end-users who lack bioinformatics training, but wish to analyse the data [4]. Hence, our project aims to generate a simplified diagnostic workflow in order to identify causal variants in Whole-Exome Sequencing (WES) data, a subset of NGS. Although exons only constitute about 1% of the human genome [5], they contain 85% of disease-causing mutations responsible for Mendelian disorders [5, 6]. Therefore, since WES only sequences these protein-coding regions, it is able to analyse a significantly smaller dataset where there is a higher probability of identifying causative variants in a shorter amount of time, allowing for a quicker diagnosis.

Our project aims to resolve the cause of NMDs in a given set of patients from WES data and improve the diagnostic capabilities of NGS by designing and implementing a simplified, unique and comprehensive pipeline based on selected tools to identify disease-causing variants more easily and accurately.

2 Materials and Methods

2.1 Clinical Presentation of Patients

Three unrelated patients were investigated. Patient 1 is a female presenting with symptoms of adult-onset spinal muscular atrophy (SMA4). She had muscle weakness predominantly affecting the legs and hip muscles starting about two years ago. This progressed slowly to mild weakness affecting the shoulders and arms. She manifested finger trembling and fasciculation and complained of increased muscle weakness and fatigue. Screening of the *SMN1* gene for mutations causative of SMA was negative. Patient 2 is a male with clinical presentations of hereditary neuropathy and no family history of the disease. Patient 3 is a male with suspected Ullrich congenital muscular dystrophy (UCMD). He had decreased foetal movement, arthrogryposis and weakness at birth. Facial weakness, distal joint hyperlaxity, decreased reflexes, bilateral wrist drop and ankle drop were also observed. Results of a muscle biopsy conducted were suggestive of congenital muscular dystrophy.

2.2 *NGS Workflow*

Sequencing reads were mapped onto Reference Genome GRCh37, and the files of variants were generated. Anonymised WES datasets containing genetic variants of the patients were provided by the National University of Singapore, Department of Paediatrics. The variants were then annotated using the integrative tool ANNOVAR. Using previously published protocols [7, 8], we developed a variant filtration pipeline to identify disease-causing variants correlating to the observed phenotypes (Fig. 1).

Only variants with 5 or more alternate reads and a quality of PASS were retained. Variants with more than 5% population frequency in population databases of EXAC, ESP and 1000 Genomes were directly excluded from the analysis. Next, only variants in the exonic or splicing regions were retained, and synonymous variants were removed. Variant prioritisation was then conducted. The variants were filtered for

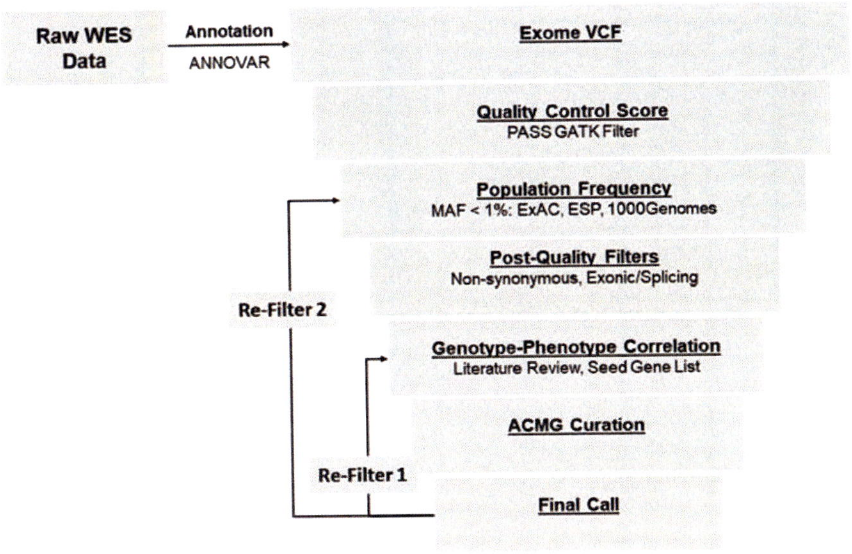

Fig. 1 NGS analysis workflow. Generated based on published protocols

those occurring in genes most closely associated with the patients' disease phenotype. These genes were based on 3 gene lists created for each patient. The first list was based on literature review. The second list was a SeedGene list generated by Phen2Gene, an online tool that generates a list of associated genes based on disease terms, comprising genes with strong associations with the phenotypes inputted, and the third is an expanded list of all predicted genes identified by Phen2Gene. The SeedGene lists generated for each case were sorted according to SeedGene score, a measure of the candidate gene's association with the inputted phenotypes [9]. This allowed for the highest-ranking candidate genes to be shortlisted (Table 1).

Associated disease phenotypes were then compared with patient phenotypes to shortlist candidate variants. Variants were then curated following the American College of Medical Genetics and Genomics (ACMG) guidelines for interpreting variants into 1 of 5 classes: benign, likely benign, variant of uncertain significance (VUS), likely pathogenic and pathogenic. After curation, should no variant be a suitable candidate, the variant filter parameters are expanded and the data is re-filtered.

Table 1 Most probable causative gene for each case with SeedGene list score and rank

Gene	Effect of variant (NCBI)	Associated disease (OMIM)
Case 1: *FUS* Rank: 62 Score: 0.138420	This gene encodes a multifunctional protein component of the heterogeneous nuclear ribonucleoprotein (hnRNP) complex. The hnRNP complex is involved in pre-mRNA splicing and the export of fully processed mRNA to the cytoplasm. This protein belongs to the FET family of RNA-binding proteins which have been implicated in cellular processes that include regulation of gene expression, maintenance of genomic integrity and mRNA/microRNA processing. Alternative splicing results in multiple transcript variants. Defects in this gene result in Amyotrophic Lateral Sclerosis type 6	Amyotrophic Lateral Sclerosis 6 (ALS6), Hereditary Essential Tremor 4 (ETM4)
Case 2: *KIF1B* Rank: 47 Score: 0.18565	This gene encodes a motor protein that transports mitochondria and synaptic vesicle precursors. Mutations in this gene cause Charcot-Marie-Tooth disease, type 2A1	Charcot-Marie-Tooth disease, type 2A1 (CMT2A1), Pheochromocytoma
Case 3: *COL6A2* Rank: 4 Score: 0.97539	Mutations in this gene are associated with Bethlem myopathy and Ullrich scleroatonic muscular dystrophy. Three transcript variants have been identified for this gene	Bethlem Myopathy 1 (BTHLM1), Ullrich Congenital Muscular Dystrophy 1 (UCMD), Congenital Myosclerosis

The maximum score is 1

A maximum of 3 rounds of variant filtration and curation were conducted for each set of WES data. The first round of prioritisation and curation followed the strictest criteria, with the cut off for population frequency set at 1% and the variants filtered for genes obtained from literature review only. Subsequent rounds of curation followed broader criteria to include more variants. In the second round, the filter was expanded to include variants from genes obtained from the SeedGene list and literature review. For the final round of filtration, the cut off for population frequency was increased to 5%.

3 Results and Discussion

3.1 Case 1

In Patient 1, 209,406 variants were initially identified. After the filters for alternate read depth, quality control score and population frequency were applied, 898 variants (MAF $< 1\%$) were present in exonic and splicing regions in the subject. After filtering for genes from the SeedGene list and literature review, 16 missense mutations, 1 in-frame insertion, 2 in-frame deletions and 1 unknown variant were identified as potential candidate variants. From these variants, curation using the ACMG guidelines revealed only 1 novel, heterozygous missense mutation in the disease-associated gene Fused in Sarcoma (*FUS*) to be the most probable candidate variant with its determined clinical significance as likely pathogenic based on genotype analysis (Appendix).

The association between the *FUS* gene and ALS6 is supported by Online Mendelian Inheritance in Man (OMIM). Based on its clinical features reported in OMIM, the symptoms of ALS6 bear some overlaps with the clinical presentation of Patient 1, such as proximal fasciculations and mobility loss at all spinal levels [10].

FUS c.1561A > T:p.Arg521Trp, a novel missense variant in a methylation site of *FUS*, was curated and predicted as likely pathogenic, fulfilling 5 pathogenic ACMG criteria: PM1, PM2, PM5, PP2, PP4 (Appendix), with no benign criteria, presenting a strong case for its pathogenicity. The *FUS* gene codes for the RNA-binding *FUS* protein and mutations in the gene have been reported to result in moderate accumulation of the FUS protein in the cytoplasm and various downstream effects on transcription, protein translation, RNA splicing and transport [11], and this variant was predicted pathogenic by 4 in 5 predictive tools. Moreover, *FUS* c.1561A > T:p.Arg521Trp is located in a well-established mutational hotspot. A mutation analysis of the *FUS* gene in a Chinese-Canadian ALS patient revealed a novel heterozygous double point mutation (p.Arg514Ser/p.Glu516Val), confirming that exon 15 is a mutation hotspot [12]. The identified variant was also found to occur in a hotspot of pathogenic and likely pathogenic variants on ClinVar (Appendix).

However, further investigations showed the patient's age of diagnosis as 24, whereas the mean age of symptom onset for ALS is 44.5 years [13]. Nevertheless, the results of predictive tools, presence of overlapping ALS features and the presence of the variant in a well-established mutation hotspot suggest this variant to be pathogenic. However, further clinical evaluation is needed to ascertain if this variant is underlying or incidental for the primary clinical presentation. Moreover, *FUS* can lead to two possible phenotypes. OMIM shows that the *FUS* gene can be associated with Tremor, Hereditary Essential, 4 (ETM4) [10]. Individuals with ETM4 are largely asymptomatic, characterised occasionally by postural and action tremors which are inconsistent with the patient's phenotype. Because of this, the likelihood that Patient 1 suffers from ETM4 is significantly lower. Hence, the clinical focus on ALS6 associated with *FUS* gene should be further explored. If further clinical assessment of the patient aligns to this, it can potentially lead to better treatment and management strategies for the patient.

3.2 Case 2

In Patient 2, WES identified 206,638 initial variants. After the filters for alternate read depth, quality control score and population frequency were applied, 929 variants (MAF < 1%) were present in exonic and splicing regions in the NMD subject. After filtering for genes obtained from literature review and the SeedGene list, 13 missense mutations and 1 in-frame insertion were identified as potential candidate variants. Curation using the ACMG guidelines revealed no variants that fulfilled the criteria for the clinical significance of likely pathogenic or pathogenic. However, 1 variant in the disease-associated gene *KIF1B* was classified as a variant of uncertain significance (VUS). This variant could be a potential causative variant but needs to be investigated further before its effects can be determined. The variant fulfilled 2 pathogenic ACMG criteria: PS3 and PP3, but it also fulfils 2 benign criteria: BS2 and BP6 (Appendix).

KIF1B c.3260A > G:p.Tyr1087Cys is a missense variant identified and predicted pathogenic by 5 in 5 computational tools. The *KIF1B* gene encodes for a protein, Kinesin-like protein KIF1B, which has 2 main isoforms; the longer beta isoform and the shorter alpha isoform. This particular variant has been identified in 2 unrelated families with CMT2A1 [14], and in vitro studies showed that the mutated protein was unable to fully complement the defects of axonal outgrowth in *KIF1B*-null mouse primary hippocampal neurons, thus appearing to demonstrate a pathogenic effect. Moreover, the phenotype of CMT2A1 is consistent with that of hereditary neuropathies, and the patient's age of diagnosis lies within the range of ages for disease development, which ranges from childhood to 50 years of age, with the peak ages of onset being from 11 to 20 years old, as described in OMIM [15]. The predictions of computational tools, similarities between the phenotypes for CMT2A1 and the reported phenotype of the patient, as well as functional evidence highlighting the effect of this variant on axonal outgrowth suggest that this could be a probable causative variant which would suggest that the patient has CMT2A1. However, this variant has also been observed in the homozygous condition in 183 healthy adults reported in gnomAD, indicating that it is likely not disease-causing. Multiple reputable studies by Illumina and other laboratories reported on ClinVar have also previously classified it as benign [16]. Thus, more information is needed to make a definite conclusion on the effect of this variant, but this variant was the highest-ranked SeedGene candidate found in Case 2.

3.3 Case 3

WES identified 214,253 variants in Patient 3. After alternate read depth, quality control and population frequency filters were applied, 527 variants (MAF < 1%) were present in exonic and splicing regions in the NMD subject. After filtering for genes obtained from literature review and the SeedGene list, 16 distinct missense mutations, 1 in-frame insertion, 1 in-frame deletion and 1 unknown variant were identified as potential candidate variants. Curation using the ACMG guidelines revealed no variants that fulfilled the criteria for the clinical significance of likely pathogenic or pathogenic. The identified candidate gene Collagen Type VI, Alpha-2 (*COL6A2*), was found to directly cause Ullrich Congenital Muscular Dystrophy 1 (UCMD1), the same disorder as reported in Patient 3. However, the 2 associated variants *COL6A2* c.2528C > G:p.Thr843Ser and *COL6A2* c.2600G > A:p.Arg867Gln were classified as VUS and likely benign, respectively, and hence were not considered as a high-ranking causative variant. *COL6A2* c.2528C > G:p.Thr843Ser fulfilled 1 benign ACMG criteria: BP1, with 358 out of 484 pathogenic variants in the *COL6A2* gene being truncating variants, and only 13 being missense variants. *COL6A2* c.2600G > A:p.Arg867Gln fulfilled BP1, as well as BP6, with multiple reputable sources such as Illumina reporting the variant on ClinVar as likely benign or benign [17], but with no accessible evidence (Appendix). It may be possible that future new supporting evidence may help in elevating the ranking of these two variants or conclusively prove non-pathogenicity.

3.4 Limitations

While candidate variants were identified in all cases, there was insufficient information to conclusively determine the pathogenicity of many variants. Information regarding the proband's parental genotypes and phenotypes were unavailable in the cases. Moreover, there was a lack of functional evidence in literature that could be utilised for most variants, since there was usually no case–control studies or functional evidence available regarding the variants. As a result, the proposed diagnostic approach could not diagnose the exact NMDs with complete certainty and thus was limited in its ability to shorten a patient's diagnostic odyssey.

4 Conclusion

Our novel NGS workflow pipeline has allowed us to filter and narrow the list of candidate variants in order to identify causative candidates in undiagnosed patients. For Patient 1, a strong candidate variant in the *FUS* gene was found, while for Patient 2, a possible causative variant in the *KIF1B* gene causing CMT2A1 was found. However, for the third patient, no variants were found to be likely causative. Nevertheless, the results of multiple predictive tools and the correlations of associated phenotypes of the candidate gene to the patient's clinical presentation as determined by the ACMG criteria in Patients 1 and 2 suggest the pipeline's effectiveness in shortlisting causal variants. Although the proposed diagnostic approach is not conclusive, it is able to reduce efforts in clinical validation by providing clinicians with a more refined dataset to investigate patients more accurately, thus shortening the diagnostic odyssey by preventing false diagnoses and repeated consultations.

To confirm the diagnosis for Patients 1 and 2, in vitro and in vivo functional studies could additionally be conducted in conjunction with the proposed NGS approach to determine the effect of the variant on the candidate genes. For Patient 3, further studies are needed to investigate the possible causative variants and determine the pathogenicity of those variants. Segregation analysis of variants in candidate genes should also be subsequently conducted in the patients' families, before phenotype-genotype correlation and clinical verification are carried out in order to produce conclusive diagnoses. Furthermore, future studies can be conducted using a larger sample size to further validate the utility of the workflow process across different samples, as the sample size investigated in the study was rather small. Further analysis could also widen the filtration criteria to include variants in genes not previously shortlisted.

Acknowledgements This project has only been possible due to the help and guidance of several individuals, who have extended their invaluable assistance in the preparation and completion of this research project. We are very grateful to Prof Lai Poh San and Ms. Tan Li Xuan, Grace from the Department of Paediatrics, Yong Loo Lin School of Medicine, National University of Singapore (NUS), for mentoring us and guiding us through our research journey. We would also like to thank our teacher-mentor Dr. Abigayle Ng from Raffles Institution (RI) for her support, as well as NUS and RI for giving us this opportunity to participate in the Science Research Programme (SRP).

Appendix: ACMG Curation and Pathogenicity Calls

Case 1		
ACMG pathogenicity: Likely pathogenic		**Variant**: *FUS* c.1561A > T:p.Arg521Trp
Criteria	**Justification**	

(continued)

(continued)

PM1	Variant in exon 15, a reported mutational hotspot, and is in a region of pathogenic/likely pathogenic variants
PM2	The variant is absent from the 3 population databases used. (EXAC, ESP, 1000 Genomes)
PM5	A highly similar missense variant NM_001170634.1:c.A1561G: p.Arg521Gly at the same nucleotide was reported as pathogenic [18]
PP2	*FUS* variants have a low rate of benign variation (ClinVar). Benign (1), likely benign (4), VUS (41), likely pathogenic (5), pathogenic (11)
PP4	ALS6 is characterised by severe lower motor neuron loss in the spinal cord and less so in the brain stem region, resulting in decreased mobility, affecting the spinal area similar to SMA4, causing proximal fasciculations and muscle weakness of the lower limbs
Case 2	
ACMG pathogenicity: Variant of uncertain significance (VUS)	**Variant**: *KIF1B* c.3260A > G:p.Tyr1087Cys
Criteria	**Justification**
PS3	Variant reported as pathogenic in a study cited in OMIM [19]. Variant reduces the binding capacity of KIF1Bβ to IGF1R, enhancing the pathogenesis of neuropathic symptoms for hereditary neuropathies like CMT2A1 [14]
PP3	All 5 in silico tools used show that this variant has a deleterious effect
BS2	183 homozygotes for this variant were healthy (gnomAD), contradicting the expected autosomal dominant inheritance pattern for CMT2A1
BP6	Multiple reputable sources on ClinVar reported this variant as benign, but the evidence is not provided
Case 3	
ACMG pathogenicity: Variant of uncertain significance (VUS)	**Variant**: *COL6A2* c.2528C > G:p.Thr843Ser
Criteria	**Justification**
BP1	Of the 484 pathogenic variants in *COL6A2*, 358 variants are truncating variants, while only 13 are missense variants
ACMG pathogenicity: Likely benign	**Variant**: *COL6A2* c.2600G > A:p.Arg867Gln
BP1	Of the 484 pathogenic variants in *COL6A2*, 358 variants are truncating variants, while only 13 are missense variants
BP6	Multiple reputable sources on ClinVar have reported this variant as benign, but the evidence is not provided

References

1. Ankala, A., da Silva, C., Gualandi, F., Ferlini, A., Bean, L. J., Collins, C., Tanner, A. K., & Hegde, M. R. (2015). A comprehensive genomic approach for neuromuscular diseases gives a high diagnostic yield. *Annals of Neurology, 77*(2), 206–214. https://doi.org/10.1002/ana.24303
2. Turnbull, C., Scott, R. H., Thomas, E., Jones, L., Murugaesu, N., Pretty, F. B., Halai, D., Baple, E., Craig, C., Hamblin, A., Henderson, S., Patch, C., O'Neill, A., Devereau, A., Smith, K., Martin, A. R., Sosinsky, A., McDonagh, E. M., Sultana, R., & Mueller, M., et al. (2018). The 100 000 genomes project: Bringing whole genome sequencing to the NHS. *BMJ (Clinical Research Ed.), 361*, k1687. https://doi.org/10.1136/bmj.k1687
3. Meregalli, M., Maciotta, S., Angeloni, V., & Torrente, Y. (2016). Duchenne muscular dystrophy caused by a frame-shift mutation in the acceptor splice site of intron 26. *BMC Medical Genetics, 17*(1). https://doi.org/10.1186/s12881-016-0318-y
4. Efthymiou, S., Manole, A., & Houlden, H. (2016). Next-generation sequencing in neuromuscular diseases. *Current Opinion in Neurology, 29*(5), 527–536. https://doi.org/10.1097/WCO.0000000000000374
5. Choi, M., Scholl, U. I., Ji, W., Liu, T., Tikhonova, I. R., Zumbo, P., Nayir, A., Bakkaloğlu, A., Özen, S., Sanjad, S., Nelson-Williams, C., Farhi, A., Mane, S., & Lifton, R. P. (2009). Genetic diagnosis by whole exome capture and massively parallel DNA sequencing. *Proceedings of the National Academy of Sciences, 106*(45), 19096–19101. https://doi.org/10.1073/pnas.0910672106
6. Rabbani, B., Tekin, M., & Mahdieh, N. (2013). The promise of whole-exome sequencing in medical genetics. *Journal of Human Genetics, 59*(1), 5–15. https://doi.org/10.1038/jhg.2013.114
7. Grada, A., & Weinbrecht, K. (2013). Next-generation sequencing: Methodology and application. *Journal of Investigative Dermatology, 133*(8), 1–4. https://doi.org/10.1038/jid.2013.248
8. Illumina. (n.d.). Understanding the NGS Workflow. https://www.illumina.com/science/technology/next-generation-sequencing/beginners/ngs-workflow.html
9. Zhao, M., Havrilla, J. M., Fang, L., Chen, Y., Peng, J., Liu, C., Wu, C., Sarmady, M., Botas, P., Isla, J., Lyon, G. J., Weng, C., & Wang, K. (2020). Phen2Gene: Rapid phenotype-driven gene prioritization for rare diseases. *NAR Genomics and Bioinformatics*, *2*(2), lqaa032. https://doi.org/10.1093/nargab/lqaa032
10. OMIM. (n.d.). Fused in sarcoma; FUS. https://www.omim.org/entry/137070
11. Dormann, D., Rodde, R., Edbauer, D., Bentmann, E., Fischer, I., Hruscha, A., Than, M. E., Mackenzie, I. R., Capell, A., Schmid, B., Neumann, M., & Haass, C. (2010). ALS-associated fused in sarcoma (FUS) mutations disrupt Transportin-mediated nuclear import. *The EMBO Journal, 29*(16), 2841–2857. https://doi.org/10.1038/emboj.2010.143
12. Robertson J., Bilbao, J., Zinman, L., Hazrati, L.-N., Tokuhiro, S., Sato, C., Moreno, D., Strome, R., Mackenzie, I. R., & Rogaeva, E. (2011). A novel double mutation in FUS gene causing sporadic ALS. *Neurobiology of Aging, 32*(3). https://doi.org/10.1016/j.neurobiolaging.2010.05.015
13. Vance, C., Rogelj, B., Hortobágyi, T., De Vos, K. J., Nishimura, A. L., Sreedharan, J., Hu, X., Smith, B., Ruddy, D., Wright, P., Ganesalingam, J., Williams, K. L., Tripathi, V., Al-Saraj, S., Al-Chalabi, A., Leigh, P. N., Blair, I. P., Nicholson, G., de Belleroche, J., Gallo, J. M., et al. (2009). Mutations in FUS, an RNA processing protein, cause familial amyotrophic lateral sclerosis type 6. *Science (New York, N.Y.), 323*(5918), 1208–1211. https://doi.org/10.1126/science.1165942
14. Xu, F., Takahashi, H., Tanaka, Y., Ichinose, S., Niwa, S., Wicklund, M. P., & Hirokawa, N. (2018). KIF1Bβ mutations detected in hereditary neuropathy impair IGF1R transport and axon growth. *The Journal of Cell Biology, 217*(10), 3480–3496. https://doi.org/10.1083/jcb.201801085
15. OMIM (n.d.). Charcot-Marie-tooth disease, axonal, type 2A1; CMT2A1. https://www.omim.org/entry/118220

16. National Center for Biotechnology Information. (2020, February 4). NM_001365951.3(KIF1B):c.3398A>G (p.Tyr1133Cys). *ClinVar*. https://www.ncbi.nlm.nih.gov/clinvar/variation/VCV000291567.12
17. National Center for Biotechnology Information. (2022, January 8). NM_001849.4(COL6A2):c.2600G>A (p.Arg867Gln). *ClinVar.* https://www.ncbi.nlm.nih.gov/clinvar/variation/VCV000258330.8
18. Kwiatkowski, T. J., Jr., Bosco, D. A., Leclerc, A. L., Tamrazian, E., Vanderburg, C. R., Russ, C., Davis, A., Gilchrist, J., Kasarskis, E. J., Munsat, T., Valdmanis, P., Rouleau, G. A., Hosler, B. A., Cortelli, P., de Jong, P. J., Yoshinaga, Y., Haines, J. L., Pericak-Vance, M. A., Yan, J., & Ticozzi, N., et al. (2009). Mutations in the FUS/TLS gene on chromosome 16 cause familial amyotrophic lateral sclerosis. *Science (New York, N.Y.)*, *323*(5918), 1205–1208. https://doi.org/10.1126/science.1166066
19. OMIM. (n.d.). Kinesin family member 1B; KIF1B. https://www.omim.org/entry/605995

Cost-Free Structural Elucidation of Ctriporin in Sodium Dodecyl Sulphate (SDS) and Dodecylphosphocholine (DPC) Micelles

Colin Chow, Justin Au, Khan Mohamed Mohamed Dhasil, and Chiradip Chatterjee

Abstract *Ctriporin* peptide (Ctr), an antimicrobial peptide (AMP) procured from the venom of the scorpion *Chaerilus Tricostatus*, shows promising results in neutralizing methicillin resistant coagulase-negative and penicillin resistant *Staphylococcus* epidermis strains. To evaluate the conformational change in the presence of eukaryotic cell membrane compared to prokaryotic cell membrane, we have utilized TOCSY-NOSY 1H Nuclear Magnetic Resonance (NMR) data obtained from (Bandyopadhyay et al. in Biopolymers 101:1143–1153 [1]) to analyse the active conformation of Ctr in negatively charged *Sodium Dodecyl Sulfate (SDS)* and Zwitterionic *n-Dodecylphosphocholine (DPC)* micelles. The strength of interactions between H^N, H^α, H^β, H^γ and H^δ are quantified in the form of Volume. This allows for the performance of sructure prediction and analysis. To carry out data analysis, we utilize SPARKY (Lee et al. in Bioinformatics 31(8):1325–1327 [2]) to interpret our NMR data and various Python data representation libraries such as Plot.ly, Seaborn, and Matplotlib. After careful deliberation, diagrams such as 3D surface diagrams, radar charts, chord diagrams, and bar graphs were used as they provide optimal visualization. We then modelled the protein for further comparison with our diagrams. The diagrams were successful in pinpointing the active component of Ctr, the Alpha Helix, and displaying the regions of strongest interactions within itself. This helps infer the configuration of the protein within the mimetic micelles.

Keywords Antimicrobial peptide · TOCSY-NOSY · Sodium dodecyl sulphate · Zwitterionic · n-Dodecylphosphocholine

C. Chow (✉) · J. Au · K. M. M. Dhasil
NUS High School of Mathematics and Science, Singapore, Singapore
e-mail: h1810029@nushigh.edu.sg

J. Au
e-mail: h1810008@nushigh.edu.sg

K. M. M. Dhasil
e-mail: h1810072@nushigh.edu.sg

C. Chatterjee
Republic Polytechnic, Singapore, Singapore
e-mail: chiradip_chatterjee@rp.edu.sg

H. Guo et al. (eds.), *IRC-SET 2022*,
https://doi.org/10.1007/978-981-19-7222-5_39

1 Introduction

1.1 Background

In the last decades, following concerted efforts towards the development of omics technologies, there has been increasing demand for structural prediction and modelling software. The influx of genomic sequence information has led to the concept of structural proteomics, the determination of protein structures on a genome-wide scale [1]. Numerous methods such as Multidimensional NMR, Florescence Spectroscopy and Gel Retardation have been established [3]. However, they largely remain unaffordable for small startups and students embarking on independent research projects. Therefore, we propose a novel and cost-effective method to visually represent NMR data in a meaningful and unconvoluted manner without compromising the quality of analysis.

1.2 Literature Review

In this venture, Ctr was chosen as the prime candidate for analysis due its short amino acid length, simplicity, and useful applications. Ctr is a 19-amino acid peptide which contains a single alpha helix in posterior region. Ctr remains an indispensable tool in our arsenal to combat against the rise of "super bugs". Various experiments [4] have shown the effectiveness of Ctr in combating Methicillin-Resistant *Staphylococcus aureus* (MRSA) which has been linked to a wide plethora of human infections, including but not limited to carbuncles, meningitis, and impetigo. It shows promising growth-inhibitory activity against Gram-positive bacteria at low concentrations [4].

Previous studies [1] have shown that when bound to micelles, Ctr displays a flexible N-terminal region with hydrophobic residues and a well-defined amphipathic α-helix at the C-terminal with positively charged Lys18 and Lys19 remaining flexible. Ctr is known to use Phe1 and Trp3 at the flexible N-terminal to anchor and attach itself to the surface of SDS and DPC. The amphipathic helical part of the peptide is directly involved in peptide binding to SDS and DPC.

As shown in Fig. 1 [1], Ctr has a longer α-helix at the C terminus spanning from residues 5–18 in SDS, while DPC only adopts a partial α-helix, spanning from 9 to 18. This results in Ctr having a more flexible and linear structure in SDS and a more angular and rigid structure in DPC. These structural differences in SDS and DPC confers that Ctr is more energetically stable in negatively charged membranes as compared to the zwitterionic nature of eukaryotic membranes. This plays a huge part in the selectivity of Ctr.

Most past studies on Ctr was conducted using exorbitant protein modelling softwares such as CYANA-2.1. However, due to its high cost, its use was is limited to all but the most well funded of laboratories. Despite this, there has been little push

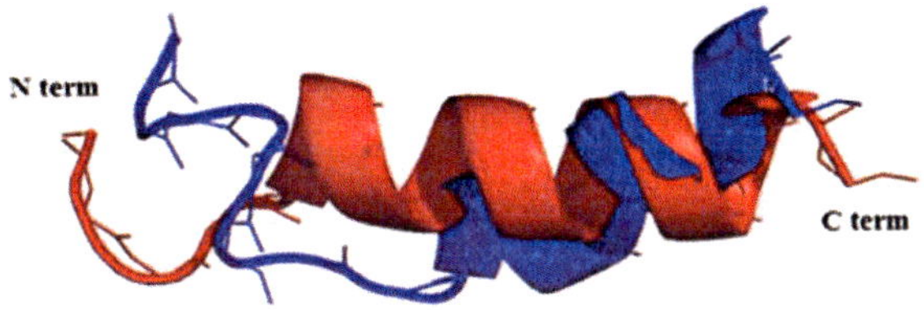

Fig. 1 Ribbon diagram of overlapped representative average structure of Ctr in SDS (red) and DPC (blue) micelles

towards the development of inexpensive tools for the discernment of protein structures. Thus, this places the review of protein structures firmly out of reach of startups or poorly funded laboratories, impeding the scientific contributions they can make.

1.3 Project Aim

The goal of this report is to study the differences in the active conformation of Ctr in SDS and DPC, which play a huge role in the peptide's mode of action, using free data analysis tools. We have drafted several data representation tools that can provide invaluable understanding of the numerous physicochemical traits of Ctr which includes the assignment of charge on the surface of Ctr, amphipathicity, and active structural conformation which grants Ctr its functions.

2 Materials and Methods

2.1 NMRFAM-Sparky

TOCSY-NOSY 1H NMR data, obtained from [1] were performed on a Bruker Avance DRX-800 operating at 800 MHz with cryo-probe and Bruker Avance DRX-400 operating at 400 MHz. The data was then uploaded to SPARKY, a software specialized to analyze 2D NMR. Using the chemical shift values provided in the supplementary materials of [1] and Fig. 2, we assigned the peak interactions accordingly. The basic approach to peak identification in 2D biomolecular NMR spectra is to search for local maxima above a chosen contour level.

Peak identification can also be automated using the tools APES and PONDEROSA. Following which, we integrated the assigned peaks using the following setting (In appendix) to obtain the volume of the peaks. The settings used were obtained from Dave.H from the University of Texas. These settings allowed us to obtain the most accurate and reproducible results.

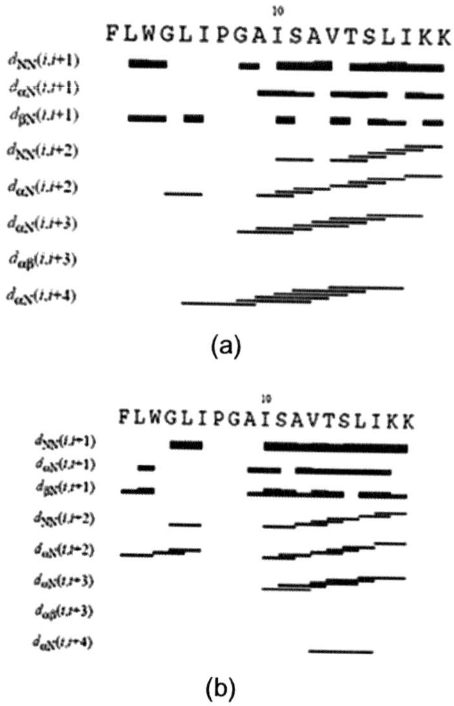

Fig. 2 Relative connectivity of NOEs of Ctr in **a** SDS micelles and **b** in DPC micelles in 298 K pH 4.5 obtained from [1]

3 Results

3.1 Bar Graph

Cross peaks result from magnetization transfer. Their presences are indicative of two nuclei that despite having two different chemical shifts are coupled together.

The greatest number of cross peaks are observed in the posterior section of Ctr in both SDS and DPC, indicating the location of the α-helix. From Fig. 3a, we observe that the anterior section of Ctr has a higher mean cross peak in SDS, indicating that the alpha helix starts earlier as compared to in DPC in Fig. 3b.

There is an absence of cross peaks at P7. This is due to the lack of H^N in Proline residues. However, this bar graph does not consider the strength of interactions. Therefore, Fig. 3 gives a general overview of the peptide, but if more conclusive analysis is needed, volume data should be considered.

The nuclear overhauser effect (NOE) is an important parameter used in NMR for conformational analysis because the magnitude of the NOE is inversely proportional to the sixth power of the interproton distance in space (INOE $\propto r^{-6}$). The volume obtained through integration in SPARKY is proportional to NOE, hence volume can be deduced to be proportional to the interproton distance. Thus, volume

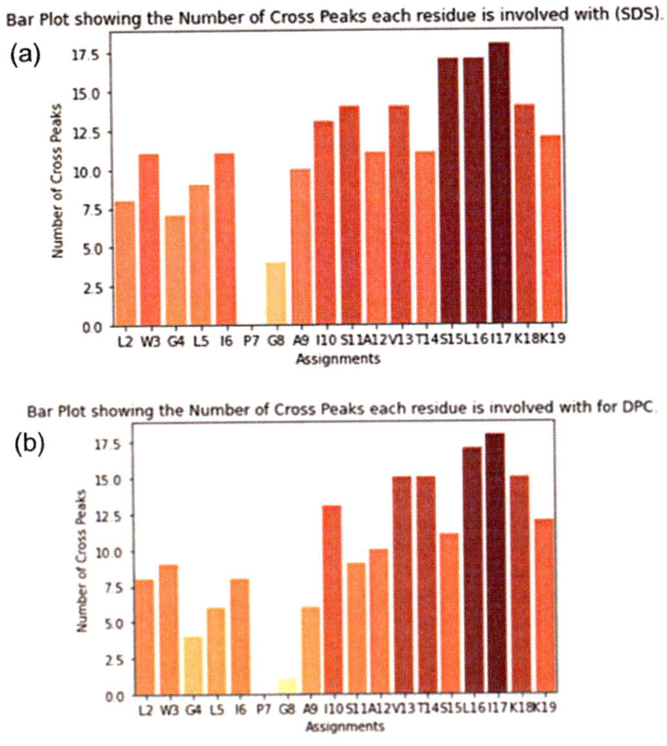

Fig. 3 Bar graph indicating the number of cross peaks of Ctr in **a** SDS and **b** DPC

is proportional to the strength of interaction between hydrogens, with a higher volume representing stronger interactions and vice versa.

3.2 3D Surface Graph

A diagonal line is formed across the *X*–*Y* plane in Fig. 4a, b. This relatively linear shape is a result of the lack of interaction between far flung amino acids. Instead, most of the interaction is between adjacent amino acids. Figure 4b shows that the interaction amongst amino acids in DPC is stronger than SDS in Fig. 4a, with DPC's interaction capping out $118 * 10^6$ ga and SDS's capping out $85.8 * 10^6$. Owing to the alpha helix of the Ctr in SDS and DPC, the start of the cluster of significantly strong interaction in SDS starts at I6 and G8 for DPC.

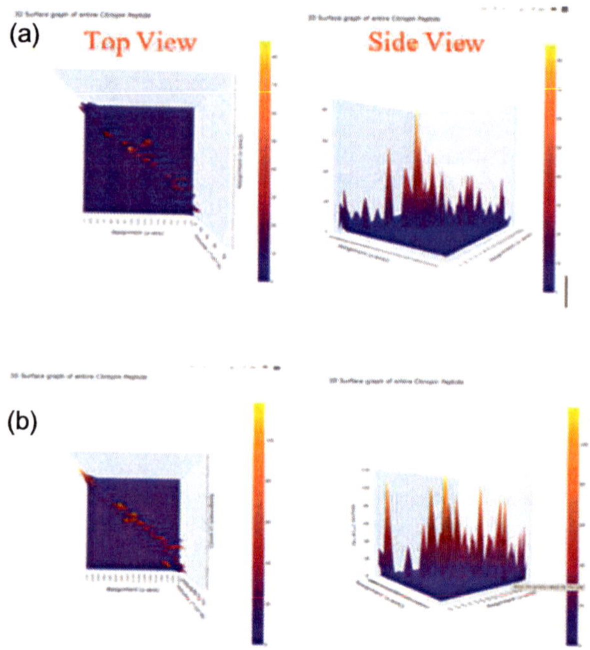

Fig. 4 3D surface diagram indicating the relative strength of interaction through volume of Ctr in **a** SDS and **b** DPC. Inset: The *X* axis includes all the assignments. The *Y* axis only includes the H^N for the residues. The *Z* axis represents the volume. (To view a clearer and interactive version of the graph, please refer to the appendix)

3.3 Chord Diagram

Figure 5 shows the connectivity of various amino acids through the relative distance occupied by the H^N of the respective residues on the circumference of the chord diagram. As an example, in Fig. 5a L16H^N occupies a relatively large distance on the circumference while G4H^N occupies a relatively minimal distance on the circumference. Figure 5 shows two different types of connections. The first is a group of residues which experience interactions only with adjacent residues, resulting in compact arches in the chord diagram. An example of this would be the residues G4 to I6 in Fig. 5b. The second is a group of residues which experiences interaction between i, $i + 2$ and i, $i + 3$ in addition to its i, $i + 1$ interaction, resulting in loose and scattered arches in the chord diagram. An example would be the residues G8 and A12 in Fig. 5a.

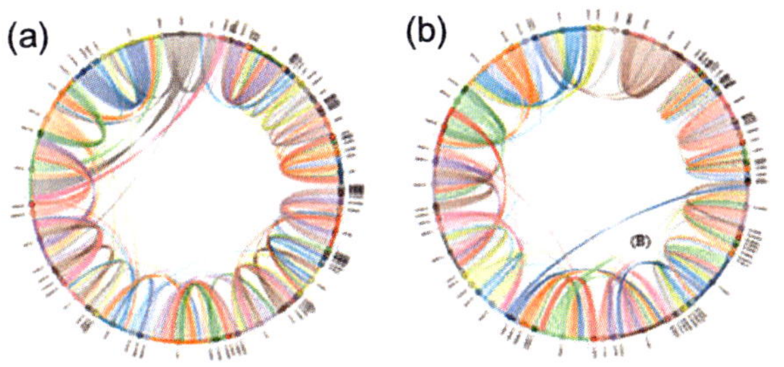

Fig. 5 Chord diagram indicating the relative strength of interaction through volume of Ctr in **a** SDS and **b** DPC. Inset: The thicker the lines in the chord diagram, the stronger the interaction (To view a clearer and interactive version of the graph, please refer to the appendix)

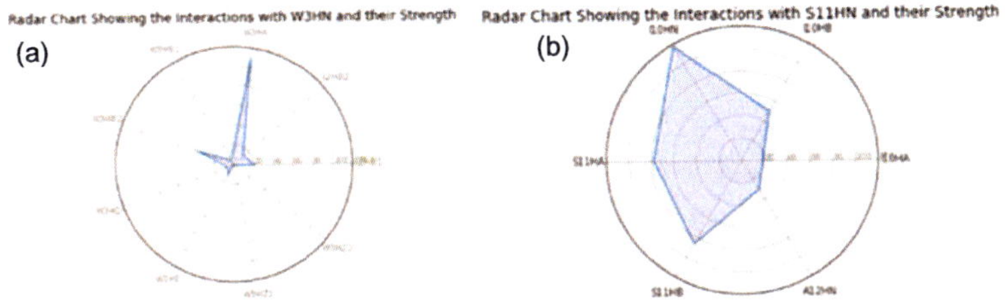

Fig. 6 Radar chart indicating the relative of strength of interactions through volume of Ctr in **a** W3H^N and **b** S11H^N in DPC

3.4 Radar Chart

As shown in Fig. 6, amino acid interactions in DPC and SDS micelles can be divided into interactions between amino acids involved in the alpha helix and amino acids that are not. Generally, amino acids involved in the alpha helix show strong interaction with other distant amino acids in the secondary structure ($i \pm 1, 2, 3$), while amino acids that are not show strong interactions limited to intra-interactions and interactions with their neighbouring amino acids ($i \pm 1$).

3.5 Secondary Structure Analysis

Figure 7 shows a high probability of α-helix in Ctr and a negligible probability of β-pleated sheets. This rule out the presence of β-pleated sheets within Ctr in DPC, which is consistent with pre-existing knowledge [1].

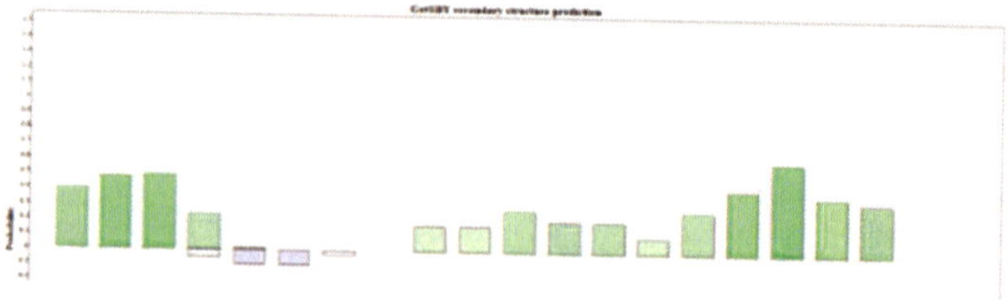

Fig. 7 Bar graph showing the probability; of secondary structure in Ctr in DPC. Inset: Green represents α-helix. Blue represents β-pleated sheets

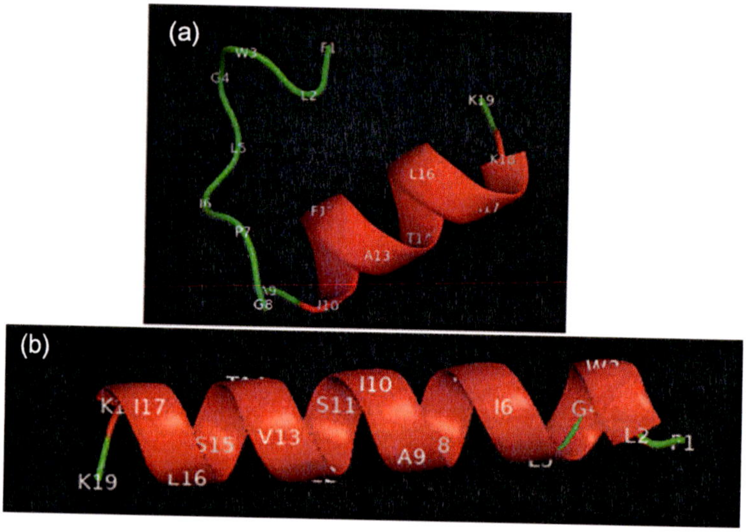

Fig. 8 Ribbon diagram of representative average structure of Ctr in **a** DPC and **b** SDS

3.6 *3D Model*

See Fig. 8.

4 Discussion

4.1 *3D Surface Graph*

From Fig. 4b, it can be noted that Ctr has a higher mean peak Volume and number of significant interactions in the presence of DPC as compared to SDS which is seen from Fig. 4a. Stronger interactions suggests that residues are closer together in space and unable to move freely, thus Ctr can be said to be more constrained in DPC,

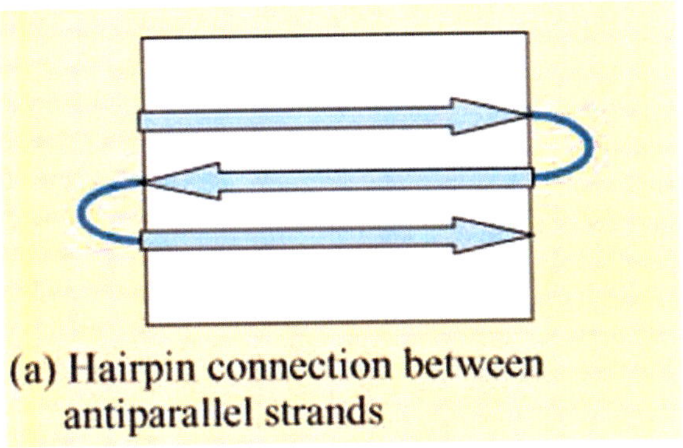

Fig. 9 Diagram showing the antiparallel arrangement of the β-pleated sheet [5]

resulting in an angular and rigid structure. On the other hand, it can be seen from Fig. 4a that residues are more flexible in Ctr and SDS due to its lower mean peak Volume and number of significant interactions. These observations are consistent with [1].

From Fig. 4, we can see a dip followed by a sharp increase in Volume. In Fig. 4a, the spike starts at I6, whereas in Fig. 4b, it starts at G8, suggesting the Ctr in SDS has a longer α-helix spanning from I6 to K18. Moreover, Ctr in DPC can be seen to have a shorter α-helix spanning G8 to K18. These observations are substantiated by Fig. 9.

4.2 *Chord Diagram*

We observe a larger number of scattered and dispersed arches starting from I6 in Fig. 5a and G8 from Fig. 5b. This suggests that interactions between i, $i + 2$ and i, $i + 3$ in addition to its i, $i + 1$ increase significantly. Moreover, the relative distance occupied by the H^N of the respective residues on the circumference increases from I6 in Fig. 5a and G8 in Fig. 5b. Since relative distance is indicative of connectivity of residues, residues with a larger connectivity can be seen more prominently from I6 in Fig. 5a and G8 in Fig. 5b onwards. These observations are indicative of a transition from a linear shape to a secondary structure where peptide backbone hydrogen bond comes into more significant play.

We observe from the larger amounts of gradual and gentle arches that span larger distances in Fig. 5b that there are more interactions between residues that are far apart along Ctr in DPC which contrasts with what Fig. 5a shows. Thus, this suggests that Ctr in DPC exhibits more numerous interactions, suggesting that residues are close together. Thus, Ctr in DPC is more constrained and rigid, as compared to Ctr in SDS which has fewer interactions, allowing it to be more flexible.

We observe from Fig. 5b that L2 and W3, as well as G4, L5 and I6 interact with one another almost exclusively. This can be largely attributed to the more angular and rigid structure of Ctr in DPC as seen in Fig. 9, where a bend is located between L2 and W3 as well as G4 and L5. This brings the residues closer in contact with one another, resulting in more significant interactions. From the lack of interactions between W3 and G4, we can conclude that they are connected by a linear segment. We also observe from Fig. 5a, that the interactions between amino acids lack exclusivity as they interact with their adjacent partners almost equally. Thus, we can surmise that Ctr in SDS is relatively linear, with gentle curves that allows for more prominent yet equal interactions between adjacent amino acids as seen in Fig. 9.

4.3 Radar Chart

Residues from Alanine-9 to Lysine-18 are involved in the alpha helix, while the others are not. Hence, a comparison will be made between Serine-11 (Fig. 6b) and Tryptophan-3 Fig. 6a, without the loss of generality (other graphs are included in the appendix). In Fig. 6a, W3HN shows strong intra-interactions with W3HB2, W3HA, and weak to moderate inter-interactions with its neighbouring residue, L2HB1. This concurs with our initial idea, that due to its non-involvement in the alpha helix, it exhibits little to no interactions with distant residues, while demonstrating moderate to string interactions with itself and immediate neighbours. Similarly in Fig. 6b, due to residue S11's participation in the alpha helix, it shows strong interactions with other distantly situated residues (I10, A12) on the sequence, since the helix brings the residues closer to each other.

Referring to the residues (Appendix, Figs. 11 and 12), residues in DPC typically have higher values (~20 * 10^6 Ga typical), whereas residues in SDS have lower values (~3 * 10^6 Ga typical). This indicates greater interactions between the residues in DPC environments, resulting in a more rigid and packed structure as compared to SDS.

It can be observed that magnitude of inter-interactions increases from L5 for residues in the SDS environment whereas, it only increases from A9 for residues in DPC environment. This is due to the longer alpha helix that Ctr forms in SDS (L5–K18) as compared to DPC (A9–K18), resulting in the earlier increase in magnitude for inter-interactions due to the proximity established between neighbouring residues because of the alpha helix.

4.4 Secondary Structure Analysis

We observe from Fig. 7 that there is a 30–50% chance of an α-helix from F1 to W3 and a 20–70% chance of an α-helix from I10 to K18, with the highest probability of 70% I17. However, from previous analysis done with 3D surface plot and chord diagram, we can conclusively state that the α-helix is between I10 to K18. The significant probability of an α-helix from F1 to W3 can thus be attributed to the coil-like structure similar to an α-helix from F1 to W3, as observed from Fig. 9b.

Therefore, while this extension from NMRFAM-SPARKY may not be 100% accurate, it does gives us a rough gauge of the presence of secondary structures in Ctr.

4.5 3D Model

While Fig. 8a shows a close resemblance in terms of its linearity and secondary structure to Fig. 1, the length of its α-helix is longer, extending from L2 to K18, instead of L5 as depicted in Fig. 1. This difference may be due to the lower accuracy of the CS-Rosetta application as compared to more well-established paid software used in [1] since the 3D surface graph, radar chart, and chord diagram agrees with Fig. 1. However, we can rectify these errors by counter checking the structure against the above graphs to refine the proposed 3D model of Ctr in SDS. However, the model provided by CS-Rosetta is still of use as the generalized properties of Ctr in SDS remains valid.

Figure 8b on the other hand, closer resembles that of [1] and closer aligns itself with the information obtained from the above graphs. This may be in part due to the more angular shape of the protein, allowing for CS-Rosetta to more clearly distinguish between the unstructured head and the α-helix as compared to the more linear and coiled shape of Ctr in SDS, contributing to the inaccuracies as seen in Fig. 8a.

4.6 Identification of Key Secondary Structures

To differentiate between α-helices and β-pleated sheets, it is important to identify defining characteristics of key interactions for both these secondary structures.

Interactions in α-helices typically include $(i, i + 2)$, $(i, i + 3)$, and $(i, i + 4)$ in the central turn of the helix, with significant interactions - both stabilizing and destabilizing–found to occur at spacings of $(i, i + 3)$ and $(i, i + 4)$ [6]. Additionally, α-helices are stabilized by exclusion of water from the hydrophobic protein core and/or by specific electrostatic interactions with neighbouring secondary structures [7]. This is due to the destabilizing nature of α-helices in the absence of tertiary interactions. This is thought to be a consequence of competition for hydrogen bonding between the amide hydrogen in the α-helical backbone with either oxygen found in water molecules in aqueous solutions or carbonyl oxygen in the backbone [8]. Thus, the α-helix in Ctr in both SDS and DPC are likely to be found within the cell membrane, providing anchorage to the peptide.

Certain amino acids are more or less likely to be found in α-helices or β pleated sheets. For instance, the amino acid proline is sometimes called a "helix breaker" because its unusual R group (which bonds to the amino group to form a ring) creates a bend in the chain and is not compatible with helix formation. Proline is typically found in bends, unstructured regions between secondary structures. Similarly, amino acids such as tryptophan, tyrosine, and phenylalanine, which have large ring structures in their R groups, are often found in β pleated sheets, perhaps because the β pleated sheet structure provides plenty of space for the side chains [5]. Thus, Ctr in DPC is unlikely to form an α-helix at F1, W3, P7 due to the presence of these residues despite what Fig. 7 predicts.

Interactions in β-pleated sheet, on the other hand, can be formed in two different ways: either residue $i + 2$ in strand 1 binds to $j + 2$ in strand 2 (referred to as a parallel β-sheet), or residue $i + 2$ in strand 1 binds to $j - 2$ in strand 2 (referred to as anti-parallel β-sheet). In contrast to the situation for helices, the hydrogen bonding residues in β-sheets might be far away in sequence [9].

Hence, we can conclude that while the alpha helix exhibits characteristic interaction as shown in Fig. 4, antiparallel β-sheets which is shown in Fig. 9, which are the dominant type of β-sheets due to their stability, will not exhibit a diagonal line across the *X*–*Y* plane. Instead, it exhibits characteristic interactions as shown in Fig. 10 due to interactions between residues which are further apart and on different sheets as described above.

In conclusion, the combined use of protein modelling by free software and the generation of 3D surface graphs, radar chart and chord diagram proves to be a viable alternative method as compared to the high-cost high effort 3D modelling application as used in [1]. This can be seen from its ability to differentiate between the key structure of Ctr in SDS and DPC, namely, its linearity, secondary structure, and rigidity, all of which are key in determining the active conformation and plausible mode of action of Ctr. The structural prediction methods provided in this report may help to cut down costs in protein research and provide an inexpensive yet accurate way of depicting the protein structure.

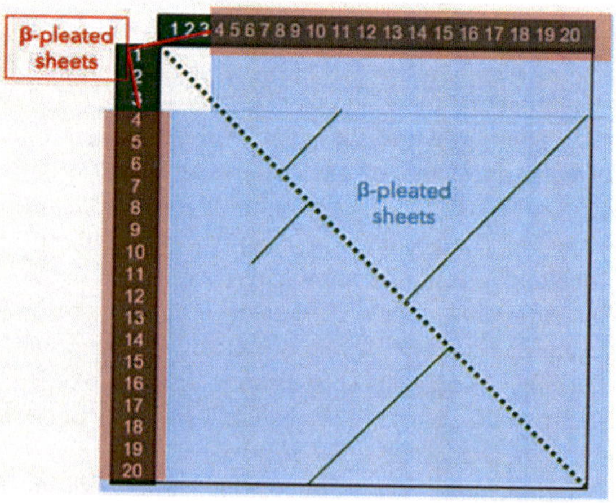

Fig. 10 Suggested (top view) of 3D surface graph for β-pleated sheet

Further studies can be conducted to improve upon these methods based on the few limitations which we have identified. Firstly, due to lack of NMR data available on peptides with β-sheets, further experimentation specifically on β-sheets may be conducted to verify the accuracy of our proposed model. Similarly, NMR data of longer and more complex proteins may be used to verify the effectiveness of the proposed methods on larger, more complex proteins.

Acknowledgements The authors would like to thank School of Applied Science, Republic Polytechnic. Help from Dr. Robin Seoh, NUS High mentor, is greatly appreciated.

Appendix

Integration Method: Gaussian Fit, Allow peak motion: No, Adjust Linewidth = Yes, Fit Baseline = Yes, Subtract off fit peaks: No, Use data above the lowest contour: No, Use data within rectangle: No, Group peaks in contour boundary: Yes, Group close peaks: Yes, Maximum motion (ppm):0,0, Linewidth Minimum (Hz): 2.0, 2.0, linewidth maximum (Hz): 80.0, 80.0, Grouping Range (Hz): 30.0, 30.0, Maximum minimization steps: 10,000, Minimization tolerance: 0.1%

Radar Chart

SDS

See Fig. 11.

DPC

See Fig. 12.

Chord Diagram

DPC

See Fig. 13.

SDS

See Fig. 14.

Bar Graph

See Fig. 15.

3D Surface Graphs

See Figs. 16 and 17.

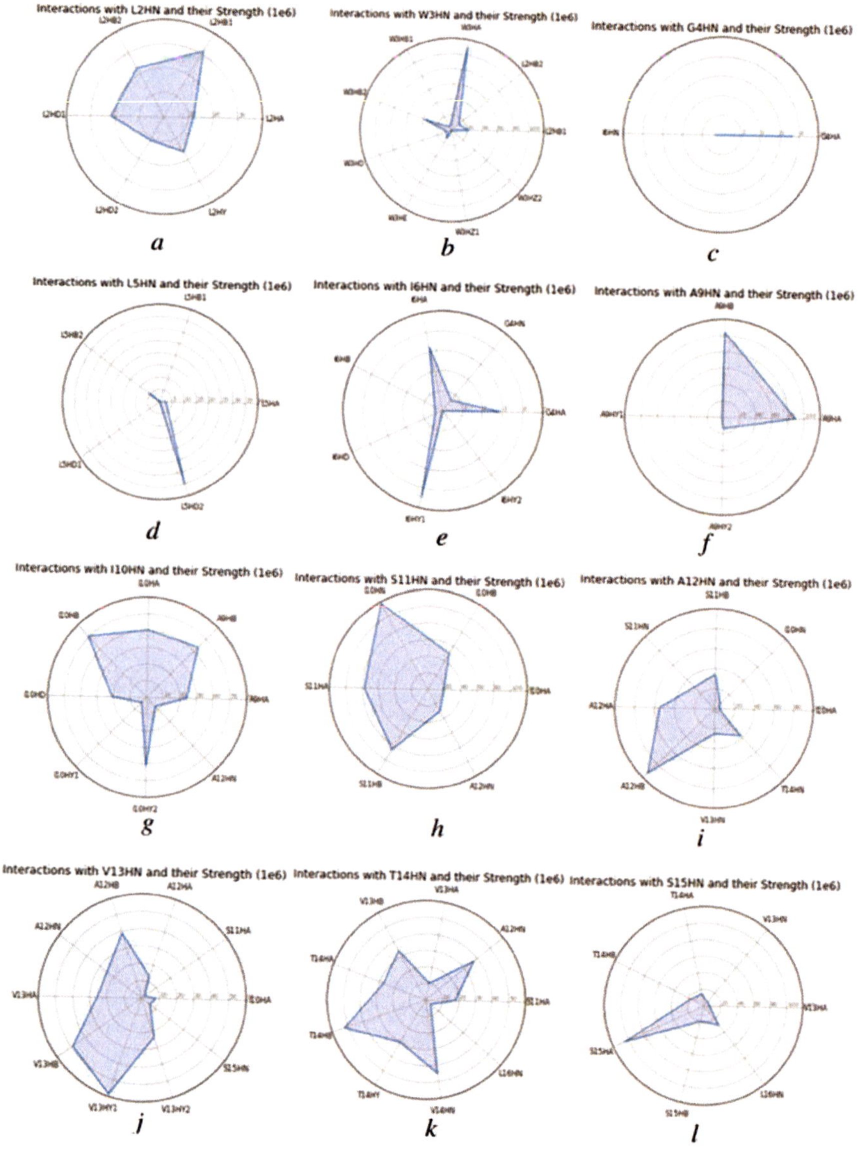

Fig. 11 Radar chart indicating the relative of strength of interactions through volume of Ctr in **a** L2H^N, **b** W3H^N, **c** G4H^N, **d** L5H^N, **e** I6H^N, **f** A9H^N, **g** I10H^N, **h** S11H^N, **i** A12H^N, **j** V13H^N, **k** T14H^N, **l** S15H^N, **m** I17H^N, **n** K18H^N, **o** K19H^N in DPC

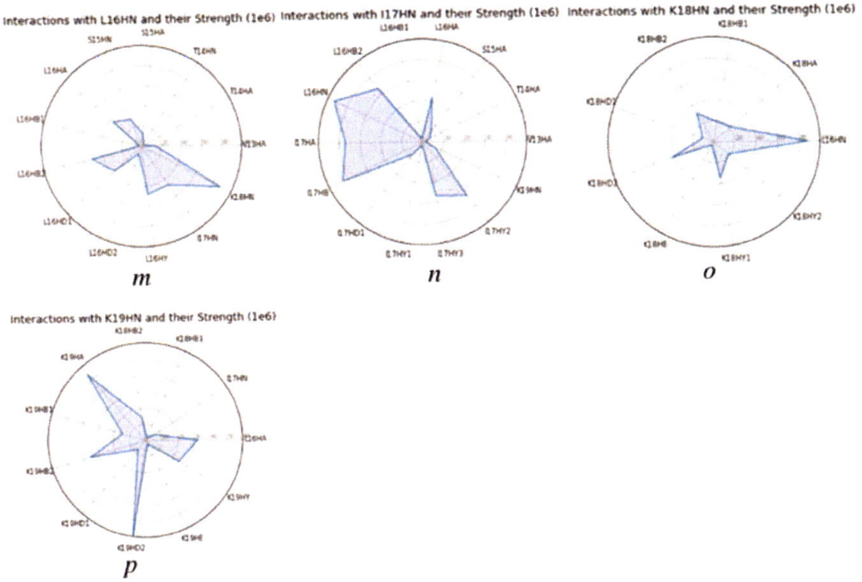

Fig. 11 (continued)

SDS

See Figs. 18, 19 and 20.

DPC

See Figs. 21, 22, 23, 24 and 25.

https://tinyurl.com/SMPInteractiveGraphDrive

Please refer to this QR code/link for the interactive version of the 3D Surface Graph and Chord Diagram

3D Model

See Figs. 26 and 27.

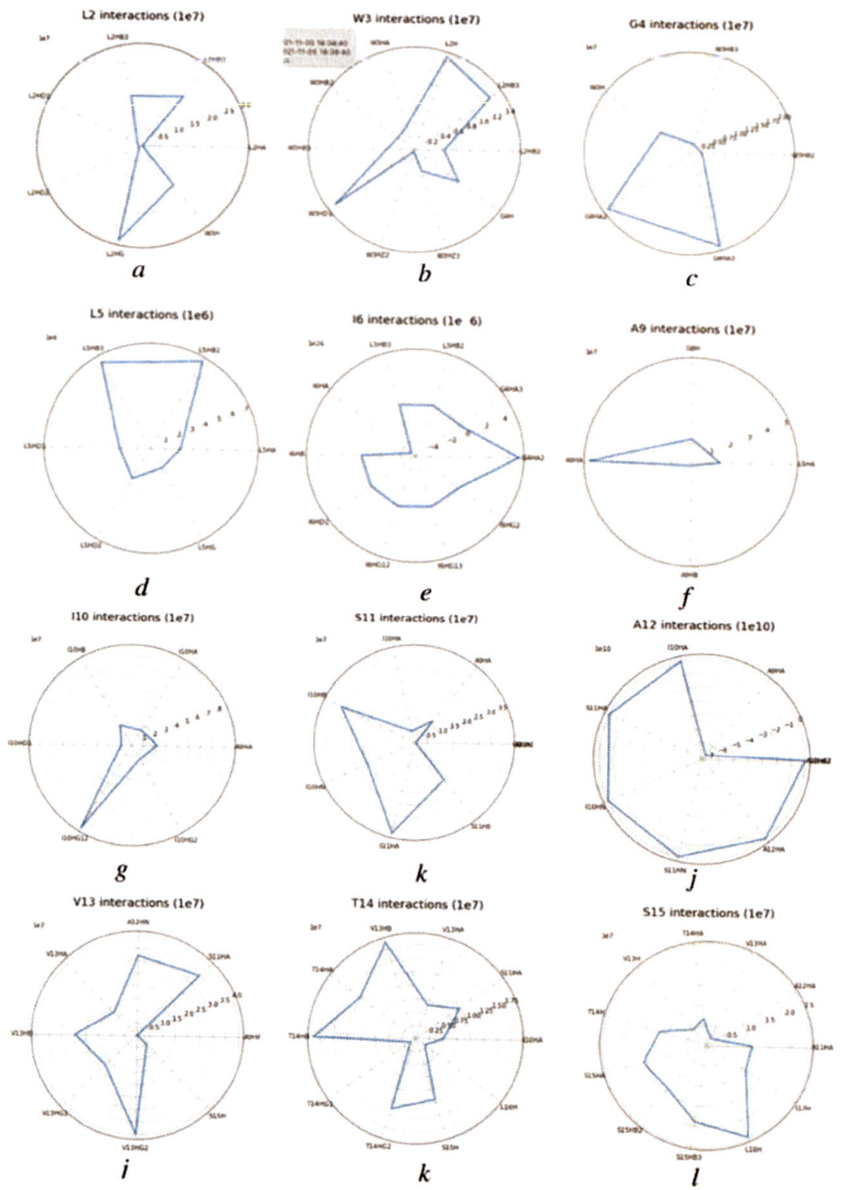

Fig. 12 Radar chart indicating the relative of strength of interactions through volume of Ctr in **a** L2H^N, **b** W3H^N, **c** G4H^N, **d** L5H^N, **e** I6H^N, **f** A9H^N, **g** I10H^N, **h** S11H^N, **i** A12H^N, **j** V13H^N, **k** T14H^N, **l** S15H^N, **m** I17H^N, **n** K18H^N, **o** K19H^N in DPC

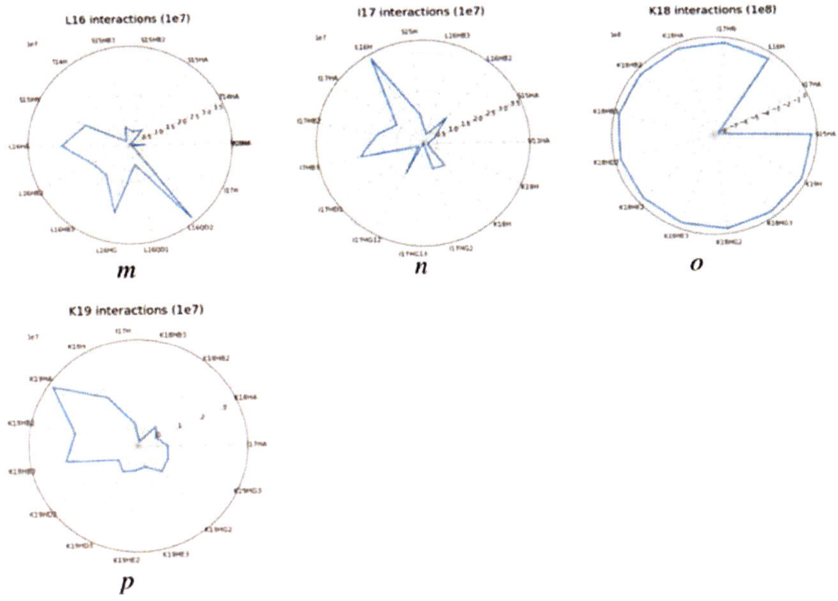

Fig. 12 (continued)

Predictive Secondary Structure Bar Graph

See Fig. 28.

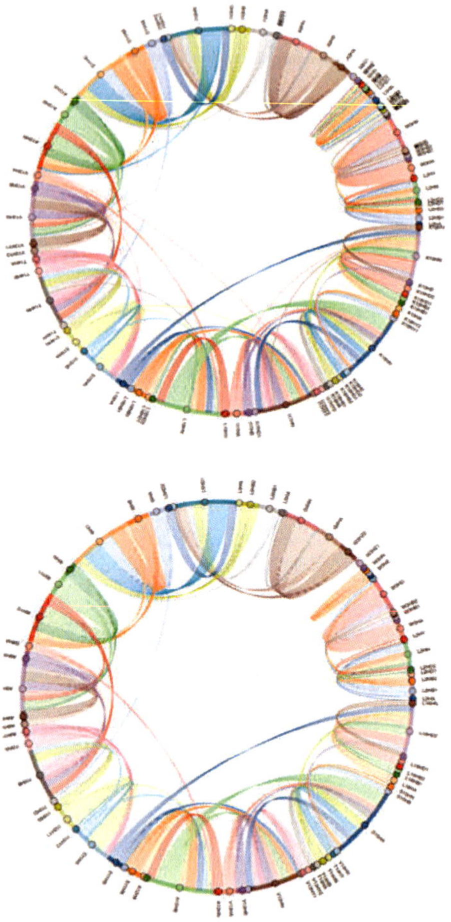

Fig. 13 Chord diagram indicating the relative of strength of interactions through volume in **a** all interactions **b** only major interactions (volume greater than 10^6) of Ctr in DPC. Access html below for interactive version of graph

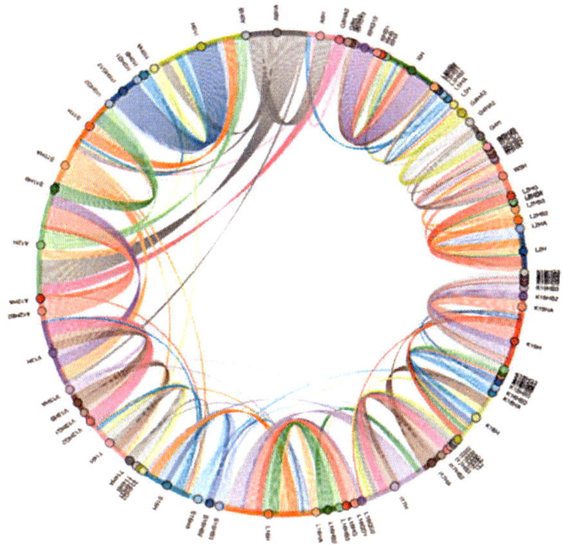

Fig. 14 Chord diagram indicating the relative of strength of interactions through volume in **a** all interactions **b** only major interactions (volume greater than 10^6) of Ctr in SDS

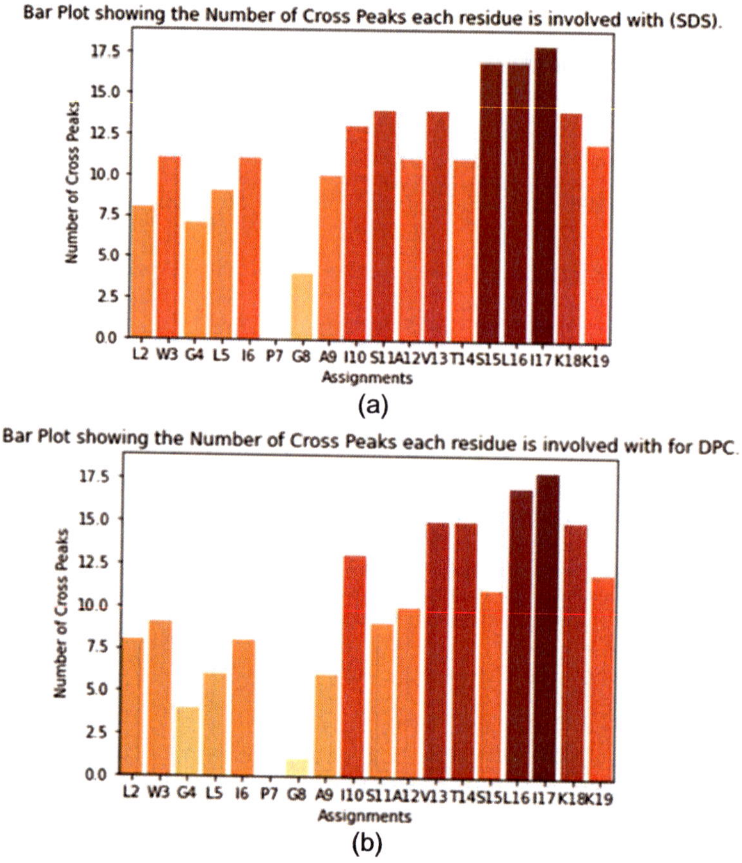

Fig. 15 Bar graph indicating the number of cross peaks of Ctr in **a** SDS and **b** DPC

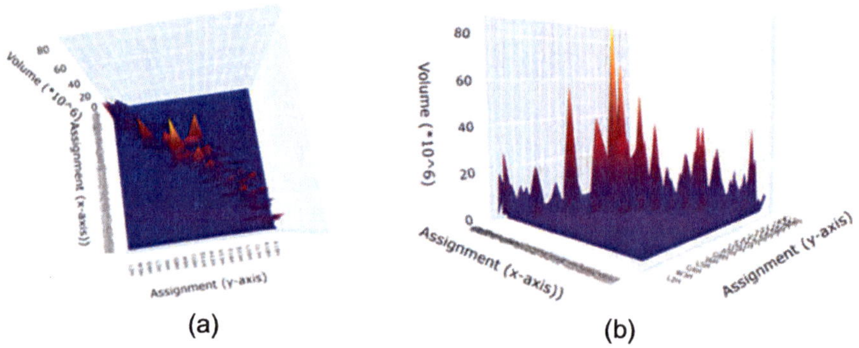

Fig. 16 3D surface diagram including the relative strength of interaction through volume of the entire Ctr **a** top view and **b** side view in SDS. Inset: The *X* axis includes all the assignments. The *Y* axis only includes the H_N for the residues. The *Z* axis represents the volume

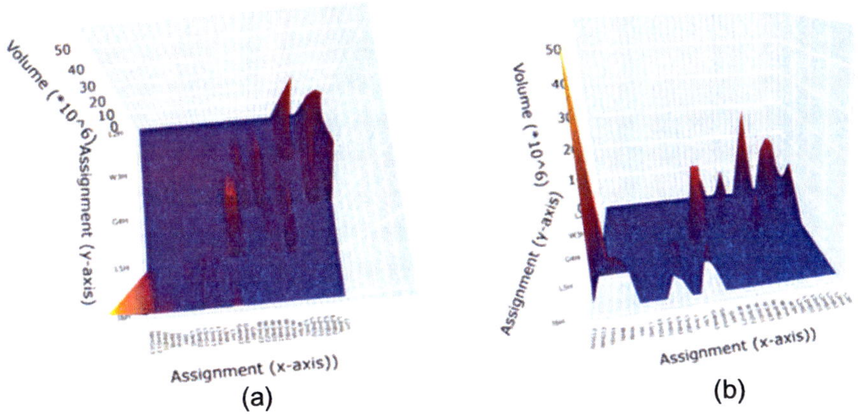

Fig. 17 3D surface diagram including the relative strength of interaction through volume of L2H to 16H of Ctr **a** top view and **b** side view in SDS. Inset: The *X* axis includes all the assignments. The *Y* axis only includes the H_N for the residues. The *Z* axis represents the volume

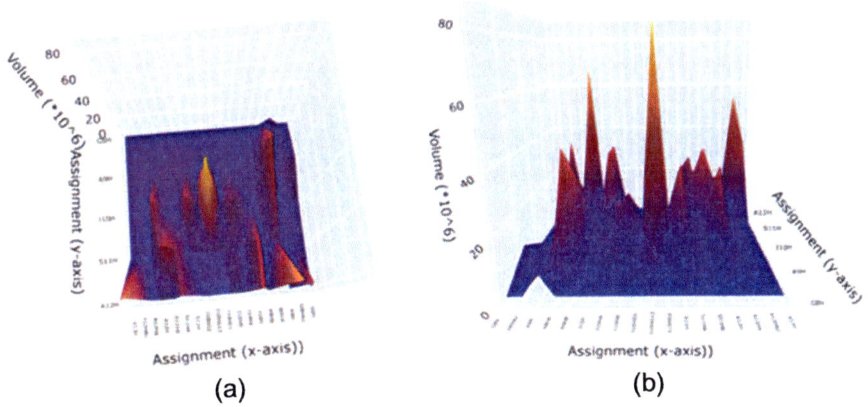

Fig. 18 3D surface diagram including the relative strength of interaction through volume of G8H to A12H of Ctr **a** top view and **b** side view in SDS. Inset: The *X* axis includes all the assignments. The *Y* axis only includes the H_N for the residues. The *Z* axis represents the volume

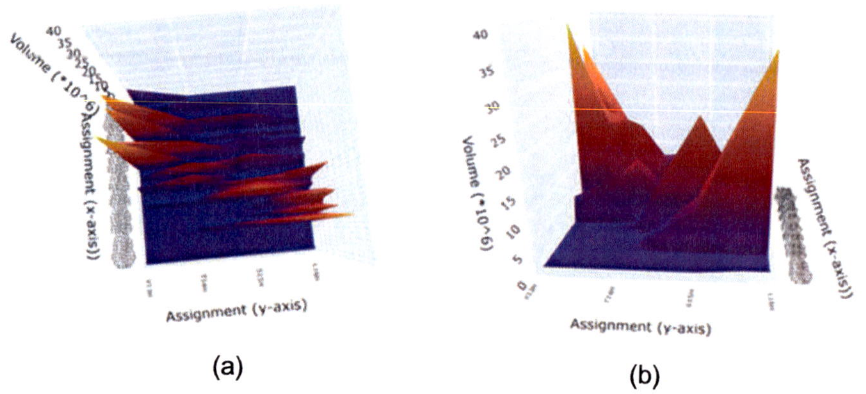

Fig. 19 3D surface diagram including the relative strength of interaction through volume of V13H to L16H of Ctr **a** top view and **b** side view in SDS. Inset: The *X* axis includes all the assignments. The *Y* axis only includes the H_N for the residues. The *Z* axis represents the volume

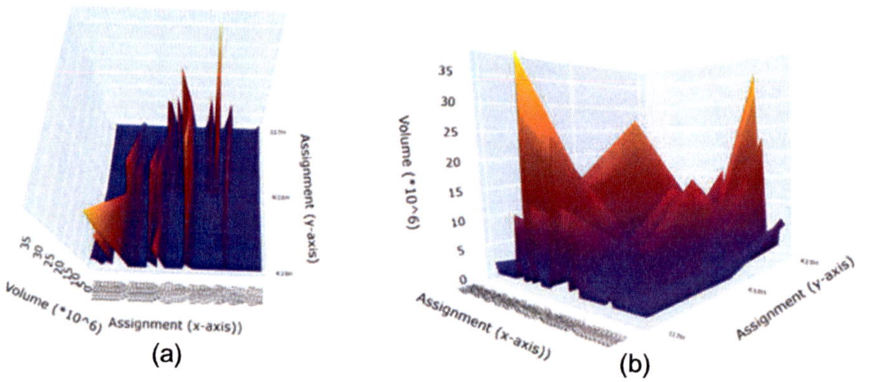

Fig. 20 3D surface diagram including the relative strength of interaction through volume of I17H to K19H of Ctr **a** top view and **b** side view in SDS. Inset: The *X* axis includes all the assignments. The *Y* axis only includes the H_N for the residues. The *Z* axis represents the volume

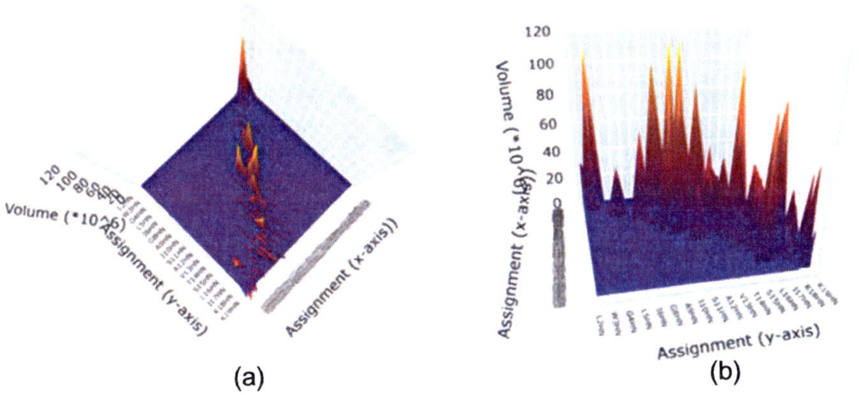

Fig. 21 3D surface diagram including the relative strength of interaction through volume of the entire Ctr **a** top view and **b** side view in SDS. Inset: The *X* axis includes all the assignments. The *Y* axis only includes the H_N for the residues. The *Z* axis represents the volume

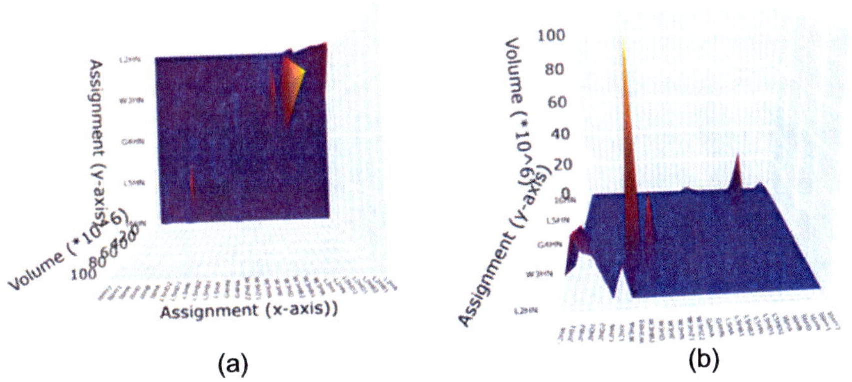

Fig. 22 3D surface diagram including the relative strength of interaction through volume of L2H to 16H of Ctr **a** top view and **b** side view in SDS. Inset: The *X* axis includes all the assignments. The *Y* axis only includes the H_N for the residues. The *Z* axis represents the volume

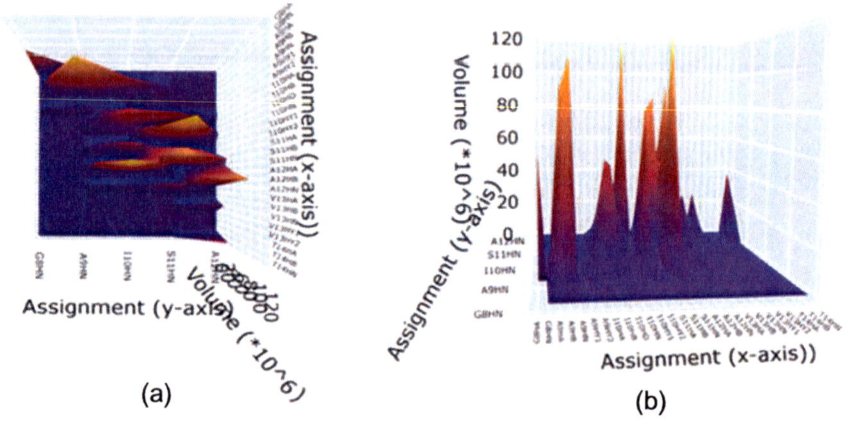

Fig. 23 3D surface diagram including the relative strength of interaction through volume of G8H to A12H of Ctr **a** top view and **b** side view in SDS. Inset: The *X* axis includes all the assignments. The *Y* axis only includes the H_N for the residues. The *Z* axis represents the volume

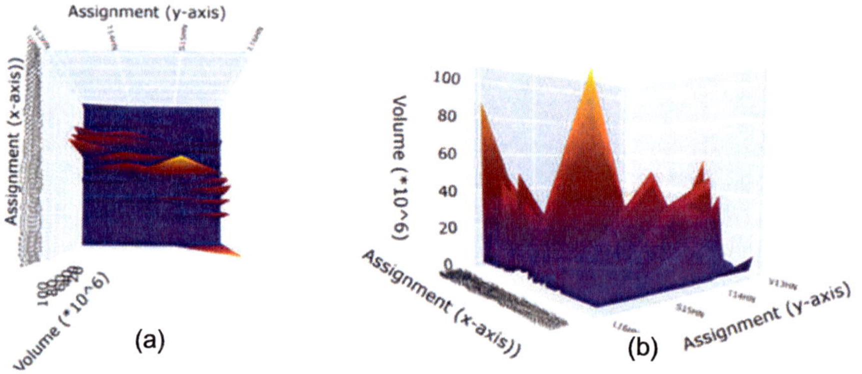

Fig. 24 3D surface diagram including the relative strength of interaction through volume of V13H to L16H of Ctr **a** top view and **b** side view in SDS. Inset: The *X* axis includes all the assignments. The *Y* axis only includes the H_N for the residues. The *Z* axis represents the volume

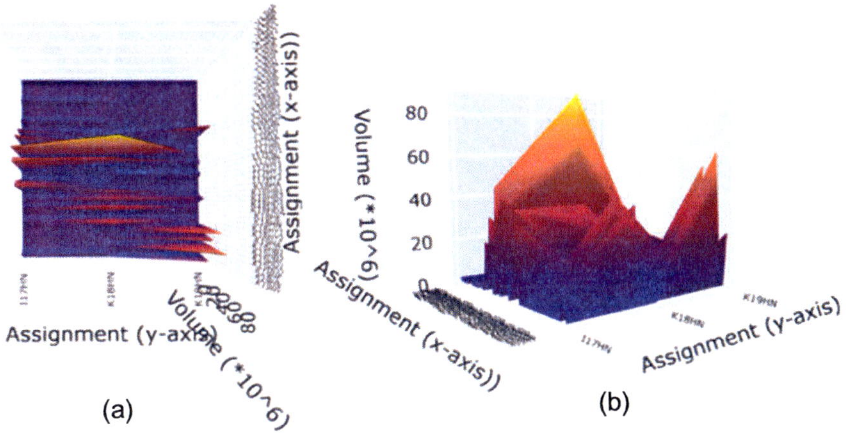

Fig. 25 3D surface diagram including the relative strength of interaction through volume of I17H to K19H of Ctr **a** top view and **b** side view in SDS. Inset: The *X* axis includes all the assignments. The *Y* axis only includes the H_N for the residues. The *Z* axis represents the volume

3D Model

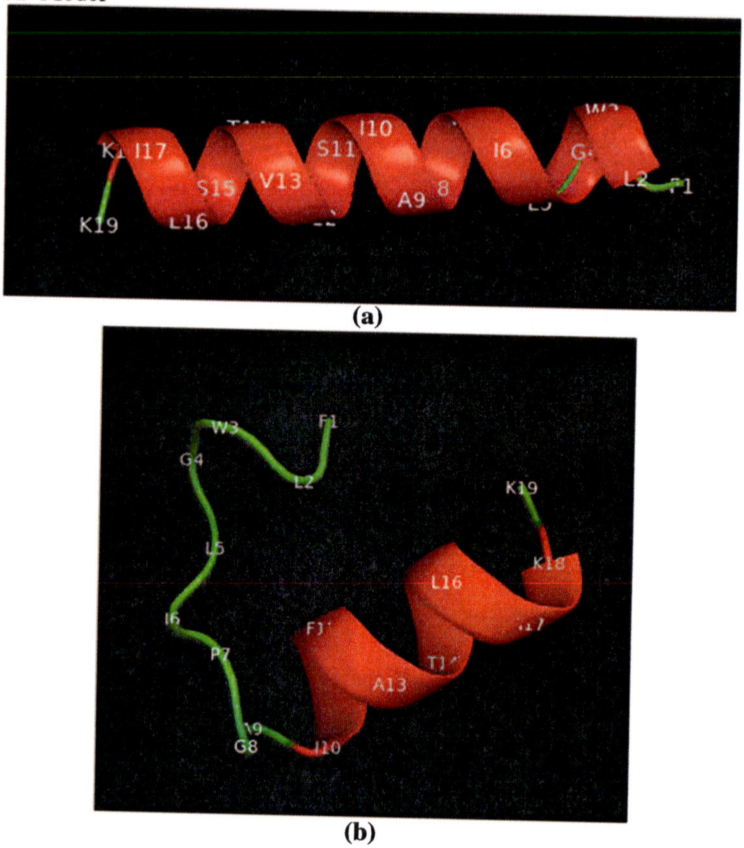

Fig. 26 Ribbon diagram of representative average structure of Ctr in **a** DPC and **b** SDS

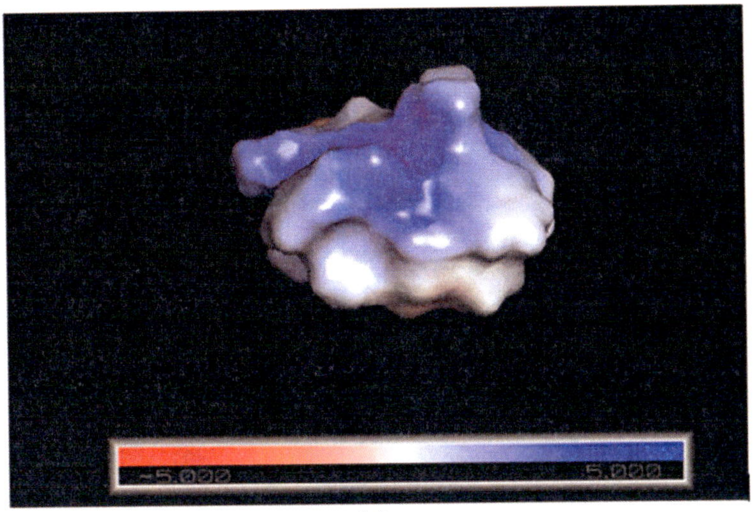

(a)

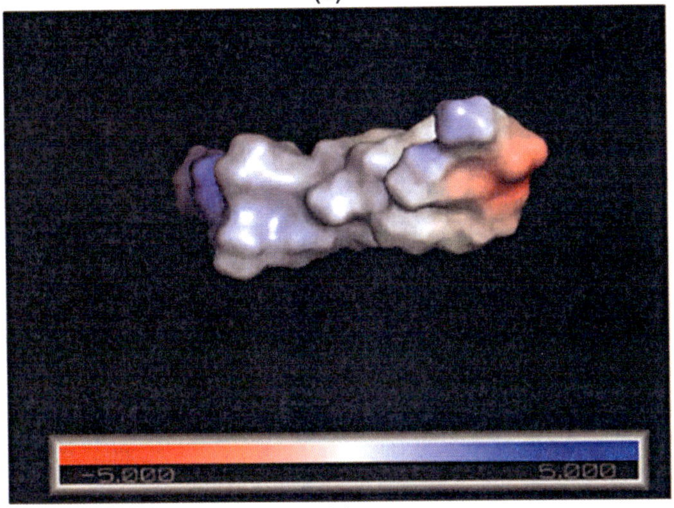

(b)

Fig. 27 Charge topography of Ctr in the presence of **a** SDS and **b** DPC. Red colour denotes negative charge and blue denotes positive charge

Predictive Secondary Structure Bar Graph

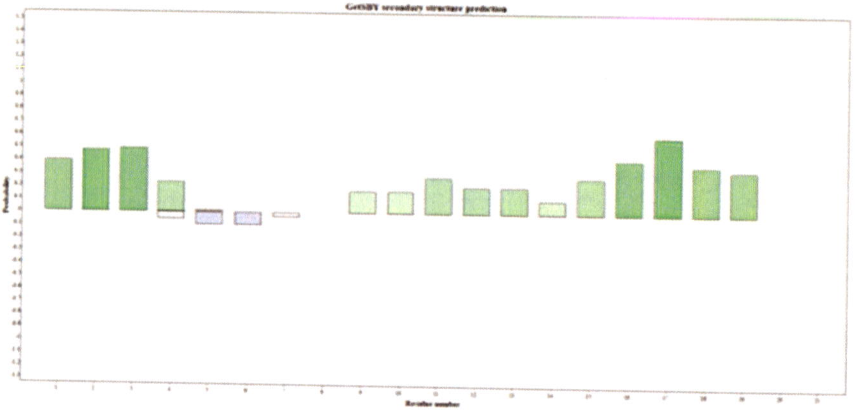

Fig. 28 Bar graph showing the probability; of secondary structure in Ctr in DPC. Inset: Green represents α-helix. Blue represents β-pleated sheets

References

1. Bandyopadhyay, S., Junjie, R. L., Lim, B., Sanjeev, R., Xin, W. Y., Yee, C. K., Hui Melodies, S. M., Yow, N., Sivaraman, J., & Chatterjee, C. (2014). Solution structures and model membrane interactions of Ctriporin, an anti-methicillin-resistant *Staphylococcus aureus* Peptide from Scorpion Venom. *Biopolymers, 101*(12), 1143–1153. https://doi.org/10.1002/bip.22519
2. Lee, W., Tonelli, M., & Markley, J. L. (2015). NMRFAM-SPARKY: Enhanced software for biomolecular NMR spectroscopy. *Bioinformatics, 31*(8):1325–1327.
3. Bogachev, M. I., Kayumov, A. R., Markelov, O. A., & Bunde, A. (2016). Statistical prediction of protein structural, localization and functional properties by the analysis of its fragment mass distributions after proteolytic cleavage. *Scientific Reports, 6*(1). https://doi.org/10.1038/srep22286
4. Fan, Z., Cao, L., He, Y., Hu, J., Di, Z., Wu, Y., Li, W., & Cao, Z. (2011). Ctriporin, a new anti-methicillin-resistant *Staphylococcus aureus* Peptide from the Venom of the Scorpion Chaerilus tricostatus. *Antimicrobial Agents and Chemotherapy, 55*(11), 5220–5229. https://doi.org/10.1128/aac.00369-11
5. 3.4 Proteins—Biology. *OpenStax* (2016). Proteins. https://openstax.org/books/biology/pages/3-4-proteins
6. Creamer, T. P., & Rose, G. D. (1995). Interactions between hydrophobic side chains within α-helices. *Protein Science, 4*(7), 1305–1314. https://doi.org/10.1002/pro.5560040706
7. Daggett, V., & Fersht, A. (2003). The present view of the mechanism of protein folding. *Nature Reviews Molecular Cell Biology, 4*(6), 497–502. https://doi.org/10.1038/nrm1126
8. Cammers-Goodwin, A., Allen, T. J., Oslick, S. L., McClure, K. F., Lee, J. H., & Kemp, D. S. (1996). Mechanism of stabilization of helical conformations of polypeptides by water containing trifluoroethanol. *Journal of the American Chemical Society, 118*(13), 3082–3090. https://doi.org/10.1021/ja952900z
9. Reeb, J., & Rost, B. (2019). Secondary structure prediction. *Encyclopedia of Bioinformatics and Computational Biology*, 488–496.https://doi.org/10.1016/b978-0-12-809633-8.20267-7

Heterotrophic Cultivation of Microalgae in Food Waste Digestate for Simultaneous Biomass Production and Nutrient Remediation

Zou Jiayue, Sheetal Kishor Parakh, and Yen Wah Tong

Abstract As food wastage becomes an increasingly dire problem, research has been conducted on food waste management. Anaerobic digestion has been presented as an alternative to traditional methods of food waste disposal, as it is able to produce biogas as a renewable energy source. However, its byproduct, liquid food waste digestate, needs to be treated appropriately before disposal. Food waste digestate usually contains a high percentage of ammonia, phosphorus, and other organic compounds. Hence, microalgae cultivation in food waste digestate has been suggested as a treatment method, as various strains of microalgae are effective in removal of nitrogen, phosphorus, heavy metals, and toxins from wastewater, while using compounds present to synthesize valuable biomass. Thus, microalgae as a treatment for food waste digestate is promising. In this study, food waste was taken from a local anaerobic digester, filtered and used for the heterotrophic cultivation of three different strains of microalgae. Hetero-trophic cultivation was carried out as it does not require light and is easier to incorporate into biorefineries. From the three strains, *Chlorella sorokiniana* has been found to have the best growth rate, reaching a final dry cell weight of 0.144 g/L, and was thus used for subsequent experiments. *C. sorokiniana* was then cultivated in different glucose and food waste digestate concentrations to investigate microalgal performance and potential products. Results found that *C. sorokiniana* was effective in ammonia removal, with the highest removal at 250.8 ppm and had the highest protein and carbohydrate percentage contents at 47% and 57.1% respectively.

Keywords Microalgae · Wastewater treatment · Heterotrophic cultivation · Protein production · Carbohydrate production · Ammonium ion removal · Food waste digestate

Z. Jiayue (✉)
Raffles Institution, 1 Raffles Institution Ln, Singapore, Singapore
e-mail: jiayue.zou1030@gmail.com

S. K. Parakh · Y. W. Tong
Department of Chemical and Biomolecular Engineering, National University of Singapore, 21 Lower Kent Ridge Rd, Singapore, Singapore

H. Guo et al. (eds.), *IRC-SET 2022*,
https://doi.org/10.1007/978-981-19-7222-5_40

1 Introduction

In 2021, about 17% of global food production was wasted, according to a report by the United Nations [1]. Food waste (FW) refers to materials intended for human consumption which are discarded, lost, degraded, or contaminated [2]. In America alone, avoidable FW exceeds 55,000,000 tonnes per year, producing greenhouse gas emissions of CO_2equivalent to 2% of the nation's total annual emission [3]. China generates around 90,000,000 tonnes of FW yearly [4], and locally in Singapore, 665,000 tonnes of FW was generated just in the year of 2020 itself [5]. With the increasing amount of FW generated globally, the need for proper FW treatment and disposal is on the rise.

FW, if disposed untreated, can cause various environmental problems of detrimental consequences. The traditional method of disposing of FW in landfills results in the release of methane into the atmosphere during organic decomposition, which is 80 times more powerful than CO_2 as a greenhouse gas [6]. Another conventional method of food waste management is incineration at Waste-To-Energy (WTE) plants. It makes use of burning of waste to generate electricity, causing airborne pollution [7]. WTE plants also do not fully utilize the nutrients still present in the FW such as nitrogen, potassium, and phosphorus [8]. Hence currently, new and innovative ways of FW treatment are being investigated, one of which is anaerobic digestion of FW, using bacterial breakdown [9]. Compared to traditional methods of FW disposal, anaerobic digestion is seen as a promising method of generating methane, which is captured as biofuel as a form of renewable energy. Food waste as a substrate for anaerobic digestion may also have the potential to provide a higher biogas yield compared to other forms of widely used substrate, such as manure and corn silage [10].

However, anaerobic digestion also produces a byproduct, a sludge digestate consisting of the end products of anaerobic digestion. FW digestate often contains macronutrients and micronutrients, and is generally rich in nitrogen, potassium and sodium [11]. Due to its high organic ion content, direct disposal of it without treatment can cause environmental pollution. The most common treatment method is using it as agricultural fertiliser [12], but liquid digestate fertilizer might cause run-off and eventual eutrophication of water systems [13]. Using liquid FW digestate for land-based agriculture also requires a large land size and its suitability depends on the type of crops and time of the year [14]. For countries which do not have large land spaces dedicated to agricultural use, using FW digestate as fertilizer may not be as effective. However, it is a highly viable medium for the cultivation of different strains of microalgae simultaneously with the treatment of digestate, a method which requires only a small land size [15].

Microalgae cultivation is a novel and promising way of utilizing FW digestate [16, 17]. Various strains of microalgae have been found to be effective in removing nitrogen, phosphorus and other compounds in waste, utilizing these compounds for growth and producing high yield of valuable products such as carbohydrates which can be converted into bioethanol, lipids which can be converted into biofuel, proteins as animal feed, and various vitamins [18]. Generally, the large-scale commercialisation of microalgae production has been hindered by the expensive biomass production procedure and product extraction steps [19]. However, using FW digestate can reduce the cost price of producing a viable medium for microalgae cultivation. As such, cultivation of microalgae using FW digestate could be beneficial, since high value products can be obtained with the concurrent removal of compounds in food waste.

Several studies have investigated the possibility of using FW digestate as a medium to cultivate microalgae, but most of them researched on the autotrophic or mixotrophic method of cultivation [15, 20, 21], which requires the presence of light. Despite various added benefits of heterotrophic cultivation method, it is rarely studied in FW digestate treatment. Studies which investigated the heterotrophic cultivation method [22, 23] only took into consideration one specific strain of microalgae. Some benefits of heterotrophic methods include its ability to be integrated into current biorefinery systems due to it occuring in the absence of light, as large-scale operations in fermenters are carried out in the dark. Heterotrophic cultivation also requires less manpower in terms of operation and daily maintenance, and has been reported to be able to enhance the algal biomass yield up to 25 fold compared to the autotrophic method [24]. Hence, heterotrophic cultivation was chosen in this study to investigate its suitability as a cultivation method for microalgae in FW digestate.

This study aims to investigate the suitability of heterotrophic cultivation of microalgae in FW digestate for nutrient removal in digestate and nutrient remediation by microalgae [15, 21, 25, 26]. Three different strains of freshwater microalgae strains were cultivated and their ability to grow heterotrophically in a medium of local FW digestate was compared. *Chlorella sorokiniana* was found to have the highest growth rate out of the three strains and used for the subsequent experiments, where mediums of different digestate and glucose concentrations were used to cultivate *C. sorokiniana* in order to investigate microalgal performance in both ammonium ion removal and biomass production. The hypothesis is that with an increasing percentage of FW digestate in the medium, the growth rate of microalgae will increase initially, but when the percentage concentration of digestate increases beyond the tolerable ammonium ion concentration of the microalgae, the cells will stop growing. With an increasing concentration of glucose, cell growth is also expected to increase, as carbon will no longer be a limiting factor. Biomass production and nutrient remediation is also expected to increase with increasing digestate and glucose, as the cell growth increases.

2 Materials and Methods

2.1 *Filtration of Food Waste Digestate*

Food waste digestate was collected from a local anaerobic digester. Digestate was then centrifuged at 4 °C and 10,000 rpm for 10 min, then the supernatant was recovered and vacuum filtered using 90, 45, and 25 mm filter papers in order to remove solids. A sample was taken from the resultant liquid food waste digestate to determine its cation and anion concentration by ionic chromatography. By composition, the digestate contains Na+ (1627 ppm), K+ (1604 ppm), Cl− (1709 ppm), SO_4^{2-} (184 ppm), PO_4 (70 ppm), and NH_4^+ (2509 ppm) Subsequently, the FW digestate was refrigerated until the commencement of the experiment.

2.2 *Pre-cultivation of Microalgal Strains*

Three strains of freshwater microalgae *Chlorella vulgaris* (UTEX259), *Chlorella sorokiniana* (UTEX1230) and *Scenedesmus obliquus* (UTEX393) were purchased from the University of Texas, Austin. For inoculation, each of the 3 strains were pre-cultivated in Bold's Basal Medium (BBM) prepared from 0.01 g/L (v/v) NaCl, $NaNO_3$, $CaCl_2{\cdot}2H_2O$, $MgSO_4{\cdot}7H_2O$, K_2HPO_4, KH_2PO_4, and 0.01 g/L (v/v)

$Na_2EDTA{\cdot}2H_2O$, $FeSO_4{\cdot}7H_2O$, H_3BO_3, and trace metal solution. A 2.5 g/L glucose was added as an organic carbon source. The medium was inoculated using 40 ml cell culture flasks wrapped in aluminium foil, shaken at 115–120 rpm and maintained at 24–25 °C for the next 4 days.

2.3 *Cultivation of Different Microalgal Strains in Filtered Food Waste Digestate*

From the pre-cultivated strains, the optical density (OD) was measured to determine the cell concentration (see Sect. 2.5), and 0.05 g/L of cells were collected and added to 150 ml baffled flasks in duplicates. A 15% of digestate (v/v) was added to each flask. BBM used is similar to that in Sect. 2.2 but omits $NaNO_3$ and NaCl as the ions in these solutions are already present in the digestate. The baffled flasks were wrapped in aluminium foil, shaken at 115–120 rpm and maintained at room temperature. Medium was inoculated for the next 8 days.

2.4 Cultivation of *Chlorella sorokiniana* in Different Concentrations of Glucose and Digestate

Chlorella sorokiniana cultivated in Sect. 2.3 had the highest growth rate out of the three strains, and was thus used for this experiment. Initial cultivation process is similar to that in Sect. 2.3, but for three flasks, filtered digestate was added in 2.5%, 5% and 10% (v/v) respectively with 2.5 g/L glucose. For the other three flasks, filtered digestate was added in 2.5%, 5%, and 10%, respectively, with 10 g/L glucose. The pH of the culture medium was kept at 7–7.5 through 1M NaOH addition. The medium was inoculated in the same conditions as in Sect. 2.3 for the next 8 days.

2.5 Analytical Methods

For all above experiments, 1ml of samples were taken from each flask everyday. Microalgal growth was determined by diluting samples with distilled water to an optical density (OD) below 0.8, and the OD was measured at wavelength of 540 nm (OD540).

Dry cell weight (DCW) was calculated using the correlations developed in previous studies.

$$\text{DCW of C.V.} = 0.444 * \text{OD540}$$
$$\text{DCW of C.S.} = 0.24 * \text{OD540}$$
$$\text{DCW of S.O.} = 0.314 * \text{OD540}$$

pH of the medium was measured using a pH metre (Orion 4-Star, Thermo Scientific, USA).

At the end of the experiment in Sect. 2.4, *C. sorokiniana* was harvested through centrifugation, washed and freeze dried at − 80 °C for determination of biomass composition. The cation concentration in the medium was determined using ionic chromatography. Protein was extracted in duplicates from all the dried cell samples by the hot alkali method. The protein content was determined using the Lowry's Method of Protein Analysis. Carbohydrate was extracted via acid hydrolysis, and the carbohydrate content was determined using the Calorimetric Dubois Method for determination of sugar and related substances.

3 Results and Discussions

3.1 Growth of Different Microalgal Strains in FW Digestate

Chlorella vulgaris, *S. obliquus* and *C. sorokiniana* were cultivated in 15% (v/v) FW digestate for 8 days. As presented in Fig. 1, *C. sorokiniana* reached the highest DCW concentration, with its DCW reaching 0.144 g/L. For *C. sorokiniana*, the DCW increased by almost 3 fold from its initial cell concentration on Day 0. Both *C. vulgaris* and *S. obliquus* had a peak in growth on Day 2, where their DCW reached 0.112 g/L and 0.0702 g/L. However, both the growth of *C. vulgaris* and *S. obliquus* experienced a decline in DCW from Day 2 to Day 7, with final DCW at 0.07546 g/L and 0.05406 g/L respectively. In the three strains of microalgae cultivated, only *C. sorokiniana* had an increasing trend of DCW all throughout the experiment duration, with *C. vulgaris* and *S. obliquus* both experiencing a decrease in DCW from Day 2 onwards. This might be due to the high ammonium ion content in the FW digestate, which might have exceeded the ammonia tolerance limit in *C. vulgaris* and *S. obliquus* [27], inhibiting further growth, as ammonia is toxic to microalgae in high concentrations. Cultivation with digestate also might not be the most optimal method of cultivation of those 2 strains [28, 29], although the ammonium ion concentration was suitable for *C. sorokiniana* [30]. This corroborates our results and thus, *C. sorokiniana* was determined to have the most optimal growth rate.

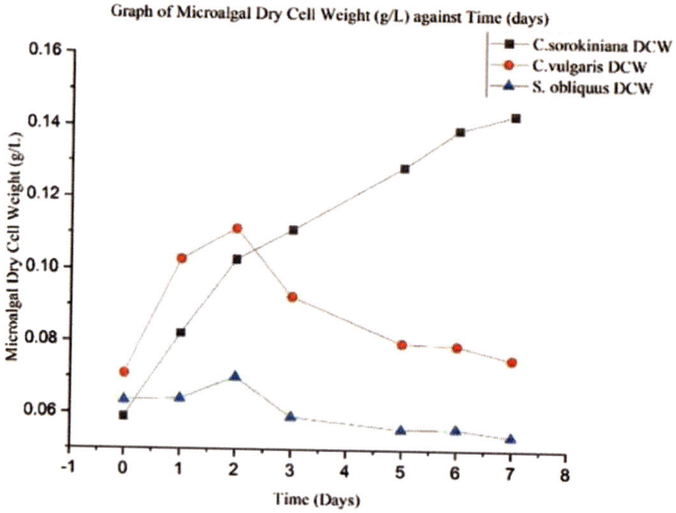

Fig. 1 Three strains of microalgae cultivated in FW digestate for 8 days

3.2 *Growth of Chlorella sorokiniana in Different Glucose Concentrations*

Since the growth rates of microalgae in 15% digestate and 2.5 g/L glucose were not significant, for subsequent investigations, this study varied the glucose and digestate concentration to investigate their effects on cell growth. Three percentages of digestate (v/v) 2.5%, 5%, and 10% were chosen. Two concentrations of glucose, 2.5 and 10g/L were used. Comparing Fig. 2 (left) and Fig. 2 (right), *C. sorokiniana* reached a generally higher DCW in 10 g/L glucose than 2.5 g/L glucose. In 2.5 g/L glucose concentration, culture with 2.5% FW digestate reached a final DCW of 0.961 g/L, but with 10 g/L glucose concentration, culture with 2.5% FW digestate reached a final DCW of 1.22 g/L. For cultures with 5% digestate, final DCW is 0.523 g/L in 2.5 g/L glucose concentration, compared to 2.00 g/L in 10 g/L glucose concentration. For cultures with 10% digestate, the final DCW is 1.24 g/L in 2.5 g/L glucose concentration and 3.05 g/L in 10 g/L glucose, with an almost 3 times difference between them.

Glucose was utilized for this experiment as it has been found to be amongst the organic carbon sources to support growth of microalgal strains under heterotrophic conditions [31, 32]. It is also easier to obtain than other organic carbon sources as it is produced on a large scale and yields a high level of substrate [33]. Increasing the glucose concentration of the culture increases the growth rate of *C. sorokiniana* significantly under heterotrophic conditions [34], as organic carbon will no longer be a limiting factor in glucose catabolic pathways, which are crucial for microalgal growth. This corroborates with the results obtained above.

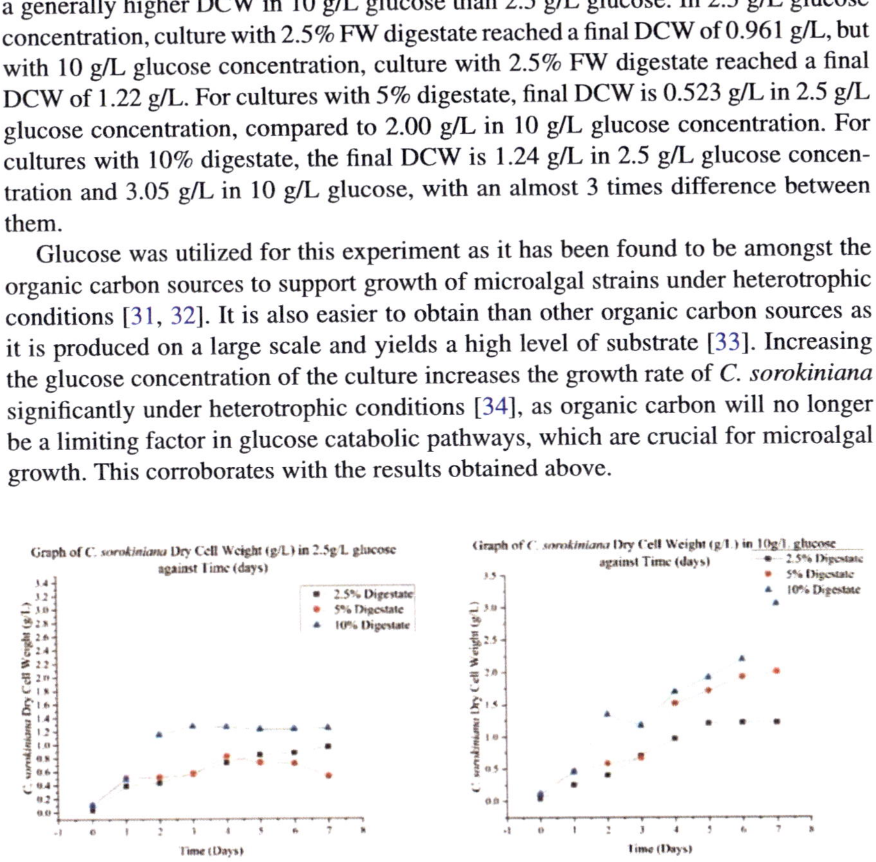

Fig. 2 (Left) and (right) showing the DCW change of *C. sorokiniana* over time, in 2.5 g/L glucose and 10 g/L glucose, respectively

3.3 Growth of *Chlorella sorokiniana* in Different Digestate Concentrations

Growth of *C. sorokiniana* in different FW digestate percentages was also compared. In Fig. 2 (left), it was observed that the culture with 10% digestate had a higher trend of DCW increase compared to the culture with 5% digestate and 2.5% digestate, reaching a DCW of 1.24 g/L at the end of 8 days. Culture with 2.5% digestate and 5% digestate reached a final DCW of 0.523 g/L and 0.961 g/L, respectively. In Fig. 2 (right), a similar trend is observed. Culture with 2.5% digestate reached 1.23 g/L of final DCW, culture with 5% digestate reached 2.00 g/L of final DCW, and culture with 10% digestate reached a final DCW of 3.05 g/L. An increase in the percentage of digestate is found to have generally increased the growth rate of *C. sorokiniana*, with culture of 10% digestate having the most DCW in both 2.5 and 10 g/L glucose environments. This is likely because the digestate consists of ammonium ion, the sole nitrogen-containing nutrient in the medium. Nitrogen is required by microalgae as a vital macronutrient which regulates the metabolism of microalgae [35], thus controlling its growth. With an optimum concentration of ammonium ion, cell growth can be enhanced.

3.4 Biomass Production and Nutrient Remediation of *Chlorella sorokiniana* in Different Glucose and Digestate Concentrations

Table 1 shows the comparison of protein and carbohydrate content in each of the cultures after the experiment, and the change in ammonium ion content in the FW digestate medium before and after the experiment for each cell culture. The carbohydrate content for the culture with 10% digestate and 2.5 g/L glucose is vastly different from other carbohydrate content. It is considered as an experimental error and not taken into account for subsequent data analysis. The protein and carbohydrate production of each flask is more clearly represented in Fig. 3 (left) and Fig. 3 (right).

Comparing different digestate concentrations, it was observed that as the digestate percentage supplied was increased, the protein content in the microalgal cells increases. The highest final protein content is found in cells cultivated in 10% digestate and 2.5 g/L glucose, of 47.3%. As the digestate percentage increases, the carbohydrate percentage decreases, with the highest carbohydrate percentage and the lowest carbohydrate percentage present in culture with 2.5% digestate, and culture with 5% digestate, respectively. The increase in protein and decrease in carbohydrate percentages may be due to the decreasing Carbon/Nitrogen (C/N) ratio as the digestate increases. It has been reported that microalgae utilize carbon in medium 25–30 times faster than nitrogen, hence a low C/N ratio may result in more protein being produced [36, 37].Studies have also observed that the highest rate of carbohydrate

Table 1 The biomass production and nutrient remediation of each culture

Samples	1A	2A	3A	1B	2B	3B
Digestate percentage (v/v) (%)	2.5	5	10	2.5	5	10
Glucose concentration (g/L)	2.5	2.5	2.5	10	10	10
Initial ammonium ion concentration (ppm)	62.7	125	251	62.7	125	251
Final ammonium ion concentration (ppm)	0.2	0.2	33	0.2	0.1	0.2
Change in ammonium ion concentration (ppm)	62.5	124.8	218	62.5	124.9	250.8
Final protein percentage (%)	17.7	29.5	47.3	13.9	14.2	24.2
Final carbohydrate percentage (%)	56.3	42.6	2.87	57.1	50.1	47.6

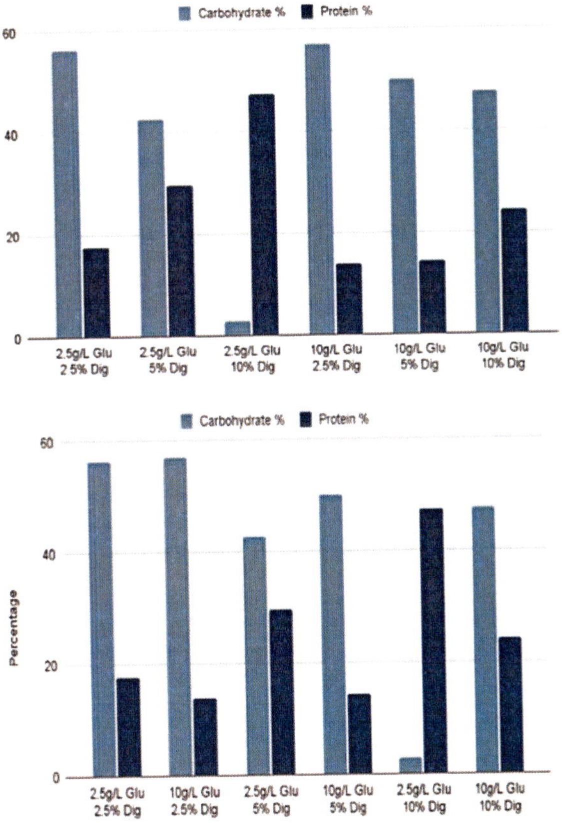

Fig. 3 (Top) and (bottom) showing comparison of biomass production between different digestate concentrations and different glucose concentrations respectively

synthesis in microalgae was present in the culture cultivated under nutrient limitation [38], where the highest carbohydrate concentrations were also found in cultures with 2.5% of digestate, limiting the nitrogen content. For nutrient remediation, an increase in digestate increases the ammonium ion absorbed by microalgal cells, with the greatest change being 250.8 ppm of ammonium ion absorbed, which is comparable to other methods of nutrient remediation.

Comparing the cultures with the same digestate percentage but different glucose concentrations, an increase in glucose caused a decrease in protein percentage, with the biggest difference being 47.3 and 24.2%. However, an increase in glucose was observed to have increased the carbohydrate percentage with the highest found from the culture with 10 g/L glucose, at 57.1%. An increase in glucose concentration was not found to have a major impact on the ammonium ion taken in by the cells, although a difference of 32.8 ppm was observed at 10% digestate content. A higher glucose concentration increases the C/N ratio in the cultivation medium, and thus results in a lower protein percentage but a higher carbohydrate percentage. The high C/N ratio causes more carbon nutrient uptake, and this may induce the storage of carbohydrates and lipids in some microalgal species [39], thus decreasing protein concentration overall.

4 Conclusion

This study found that using FW digestate for microalgal cultivation holds potential in terms of ammonium ion removal and production of protein and carbohydrates. Depending on the method of utilization of microalgae after cultivation, different C/N ratios should be chosen to optimize protein or carbohydrate content. Growth was not found to be limited at high FW digestate percentages, suggesting that *C. sorokiniana* has not yet reached its ammonia tolerance capacity. However, this method requires more research before it can be applied large scale, as dilution of FW digestate is not cost effective [26]. Another cost factor of heterotrophic cultivation is that it requires an organic carbon source, as of now, the most commonly used is glucose. However, further research can be conducted to obtain carbon sources suitable for microalgae cultivation from other waste sources.

References

1. UNEP food waste index report 2021 [Internet]. UNEP—UN Environment Programme [cited 21 January 2022]. https://www.unep.org/resources/report/unep-food-waste-index-report-2021
2. Girotto, F., Alibardi, L., & Cossu, R. (2015). Food waste generation and industrial uses: A review. *Waste Management, 45*, 32–41.
3. Venkat, K. (2022). The climate change and economic impacts of food waste in the United States [Internet]. Centmapress.ilb.uni-bonn.de [cited 21 January 2022]. http://centmapress.ilb.uni-bonn.de/ojs/index.php/fsd/article/view/198
4. Zhang, C., Su, H., Baeyens, J., & Tan, T. (2014). Reviewing the anaerobic digestion of food waste for biogas production. *Renewable and Sustainable Energy Reviews, 38*, 383–392.
5. (2022). Commentary: Singapore's festive indulgence creates enormous food waste [Internet]. CAN [cited 21 January 2022]. https://www.chan-nelnewsasia.com/commentary/food-waste-christmas-festive-indulgence-2397511

6. (2022). [Internet]. Environmental defence fund [cited 21 January 2022]. https://www.edf.org/climate/methane-crucial-opportunity-climate-fight
7. Bosello, F. Energy from waste
8. Kuppusamy, S., Venkateswarlu, K., & Megharaj, M. (2017). Evaluation of nineteen food wastes for essential and toxic elements. *International Journal of Recycling of Organic Waste in Agriculture* [Internet]. SpringerLink. Springer [cited 2022 Jan 21]. https://doi.org/10.1007/s40093-017-0178-2
9. Premo, B. (2022). Anaerobic digestion: A solution to diverting wasted food [Internet]. Center for EcoTechnology [cited 21 January 2022]. https://www.centerfore-cotechnology.org/anaerobic-digestion-a-solution-to-diverting-food-waste/#:~:text=Anaero-bic%20Digestion%20is%20a%20process,biogas%20composed%20mostly%20of%20me-thane
10. Curry, N., & Pillay, P. (2012). Biogas prediction and design of a food waste to energy system for the urban environment. *Renewable Energy, 41*, 200–209.
11. Logan, M., & Visvanathan, C. (2019). Management strategies for anaerobic digestate of organic fraction of municipal solid waste: Current status and future prospects. *Waste Management & Research: The Journal for a Sustainable Circular Economy, 37*(1_suppl), 27–39.
12. Sayedin, F., Kermanshahi-pour, A., He, Q., Tibbetts, S., Lalonde, C., & Brar, S. (2020). Microalgae cultivation in thin stillage anaerobic digestate for nutrient recovery and bioproduct production. *Algal Research, 47*, 101867.
13. Xia, A., & Murphy, J. (2016). Microalgal cultivation in treating liquid digestate from biogas systems. *Trends in Biotechnology, 34*(4), 264–275.
14. Praveen, P., Guo, Y., Kang, H., Lefebvre, C., & Loh, K. (2018). Enhancing microalgae cultivation in anaerobic digestate through nitrification. *Chemical Engineering Journal, 354*, 905–912.
15. Chuka-ogwude, D., Ogbonna, J., Borowitzka, M., & Moheimani, N. (2020). Screening, acclimation and ammonia tolerance of microalgae grown in food waste digestate. *Journal of Applied Phycology, 32*(6), 3775–3785.
16. Prajapati, S., Kumar, P., Malik, A., & Vijay, V. (2014). Bioconversion of algae to methane and subsequent utilization of digestate for algae cultivation: A closed loop bioenergy generation process. *Bioresource Technology, 158*, 174–180.
17. Bjornsson, W., Nicol, R., Dickinson, K., & McGinn, P. (2013). Anaerobic digestates are useful nutrient sources for microalgae cultivation: Functional coupling of energy and biomass production. *Journal of Applied Phycology, 25*(5), 1523–1528.
18. Dourou, M., Dritsas, P., Baeshen, M., Elazzazy, A., Al-Farga, A., & Aggelis, G. (2020). High-added value products from microalgae and prospects of aquaculture wastewaters as microalgae growth media. *FEMS Microbiology Letters, 367*(12).
19. Tan, J., Lee, S., Chew, K., Lam, M., Lim, J., Ho, S., et al. (2020). A review on microalgae cultivation and harvesting, and their biomass extraction processing using ionic liquids. *Bioengineered, 11*(1), 116–129.
20. Stiles, W., Styles, D., Chapman, S., Esteves, S., Bywater, A., Melville, L., et al. (2018). Using microalgae in the circular economy to valorise anaerobic digestate: Challenges and opportunities. *Bio-resource Technology., 267*, 732–742.
21. Koutra, E., Economou, C., Tsafrakidou, P., & Kornaros, M. (2018). Bio-based products from micro-algae cultivated in digestates. *Trends in Biotechnology, 36*(8), 819–833.
22. Sobhi, M., Guo, J., Cui, X., Sun, H., Li, B., Aboagye, D., et al. (2019). A promising strategy for nutrient recovery using heterotrophic indigenous microflora from liquid biogas digestate. *Science of the Total Environment, 690*, 492–501.
23. Pleissner, D., Lindner, A., & Händel, N. (2021). Heterotrophic cultivation of Galdieria sulphuraria under non-sterile conditions in digestate and hydrolyzed straw. *Bioresource Technology., 337*, 125477.
24. Morales-Sánchez, D., Martinez-Rodriguez, O., & Martinez, A. (2016). Heterotrophic cultivation of microalgae: Production of metabolites of commercial interest. *Journal of Chemical Technology and Biotechnology, 92*(5), 925–936.

25. Nwoba, E., Mickan, B., & Moheimani, N. (2019). Chlorella sp. growth under batch and fed-batch conditions with effluent recycling when treating the effluent of food waste anaerobic digestate. *Journal of Applied Phycology, 31*(6), 3545–3556.
26. Torres Franco, A., da Encarnação, A. S., Passos, F., de Lemos, C. C., Mota Filho, C., & Cunha, F. C. (2018). Treatment of food waste digestate using microalgae-based systems with low-intensity light-emitting diodes. *Water Science and Technology., 78*(1), 225–234.
27. Cai, T., Park, S., & Li, Y. (2013). Nutrient recovery from wastewater streams by microalgae: Status and prospects. *Renewable and Sustainable Energy Reviews., 19*, 360–369.
28. Safi, C., Zebib, B., Merah, O., Pontalier, P., & Vaca-Garcia, C. (2014). Morphology, composition, production, processing and applications of *Chlorella vulgaris*: A review. *Renewable and Sustainable Energy Reviews, 35*, 265–278.
29. Mandal, S., & Mallick, N. (2009). Microalga *Scenedesmus obliquus* as a potential source for bio-diesel production. *Applied Microbiology and Biotechnology., 84*(2), 281–291.
30. Kim, S., Park, J., Cho, Y., & Hwang, S. (2013). Growth rate, organic carbon and nutrient removal rates of *Chlorella sorokiniana* in autotrophic, heterotrophic and mixotrophic conditions. *Bioresource Technology., 144*, 8–13.
31. Theriault, R. (1965). Heterotrophic growth and production of Xanthophylls by Chlorella pyrenoidosa. *Applied Microbiology., 13*(3), 402–416.
32. Samejima, H., & Myers, J. (1958). On the heterotrophic growth of Chlorella pyrenoidosa. *Journal of General Microbiology., 18*(1), 107–117.
33. Cheirsilp, B., & Torpee, S. (2012). Enhanced growth and lipid production of microalgae under mixotrophic culture condition: Effect of light intensity, glucose concentration and fed-batch cultivation. *Bioresource Technology., 110*, 510–516.
34. Kim, H., Park, W., Lee, B., Seon, G., Suh, W., & Moon, M., et al. (2019). Optimization of heterotrophic cultivation of Chlorella sp. HS2 using screening, statistical assessment, and validation. *Scientific Reports, 9*(1).
35. Zarrinmehr, M., Farhadian, O., Heyrati, F., Keramat, J., Koutra, E., Kornaros, M., et al. (2020). Effect of nitrogen concentration on the growth rate and biochemical composition of the microalga, Isochrysis galbana. *The Egyptian Journal of Aquatic Research., 46*(2), 153–158.
36. Barik, D. (2019). *Energy from toxic organic waste for heat and power generation*. Woodhead Publishing.
37. (2004). Handbook of water and wastewater microbiology. *Choice Reviews Online, 41*(07):41-4037–41-4037.
38. Recht, L., Zarka, A., & Boussiba, S. (2012). Patterns of carbohydrate and fatty acid changes under nitrogen starvation in the microalgae Haematococcus pluvialis and Nannochloropsis sp. *Applied Microbiology and Biotechnology., 94*(6), 1495–1503.
39. Hannon, M., Gimpel, J., Tran, M., Rasala, B., & Mayfield, S. (2010). Biofuels from algae: Challenges and potential. *Biofuels, 1*(5), 763–784.

Automatic Detection of Vaccination and Covid-19 Falsehoods in Social Media

Yi Lin Zhao and Guoxian Tan

Abstract As COVID-19 spread on an unprecedented global scale, vaccinations offered a way out, lowering not just transmission rates, but also fatality rates for vaccinated individuals. However, misinformation on both the virus and its vaccine also spread online, from social media posts, to public chat rooms. Thus, many rejected these vaccines for untrue reasons when they might not have otherwise, or used unauthorised drugs as vaccine "alternatives". Hence, it is crucial to allow readers to detect vaccine misinformation, so they may make more informed decisions amidst COVID-19. To investigate how machine learning can be used, this project uses data from Kaggle datasets alongside 452 total texts scraped from Twitter, telegram group SG COVID La Kopi, sites Truth Warriors and Channel News Asia, vectorizing text with word frequency. The effectiveness of detection across information categories and models is evaluated at the character and word levels. Results indicate decreasing accuracy as the input's category gets more specific, with General, COVID-19 and COVID-19 vaccine information categories, from broadest to most specific. Results showed an increased accuracy as character counts of inputs increased. In addition, our single-layer linear support vector classification model classifies inputs as true or false with the highest accuracy for both overall detection and vaccination-related falsehood detection rates, at 98.6% and 85.6%, respectively, when compared against long short-term memory with 22 outputs classified by a sigmoid dense layer and 96.7% and 72.22%, respectively, and Naive Bayes with true or false outputs and 91.8 and 71.1%.

Keywords Covid-19 · Vaccines · Social media · Falsehoods · Fake news · Machine learning

Y. L. Zhao (✉) · G. Tan
Raffles Institution, Singapore, Singapore
e-mail: 22YZHAO557D@student.ri.edu.sg

G. Tan
e-mail: guoxian.tan@ri.edu.sg

H. Guo et al. (eds.), *IRC-SET 2022*,
https://doi.org/10.1007/978-981-19-7222-5_41

1 Introduction

1.1 Background Introduction

As COVID-19 (COVID) spread on an unprecedented global scale, vaccines offered a way out, lowering both transmission rates and fatality rates significantly for vaccinated individuals [1]. However, misinformation on COVID and its vaccines also spread throughout the Internet, falsely deterring many from scientific measures. The proliferation of these falsehoods was not just rapid, but also widespread, often through the common route of social media platforms and chat groups.

As these falsehoods spread, by May 2020, 60% of Singaporeans had received fake news about COVID [2]. Verified and unverified information mixed on social media, as Twitter collected over 33 million COVID-related posts, many being from users who are not verified authorities [3]. Similarly, falsehoods proliferated on COVID messaging groups with large followings, such as SG Covid La Kopi with ~ 15,000 members [4]. Thus, social media's widespread nature enabled the dissemination of such fallacies, which many users unfortunately came into contact with.

Even so, "fake news" can be highly subjective, there being no current consensus on a formal definition; the popular meaning of "misleading" information is greatly dependent on its context, while "malicious intent" can be hard to prove [5]. Similarly, opinions signify relative truth [6], which may differ based on perspective. However, while scientific fact may change from discoveries, it can be deemed objectively true at the moment. Thus, we shall assume these factually untrue claims to be falsehoods.

These claims may mislead people to make decisions they would not have otherwise. After all, they are highly convincing; A COVID-19 study by the World Health Organization (WHO) in Singapore found that 77.1% of respondents trust social media as an information source on infectious disease outbreaks [7]. Hence, many may mistakenly believe social media misinformation on COVID and its vaccines, potentially misplacing certain concerns. With over 60% of Singaporean respondents between ages 21 and 39 expressing concern on possible side effects and safety [8], vaccine safety is likely already a concern for many. Falsehoods on side effects may then falsely exaggerate these existing concerns or generate new ones, falsely exacerbating vaccine hesitancy [9]. Worse, claims that unauthorised substances prevent COVID risk harming those who believe them, including a grandmother who was hospitalised after consuming Ivermectin [10]. Falsehoods may hence impair readers' decision-making, misleading them to make choices they might not have otherwise, including potentially putting themselves, and those around them, at greater risk from COVID.

Currently, many social media platforms and public forums do not actively identify falsehoods, at most labelling any information related to COVID as something users should self-verify. Similarly, detection systems for public messages are even more scarce. With the detrimental effects of falsehoods, the need to effectively detect vaccine and COVID misinformation on public platforms has only become even more pressing, an invaluable component in the fight against COVID and in aiding informed decisions. Thus, we aim to address this need.

1.2 Fundamental Theory for Natural Language Processing in Fake News Detection

To do so, all models first vectorize the texts by matching indexes to each unique word.

Linear Support Vector Classifier (Linear SVC). Linear SVC [11], a type of support vector machine (SVM), constructs the hyperplane line that separates the plotted inputs into their categories and the support vectors. It then creates a decision rule that decides which "hyperplane" divided category the input lies on. X_i is the i_{th} input from the dataset, Y_i the output label, $\vec{w}$ the margin width and b the translation. [12]

$$Y_i\left(\vec{w} \cdot \vec{X}_i + b\right) \geq 1 \tag{1}$$

Long Short-Term Memory (LSTM). LSTM [11] consists of cells in Appendix 1 Fig. 1, where C_t, the "cell state", is memory passed on from the current cell, H_t is the cell input and "hidden state", U is the input, W is the weight vectors and b is the translation. Each sigmoid layer σ produces values 1 or 0 for each word index, where 0 does not pass information through the gate, and 1 does.

First, "forget gate" f_t decides to either "forget" or "remember" previous inputs in C.

$$f_t = \sigma\left(W_f * \left[H_{t-1}, x_t\right] + b_f\right) \tag{2}$$

Next, the "input gate" $\hat{C}_t$ and I_t decide what new information to include in Eq. 2.

$$i_t = \sigma\left(W_i * \left[H_{t-1}, x_t\right] + b_i\right) \tag{3}$$

$$\hat{C}_t = \sigma\left(W_c * \left[H_{t-1}, x_t\right] + b_c\right) \tag{4}$$

These gates then update C_{t-1} with C_t using

$$C_t = f_t * C_{t-1} + i_t * \widehat{C_t} \tag{5}$$

Lastly, "output gate" O_t decides what parts of the cell state to include in the output.

$$O_t = \sigma\left(W_o * \left[H_{t-1}, x_t\right] + b_o\right) \tag{6}$$

$$H_t = O_t * \tanh(C_t) \tag{7}$$

Then, it uses a dense layer with a sigmoid activation function to constrain the 22 outputs between 0 and 1. Finally, binary cross entropy classifies a result between 0 for false and 1 for true. Given y_i is the outcome to be predicted, $P(y_i)$ is the probability of all points being true, and N the number of inputs [12].

$$H_p(q) = \frac{-1}{N}\sum_{i=1}^{N} y_i * \log(P(y_i)) + (1 - y_i) * \log(1 - P(y_i)) \tag{8}$$

Naive Bayes (NB). NB assumes the features are independent, calculating the probability of whether each feature is true or false, and using it to find the overall conditional probability, where $P(y|x)$ is the probability of label y given input x [13].

$$P(y|x) = \frac{P(x_1|y) * P(x_2|y) * P(x_3|y) * \cdots * P(x_n|y) * P(y)}{P(x_1) * P(x_2) * P(x_3) * \cdots * P(x_n)} \tag{9}$$

Thus, we aim to detect falsehoods with the models above, bringing additional analysis and implementations, covering gaps in existing works.

1.3 Literature Review

Existing projects first detect general fake news articles, Wang et al. [14] finding CNNs to have the highest accuracy followed by SVMs and BiLSTM, Saleh et al. [15] similarly finding SVMs to have one of the highest accuracies, outperforming NB for 2 out of 3 datasets. Similarly, projects detect COVID falsehoods, Mohamed et al. [16] using COVID facts from fact-checking sites, finding Linear SVM to outperform NB models. However, while work was done on general knowledge and COVID facts, few look into COVID vaccines specifically. Analysis between information categories is also lacking, such as comparing results and reasons for error across different, increasingly specific information categories of general knowledge, COVID and COVID vaccines.

Additionally, the above focus on articles and claims, while extensions into social media are mostly focused on posts such as tweets. For instance, Ni et al. [17] and Sansonetti et al. [18] identify general false tweets, [17] using multi-view attention networks while [18] focuses on the reliability of social media users. Others identify false COVID tweets, Abdelminaam et al. [19] finding SVMs to generally outperform NB and a modified LSTM to outperform both, while Quershi et al. [20] analyses tweet propagation through users and communities. Thus, there is also a gap in analysing forms of social media beyond posts and tweets, such as public messages. Additionally, most works focus on one type of information, such as either articles or social media posts, with missing analysis on results across such types of information.

Investigating reasons for errors in detecting their respective falsehoods, Xu et al. [21] analysing certain features of falsehoods on social media posting platforms Delving deeper, Barua et al. [22] found their article detection model to require inputs of decent size to provide context for it to function; Poddar et al. [23] a higher number of features in news causes a higher performance; Ghinadya et al. [24] found sarcasm to be a reason for misclassification, and Yanagi et al. [25] finding special characters common in article falsehoods. Aiming to fill the previously lacking analysis across both information categories and types, we thus also aim to identify potential reasons for errors between categories that may have been missed.

1.4 Engineering Goal

Hence, we aim to effectively detect vaccine misinformation to ensure readers may make more informed decisions to maximise their welfare amidst COVID-19. To compare the effect of social media information as compared to formal texts on various models, in order to determine potential reasons for better performance. The model currently has a working website in Appendix 8, with potential implementations including a telegram bot or chrome extension.

2 Methodology

2.1 Datasets

Table 1 shows the number of documents contained in the different datasets used, their information categories and types, the maximum, minimum and average character count, and whether it is factually true or false.

General. "Fake and Real News Dataset" from Kaggle [26] provides general fake and real news. Segments of the real news had repetitive starting phrases of a location followed by "(Reuters) -", which was removed as it affected the model's overall accuracy as the model detected the starting phrase instead.

COVID-19. Kaggle Dataset "Covid-19 Fake News" [27] has COVID News, Tweets and Tweeted claims as Twitter Ids, which were converted to tweets with Tweepy and the Twitter API.

COVID-19 vaccines. The following sites were scraped using keywords "vaccine", "anti-vaxxers", "mRNA", "pfizer", "moderna", "sinovac", "sinopharm", "novavax", "booster", "vaxxed", "vaccination". The results were then researched and labelled.

Table 1 Dataset overview

Category	Dataset	Real/fake	Type	Character count			Num. of documents
				Average	Min	Max	
General	Fake and real news dataset [26]	Real	News	~ 2383	152	29,781	21,416
		Fake	News	~ 2618	5	51,794	22,851
COVID	Covid-19 fake news [27]	Real	News	~ 441	29	498	517
		Fake	News	~ 238	31	500	397
		Real	Tweets	~ 123	17	148	5606
		Fake	Tweets	~ 121	40	149	15,635
COVID vaccines	Channel News Asia (CNA) [28]	Real	News	~ 1633	228	5749	56
	Truth warriors [29]	Fake	News	~ 3700	58	32,215	78
	Scraped Twitter	Real & fake	Tweets	~ 216	20	412	207
	Sg Covid La Kopi	Real & fake	Messages	~ 291	30	3347	111

First, the Newspaper API was used to scrape Channel News Asia (CNA) [28] and Truth Warriors (TW) [29]. While CNA is a credible organisation for articles, including COVID and vaccine news, TW's articles on COVID and vaccines are often false; Their articles have been issued multiple correction directions from The Protection from Online Falsehoods and Manipulation Act (POFMA) Office due to falsehoods, Ministry of Health (MOH) calling it "misleading" [30].

Next, the Telethon API was used to scrape public Telegram Group Chat, SG Covid La Kopi. With ~ 15,000 members, misinformation in these chats often comes in messages or articles from "credible" personnel, many of which have been debunked [31]. Messages were scraped from dates 10 October and 10 July, the days after vaccination differentiated measures were announced, and on 9 and 16 December.

Finally, Tweepy and the Twitter API were used to scrape COVID vaccine tweets from Twitter for data on social media posts. Character counts of inputs include user tags in replies, which are not constrained by twitter's character limit.

3 Results and Discussion

3.1 Overall Feasibility

All three models have high overall accuracies (see Appendix 2, Table 2); The Linear SVC model has the highest accuracy of 0.9856, followed by the LSTM model with

0.9670 and the NB model with 0.9178. This order of performance aligns with dataset 1 of paper [15], but contradicts dataset 2 where NB could outperform LSTMs and SVMs. It corroborates with both datasets and paper [14] that SVMs outperform LSTMs, and [23] where SVM outperforms NB.

All models have generally low false positive rates (FPR) and false negative rates (FNR) (see Appendix 2, Figs. 2–4). For LSTM, plotted graphs of FPR and FNR intersected at the low equal error rate (EER) 3.31%. Both graphs have low areas under them, showing overall low error rates with relatively even FPR and FNR, matching the confusion matrix (see Appendix 2, Fig. 5). Precision and recall are high for all models, a large area under the curve (AUC) showing they are robust (see Appendix 2, Figs. 6–8).

3.2 *Decreasing Accuracy as Information Gets More Specific*

All models demonstrate a decreasing accuracy as information gets more specific, from general news to COVID information to vaccine information (see Appendix 3 Table 3); As information gets more specific, all models' AUC for PR curves decrease, showing slightly less robustness (see Appendix 5 Figs. 24–32). Similarly, while all models exhibit a generally low FPR and FNR for general news and COVID information, COVID and Vaccine information exhibit higher FNR than FPR, vaccine information having the highest FNR. (see Appendix 4). This higher FNR is further supported by plotting the FPR and FNR for LSTM, where the AUC increases as information gets more specific, increasing more for FNR. The equal error rate for the plotted graphs also increases from 2.72% for general news and 5.85% for COVID tweets and news to 29.15% for COVID vaccine-specific tweets and articles, aligning with the comparatively lower accuracy (see Appendix 3 Figs. 12–14), as per reasons discussed below, with examples in Appendix 7.

Decreasing accuracy as Character Count decreases. A graph of character count against accuracy was plotted by evaluating the accuracy of inputs within character counts by intervals of 200 characters, from 0 characters to maximum length for each category and overall. A smaller interval of 30 characters was used for COVID information to observe the trend due to its lower maximum character count. However, most of the character counts are clustered around the averages of their datasets, with few near the maximum. Hence, as numbers increase away from the mean, the number of datasets in the intervals can fall to single digits, even 1 or 0, causing fluctuating accuracies as each dataset's result has a bigger impact. This is visible in all accuracy against character count graphs, for all types of information and overall (see Appendix 2, Figs. 9 and 10, Appendix 6).

Before these fluctuations, all models maintained a generally high accuracy, in the same order of Linear SVC, LSTM, NB (see Appendix 2, Figs. 9–11). Prior to fluctuations, Linear SVC and LSTM had increasing accuracy across all information types as character count increases to provide context to offset reasons for errors.

Beyond this, it could also be as Linear SVC works better when more features are present, thus a higher character count provides more features for it to analyse.

Accuracy for NB increases slightly as character count increases before starting to decrease and fluctuates at a greater magnitude. This decrease is due to the general news category, while COVID and COVID vaccine categories maintain an increasing trend similar to Linear SVC and LSTM.

Ultimately, all three models generally have an increase in accuracy as the character count increases for most categories of information. It is also worth noting that a longer character count might provide greater context to mitigate other errors, such as sarcasm and special characters mentioned below.

Despite this, the models are still generally able to achieve accuracy for inputs of short lengths, especially for Linear SVC. The model's ability to still detect falsehoods in shorter inputs contradicts that of the model proposed in [22], which needs articles of "decent size" and may not function properly on sites, e.g. Twitter.

Sarcasm. What the text states may not be what it means, while the model may take it literally. This largely applies to social media posts and messages as they tend to be more informal, corroborating with [24].

Special Characters. Unlike fake article detection, the model cannot rely as much on common informal characters in false articles such as "@", "!" [25] for social media and message detection as social media itself takes an informal tone.

While all three of these aspects: shorter lengths, sarcasm and special characters can affect all types of information, they are most likely to affect social media information and messages, which have lower average character counts. Similarly, they tend to be more informal compared to articles, with even factually true information being more likely to contain characters and sarcasm. Thus, social media and message information may be more prone to being falsely marked negative and having a higher FNR, and since COVID and COVID vaccine information have more social media information, this could explain these information categories having higher FNRs (see Appendix 3) and comparatively lower accuracies. Consequently, this could explain the lower and decreasing accuracy for COVID and COVID vaccine information, as information gets more specific.

3.3 Across Models

For all models, Linear SVC continues to maintain the highest accuracy across all categories and overall, followed by LSTM and NB, corroborating with [15]. SVM outperforming NB for COVID information corroborates with [18] and [19], but contradicts LSTMs outperforming both models in [19].

Interestingly, [23] suggests that SVM performs better due to a higher number of features in text classification, which could be the case in our data due to its high number of features. The paper also states that NB tends to outperform other models when the sample size is small, which could be a reason for underperformance in this case as the sample size is larger. This paper suggests Bi-LSTMs could have performed worse due to overfitting, which could also be the case for LSTM.

4 Implications

Hence, this project could provide readers with a gauge on whether the information they have received is true, so they can make more informed decisions. It can provide an add-on to public messaging channels as an automated fact-checker, or complement existing measures against COVID "misinformation", such as Instagram's reminder to check the validity of COVID information. Detection of these falsehoods may also allow more efficient evaluation of when POFMA is necessary, prompting careful consideration before believing information. Thus, the model can come in many forms to be as accessible as possible. Currently, a website exists for detection on text inputs (see Appendix 8, Figs. 42 and 43).

5 Limitations

For the models, more strategies such as combining content with user credibility could be used. Further investigations can be done by varying the amount of training data.

6 Future Works

Beyond the website, other potential implementations include a telegram bot for easy forwarding of messages for detection, or a chrome extension that can detect falsehoods for any highlighted text. The implementation could also suggest factually true information based on detected keywords, to clarify false facts and provide convenience in making informed decisions. The model could also be tested to detect fake news in videos or audios, from visual text or spoken text.

7 Conclusion

With falsehoods falsely increasing vaccine hesitancy rates when they may not have otherwise [9], preventing misinformation becomes even more crucial in the fight against COVID. After all, vaccines are vital in curbing the spread of COVID and preventing loss of life. Being able to determine misinformation may prevent these concerns from mistakenly arising, helping people to make informed choices.

Knowing the reasons for model inaccuracies, such as accuracy decreasing as the information gets more specific, gives the project a direction for improvement. Overall, this project presents the ability to use Linear SVC, LSTM and NB models to help distinguish and limit the spread of falsehoods, and hence, the realistic prospect of helping users make educated decisions and stay safe amidst COVID.

Appendices

Appendix 1—LSTM Diagram

See Fig. 1.

Appendix 2—Overall Results

See Figs. 2, 3, 4, 5, 6, 7, 8, 9, 10, 11 and Table 2.

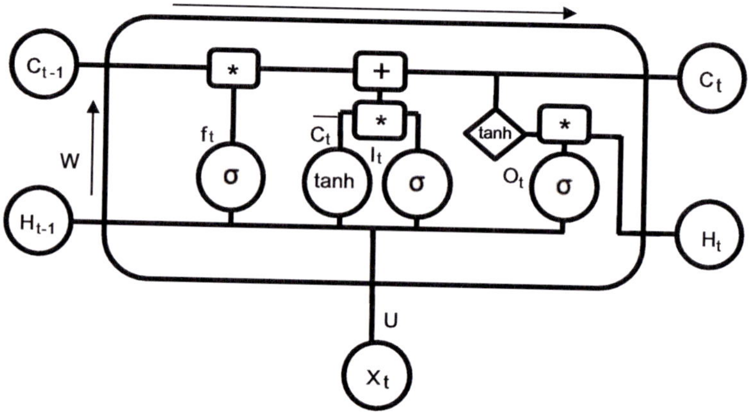

Fig. 1 LSTM diagram

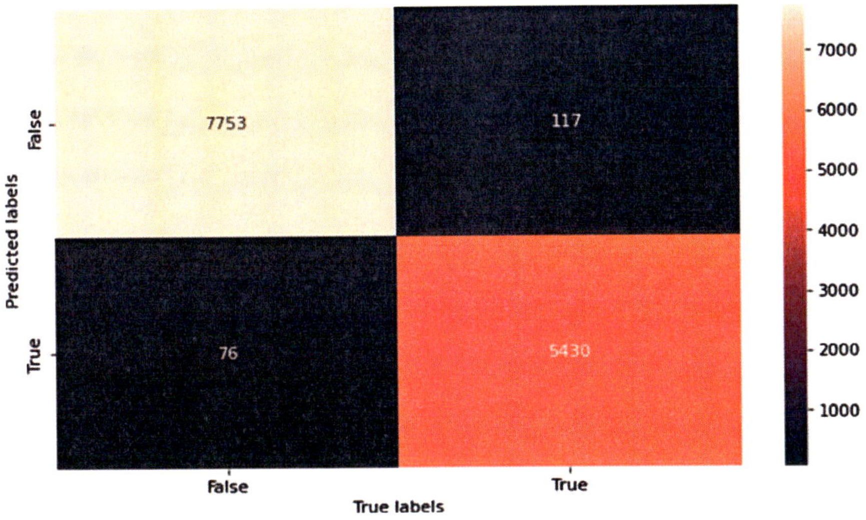

Fig. 2 Confusion matrix, overall Linear SVC performance

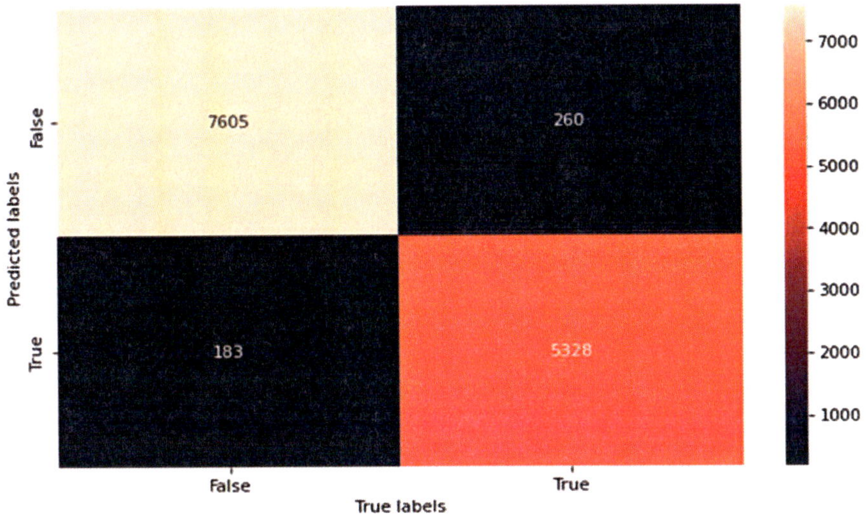

Fig. 3 Confusion matrix, overall LSTM performance

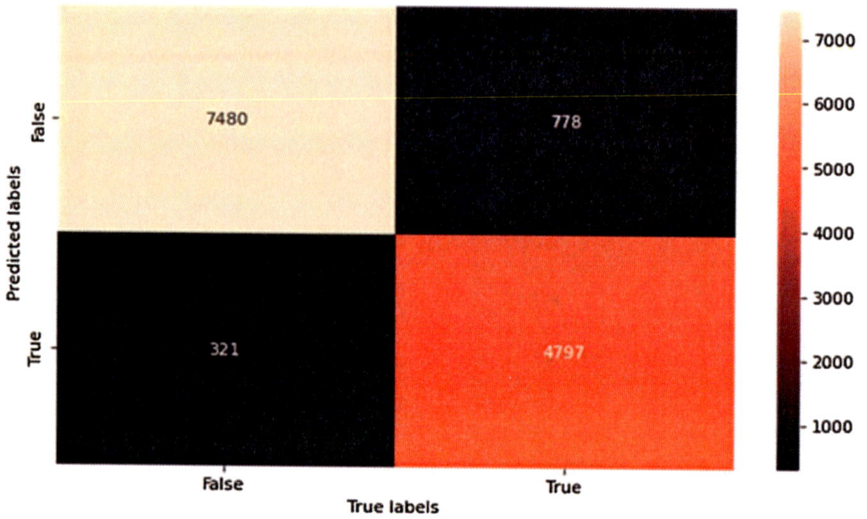

Fig. 4 Confusion matrix, overall NB performance

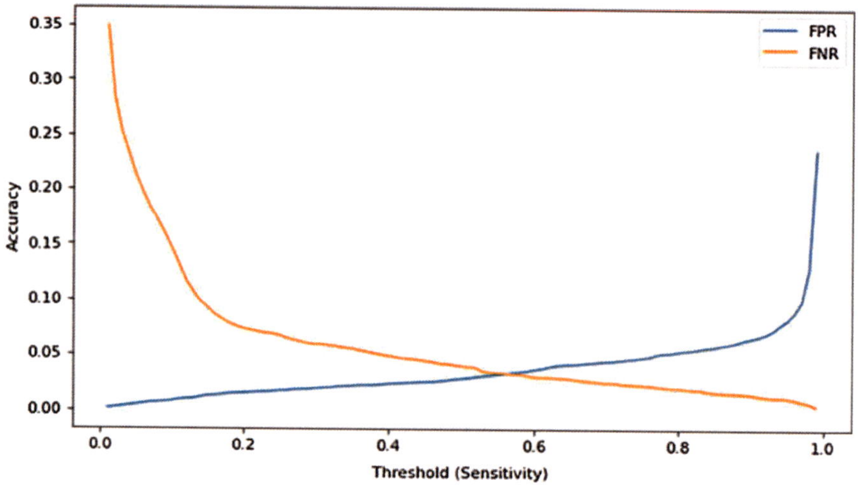

Fig. 5 EER curve, overall LSTM performance

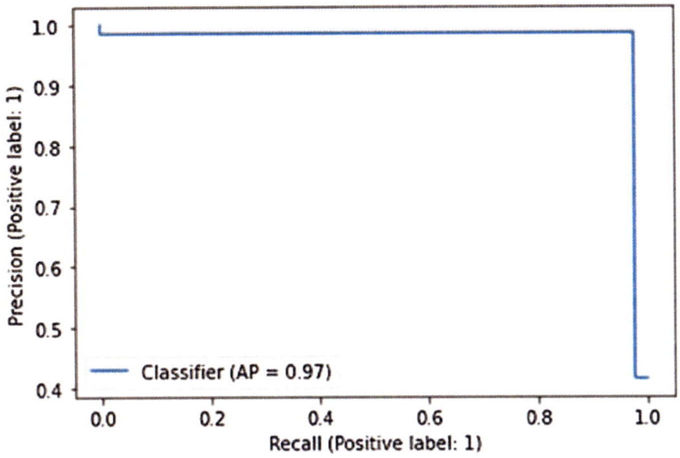

Fig. 6 PR curve, overall linear SVC performance

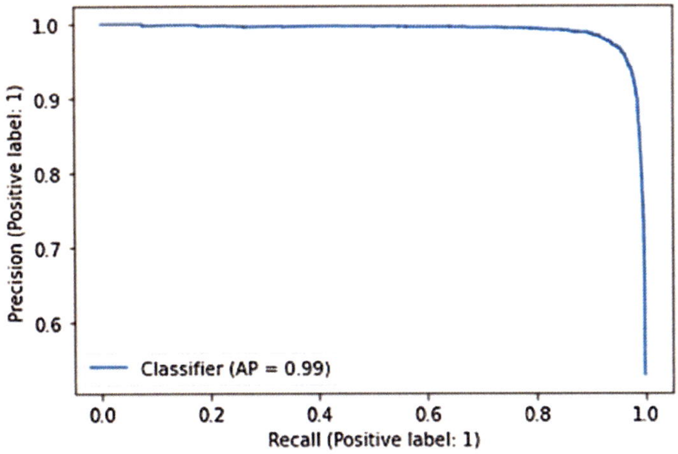

Fig. 7 PR curve, overall LSTM performance

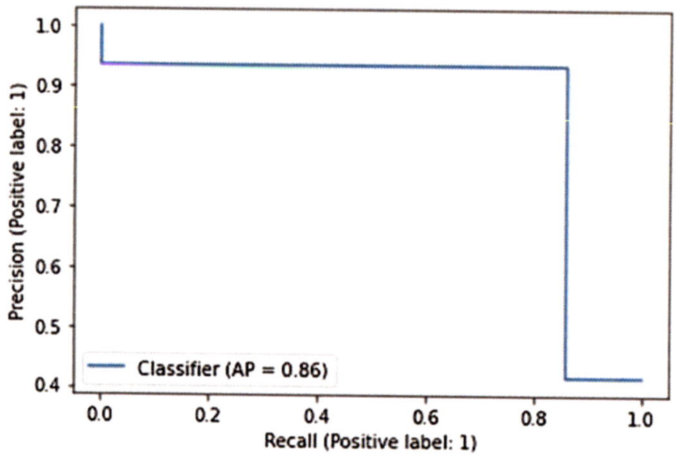

Fig. 8 PR curve, overall NB performance

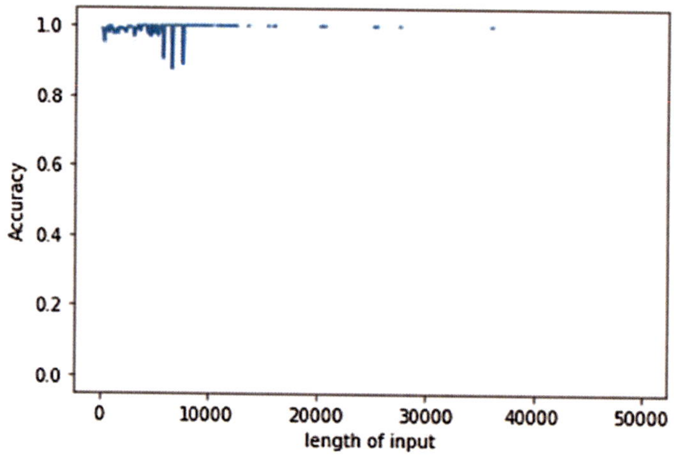

Fig. 9 Accuracy against character count, overall linear SVC performance

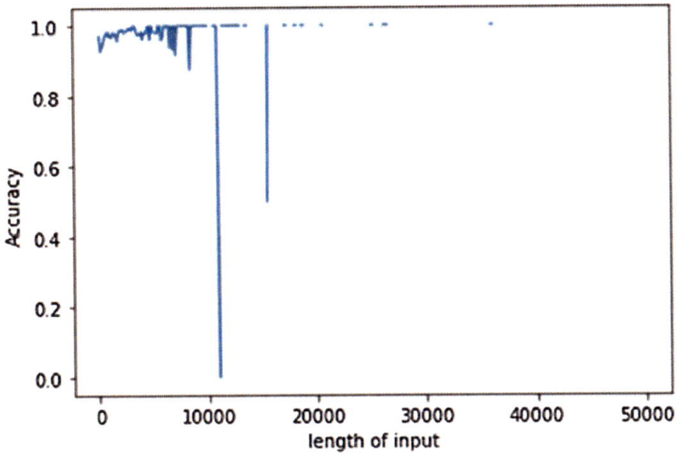

Fig. 10 Accuracy against character count, overall LSTM performance

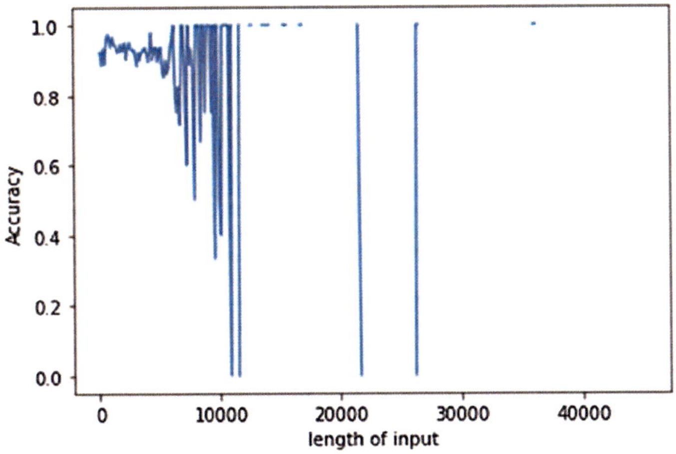

Fig. 11 Accuracy against character count, overall NB performance

Table 2 Overall accuracy across models

Model	General news accuracy
Linear SVC	0.9855711722
LSTM	0.9670000000
NB	0.9178379187

Appendix 3—Accuracy Across Categories

See Figs. 12, 13, 14 and Table 3.

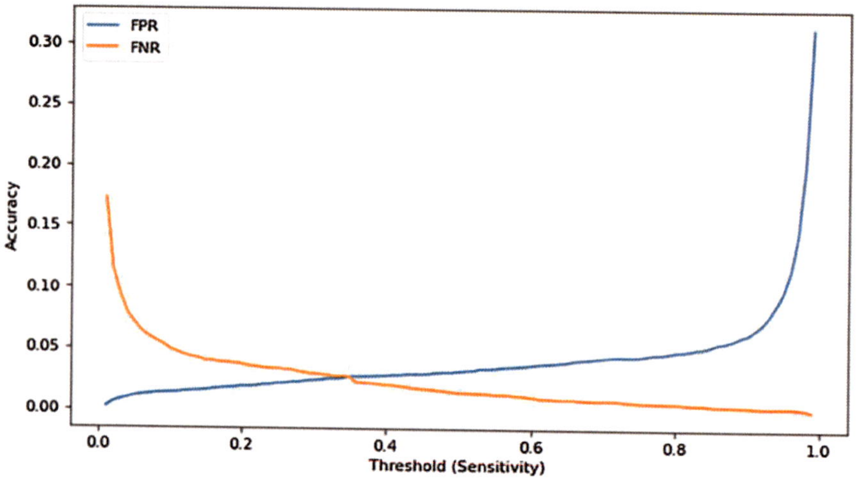

Fig. 12 EER curve, general news LSTM performance

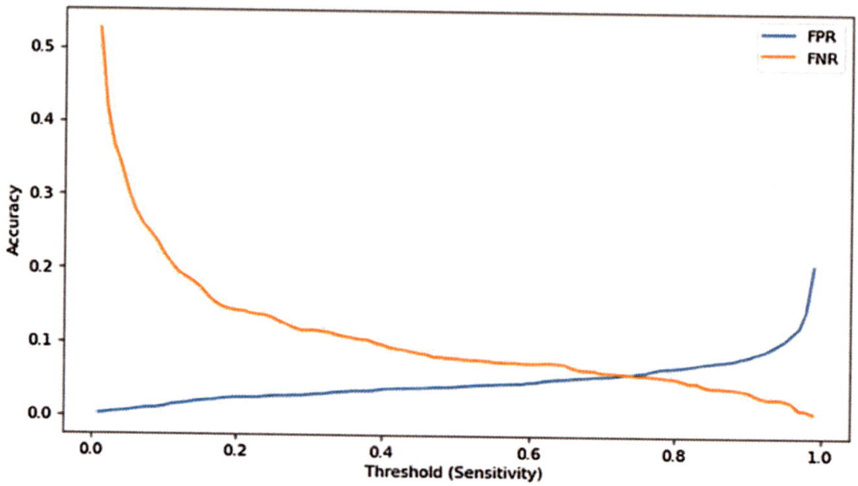

Fig. 13 EER curve, COVID information LSTM performance

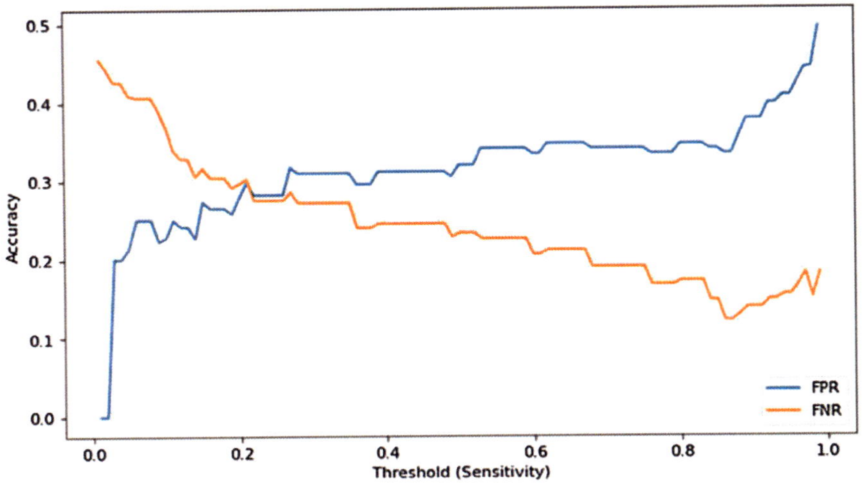

Fig. 14 EER curve, COVID vaccination information LSTM performance

Table 3 Accuracy across models and categories

Model	General news accuracy	COVID information accuracy	COVID vaccine information accuracy
Linear SVC	0.9885927265	0.9821750903	0.8555555556
LSTM	0.9787	0.9488	0.7222
NB	0.9333634515	0.8910198556	0.7111111111

Appendix 4—Confusion Matrices Across Categories

See Figs. 15, 16, 17, 18, 19, 20, 21, 22 and 23.

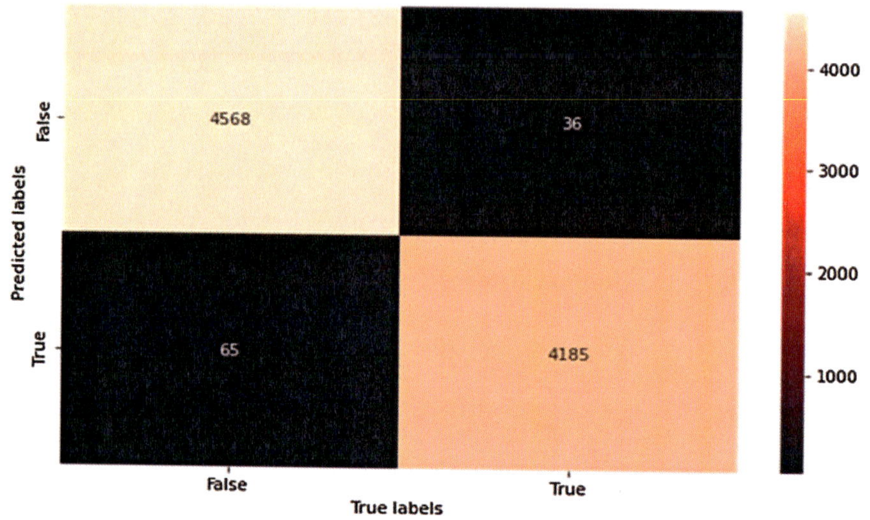

Fig. 15 Confusion matrix, general news linear SVC performance

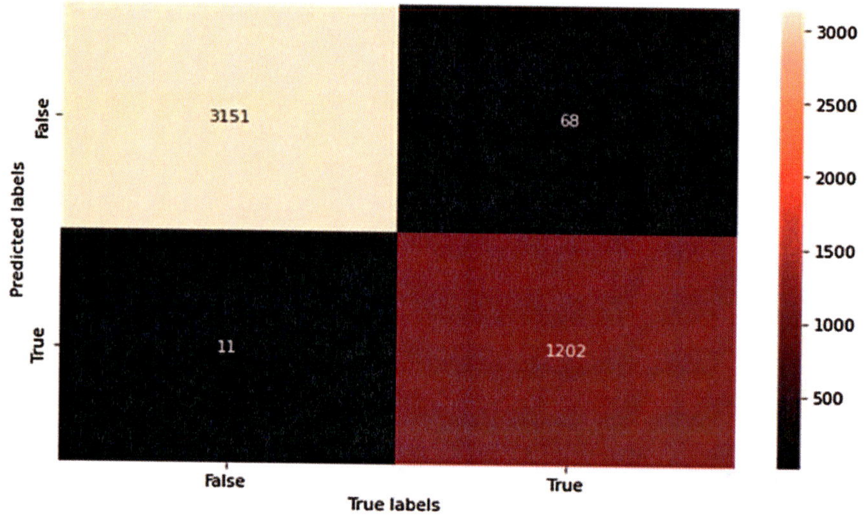

Fig. 16 Confusion matrix, COVID information linear SVC performance

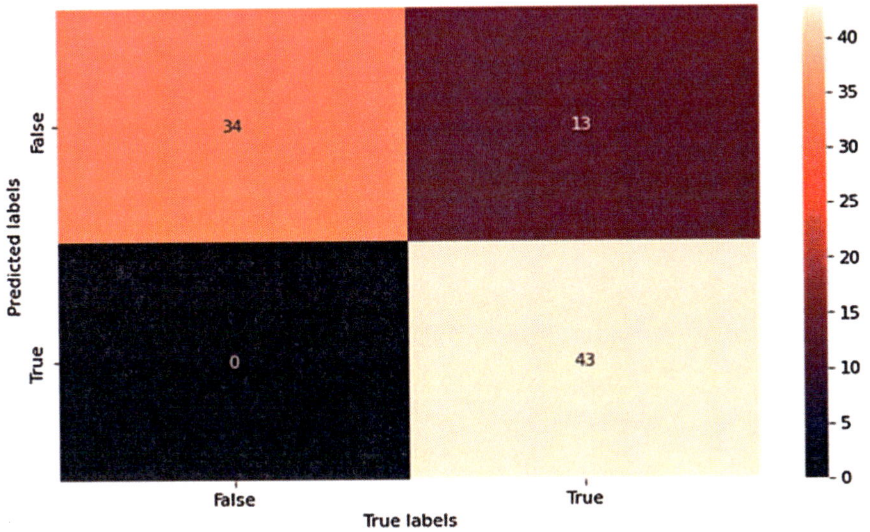

Fig. 17 Confusion matrix, COVID vaccine information linear SVC performance

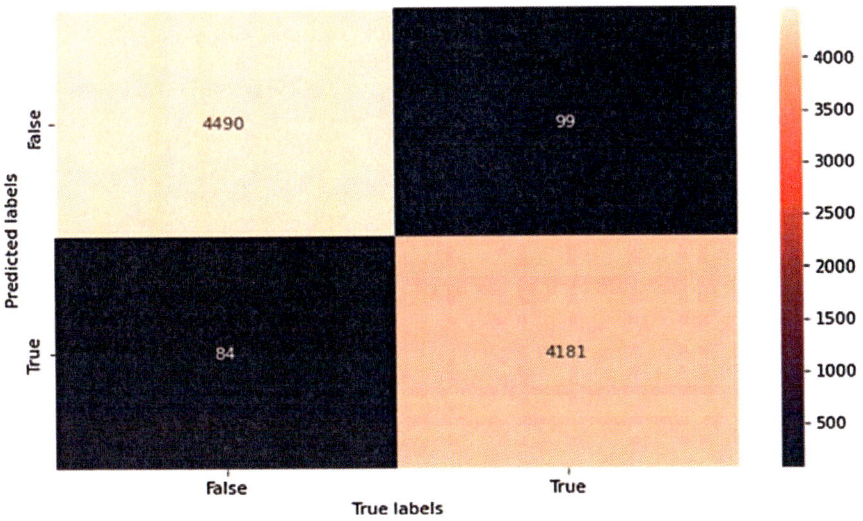

Fig. 18 Confusion matrix, general news LSTM performance

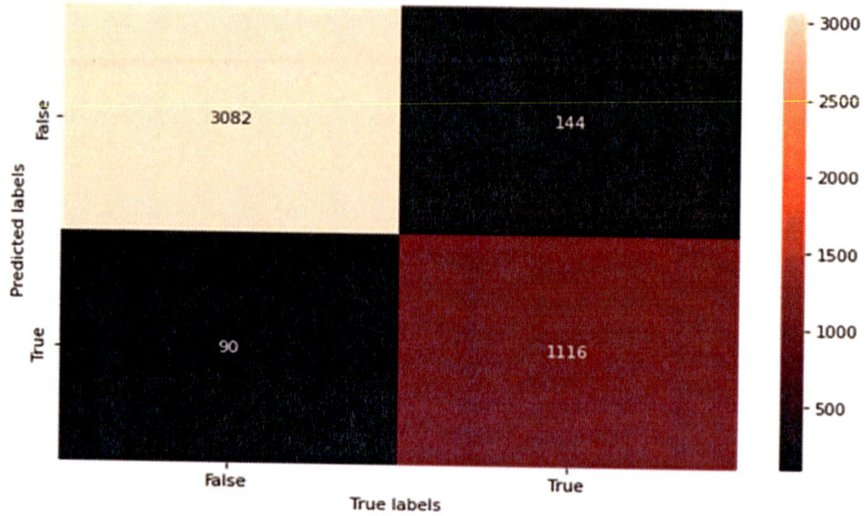

Fig. 19 Confusion matrix, COVID information LSTM performance

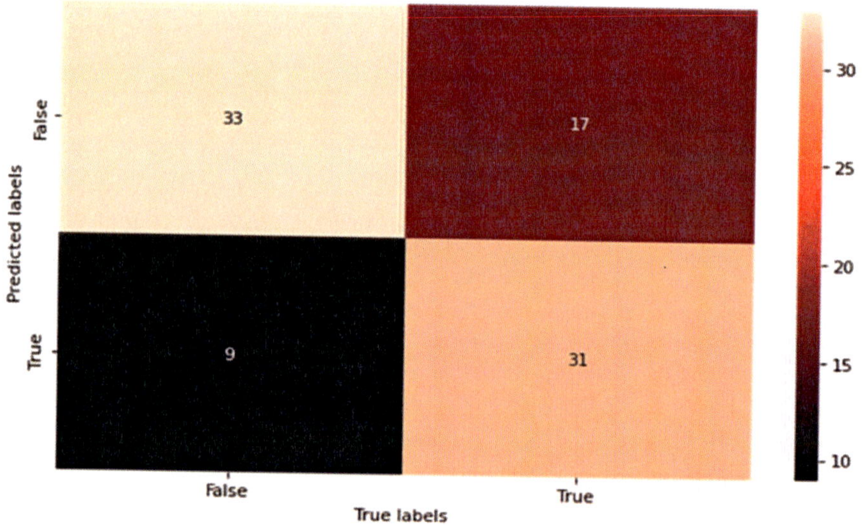

Fig. 20 Confusion matrix, COVID vaccine information LSTM performance

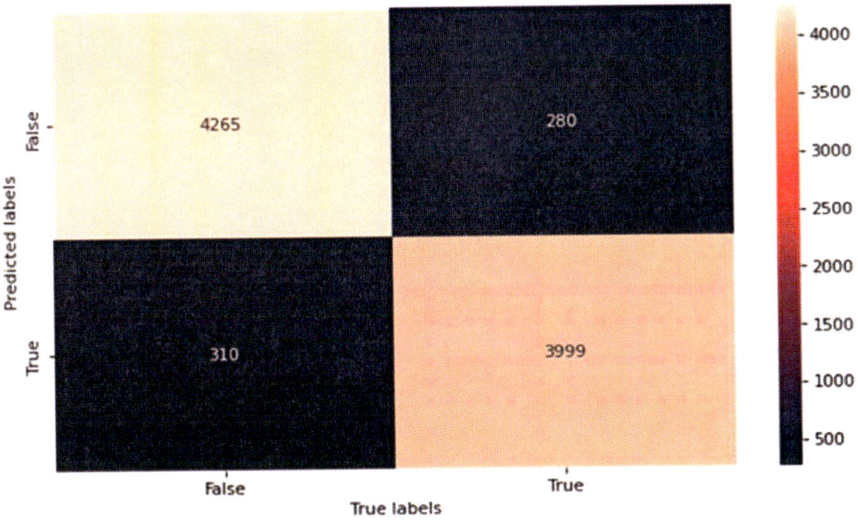

Fig. 21 Confusion matrix, general news NB performance

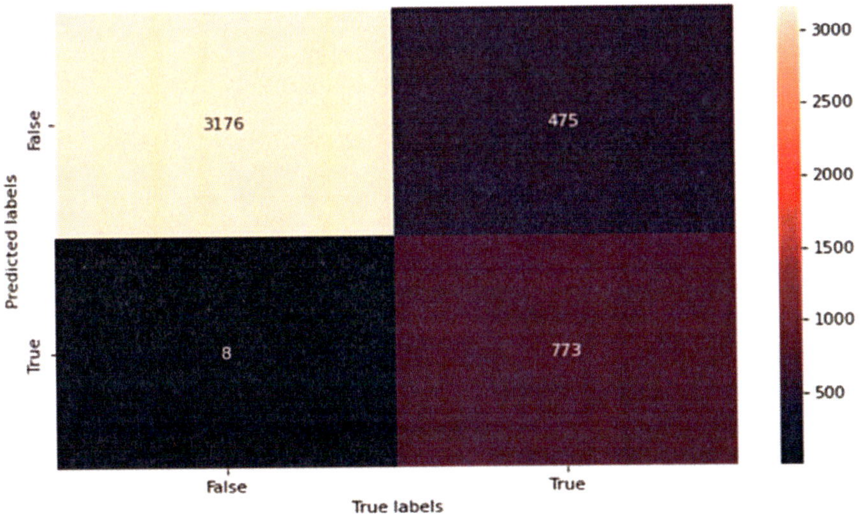

Fig. 22 Confusion matrix, COVID information NB performance

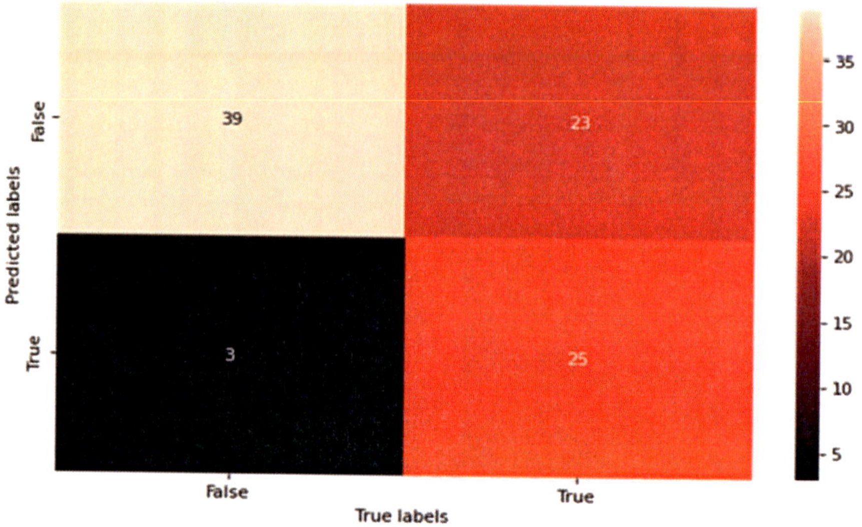

Fig. 23 Confusion matrix, COVID vaccine information NB performance

Appendix 5—PR Curve Across Categories

See Figs. 24, 25, 26, 27, 28, 29, 30, 31 and 32

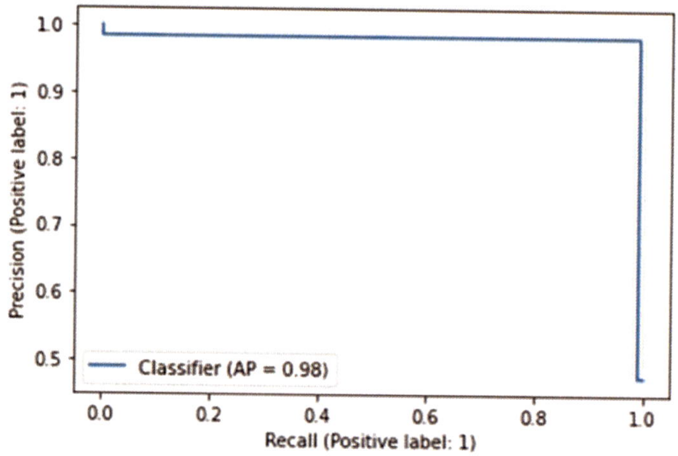

Fig. 24 PR curve, general news linear SVC performance

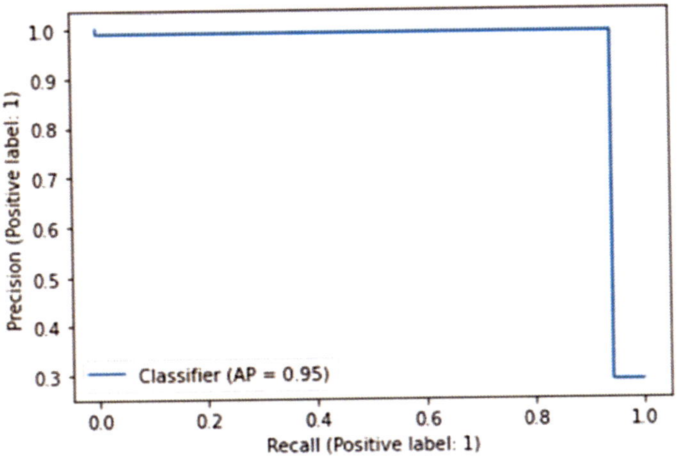

Fig. 25 PR curve, COVID information linear SVC performance

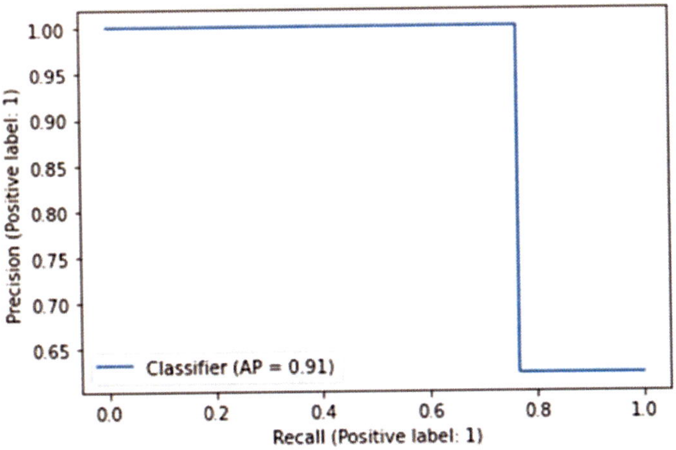

Fig. 26 PR curve, COVID vaccine information linear SVC performance

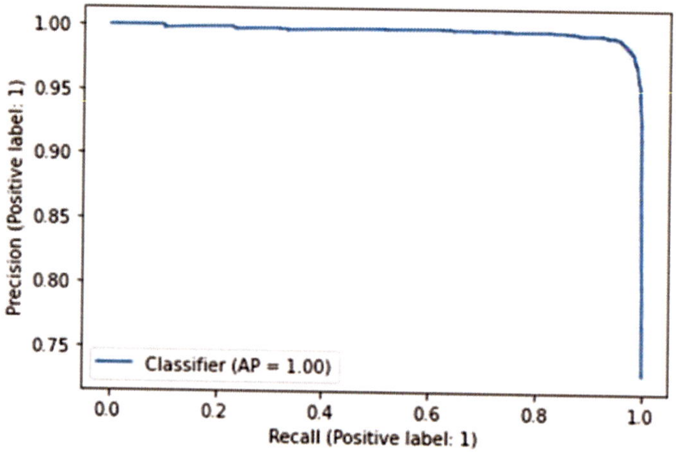

Fig. 27 PR curve, general news LSTM performance

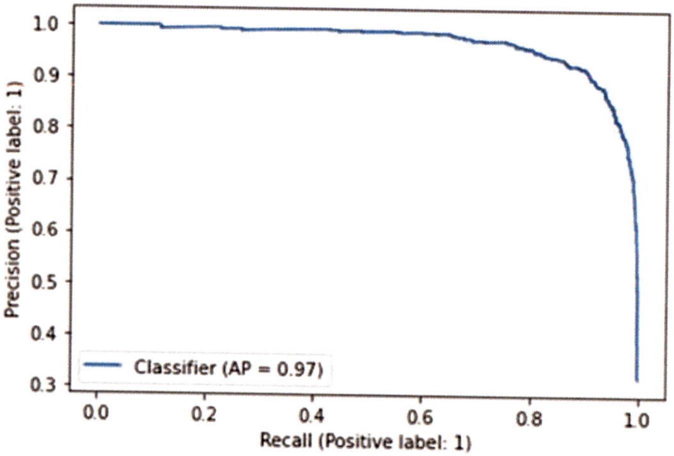

Fig. 28 PR curve, COVID information LSTM performance

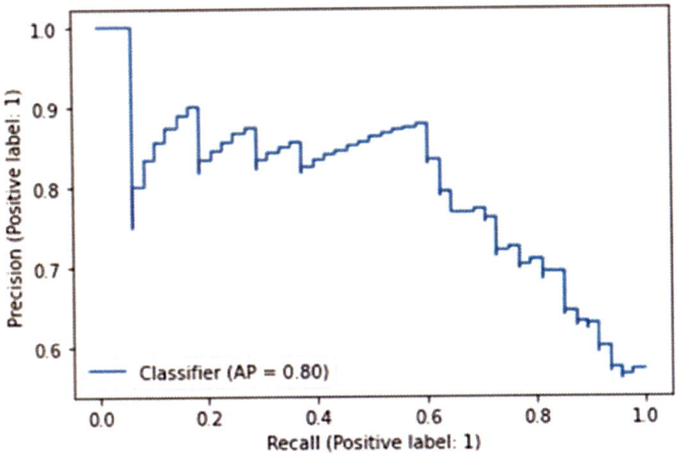

Fig. 29 PR curve, COVID vaccine information LSTM performance

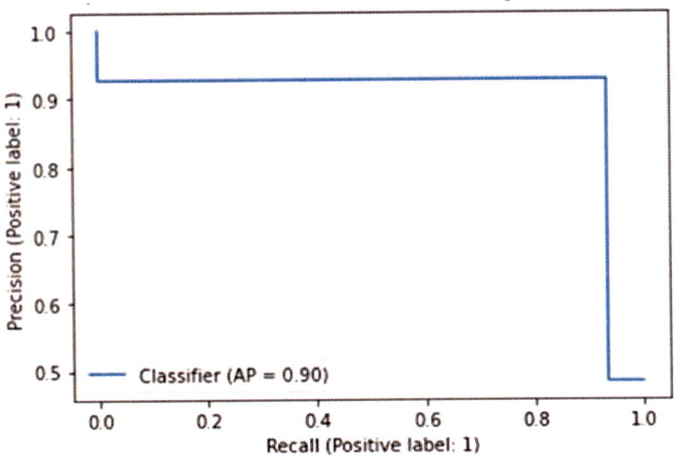

Fig. 30 PR curve, general news NB performance

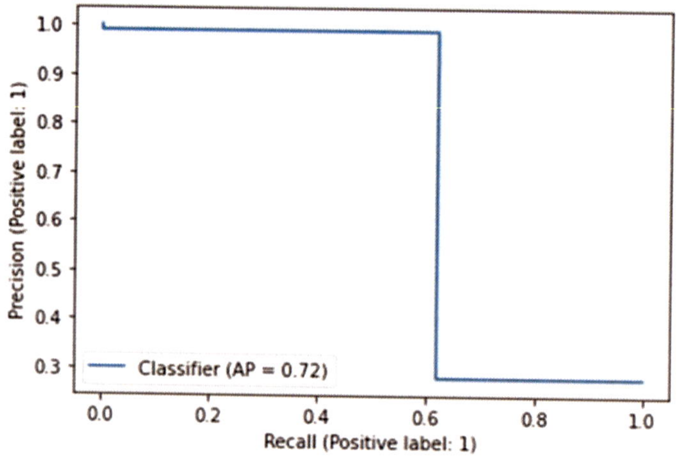

Fig. 31 PR curve, COVID information NB performance

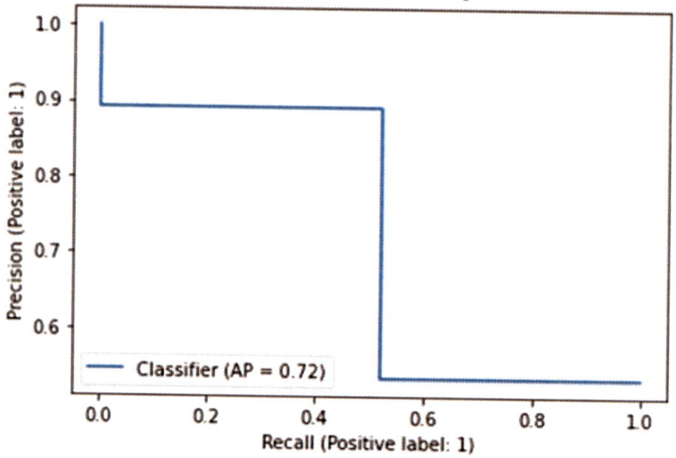

Fig. 32 PR curve, COVID vaccine information NB performance

Appendix 6—Accuracy Against Character Count Across Categories

See Figs. 33, 34, 35, 36, 37, 38, 39, 40 and 41.

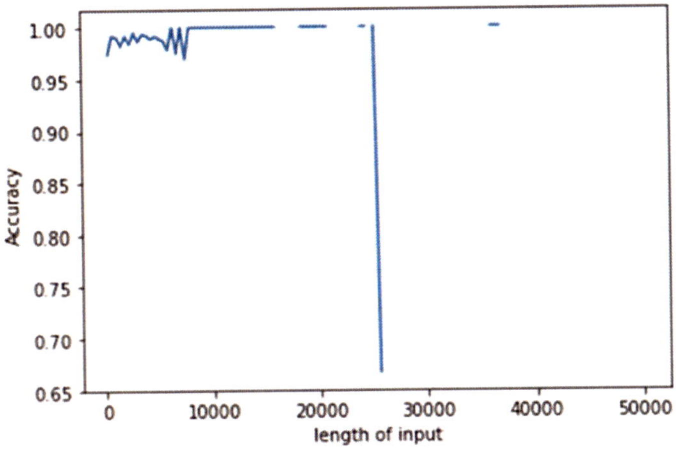

Fig. 33 Accuracy against character count, general news linear SVC performance

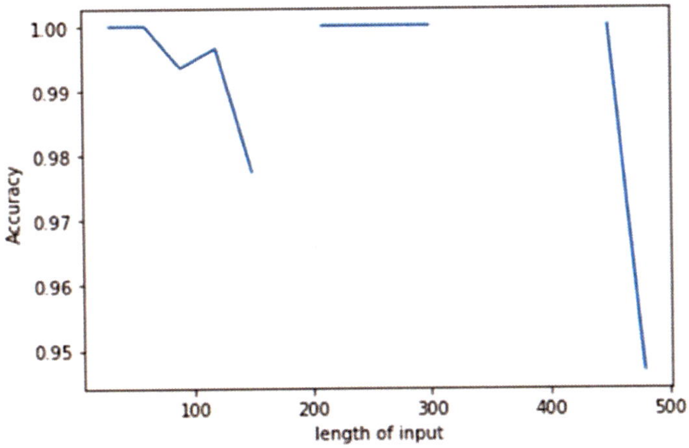

Fig. 34 Accuracy against character count, COVID information linear SVC performance

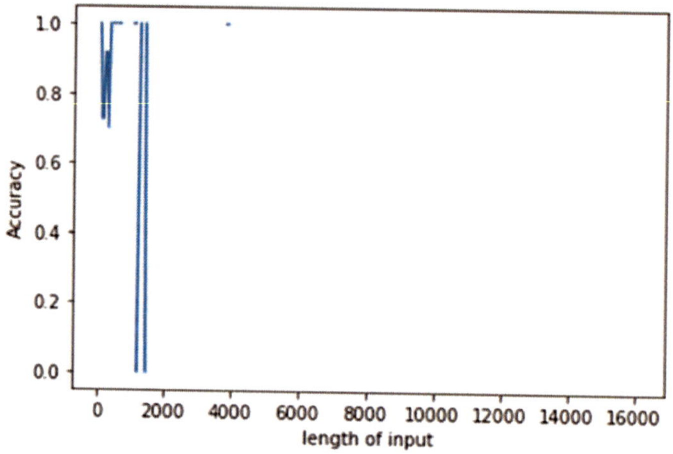

Fig. 35 Accuracy against character count, COVID vaccine information linear SVC performance

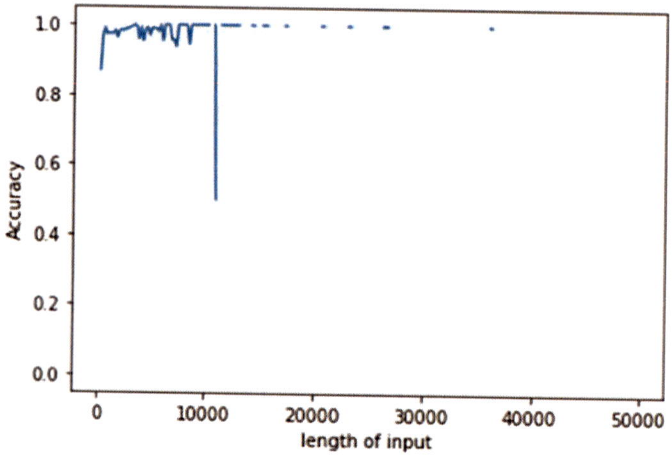

Fig. 36 Accuracy against character count, general news LSTM performance

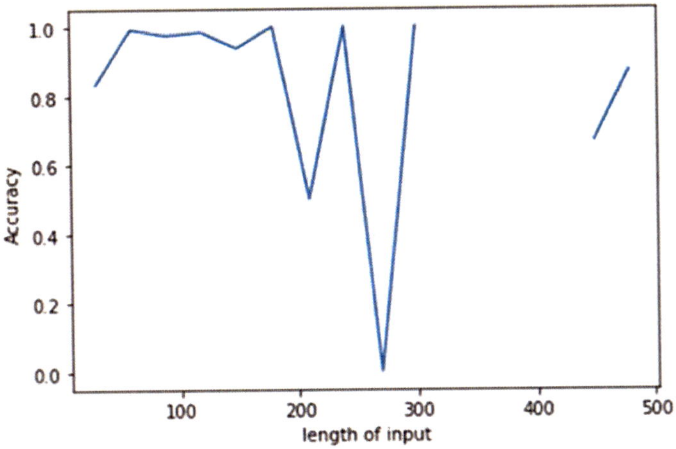

Fig. 37 Accuracy against character count, COVID information LSTM performance

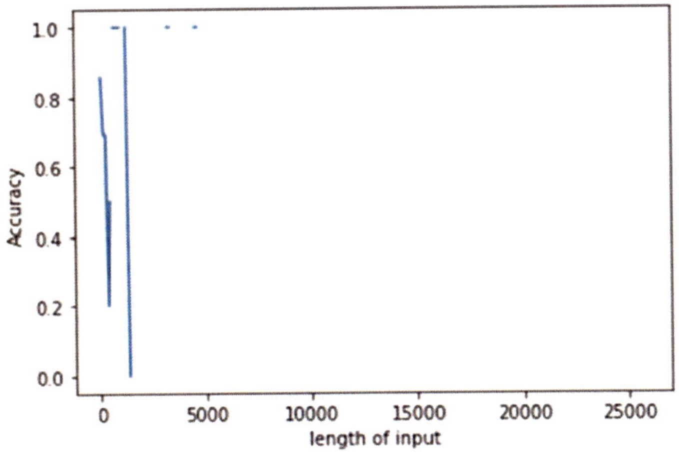

Fig. 38 Accuracy against character count, COVID vaccine information LSTM performance

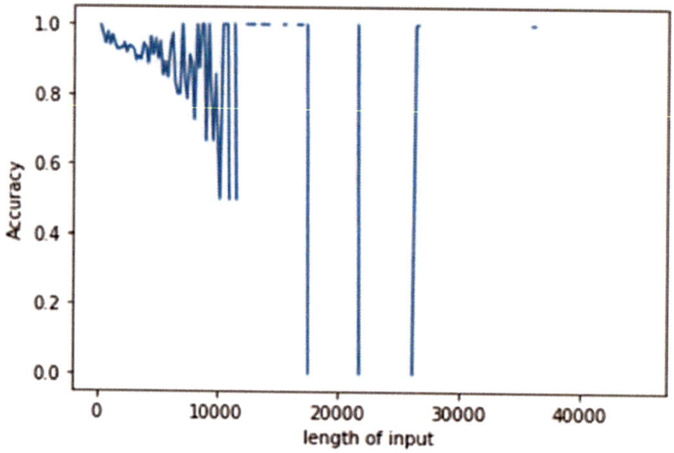

Fig. 39 Accuracy against character count, general news NB performance

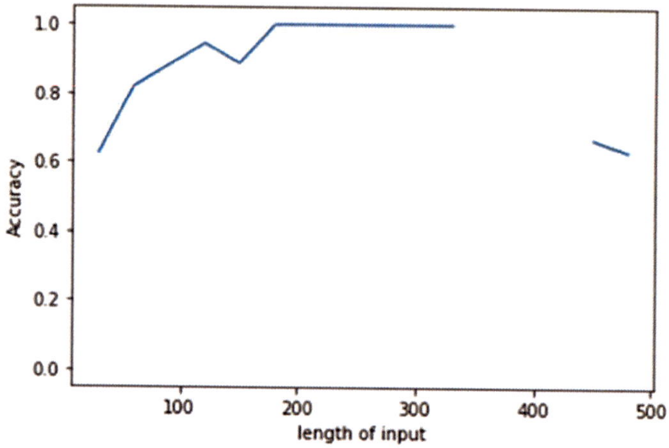

Fig. 40 Accuracy against character count, COVID information NB performance

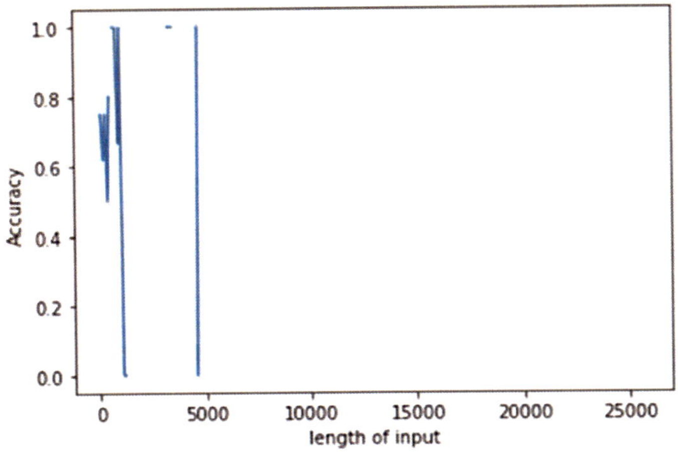

Fig. 41 Accuracy against character count, COVID vaccine information NB performance

Appendix 7—Misclassification Examples

Italics text is classified as true, while Bolded text is classified as false.

Misclassification due to lack of context.

Source: Channel News Asia (segment).

Label: True.

Contents:

SINGAPORE: My colleagues and I are seeing an increase in cases of intractable family conflicts over vaccines. While many of these clients also report frequent and intense conflicts compared to other families, COVID-19 has opened up some of these fissures even more.

Take Sarah for instance. She is the main caregiver of her elderly mother. Besides the mental strain of caregiving, Sarah experiences anxieties over her unvaccinated siblings visiting their mother.*

She tried explaining to her siblings that they might endanger the health of their elderly mother who has pre-existing allergies, but they are adamant the vaccines are unsafe.

Pleas for them to get vaccinated often turned into arguments, causing a rift in their relationship.

Then there's Mike* who has been talking to his parents for months trying to persuade them to get their jabs. But these conversations often become heated, creating tension.

His parents are religious and believe that vaccines are harmful. As a result of their arguments, Mike's relationship with his parents has become very strained.

Tension and conflict are unavoidable realities in every family setting. Family members can hold strong but opposing views on many issues: Money, politics, chores, romantic partners, tastes and preferences.

Misclassification due to "sarcasm" or "tone".

Source: Twitter

Label: False (implications of the vaccines not being "safe and effective" and leading to "the healthiest people among us" dropping "dead").

Contents:

The New Normal means that the healthiest people among us drop dead... the "safe and effective" vaccines couldnt possibly be the cause says the **#COVIDIDIOTS**.

#vaccine #vaccination

#COVID19 #Covid

#VaccinePassport

#VaccineMandate

#NoVaccinePassports

Source: Twitter

Label: True (implications of vaccine not causing glowing).

Contents:

@Diana011962 @chipfranklin **We all glow now...well, that was the promise.** *The anti-vaxxers* **lied. Only the rest of the animal kingdom can utilise bioluminescence. I'm very disappointed.**

Misclassification due to characters such as "@", "!" and emojis.

Source: Twitter

Label: True

Contents:

BOOK REVIEW: *Please Explain Vaccines to Me: Because* **I HATE SHOTS!** *by Dr. Laurie Zelinger, from the ever popular* **Please Explain series.**

Enjoy the colorful illustrations and happy story ending.

https://t.co/RhHwRJ51vr @vvolkman.

#childrensbook #vaccine.

Appendix 8—Website with Prototype UI

See Figs. 42 and 43.

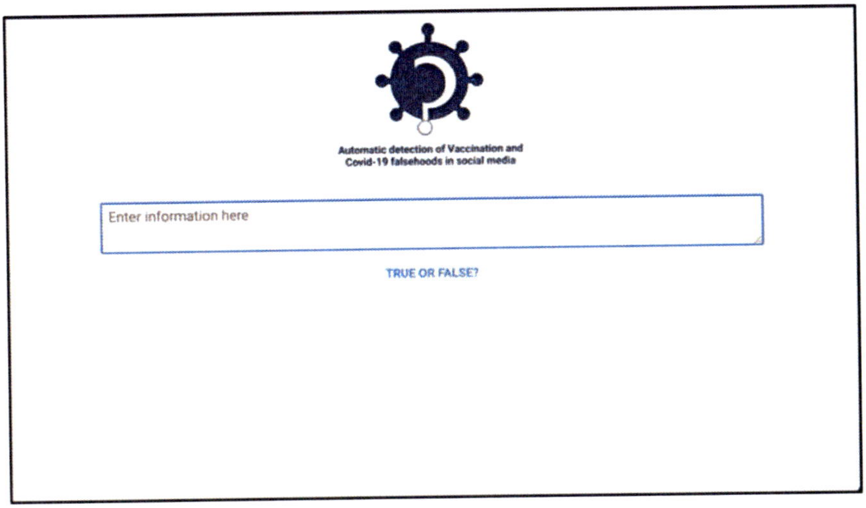

Fig. 42 Website before input

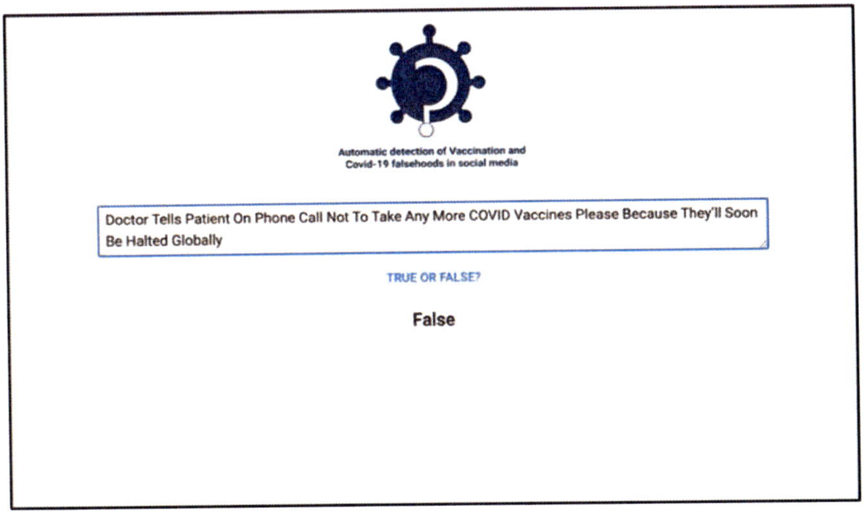

Fig. 43 Website after input

References

1. Straits Times. (2021). *Vaccines reduce Covid-19 transmission by 40%, other precautions should still be taken: WHO*. Retrieved from https://www.straitstimes.com/world/vaccines-reduce-covid-19-transmission-by-40-per-cent-who
2. National Center for Infectious Diseases, National University of Singapore, Nanyang Technological University. (2020). *NCID, NUS and NTU studies highlight the role of socio-behavioural factors in managing COVID-19*. Retrieved from https://www.ncid.sg/Documents/Media%20Release_The%20role%20of%20socio-behavioural%20factors%20in%20outbreak%20control%2021%20May.pdf
3. National Center for Infectious Diseases (NCID). (2020). *NCID, NUS AND NTU studies highlight the role of socio-behavioural factors in managing COVID-19*. Retrieved from https://www.ncid.sg/News-Events/News/Pages/NCID,-NUS-and-NTU-Studies-Highlight-the-Role-of-Socio-Behavioural-Factors-in-Managing-COVID-19-.aspx&sa=D&source=docs&ust=1642127416593436&usg=AOvVaw2Qe561lLPcEGavkTT-_v6c
4. Linette, L., & Clara, C. (2021). *Half-truths and lies: How Covid-19 misinformation spreads in S'pore*. Retrieved from https://www.straitstimes.com/singapore/pandemic-of-online-misinformation-on-covid-19-takes-its-toll
5. Weth, C. v. d., & Kankanhalli, M. (2019). *Fake news and social media*. Retrieved from https://news.nus.edu.sg/research/fake-news-and-social-media
6. Peter Lasersohn, P. (2015). *Subjectivity and perspective in truth-theoretic semantics*. Retrieved from https://semanticsarchive.net/Archive/zEzNWI3O/subjectivityandperspective.pdf
7. Vanessa, W. L., Rachel, L. L., Yi Roe, T., Alexis, S. E. S., Mei Xuan, T., Norhudah, B. O., Sue, B. D., Tun-Linn, T., May, O. L., Rick, T. O., Yee-Sin, L., Vernon, J. L., & Mark I. C. C. (2020). *Government trust, perceptions of COVID-19 and behaviour change: cohort surveys, Singapore*. Retrieved from https://www.who.int/bulletin/online_first/BLT.20.269142.pdf
8. Justin, O. (2021). *67% of S'poreans willing to take Covid-19 vaccine, 20% neutral; younger ones more likely to be concerned: IPS study*. Retrieved from https://www.straitstimes.com/singapore/health/67-per-cent-willing-to-take-vaccine-20-per-cent-neutral-younger-sporeans-more
9. Ng, E. (2021). *How governments are tackling anti Covid-19 vaccine misinformation*. Retrieved from https://www.straitstimes.com/asia/how-governments-are-tackling-anti-covid-19-vaccine-misinformation
10. Koh, V. (2021). *Grandmother hospitalised after taking ivermectin to 'protect herself' against Covid-19*. Retrieved from https://www.straitstimes.com/singapore/grandmother-hospitalised-after-taking-ivermectin-to-protect-herself-against-covid-19
11. Raschka, S., & Mirjalili, V. (2019). *Applied machine learning with a solid foundation in theory. Revised and expanded for TensorFlow 2, GANs, and reinforcement learning*. Retrieved from https://books.google.com.sg/books?id=sKXIDwAAQBAJ&printsec=frontcover&dq=Python+Machine+Learning+Packt&hl=en&sa=X&redir_esc=y#v=onepage&q=Python%20Machine%20Learning%20Packt&f=false
12. Duda, R. O., Hart, P. E., & Stork, D. G. (2012). *Pattern classification*. Retrieved from https://books.google.com.sg/books?id=Br33IRC3PkQC&pg=PA2&dq=Pattern+Recognition&source=gbs_selected_pages&cad=3#v=snippet&q=Naive%20Bayes&f=false
13. Bishop, C. M. (2012). *Pattern recognition and machine learning*. https://www.goodreads.com/review/show/390431655?book_show_action=true&page=1
14. Wang, W. Y. (2017). *"Liar, Liar Pants on Fire": A new benchmark dataset for fake news detection*. Retrieved from https://sites.cs.ucsb.edu/~william/papers/acl2017.pdf
15. Saleh, H., Alharbi, A., & Alsamhi S. H. (2021). *OPCNN-FAKE: Optimized convolutional neural network for fake news detection*. Retrieved from https://ieeexplore.ieee.org/stamp/stamp.jsp?tp=&arnumber=9537782
16. Mohamed. K. E., Li, K. F., & Gebali, F. (2020). *Detecting misleading information on COVID-19*. Retrieved from https://ieeexplore.ieee.org/document/9189767

17. Ni, S., Li, J., & Kao, H. -Y. (2021). MVAN: Multi-view attention networks for fake news detection on social media. Retrieved from https://ieeexplore.ieee.org/document/9497048
18. Sansonetti, G., Gasparetti, F., D'aniello, G., & Micarelli, A. (2020). *Unreliable users detection in social media: Deep learning techniques for automatic detection.* Retrieved from https://ieeexplore.ieee.org/document/9269985
19. Abdelminaam, D. S., Ismail, F. H., Taha, M., Tah, A., Houssein, E. H., & Nabil, A. (2021). *CoAID-DEEP: An optimized intelligent framework for automated detecting COVID-19 misleading information on Twitter.* Retrieved from https://ieeexplore.ieee.org/document/9350542
20. Qureshi, K. A., Malick, R. A. S., Sabih, M., & Cherifi, H. (2021). *Complex network and source inspired COVID-19 fake news classification on Twitter.* Retrieved from https://ieeexplore.ieee.org/document/9567692
21. Xu, K., Wang, F., Wang, H., & Yang, B. (2020). *Detecting fake news over online social media via domain reputations and content understanding.* Retrieved from https://ieeexplore.ieee.org/stamp/stamp.jsp?tp=&arnumber=8768083
22. Barua, R., Maity, R., Minj, D., Barua, T., & Layek, A. K. (2019). *F-NAD: An application for fake news article detection using machine learning techniques.* Retrieved from https://ieeexplore.ieee.org/document/8973059
23. Poddar, K., Amali, G. B., & Umadevi, K. S. (2019). *Comparison of various machine learning models for accurate detection of fake news.* Retrieved from https://ieeexplore.ieee.org/document/8960044
24. Ghinadya & Suyanto, S. (2020). *Synonyms-based augmentation to improve fake news detection using bidirectional LSTM.* Retrieved from https://ieeexplore.ieee.org/document/9166230
25. Yanagi, Y., Orihara, R., Sei, Y., Tahara, Y., & Ohsuga, A. (2020). *Fake news detection with generated comments for news articles.* Retrieved from https://ieeexplore.ieee.org/document/9147195
26. Bisaillon, C. (2020). *Fake and real news dataset.* Retrieved from https://www.kaggle.com/clmentbisaillon/fake-and-real-news-dataset
27. Möbius. (2020). *COVID-19 fake news dataset.* Retrieved from https://www.kaggle.com/arashnic/covid19-fake-news
28. Channel News Asia. (n.d.). Retrieved from https://www.channelnewsasia.com/
29. Truth Warriors. (n.d.). Retrieved from https://www.truthwarriors.info/articles/
30. Lim, Y. L. (2021) *Pofma correction direction issued to Truth Warriors website over Covid-19 false claims.* Retrieved from https://www.straitstimes.com/singapore/correction-direction-issued-to-truth-warriors-website
31. Guest, P., Firdaus, F., & Danan, T. (2021). *"Fake news" laws are failing to stem Covid-19 misinformation in Southeast Asia.* Retrieved from https://restofworld.org/2021/fake-news-laws-are-failing-to-stem-covid-19-misinformation-in-southeast-asia/

Modelling Singapore's Covid-19 Pandemic Using SEIRQV and Hybrid Epidemiological Models

Chuping Mu, Jovern Teo, and Joshua Cheong

Abstract In this paper, we aim to simulate the Covid-19 spread in Singapore using a SEIRQV model. It is a modified SEIR model that includes quarantining, provision of vaccines (with waning efficacy), and booster shots. To address issues caused by the many assumptions made to run a numerical simulation, we decided to utilise Kalman Filter (KF) for a simpler SIR model for data assimilation as it is less sensitive to imperfections in input data and model assumptions. It is coupled with the hybrid AI model (using LSTM, or Long Short-Term Memory, a recurrent neural network). The 3 models show how Singapore's Safe Management Measures have delayed and greatly ameliorated the spread of Covid-19 while allowing our economy to reopen. Moreover, we have attempted to use these hybridised models to benchmark the accuracy of the pure mathematical model. Both hybrid models are much better than the non-hybrid SEIRQV model, and the latter shown to achieve a significantly better performance than that of SEIR.

Keywords Covid-19 · Epidemiological compartmental models · Hybrid models · Kalman filter · LSTM · Intervention · Vaccination · Quarantine

1 Introduction

As the pandemic situation evolved, an increase in Covid-19 variants and information regarding waning vaccine effectiveness [1, 2] have surfaced. These variants have brought greater concerns of vaccine breakthrough infections. Singapore's unprecedentedly high vaccination rate and strict Safe Management Measure (SMM) compliance has led to the pressing need for a specialised epidemiological model. We propose a mathematical SEIRQV model based on the SEIR model that represents Singapore's government's major measures such as lockdown, vaccinations, boosters, and quarantine. We also include waning vaccine efficacy and provision of booster shots.

C. Mu (✉) · J. Teo · J. Cheong
School of Math and Science, NUS High, Singapore, Singapore
e-mail: h1810107@nushigh.edu.sg

H. Guo et al. (eds.), *IRC-SET 2022*,
https://doi.org/10.1007/978-981-19-7222-5_42

To address issues caused numerous assumptions made in a numerical simulation, we decided to utilise Kalman Filter (KF) with a simpler SIR model for data assimilation as it is less sensitive to imperfections in input data and model assumptions. It is coupled with the hybrid AI model (using Long Short-Term Memory (LSTM), a recurrent neural network). The three models show how Singapore's SMM have greatly ameliorated the spread of Covid-19 while reopening our economy.

2 Literature Review

Ever since the start of the Covid-19 pandemic, researchers around the world have taken to model the spread of the virus. Using these models, they forecast the epidemic's spread, magnitude, and duration, via estimating various epidemiological parameters such as the reproductive number. Most Covid-19 prediction models build upon the basic Susceptible-Infectious-Removed (SIR) [3] or SEIR compartmental model with the addition of the "Exposed" compartment [4], which is more useful due to most diseases, including Covid-19 having a dormant period before becoming infectious.

2.1 *Hybrid Modelling*

Compartmental models are often used by epidemiologists, especially during the current Covid-19 pandemic, to model disease transmission in a population under certain scenarios and are frequently used to carry out counterfactual analysis. Thus, hybridising it with a good data assimilation method allows for better visualisation of the positive impact brought by Covid-19 mitigation measures (CMMs). These measures span mobility restrictions, mask donning, vaccination, quarantining of travellers, contact tracing, and social distancing.

2.2 *AI Model*

The use of AI is also being actively explored to improve models [5, 6]. AI is mainly used to approximate unknown values [7, 8], extrapolate a time series [9, 10], or consider other environmental conditions [10]. In this paper, we utilise an approach from 15 Feb 2020 to 17 Dec 2021. Model is adapted from Ossa et al. 2021's Neuro-hybrid Covid-19 model [11].

2.3 KF Model

The Kalman Filter is a recursive algorithm that specialises in finding unknown parameters of a model. In our case, we hybridised an SIR model with a Kalman Filter to determine unknown variables β, γ, and μ. We opted for a simpler SIR model as opposed to one with more compartments [12, 13] as limited data is available for Singapore's Covid-19 spread.

3 The SEIRQV Mathematical Model

Based on the susceptible-exposed-infected-removed (SEIR) compartmental model, we included quarantined (Q) and vaccinated (V) compartments to improve the accuracy of pandemic modelling (Fig. 1).

Mathematically, the dynamic process of SEIRQV can be described as such:

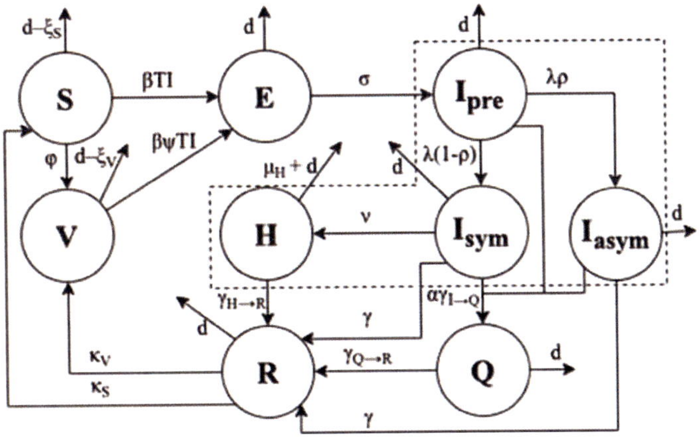

Fig. 1 SEIRQV model. An individual can be in one of each state, susceptible **(S)**, exposed **(E)**, infectious **(I)**, recovered **(R)**, dead **(D)**, vaccinated **(V)**, and quarantined **(Q)**. Below shows a flow chart summarising the model. Due to the high rate of asymptomatic Covid-19 cases, the **I** compartment includes the hospitalised **(H)** subcategory, pre-symptomatic **(I_{pre})**, asymptomatic **(I_{asym})**, and symptomatic infected **(I_{sym})**. Each compartment represents a fraction of the total population, N. . The parameters on the arrows in the figure indicate the proportion of individuals from one compartment entering another (they are defined in Table 1)

$$
\begin{aligned}
\dot{S} &= -\beta T S I + \kappa_S R - (\phi + d - \xi_S)S \\
\dot{E} &= \beta T(S + \psi V)I - (\sigma + d)E \\
\dot{I} &= \sigma E - \left(\alpha\gamma_{I\to Q} + \mu + (1-\alpha)\gamma + d\right)I = \sigma E - C_I I \\
\dot{Q} &= \alpha\gamma_{I\to Q} I - \left(\gamma_{Q\to R} + d\right)Q = \alpha I - C_Q Q \\
\dot{R} &= \gamma_{Q\to R} Q + (1-\alpha)\gamma I - (\kappa + d)R \\
\dot{V} &= \phi S - \psi\beta T V I + \kappa_V R - (d - \xi_V)V
\end{aligned}
$$

$(1 - \psi_V)$ denotes the protection against infection after being doubly vaccinated. Every individual that is newly vaccinated/received a booster shot is assumed to have a vaccine efficacy "half-life" of 300 and 500 days, respectively, such that $\psi_{V_i} = 1 - (1 - \psi_{V,i})\left(\frac{1}{2}^{\frac{t}{300}}\right)$. To take booster shots into account, we use ψ to denote net vaccine non-protection, where: $\psi = \sum_{i=t_V}^{t \geq t_V} \left(\psi_{V,i}(V(i) - B(i)) + \psi_{B,i} B(i)\right)$ and $B = \phi_B \max\left(0, \min\left(t_B + \frac{V}{\phi_B}, t\right) - t_B\right)$.

4 Mathematical Approach to Predicting Pandemic Trajectory

As the SEIRQV model is a system of non-homogenous differential equations, where:

$$
\boldsymbol{y}' = \boldsymbol{U}(t)\boldsymbol{y}(t)
$$

We can express the system (excluding deaths) as such:

$$
\begin{pmatrix} \dot{S} \\ \dot{E} \\ \dot{I} \\ \dot{Q} \\ \dot{R} \\ \dot{V} \end{pmatrix} = \left(\begin{pmatrix} -\phi - d + \xi_S & 0 & -\beta T S & 0 & \kappa_S & 0 \\ 0 & -\sigma - d & \beta T S & 0 & 0 & \beta T I \psi \\ 0 & \sigma & -C_I & 0 & 0 & 0 \\ 0 & 0 & \alpha\gamma_{I\to Q} & -C_Q & 0 & 0 \\ 0 & 0 & (1-\alpha)\gamma & \gamma_{Q\to R} & -\kappa - d & 0 \\ \phi & 0 & 0 & 0 & \kappa_V & -\beta T I \psi - d + \xi_V \end{pmatrix} \right) \begin{pmatrix} S \\ E \\ I \\ Q \\ R \\ V \end{pmatrix}
$$

Unfortunately, since $\boldsymbol{U}$'s eigenvalues are extremely complex (refer to Appendix), we decided to simplify the matrix to greatly reduce its complexity and obtain more useful expressions.

4.1 Observing Dynamics of Extended SEIQR Model

Thus, we can approximate the solution when $\dot{V} \approx 0$, (valid for short-term predictions)

$$\begin{pmatrix} \dot{S} \\ \dot{E} \\ \dot{I} \\ \dot{Q} \\ \dot{R} \end{pmatrix} = \begin{pmatrix} -\phi - d + \xi_S & 0 & -\beta T S & 0 & \kappa_S \\ 0 & -\sigma - d & \beta T(S + V\psi) & 0 & 0 \\ 0 & \sigma & -C_I & 0 & 0 \\ 0 & 0 & \alpha\gamma_{I\to Q} & -C_Q & 0 \\ 0 & 0 & (1-\alpha)\gamma & \gamma_{Q\to R} & -\kappa - d \end{pmatrix} \begin{pmatrix} S \\ E \\ I \\ Q \\ R \end{pmatrix}$$

The eigenvalues are $-d - \phi$, $-C_Q$, $-d - \kappa$, $\frac{-(\sigma+C_I+d)-\sqrt{\Delta}}{2}$, $\frac{-(\sigma+C_I+d)+\sqrt{\Delta}}{2}$, assuming $\xi_S = 0$. The determinant, $\Delta = (\sigma + C_I + d)^2 - 4((d+\sigma)C_I - \sigma\beta T(\psi V + S))$ is always positive (*solution in A3*), hence only the lattermost term, λ_5, can be positive (others are always negative regardless of the population fractions in any compartment). Finally, we arrive at the solution (all eigenvectors, X_n, are positive):

$$\begin{pmatrix} S \\ E \\ I \\ Q \\ R \end{pmatrix} = \sum_{n=1}^{5} c_n e^{\lambda_n t} X_n, \text{ for } c_n \text{such that} \sum_{n=1}^{5} c_n X_n = \begin{pmatrix} S(0) \\ E(0) \\ I(0) \\ Q(0) \\ R(0) \end{pmatrix}$$

Currently, $\gamma_{I\to Q} \approx \frac{1}{4+1} = \frac{1}{5}$ [17–19], and assuming $\alpha = 78.4\%$ as rationalised in the appendix, we obtain $C_I \approx 0.8056$. From this, we can estimate the boundary conditions needed to for a growth or decline in caseload, where $R_0 = 1$. As the largest eigenvalue (and thus rate-determining exponent) is λ_5 Covid-19 trends can be predicted by checking its sign. This criterion (given in Table 2) is only valid when quarantine is a very commonly followed procedure, especially when it is enforced by quarantine orders for close contacts, which is no longer done after October 11th, 2021.

This is further verified by Diekmann et al. [20], who proved that $\lambda + 1 = R_0$, where λ is the largest eigenvalue. Thus, this eigenvalue criterion can quickly approximate the trend of the epidemic with just one datapoint. The graphs below show the accuracies of the graphs.

From Fig. 2, it is evident that the criterion is highly insensitive. However, it still accurately showed that Singapore's Covid-19 spread was kept at bay until about 5 August 2021 (day = 560), when case numbers began growing exponentially, despite a decrease from November to December (days 648–708). This is likely due to the sharp increase in transmissibility of Covid-19 due to the Omicron variant becoming dominant (Fig. 3).

Since the criterion is theoretical and is based on many inaccurate assumptions, we must resort to using data assimilation hybrid models like KF-SIR and LSTM-SIR models to accurately predict the progress of the pandemic. In the following graph, R is determined by taking the 7-day average of daily Covid-19 cases and adjusted for changes in susceptible populations, by comparing it relative to the 7-day average of

Table 1 Table of expressions

Symbol	Represents
ϕ, ϕ_B	Those getting fully vaccinated or booster shots per day as a fraction of S and V,. respectively
ψ	Net vaccination non-protection against infection
κ	The rate needed for recovered individuals to become vulnerable to infection (assume it is very small) [Note: $\kappa_S = \frac{S}{S+\psi V}\kappa, \kappa_V = \kappa - \kappa_S$]
σ	Reciprocal of the incubation period, $\sigma_{og} = \frac{1}{\tau_{inc}}$ [14]
B	Total number of individuals who had booster shots (were previously vaccinated) as a fraction of V
α	Net proportion of infected individuals quarantined, assumed as 78.4%, (Appendix)
ρ	Proportion of asymptomatic individuals, about 45%
γ	$\frac{1}{\tau_{inf}}$, , reciprocal of infectious period (Note: $\frac{1}{\gamma_{I\to Q}} + \frac{1}{\gamma_{Q\to R}} = \frac{1}{\gamma_{I\to H}} + \frac{1}{\gamma_{H\to R}} = \frac{1}{\gamma} \approx$ 10days [15])
μ	Rate of death among infected individuals **per day**, about 4.1×10^{-6} as of January 2022
β	Contact rate, usually multiplied with **transmissibility rate**, $T = (1 + (\eta - 1)\rho)T_{Covid}$
t_V, t_B	Relative date (from t_0) when vaccine/booster shots become available
d	Daily deaths (in 2019), about 1.36×10^{-5} [16]
η	Relative transmissibility of asymptomatic individuals with respect to symptomatic, about 50%,
λ	Reciprocal of the pre-symptomatic infectious period, about $\frac{1}{5}$ days^{-1}
ξ_S, ξ_V	Sum of births (S) and overall immigration as a fraction of S or V per day, ξ totals to -3.382×10^{-5}

Table 2 Evaluating the changes of λ_5 on caseload

R_0	λ_5	Requirement (valid until Oct 11th, 2021 [12])	Daily caseload trend
$R_0 > 1$	$\lambda_5 > 0$	$\sqrt{\Delta} > 0.8056 + \sigma$	Exponential growth
$R_0 = 1$	$\lambda_5 = 0$	$\sqrt{\Delta} \approx 0.8056 + \sigma$	Approximately constant
$R_0 < 1$	$\lambda_5 < 0$	$\sqrt{\Delta} < 0.8056 + \sigma$	Exponential decrease

the previous week, as shown below: $R_i = \frac{\beta_i S_i I_{avg,i}}{\beta_{i-7} S_{i-7} I_{avg,i-7}}$. The graph below illustrates how significant government interventions have been on reducing R_t (Fig. 4).

This illustrates how government interventions are strongly associated with a sharp decrease in R_t, as seen by the effect of the progressive implementation of CMMs since 29th January, circuit breaker (4th April to 1st June), and each subsequent tightening up of CMMs in May, July, and October. These stabilise lower infection rates. We can hypothesise that government measures are highly effective at reducing Covid-19 spread.

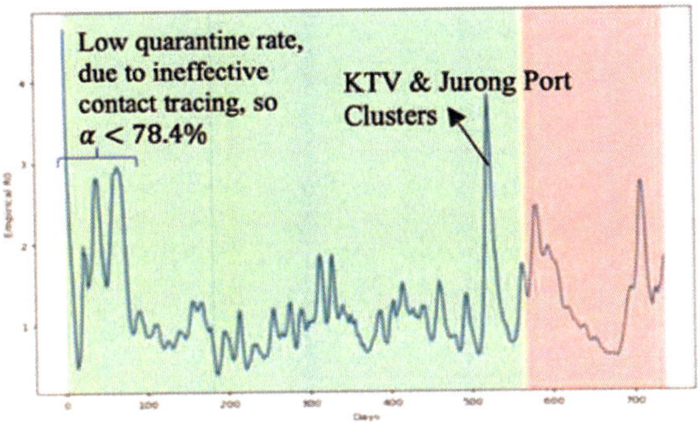

Fig. 2 Graph of empirical R_o value taken from 23 January 2020 to 15 January 2022

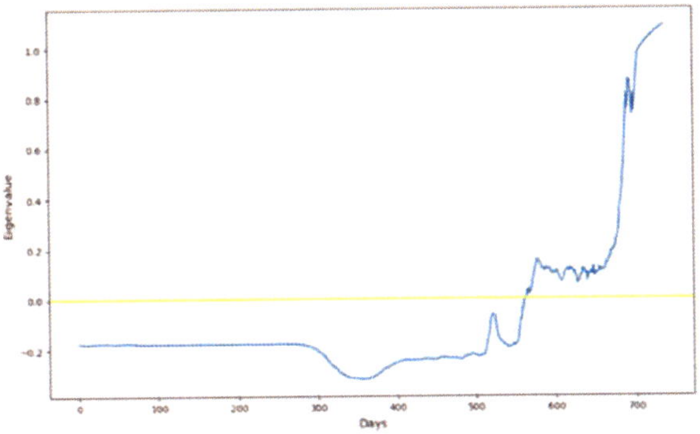

Fig. 3 Graph of determined eigenvalue from 23 January 2020 to 15 January 2022

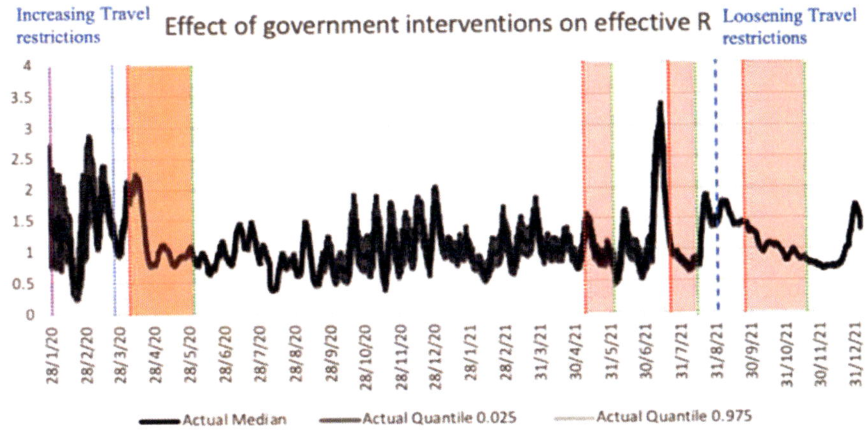

Fig. 4 Graph of estimated R_0 with 97.5% confidence intervals from 23 January 2020 to 3 January 2022

5 Methodology

5.1 *Pure Mathematical Model*

As mentioned above, we used a SEIRQV model to simulate Covid-19 spread in Singapore. We coded our simulation in Python and used the library SciPy to numerically integrate the differential equations governing the system, obtaining a time series of the SEIRD compartments.

5.2 *Hybrid AI Model*

Rather than start with an initial state, this model utilises actual data of Covid-19 spread in Singapore. The model also used other factors such as stringency index and booster shots to better model the changes in parameters β, γ,ϕ.

It is a cyclic process which progressively refines critical values β, γ, R_0. Our hybrid model is based on an SIR hybrid model done in paper [7]. We improve on their model by adding compartments for exposed (E) and vaccinated (V) populations. From the existing data, a time trend of β is determined. To extrapolate data, a Long Short-Term Memory (LSTM) neural network is trained on the time trend of β and predicts the future trend. The data source originates from Our World in Data [21] and Worldometer [22].

Figure 5 shows how the hybrid AI model is implemented. In the 2nd step, value of β on a certain day is calculated by rearranging the SEIR differential equations, as shown:

$$S_{t+1} = S_t - \frac{\beta I_t S_t}{N}$$

$$\beta = \frac{N}{I_t}\left(\frac{S_t - S_{t+1}}{S_t}\right)$$

5.3 *SEIRD + KF Model*

The KF model takes in data of active cases, recovered cases, and deaths. State vectors of SEIRD compartments and parameters β, γ and μ are used with the matrix update equations seen in KF models [39–41] to get a series of the parameters over time.

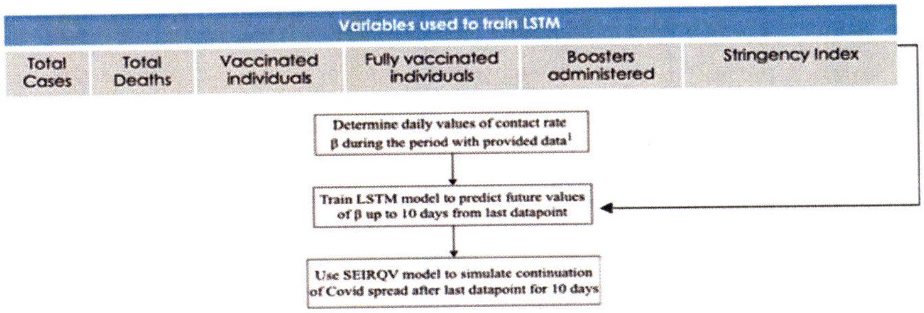

Fig. 5 Diagram of variables and process used to train AI hybrid model

6 Results and Discussion

6.1 SEIRQV Model Simulation

This SEIRQV model shows the possible results when each of the following scenarios occur:

1. No government measures (assume $\alpha \approx (1 - \rho) * 0.5 = 0.2875$)
2. No circuit breaker coupled with lax government measures (assume $\alpha \approx 0.65$)
3. No circuit breaker coupled with stricter measures—(assume $\alpha \approx 0.784$)
4. Has circuit breaker coupled with lax measures (attempt at modelling Singapore's Covid-19 response).

All of them assumes that migrant workers have a much higher chance (based on a logistic function decreasing from 5 times to 2) of getting a Covid-19 infection than the general populace (Fig. 6).

This model based on the constants given in Table 3.

6.2 SIR-LSTM Hybrid AI Model

The following graph shows the comparison of the LSTM model's prediction from day 400 to 700 compared with the actual data, showing that the initial 150 days rather accurately models the rise in cumulative cases. After 150 days, they diverge sharply (Figs. 7 and 8).

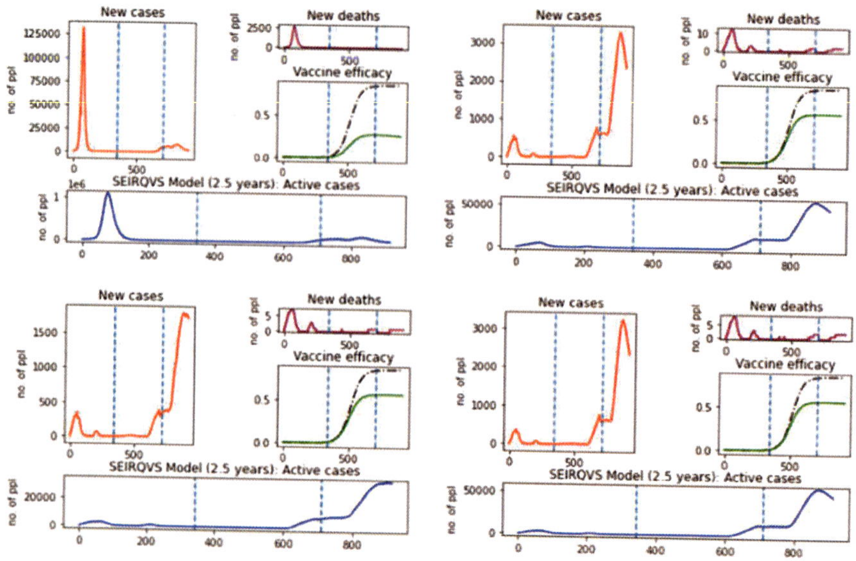

Fig. 6 Results of the SEIRQV model for each scenario (Scenarios 1, 2, 3, 4 in clockwise order)

Table 3 How transmission rate and latency rate change depending on variant

Variant name	Transmission rate (unvaccinated)	σ	Transition period
Original	0.075 [23, 24]	$\frac{1}{5.2}$ days^{-1} [25]	Mid-Feb 2021
Alpha	0.1305 [26]	$\frac{1}{3}$ days^{-1} [27]	Mid-May 2021
Delta	0.196 [28]	$\frac{1}{4}$ days^{-1} [29]	Mid-Dec 2021
Omicron	0.402 [30]	$\frac{1}{3}$ days^{-1} [31]	–

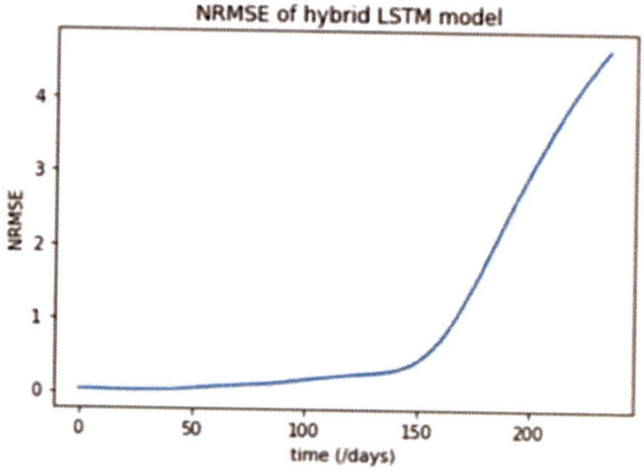

Fig. 7 Graph of normalised root mean squared error (NRMSE) of the model over time

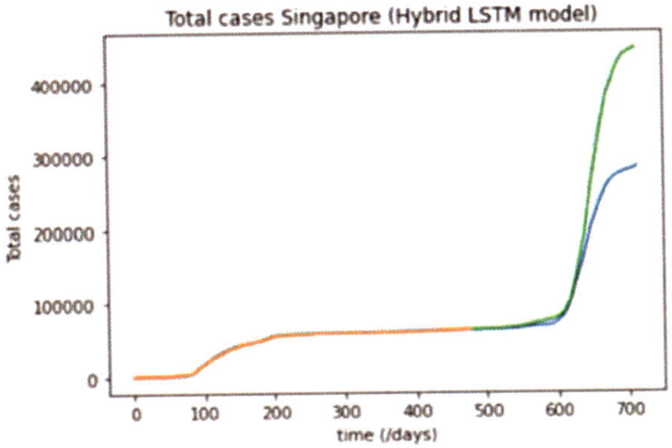

Fig. 8 Graph of total cases predicted by the hybrid AI model (green) vs the actual cases (blue) (begins prediction at Day 500 or 23 January 2022)

6.3 *SIR-KF Model*

Figure 10 shows that the uncertainty of predicted active cases increases at an exponential rate and fit the actual data well. This shows that the KF predications are moderately accurate (Fig. 9).

After interpreting this graph, we think the Singapore Government would not allow such a surge in Covid-19 cases, assuming the prediction is accurate, a tightening of restrictions during mid-February may be necessary.

In addition, we have been able to estimate the values of βT and γ throughout the pandemic using this hybrid model shown in Fig. 11.

Knowing that βT is the product of transmissibility and contact rate, by using logistic graphs with asymptotes fixed at values given in Table 3, we can find the contact rate, β, over the pandemic (Fig. 12).

This allows us to conclude that contact rate dropped to levels like that of circuit breaker during subsequent heightened alerts. This was the maximum during late

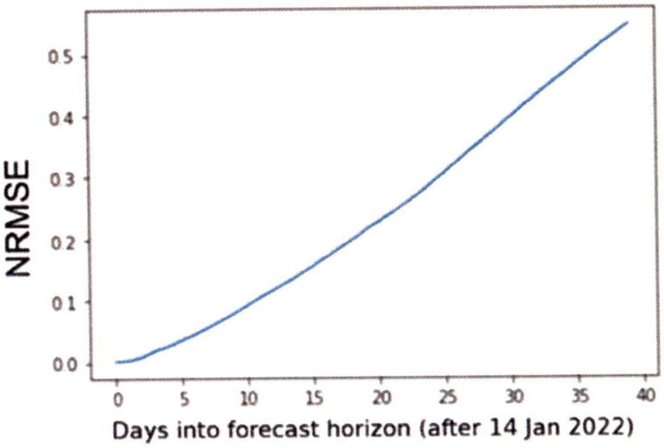

Fig. 9 Graph showing the accuracy of SIR-KF model over time by plotting its NRMSE against time/days

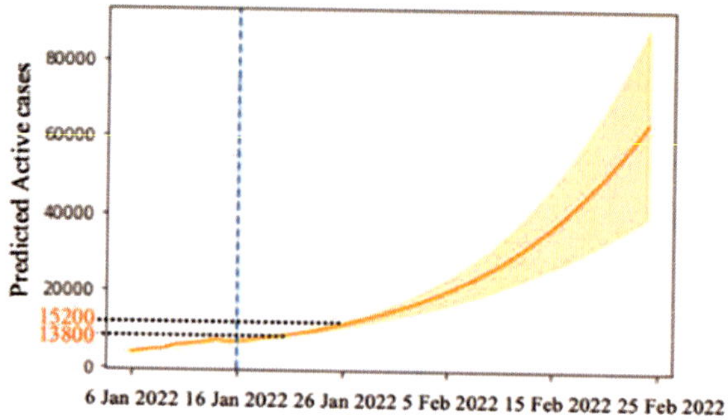

Fig. 10 Graph showing the predicted number of active cases for the 40-day prediction horizon after 16 January 2022 (assuming no change in government intervention, based on data from Jan 23rd, 2020, to Jan 16th, 2022). The numbers in orange (actual 7-day average of active cases [32] for each date 21st and 27th January 2022), fit closely to the KF predictions are rather accurate. The upper and lower bounds of this graph represent the 95% confidence interval

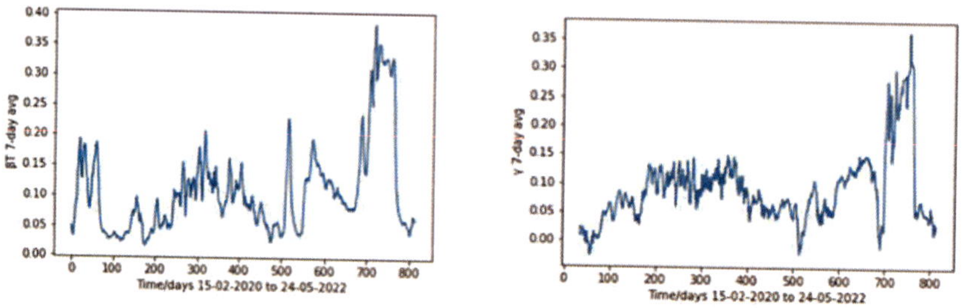

Fig. 11 Graph of βT against time/days (left) and graph of γ against time/days (right)

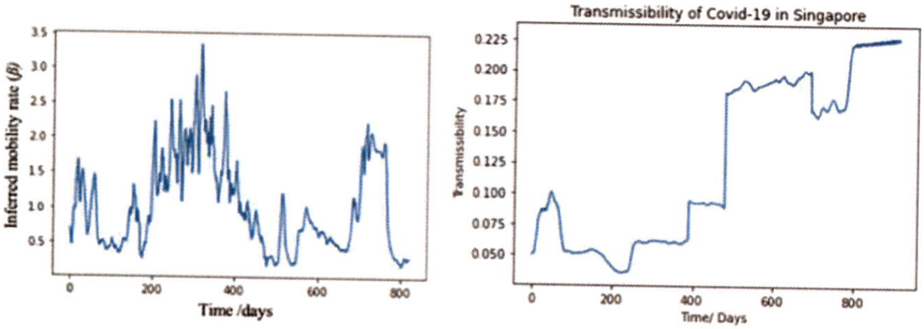

Fig. 12 Inferred mobility rate, β (left) and estimated transmissibility, T, (right) over the course of the pandemic

2020–early 2021, which was unexpected, as we expected higher than circuit breaker levels due to the relaxation of measures.

7 Limitations

7.1 *Pure Mathematical Model*

Firstly, the model overgeneralizes the movement of individuals and treats the parameters as discrete variables rather than distributions. Thus, we are unable to account for super-spreader events in our model, as β varies widely per person, as shown by the superclusters during mid to end of July 2021.

Additional depth in compartments can also account for social interactions between and within groups, by creating multiple SEIR compartments for different demographics of people [33].

Secondly, while we did take a rolling seven-day average for Covid-19 infections to smoothen weekly pattern of over and under reporting, we have no way to ensure that there was no overall underreporting in Covid-19 infections.

Finally, we merged the Infected compartment subcategories, of asymptomatic, mild, and severe [32, 34–36], to decrease the analytical complexity of the model, this has the trade-off of making the model less comprehensive.

7.2 *SIR-LSTM Hybrid AI Model*

Firstly, there was a lack of data to train the model on as it was trained solely with Singaporean data.

Secondly, the standard for being Covid-19 positive changed from being Polymerase Chain Reaction (PCR) positive to Antigen Rapid Testing (ART) positive on 11th October 2021 [37], shortly after we gathered our data. ART testing has shown to be less accurate than the PCR [38] potentially resulting in more false positives which could inflate case numbers.

7.3 *Kalman-SIR Hybrid Model*

The Kalman Filter model requires the variances of uncertainty in time series of active cases, recovered cases and deaths. No estimates are currently available and determining it results in a large error, causing all final outputs to have an innate uncertainty. Additionally, forecasting of cases into the future is difficult using the KF model due to it assuming constant coefficients during forecasting.

8 Conclusion

From the results, we have clearly showed that the presence of government restrictions has very effectively reduced the spread of Covid-19.

Due to the overgeneralisations as aforementioned, further research can be done on models where parameters are not discrete but rather distributions, where β, γ can be modelled based on a gamma distribution while the latency rate can be modelled as a logistic distribution, all of which has been shown to accurately fit real-life data for specific k, θ and μ, s, respectively. The model can be made more accurate by integrating models that account for the assumptions made by the other and processing their predictions via a weighted average which is more accurate.

In conclusion, the modelling capability of the numerical SEIRQV model can be improved by using better parameter estimation and a more comprehensive compartmentalisation of the population based on public health interventions, healthcare system capacity, population distribution and vaccination rate. Our KF model could be improved with more accurate methods to estimate parameters based off crude data.

Appendix

A1: Eigenvalues of 6×6 matrix (Assuming d, ξ_S, ξ_V are negligible): https://www.wolframalpha.com/input/?i=calculate+eigenvalues+transpose%7B%7B-z%2C0%2C0%2C0%2C0%2Cz%7D%2C%7B0%2C-u%2Cu%2C0%2C0%2C0%7D%2C%7B-b*t*s%2Cb*t*s%2C-c%2Ca*f%2C%281-a%29*g%2C0%7D%2C%7B0%2C0%2C0%2C-h-m%2Ch%2C0%7D%2C%7B%28k*s%29%2F%28s%2Bp*v%29%2C0%2C0%2C0%2C-k%2C%28k*p*v%29%2F%28s%2Bp*v%29%7D%2C%7B0%2Cb*t*p*v%2C0%2C0%2C0%2C-b*t*p*v%7D%7D

A2: We obtain $\alpha_{\text{asym}} = 60\%$ and $\alpha_{\text{sym}} = 92\%$ using the assumption that 80% of individuals would self-quarantine even without being traced. Since symptomatic individuals tend to self-quarantine, and contact-tracing effectiveness stands at about 60% [25], after an individual was traced, just 1 day after they no longer are pre-symptomatic (can be either symptomatic or asymptomatic), thus $\alpha = \alpha_{\text{sym}}(1 - \rho) + \alpha_{\text{asym}}\rho = 78.4\%$

A3: In observing dynamics of SEIRQV model, the detailed solution is as follows:

$$\det \boldsymbol{U} = (-\kappa - d - \lambda)(-\phi - d + \xi_S - \lambda)(Q(x))\left(-C_Q - \lambda\right) = 0$$

$$\textbf{Solve: } Q(x) = -\sigma\beta T(\psi V + S) + \left(d + \sigma + \lambda_{4,5}\right)\left(C_I + \lambda_{4,5}\right) = 0$$

$$\lambda_{4,5} = \frac{-(\sigma + C_I + d) \pm \sqrt{\Delta}}{2}, \Delta = (\sigma + C_I + d)^2$$

$$-4((d + \sigma)C_I - \sigma\beta T(\psi V + S)) > 0$$

$$(x + y)^2 \geq 4xy \Rightarrow (\sigma + d + C_I)^2 \geq 4(d + \sigma)C_I;$$

$$\because 4\sigma\beta T(\psi V + S) > 0 \Rightarrow \Delta > 0 \blacksquare$$

A4: Google drive folder of all codes used in this project: https://drive.google.com/drive/folders/1w7XzmSNMVff8bUgqgMK2gY5d0a736JF_?usp=sharing

References

1. Levin, E. G., Lustig, Y., Cohen, C., Fluss, R., Indenbaum, V., Amit, S., Doolman, R., Asraf, K., Mendelson, E., Ziv, A., Rubin, C., Freedman, L., Kreiss, Y., & Regev-Yochay, G. (2021). Waning immune humoral response to BNT162b2 Covid-19 vaccine over 6 months. *New England Journal of Medicine, 385*(24), e84. https://doi.org/10.1056/NEJMoa2114583
2. Chemaitelly, H., Tang, P., Hasan, M. R., AlMukdad, S., Yassine, H. M., Benslimane, F. M., Al Khatib, H. A., Coyle, P., Ayoub, H. H., Al Kanaani, Z., Al Kuwari, E Jeremijenko, A., Kaleeckal, A. H., Latif, A. N. Shaik, R. M., Abdul Rahim, H. F., et al. (2021). Waning of BNT162b2 vaccine protection against SARS-CoV-2 infection in Qatar. *New England Journal of Medicine*. https://doi.org/10.1056/NEJMoa2114114
3. Cooper, I., Mondal, A., & Antonopoulos, C. G. (2020). A SIR model assumption for the spread of COVID-19 in different communities. *Chaos, Solitons, and Fractals, 139*, 110057. https://doi.org/10.1016/j.chaos.2020.110057
4. Carcione, J. M., Santos, J. E., Bagaini, C., & Ba, J. (2020). A simulation of a COVID-19 epidemic based on a deterministic SEIR model. *Frontiers in Public Health, 8*, 230. https://doi.org/10.3389/fpubh.2020.00230
5. Musulin, J., Baressi Šegota, S., Štifanić, D., Lorencin, I., Anđelić, N., Šušteršič, T., Blagojević, A., Filipović, N., Ćabov, T., & Markova-Car, E. (2021). Application of artificial intelligence-based regression methods in the problem of COVID-19 spread prediction: A systematic review. *International Journal of Environmental*. https://doi.org/10.3390/ijerph18084287
6. Malik, Y. S., Sircar, S., Bhat, S., Ansari, M. I., Pande, T., Kumar, P., Mathapati, B., Balasubramanian, G., Kaushik, R., Natesan, S., Ezzikouri, S., El Zowalaty, M. E., & Dhama, K. (2021). How artificial intelligence may help the Covid-19 pandemic: Pitfalls and lessons for the future. *Reviews in Medical Virology, 31*(5), 1–11. https://doi.org/10.1002/rmv.2205
7. Castillo Ossa, L. F., Chamoso, P., Arango-López, J., Pinto-Santos, F., Isaza, G. A., Santa-Cruz-González, C., Ceballos-Marquez, A., Hernández, G., Corchado, J. M. (2021). A hybrid model for COVID-19 monitoring and prediction. *Electronics, 10*, 799. https://www.mdpi.com/2079-9292/10/7/799/pdf
8. Soures, N., Chambers, D., Carmichael, Z., Daram, A., Shah, D. P., Clark, K., Potter, L., & Kudithipudi, D. (2020). SIRNet: understanding social distancing measures with hybrid neural network model for COVID-19 infectious spread. *arXiv preprint* arXiv:2004.10376. https://arxiv.org/pdf/2004.10376.pdf
9. Kuvvetli, Y., Deveci, M., Paksoy, T., & Garg, H. (2021). A predictive analytics model for COVID-19 pandemic using artificial neural networks. *Decision Analytics Journal, 1*, 100007. https://www.sciencedirect.com/science/article/pii/S2772662221000060
10. Feng, S., Feng, Z., Ling, C., Chang, C., & Feng, Z. (2021). Prediction of the COVID-19 epidemic trends based on SEIR and AI models. *PLoS ONE, 16*(1), e0245101. https://doi.org/10.1371/journal.pone.0245101
11. Castillo Ossa, L. F., Chamoso, P., Arango-López, J., Pinto-Santos, F., Isaza, G. A., Santa-Cruz-González, C., Ceballos-Marquez, A., Hernández, G., & Corchado, J. M. (2021). A hybrid model for COVID-19 monitoring and prediction. *Electronics, 10*(7), 799. https://www.mdpi.com/2079-9292/10/7/799
12. Baker, J. A., & Bookmark Bookmark Share WhatsApp Telegram Face. (n.d.). *No quarantine orders for close contacts of covid-19 cases as MOH urges personal responsibility*. CNA. Retrieved January 28, 2022, from https://www.channelnewsasia.com/singapore/covid-19-no-quarantine-orders-close-contacts-health-risk-warning-isolation-2232566

13. Song, J., Xie, H., Gao, B., Zhong, Y., Gu, C., Choi, K. S. (2021). Maximum likelihood-based extended Kalman filter for COVID-19 prediction. *Chaos Solitons Fractals, 146*, 110922. https://doi.org/10.1016/j.chaos.2021.110922. Epub 2021 Apr 2. PMID: 33824550; PMCID: PMC8017556.
14. Qin, J., You, C., Lin, Q., Hu, T., Yu, S., & Zhou, X. H. (2020). Estimation of incubation period distribution of COVID-19 using disease onset forward time: a novel cross-sectional and forward follow-up study. *Science advances, 6*(33), eabc1202. https://doi.org/10.1126/sciadv.abc1202#T1
15. Lai, S. H. S., Tang, C. Q. Y., Kurup, A., & Thevendran, G. (2021). The experience of contact tracing in Singapore in the control of COVID-19: highlighting the use of digital technology. *International orthopaedics, 45*(1), 65–69. https://www.ncbi.nlm.nih.gov/pmc/articles/PMC7666400/
16. Population and population structure—Latest data. *Singstat*. (n.d.). Retrieved January 19, 2022, from https://www.singstat.gov.sg/find-data/search-by-theme/population/population-and-population-structure/latest-data
17. Cao, C., Li, Y., Fu, S., Zhang, Y., Li, N., Hou, S., & Fan, H. (2021). Patient delay in a coronavirus disease 2019 (COVID-19) outbreak in Tianjin, China from January to February 2020. *Journal of the Formosan Medical Association*. https://doi.org/10.1016/j.jfma.2021.11.003
18. Kurohi, R. (2021, August 2). Moh apologises for delays, lapses in Covid-19 quarantine process. *The Straits Times*. Retrieved January 12, 2022, from https://www.straitstimes.com/singapore/health/moh-apologises-for-delays-lapses-in-issuing-of-quarantine-orders
19. He, X., Lau, E. H., Wu, P., Deng, X., Wang, J., Hao, X., Lau, Y. C., Wong, J. Y., Guan, Y., Tan, X., Mo, X., Chen, Y., Liao, B., Chen, W., Hu, F., Zhang, Q., Zhong, M., Wu, Y., Zhao, L., Zhang, F., et al. (2020). Temporal dynamics in viral shedding and transmissibility of COVID-19. *Nature medicine, 26*(5), 672–675. https://www.nature.com/articles/s41591-020-0869-5
20. Diekmann, O., Heesterbeek, J. A. P., & Metz, J. A. (1990). On the definition and the computation of the basic reproduction ratio R 0 in models for infectious diseases in heterogeneous populations. *Journal of Mathematical Biology, 28*(4), 365–382. https://doi.org/10.1007/BF00178324
21. Ritchie, H., Mathieu, E., Rodés-Guirao, L., Appel, C., Giattino, C., Ortiz-Ospina, E., Hasell, J., Macdonald, B., Beltekian, D., & Roser, M. (2020). *Coronavirus Pandemic (COVID-19)*. Published online at OurWorldInData.org. Retrieved from: https://ourworldindata.org/coronavirus (Online Resource)
22. Worldometer. https://www.worldometers.info/coronavirus/
23. Tian, T., & Huo, X. (2020). Secondary attack rates of COVID-19 in diverse contact settings, a meta-analysis. *The Journal of Infection in Developing Countries, 14*(12), 1361–1367. https://www.jidc.org/index.php/journal/article/view/33378276/2411
24. Madewell, Z. J., Yang, Y., Longini, I. M., Halloran, M. E., & Dean, N. E. (2020). Household transmission of SARS-CoV-2: A systematic review and meta-analysis. *JAMA network open, 3*(12), e2031756–2031756. https://jamanetwork.com/journals/jamanetworkopen/fullarticle/2774102
25. Sofonea, M. T., Roquebert, B., Foulongne, V., Verdurme, L., Trombert-Paolantoni, S., Roussel, M., ... & Alizon, S. (2022). From Delta to Omicron: analysing the SARS-CoV-2 epidemic in France using variant-specific screening tests (September 1 to December 18, 2021). MedRxiv, 2021-12. https://doi.org/10.1101/2021.12.31.21268583
26. Volz, E., Mishra, S., Chand, M., Barrett, J. C., Johnson, R., Geidelberg, L., Hinsley, W. S., Laydon, D. J., Dabrera, G., O'Toole, A., Amato, R., Ragonnet-Cronin, M., Harrison, I., Jackson, B., Ariani, C. V., Boyd, O., Loman, N. J., McCrone, J. T., Gonçalves, S., Jorgensen, D., et al. (2021). Transmission of SARS-CoV-2 Lineage B. 1.1. 7 in England: Insights from linking epidemiological and genetic data. *MedRxiv*, 2020-12. https://doi.org/10.1101/2020.12.30.20249034
27. Homma, Y., Katsuta, T., Oka, H., Inoue, K., Toyoshima, C., Iwaki, H., ... & Shinomiya, H. (2021). The incubation period of the SARS-CoV-2 B1. 1.7 variant is shorter than that of other strains. Journal of Infection, 83(2), e15-e17. https://doi.org/10.1016%2Fj.jinf.2021.06.011

28. Liu, Y., & Rocklöv, J. (2021). The reproductive number of the Delta variant of SARS-CoV-2 is far higher compared to the ancestral SARS-CoV-2 virus. *Journal of Travel Medicine.* https://academic.oup.com/jtm/article/28/7/taab124/6346388
29. Kang, M., Xin, H., Yuan, J., Ali, S. T., Liang, Z., Zhang, J., Hu, T., Lau, E. H. Y., Zhang, Y., Zhang, M., Cowling, B. J., Li, Y., & Wu, P. (2021, January 1). Transmission dynamics and epidemiological characteristics of Delta variant infections in China. *medRxiv*. Retrieved January 28, 2022, from https://doi.org/10.1101/2021.08.12.21261991V1
30. Sofonea, M. T., Roquebert, B., Foulongne, V., Verdurme, L., Trombert-Paolantoni, S., Roussel, M., Haim-Boukobza, S., & Alizon, S. (2022). From delta to omicron: Analysing the SARS-COV-2 epidemic in France using variant-specific screening tests (September 1 to December 18, 2021). *MedRxiv*. https://doi.org/10.1101/2021.12.31.21268583
31. Jansen, L., Tegomoh, B., Lange, K., Showalter, K., Figliomeni, J., Abdalhamid, B., Iwen, P. C., Fauver, J., Buss, B., & Donahue, M. (2021). Investigation of a SARS-CoV-2 B.1.1.529 (Omicron) Variant Cluster — Nebraska, November–December 2021. *MMWR. Morbidity and Mortality Weekly Report*, 70(5152), 1782–1784. https://doi.org/10.15585/mmwr.mm705152e3
32. Grimm, V., Mengel, F., & Schmidt, M. (2021). Extensions of the SEIR model for the analysis of tailored social distancing and tracing approaches to cope with COVID-19. *Science and Reports, 11*, 4214. https://doi.org/10.1038/s41598-021-83540-2
33. Chung, N. N., & Chew, L. Y. (2021). Modelling Singapore COVID-19 pandemic with a SEIR multiplex network model. *Science and Reports, 11*, 10122. https://doi.org/10.1038/s41598-021-89515-7
34. Ndaïrou, F., Area, I., Nieto, J. J., & Torres, D. F. (2020). Mathematical modeling of COVID-19 transmission dynamics with a case study of Wuhan. *Chaos, Solitons & Fractals, 135*, 109846.https://www.sciencedirect.com/science/article/pii/S0960077920302460
35. Grimm, V., Mengel, F., & Schmidt, M. (2021) Extensions of the SEIR model for the analysis of tailored social distancing and tracing approaches to cope with COVID-19. *Scientific Reports, 11*, 4214. https://doi.org/10.1038/s41598-021-83540-2
36. Piccolomini, E. L., & Zama, F. *Preliminary analysis of COVID-19 spread in Italy with an adaptive SEIRD model.* arXiv:2003.09909
37. *Appendix I: Comparison of PCR, serology and ... - ministry of health.* (n.d.). Retrieved June 7, 2022, from https://www.moh.gov.sg/docs/librariesprovider5/default-document-library/appendix-i47e4ee9f78bc41d0a14c14d70143467c.pdf
38. Today. (n.d.). Covid-19 protocols streamlined: Art to be primary mode of testing, no more quarantine orders and exit swabs from Oct 11. *TODAY*. Retrieved July 7, 2022, from https://www.todayonline.com/singapore/covid-19-protocols-streamlined-art-be-primary-mode-testing-no-more-quarantine-orders-and
39. Ghostine, R., Gharamti, M., Hassrouny, S., & Hoteit, I. (2021). An extended SEIR model with vaccination for forecasting the COVID-19 pandemic in Saudi Arabia using an ensemble Kalman filter *Mathematics, 9*, 636 https://www.mdpi.com/2227-7390/9/6/636/pdf
40. Arroyo-Marioli, F., Bullano, F., Kucinskas, S., & Rondón-Moreno, C. (2021). Tracking R of COVID-19: A new real-time estimation using the Kalman filter. PloS one, 16(1), e0244474. https://doi.org/10.1371/journal.pone.0244474
41. Lal, R., Huang, W., & Li, Z. (2021). An application of the ensemble Kalman filter in epidemiological modelling. *PLoS ONE, 16*(8), e0256227. https://doi.org/10.1371/journal.pone.0256227

Determining Effects of Urea on the Structure of Dengue Virus Serotype 2 Using Fluorescence Spectroscopy

Aayush Arya, Joseph Mu Jie Poon, Kamal Kant Sharma, Xin Ee Yong, and Thorsten Wohland

Abstract Dengue viruses pose a growing public health hazard to an estimated 3.9 billion individuals at risk of infection, prioritising the development of effective vaccines and therapeutics to prevent and treat disease. While antibodies with high affinities for epitopes on the dengue virus have been developed, the dynamism of the viral envelope occludes major antigenic sites, impairing neutralisation potency. The growth of fluorescence spectroscopy, and especially time-resolved fluorescence, provides exceptional avenues for measurements at the level of single-molecule detection, and is used extensively in this study to examine viral morphologies. Through assessing Förster Resonance Energy Transfer, this study serves as a probe for elucidating changes in the antigenic surface of DENV virions and for the quantitative determination of changes in the populations of viral conformations when subjected to increasing concentrations of urea. The findings of this study can be applied to shed light on optimising antibody development.

Keywords Dengue · DENV-2 · FRET · trFRET · Urea · Fluorescence spectroscopy · Confocal microscopy

---

A. Arya · J. M. J. Poon
NUS High School of Mathematics and Science, 20 Clementi Avenue 1, Singapore 129957, Singapore

K. K. Sharma · X. E. Yong · T. Wohland (✉)
Department of Biological Sciences, National University of Singapore, Centre of Bioimaging Sciences, 14 Science Drive 4, Singapore 117543, Singapore
e-mail: twohland@nus.edu.sg

Centre for Bioimaging Sciences, National University of Singapore, 14 Science Drive 4, Singapore 117557, Singapore

T. Wohland
Department of Chemistry, National University of Singapore, 14 Science Drive 4, Singapore 117543, Singapore

H. Guo et al. (eds.), *IRC-SET 2022*,
https://doi.org/10.1007/978-981-19-7222-5_43

1 Introduction

Dengue viruses (DENV) cause the most common mosquito-borne viral disease in humans. Dengue causes an estimated 390 million cases of infection per year [1], and is endemic in the conducive environmental conditions of Singapore [2], with Asia representing ~70% of the global burden of disease [3].

DENV is a member of the flavivirus genus, encompassing a group of compact (40–60 nm in diameter), enveloped, positive-sense single-stranded RNA viruses. DENV, West Nile Virus, Yellow Fever Virus and Zika Virus are examples of mosquito-borne flaviviruses.

The ectodomain of the DENV-2 E protein is composed of three principal domains (E-DI, E-DII, E-DIII) and a stem domain that anchors to a trans-membrane region. Of these domains, studies suggest the interaction of E-DIII with target cell receptors facilitates receptor-mediated endocytosis of the virion [4–7]. Antibody development for DENV further evidences the localisation of major antigen sites to E-DIII [7, 8].

Two prominent phenomena contribute to pronounced structural heterogeneity within flavivirus populations.

First, flaviviruses are shown to exist in a series of subpopulations of related structures at equilibrium [9–12], interchanging through a process termed viral "breathing" that is pronounced at vertices on E-DIII [12]. This spectrum of viral conformations results in slightly varied epitope exposure and accessibility depending on the arrangement of surface E proteins. It is suggested that the binding of antibodies to cryptic epitopes exposed briefly through the dynamic changes of the viral envelope inhibits the previous range of viral breathing due to steric strains induced by the antibody [12], indicating that antibodies can dictate this equilibrium of morphologies.

Second, studies report inefficiency in the maturation of flaviviruses whilst cleaving the precursor-to-membrane (prM) structural protein during viral egress from a host cell [13, 14]. The cleavage is a vital step in the transition from non-infectious to infectious viral particles [15]. This results in a mosaic of partially mature infectious virions with varying protein arrangements [6, 16]. Differences in epitope exposure in a highly heterogenous viral population significantly affect the neutralisation potency of antibodies.

Primary infection by DENV provides durable immunity to subsequent re-infection by a homologous DENV serotype, but secondary infection by heterologous serotypes is frequent in endemic areas and reported as a leading cause of severe dengue [16, 17].

Antibodies generated during primary infection may show modest affinity for epitopes on heterologous serotypes, but overall binding falls below the required stoichiometry for neutralisation [9, 18, 19], thus displaying low neutralisation potencies towards heterologous serotypes. Consequently, cross-reactive antibodies facilitate antibody-dependent enhancement of DENV infection that promotes entry of virions into cells expressing Fc-γ receptors [17].

Prolonged incubation of modified West Nile Viruses showed significant increases in neutralisation potency when virions were incubated in the presence of antibodies, suggesting a time-dependent change in epitope accessibility driven by the entrapment of exposed epitopes by antibodies [9]. An analogous prolonged incubation could not be performed on DENV due to a high rate of intrinsic decay of infectivity. On the other hand, time-resolved measurements using confocal microscopy can be achieved over much shorter periods of time and provide the relative abundances of different conformations.

In our experimentation, DENV-2 is dual-labelled using tetramethyl-indocarbocynanine, 'DiI', and AlexaFluor 488-TFP Ester (AF488-TFP). The highly lipophilic C-18 alkyl tails of DiI are incorporated into the lipid bilayer of the viral envelope [20]. The bilayer is evenly stained through lateral diffusion of the DiI molecules [21]. AF488-TFP binds to lysine residues on E-DIII, proximal to the region undergoing conformational change [22]. The distance between the fluorophores thereby accurately represents the motion of E-DIII relative to the bilayer.

This study aims to elucidate the effect of urea on the structure of the DENV viral envelope, and changes to the heterogeneity of the viral population due to disruption by urea.

These conformational changes are assessed through the phenomenon of Förster Resonance Energy Transfer (FRET). FRET is premised on the non-radiative energy transfer between an excited donor fluorophore and a receiving acceptor fluorophore that proceeds without any intermediate photons [11]. FRET occurs when the emission spectrum of the donor fluorophore overlaps with the excitation spectrum of the acceptor fluorophore-conditions satisfied by AF488-TFP and DiI. The efficiency of this energy transfer (FRET efficiency), observed as a change in relative intensity of the emission peaks of the donor or acceptor fluorophores, is contingent exclusively on the distance between the fluorophores, given no changes to the external environment [23]. Steady-state and time-resolved FRET (trFRET) measurements are employed to determine changes to viral structure.

2 Methodology

2.1 Titration of DENV-2 Against Urea

An initial viral sample is prepared with 1 μL of labelled virus in 49 μL of buffer. Taken as 0 M urea, urea is then added sequentially to raise the concentration of urea in the sample in increments of 0.1 M, until 3 M. A final addition is made to increase the concentration to 4 M. Samples are excited at 488 nm to obtain emission spectra (510–610 nm).

2.2 Confocal Microscopy of DENV-2 in Urea

An initial viral sample is prepared with 2 μL of labelled virus in 48 μL of buffer, and taken as 0 M urea. Separate samples with urea concentrations of 0.5, 1, 2 and 3 M are prepared, with the same volume of labelled virus and total volume of 50 μL. To measure trFRET, time-correlated single photon counting (TCSPC) is performed with a confocal microscope. Samples are excited at 488 nm and their fluorescence decay is measured.

3 Results

3.1 Steady-State FRET Measurements Through Urea Titration

Here, we define an 'expanded' and 'unexpanded' conformation of the virus as the two principal conformations observed. Expansion of the viral envelope occurs due to urea-mediated denaturation of E-DIII.

AF488-TFP is maintained at a final virus labelling concentration of 24 μM, while DiI is adjusted from 500 to 10 μM. 0.1% v/v Triton X-100 (TX100) is added to dissolve the lipid bilayer [24] and free labelled fluorophores from their viral binding sites to test for presence of fluorophores at different DiI labelling concentrations (Fig. 1a).

Our observed emission spectra yield peak emission wavelengths of 516 nm and 567 nm for AF488-TFP and DiI, respectively. Subsequently, our trials are conducted using DENV-2 labelled with 24 μM AF488-TFP and 10 μM DiI.

To investigate the disruption of DIII, DENV-2 is subjected to varying concentrations of urea, from 0 to 4 M urea in viral sample.

To eliminate the potential scattering of light by urea molecules, especially at higher urea concentrations, blank emission spectra of samples with urea at all tested concentrations, but without any viruses, prepared in buffer solution are taken and subtracted from the corresponding emission spectra of labelled DENV-2. The intensities recorded in each emission spectrum are adjusted for the dilution of DENV-2 in sample due to sequential additions of urea.

The virions are excited at the excitation wavelength of AF488-TFP (donor fluorophore), followed by a non-radiative energy transfer from the donor fluorophore to DiI (acceptor fluorophore), producing two peaks at the emission wavelengths of the donor and acceptor fluorophores. As the fluorophores separate due to denaturation of E-DIII, the efficiency of this energy transfer decreases, thus more energy is retained and emitted directly by the donor, resulting in a higher donor peak (Fig. 1B).

Donor intensity is defined as the recorded intensity of the emission spectrum at the peak emission wavelength of AF488-TFP (517 nm), and acceptor intensity as the recorded intensity at the peak emission wavelength of DiI (567 nm). The increase

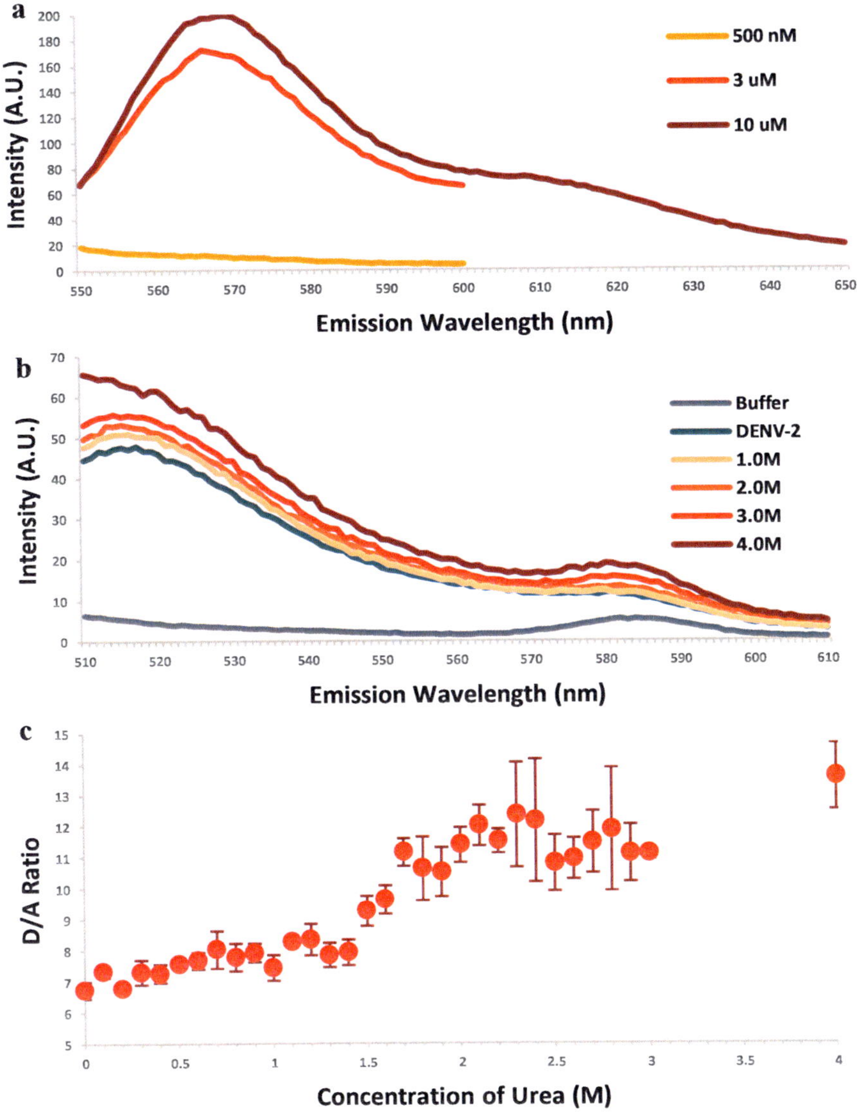

Fig. 1 FRET shows expansion of E-DIII under influence of urea. **a** Emission spectra of dual-labelled DENV-2 in buffer with 0.1% v/v TX100, excited at 530 nm, close to the excitation peak of DiI. The DiI peak at 567 nm is lacking when using 500 nm (yellow line) and best observed when using 10 μM (maroon line). **b** The emission spectra for dual-labelled DENV-2 (showing select urea concentrations), excited at 488 nm. Spectra demonstrate a noticeable increase in the donor intensity at higher urea concentrations. The acceptor peak (567 nm) overlaps with the tail of the Raman peak of the buffer solution (~582 nm). **c** The trend of the D/A ratio. Increasing urea concentration from 0 to 1.3 M shows a gradual increase in D/A ratio from 6.73 ± 0.28 to 7.86 ± 0.39. A more significant increment in *D/A* ratio is observed between 1.4 M urea and 2.3 M urea, with an increase from 7.94 ± 0.41 to 12.35 ± 1.68

of the ratio of the donor to acceptor intensity (*D*/*A* ratio) is used as an index to demonstrate a stepwise decrease in FRET efficiency.

Within the range of 1.4–2.3 M urea, the significant increase in D/A ratio likely indicates a large-scale conformational change to DIII elicited (Fig. 1c). Our findings show a 37.82–83.64% increase in the *D*/*A* ratio in this range, relative to the initial 0 M urea sample. From the steady-state measurement alone, it is unclear whether this represents a continuum of denaturation of E-DIII, or a net shift towards an expanded conformation with the same extent of E-DIII denaturation. We show using trFRET the validity of the latter mechanism.

The *D*/*A* ratio plateaus in the range of 2.5 M urea to 3 M urea. There is a notable rise of the D/A ratio from 3 to 4 M (intermediate values not sampled). This could indicate a second major conformational change to E-DIII, or an unravelling of the domain at such high urea concentrations, the latter of which is not an expected result in antibody development.

3.2 *Time-Resolved FRET Through Confocal Microscopy*

Denaturation of E-DIII creates a range of donor–acceptor (D–A) distances. Results from steady-state FRET showed an increase in *D*/*A* ratio, which provided us information on the apparent D–A distance. However, this could not determine the D–A distance distribution and could not reveal the heterogeneity of DENV-2 membrane conformation. To achieve a more qualitative analysis, a confocal microscope was used to provide more information on the trFRET.

Time-correlated single photon counting (TCSPC) was performed to measure the fluorescence decay in a time domain. The dual labelled virus is excited by a pulsed laser source with a high repetition rate. Single photon events are detected with a photomultiplier and their time of arrival with respect to the excitation pulse is measured. By counting many events a histogram of the photon distribution over time is built up.

The photon distribution is corrected with a calculated instrument response function (IRF) of ~0.2 ns, and fitted with a biexponential decay model. An average fluorescence lifetime of ~2.69 ns was calculated and best fitted with two discrete (long and short) lifetime components of ~4.01 ns and ~0.47 ns, having population of 62.6% and 37.4% respectively, indicating viral structural heterogeneity. This is consistent with literature, which showed an average fluorescence lifetime of ~2.42 ns, best fitted with two discrete (long and short) lifetime components of ~3.72 ns and ~1.18 ns [20, 22]. The discrete lifetimes remained approximately constant over the increase in urea concentration (from 0 to 3 M), and only the population of these lifetimes changed. This indicates a population interchange between an expanded and unexpanded virus conformation with respect to urea concentration. The observed conformations change, resembles the temperature-dependent conformation transition from a smooth viral surface to a bumpy viral surface above 33 °C [20, 22].

An increase in population of the "smooth" virus (α_1) will result in an equivalent decrease in the population of the unexpanded virus (α_2). Thus, we will only represent α_1 and the average fluorescence lifetime (τ).

An increase in urea concentration from 0 to 1 M in steps of 0.5 M and from 1 to 3 M in steps of 1 M, resulted in an increase in τ from ~2.69 to ~3.10 ns, and in α_1 from 62.6 to 78.0% (Fig. 2). As FRET provides a competing pathway by which excited state energy of the donor fluorophore (AF448-TFP) can decay, an increase in FRET efficiency results in a shorter fluorescence lifetime and vice versa. The increase in τ indicates a decrease in FRET efficiency, which indicates that the average distance between E-DIII and the viral membrane bilayer increased. The overall difference in α_1 at 3 M urea, indicates that ~15.4% of DENV-2 virions undergone large-scale conformational change from unexpanded to expanded state.

The findings from trFRET agree with the steady-state FRET results that suggest a large-scale conformational change elicited within the range of 1.4–2.3 M. An increase in urea concentration from 1.0 to 2.0 M, leads to an increase in α_1 from 69.3 to 74.48%. This indicates that 5.18% of virions undergone large-scale conformation change to the expanded state. This is significantly higher than the 3.56% increase in α_1 when the urea concentration increased from 2.0 to 3.0 M. This also agrees with the *D*/*A* ratio plateau from 2.5 to 3.0 M.

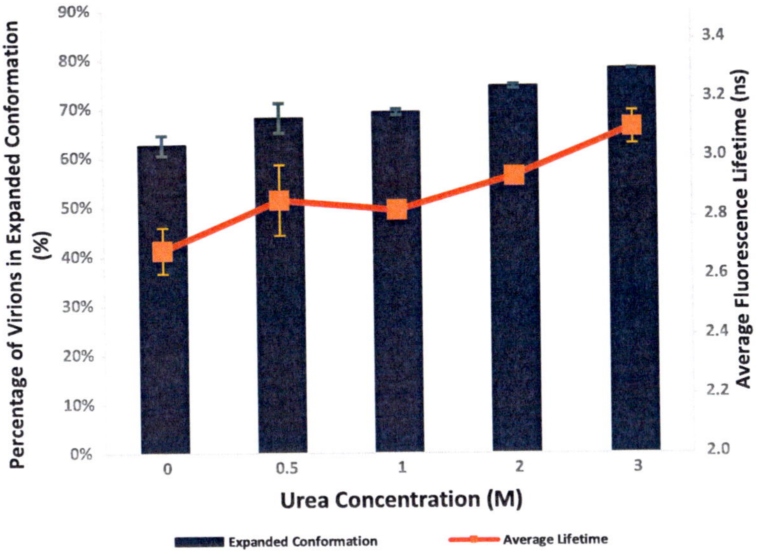

Fig. 2 Structural transition of DENV-2 under influence of urea. The lifetime traces for dual-labelled DENV-2 viruses were fitted to a biexponential decay model. In all cases, the c2 values were close to 1 and the weighted residuals as well as their autocorrelation were distributed randomly around 0, indicating a good fit. Average fluorescent lifetime of DENV-2 (orange squares) and percentage of virions in expanded conformation (blue bars)

4 Conclusion

Our findings show an expansion of the DENV-2 NGC structure correlated to the denaturation of E-DIII in the presence of urea, with a significant outward expansion of E-DIII in the range of 1.4–2.3 M urea. Our trFRET measurements indicate the persistence of two principal virus morphologies, and that urea (at 3 M) induces a shift of up to 15.4% of virions towards the expanded conformation. It is anticipated that the steric strains induced via antibody binding will similarly elicit such conformational changes to the viral envelope that can be traced through fluorescence spectroscopy and trFRET. This holds potential for the determination of virus morphologies favoured through antibody binding, and of the epitopes exposed in these specific conformations.

5 Future Work

More can be explored about the trFRET results. An increase in urea concentration from 0 to 0.5 M leads to an increase in α_1 from 62.60 to 68.13%. This indicates that 5.53% of virions undergone large scale conformation, the steepest increase within the measured increments. The average fluorescence lifetime also experienced the steepest increase from 2.69 to 2.86 ns, indicating a significant decrease in FRET efficiency. However, this contradicts the subtle decrease in overall FRET efficiency observed from steady-state FRET. To better understand the initial changes caused by the addition of urea, fluorescence lifetime imaging FRET (FLIM-FRET) microscopy can be performed, within the interval of 0–0.5 M urea, to report on the photophysical events occurring in the proximity of the FRET D-A fluorophore pair.

Appendix: Method Details

Dual-labelling of DENV-2 with AlexaFluor 488-TFP Ester and DiI

AF488-TFP and DiI are purchased from ThermoFisher Scientific, Singapore. Stock solutions of AF488-TFP and DiI are prepared in dimethylsulfoxide, and their concentrations are determined using molar extinction coefficients of 71,000 M^{-1} cm^{-1} and 148,000 M^{-1} cm^{-1}, respectively. The fluorophores are simultaneously added to 1.0×10^8 plaque forming units of the purified DENV-2 NGC viruses in a buffer containing 50 mM HEPES, 150 mM NaCl, adjusted to pH 7.4. The sample is incubated for 1.5 h at 25 °C. Free fluorophore molecules are removed via gel filtration (MicroSpinTM S-200h columns, GE Healthcare, Singapore).

Titration of DENV-2 Against Urea

Fluorescence data is collected using Cary Eclipse Spectrophotometer (Agilent) using Hellma® fluorescence cuvettes of an internal chamber volume of 45 μL. For titration, samples are excited at 488 nm, 10 nm slit width, 800 V at slow speed at 25 °C. Reported values are mean and standard error from at least three repetitions.

Confocal Microscopy of DENV-2 in Urea

trFRET measurements of the fluorescence lifetimes were taken using a commercial Olympus FV1200 laser scanning confocal microscope (IX83, Olympus, Singapore) using a time-resolved LSM upgrade kit (Microtime 200, PicoQuant, GmbH, Berlin, Germany). The samples were excited using a 485 nm pulsed diode laser with a repetition rate of 20 MHz and 29 mW of power (PDL series, Sepia II combiner module). The beam was reflected by a dichroic mirror (DM405/485/543/635 band pass, Olympus, Singapore) and the scanning unit, then focused into the sample by a water immersion objective (360, NA 1.2; Olympus, Tokyo, Japan). The same objective collects the fluorescence followed by a pinhole (120 mm) to eliminate out-of-focus light. The fluorescence signal was spectrally divided into donor (green) and acceptor (red) channels by a 560DCLP dichroic mirror. The donor fluorescence was recorded by a set of single molecule avalanche photodiodes (SPADs) (SPCM-AQR-14, PerkinElmer Optoelectronics, Quebec, Canada), through a 520/35 band pass emission filter (Omega, VT). A time correlated single photon counting card (Time-Harp 260, PicoQuant) processes this donor signal to build up the histogram of photon arrival times. The trFRET measurements were recorded for 300 s. The mean lifetime angle (τ) was calculated from the individual fluorescence lifetimes (τ_i) and their relative amplitudes (α_i) according to $(\tau) = \sum \alpha_i \tau_i)$. Donor fluorescence lifetime decay data were treated using the software SymPhoTime 64 (PicoQuant, GmbH). Reported values are mean and standard error from at least three repetitions.

References

1. Dowd, K., Mukherjee, S., Kuhn, R. J., & Pierson, T. C. (2014). Combined effects of the structural heterogeneity and dynamics of flaviviruses on antibody recognition. *Journal of Virology, 88*(20), 11726–11737.
2. Rajarethinam, J., et al. (2018). Dengue in Singapore from 2004 to 2016: cyclical epidemic patterns dominated by serotypes 1 and 2. *American Journal of Tropical Medicine and Hygiene, 99*(1), 204–210.
3. Dowd, K., & Pierson, T. (2011). Antibody-mediated neutralization of flaviviruses: A reductionist view. *Virology, 411*(2), 306–315.

4. Lok, S. M., Kostyuchenko, V., Nybakken, G. E., & Holdaway, B. (2008). Binding of a neutralizing antibody to dengue virus alters the arrangement of surface glycoproteins. *Nature Structural and Molecular Biology, 15*, 312–317.
5. Pierson, T., & Diamond, M. (2012). Degrees of maturity: The complex structure and biology of flaviviruses. *Current Opinion in Virology, 2*(2), 168–175.
6. Elshuber, S., Allison, S., Heinz, F., & Mandl, C. (2003). Cleavage of protein prM is necessary for infection of BHK-21 cells by tick-borne encephalitis virus. *Journal of General Virology, 84*(1), 183–191.
7. Chen, Y., et al. (1997). Dengue virus infectivity depends on envelope protein binding to target cell heparan sulfate. *Nature Medicine, 3*(8), 866–871.
8. Chiu, M., & Yang, Y. (2003). Blocking the dengue virus 2 infections on BHK-21 cells with purified recombinant dengue virus 2 E protein expressed in Escherichia coli. *Biochemical and Biophysical Research Communications, 309*(3), 672–678.
9. Spence, M. T. Z., & Johnson, I. D. (2010). *The molecular probes handbook: A guide to fluorescent probes and labeling technologies* (11th ed.). Life Technologies Corporation.
10. Zhang, X., Sun, L., & Rossmann, M. G. (2015). Temperature dependent conformational change of dengue virus. *Current Opinion in Virology, 12*, 109–112.
11. Lakowicz, J. R. (2006). *Principles of fluorescence spectroscopy* (3rd ed.). Springer.
12. Bhatt, S., Gething, P. W., & Brady, O. J. (2013). The global distribution and burden of dengue. *Nature, 496*(7446), 504–507.
13. Brady, O. J., Gething, P. W., & Bhatt, S. (2012). Refining the global spatial limits of dengue virus transmission by evidence-based consensus. *PLOS Neglected Tropical Diseases, 6*(8), e1760.
14. Crill, W. D., & Roehrig, J. T. (2001). Monoclonal antibodies that bind to domain III of dengue virus E glycoprotein are the most efficient blockers of virus adsorption to Vero cells. *Journal of Virology, 75*(16), 7769–7773.
15. França, A., et al. (2012). Effects caused by triton X-100 on lipid bilayers of different composition. *Biophysical Journal, 102*(3), 290.
16. Zhou, Y., et al. (2013). The mechanism of differential neutralization of dengue serotype 3 strains by monoclonal antibody 8A1. *Virology, 439*(1), 57–64.
17. Halstead, S. B., & O'Rourke, E. J. (1977). Dengue viruses and mononuclear phagocytes. I. Infection enhancement by non-neutralizing antibody. *Journal of Experimental Medicine, 146*(1), 201–217.
18. Sarathy, V., et al. (2017). A DENV-2-type-specific monoclonal antibody binds to the DENV-complex-reactive antigenic site on envelope protein domain 3. *Journal of General Virology, 98*(6), 1299–1304.
19. Junjhon, J., et al. (2020). Influence of pr-M cleavage on the heterogeneity of extracellular dengue virus particles. *Journal of Virology, 84*(16), 8353–8358.
20. Sharma, K. K., et al. (2019). Infectivity of dengue virus serotypes 1 and 2 is correlated with e-protein intrinsic dynamics but not to envelope conformations. *Structure, 27*(4), 618–630.
21. Keelapang, P., et al. (2004). Alterations of pr-M cleavage and virus export in pr-M junction chimeric dengue viruses. *Journal of Virology, 78*(5), 2367–2381.
22. Henchal, E. A., Gentry, M. K., McCrown, J. M., & Brandt, W. E. (1982). Epitope analysis of antigenic determinants on the surface of dengue-2 virions using monoclonal antibodies. *The Journal of Tropical Medicine and Hygiene, 31*(4), 830–836.
23. Della-Porta, A. J., & Westaway, E. G. (1978). A multi-hit model for the neutralization of animal viruses. *Journal of General Virology, 38*(1), 1–19.
24. Pierson, T. C., et al. (2007). The stoichiometry of antibody-mediated neutralization and enhancement of West Nile virus infection. *Cell Host & Microbe, 1*(2), 135–145.

Honeypot for Cybersecurity Threat Intelligence

Ren Rui Tan, Simon Eng, Kiam Cheng How, Yongqing Zhu, and Paul Wu Horng Jyh

Abstract Small and Medium-sized Enterprises (SMEs) have limited resources to guard their IT infrastructures because of high upfront costs and fully engaged manpower, resulting in lack of focus on cybersecurity. Though cybersecurity threat intelligence derived from honeypots can help businesses to mitigate cybersecurity risks, current implementations target huge enterprises that have in-house cybersecurity teams to digest the intelligence and implement countermeasures against emerging threats. In this paper, a honeypot mimicking SME web portals will be set up to acquire insights into cyberattacks. The honeypot aims to provide knowledge for system administrators from the SMEs to understand the current cyberthreat landscape and the threat actor's behavior, learning about the latest types of cyberthreats and identifying potential underway cyberattacks in the organization. The web activity logging feature from the honeypot allows the recording of all requests made to the honeypot and is being closely monitored. Analysis is carried out for all potential attacks and attacker details extracted from the web logs. After analyzing, the results will be referenced as enhancements to implement on the next phase of honeypot deployment to yield better performance. After three phases of the honeypot deployment is completed, a dashboard is developed to summarize the findings for all the honeypot deployment phases and provides the attack details, frequency, and count of every type of vulnerabilities detected during each deployment phase. The honeypot

R. R. Tan · S. Eng · K. C. How · Y. Zhu (✉) · P. W. H. Jyh
School of Science and Technology, Singapore University of Social Sciences,
463 Clementi Road, Singapore 599494, Singapore
e-mail: yqzhu@SUSS.edu.sg

R. R. Tan
e-mail: rrtan003@SUSS.edu.sg

S. Eng
e-mail: simoneng001@SUSS.edu.sg

K. C. How
e-mail: kchow002@SUSS.edu.sg

P. W. H. Jyh
e-mail: paulwuhj@SUSS.edu.sg

H. Guo et al. (eds.), *IRC-SET 2022*,
https://doi.org/10.1007/978-981-19-7222-5_44

can achieve its objective to inculcate the importance of cyberthreat security protection to the SMEs through the studies carried out on the honeypot.

Keywords SME · Cybersecurity threat intelligence · Web honeypot · Threat intelligence dashboard first section

1 Introduction

Cyberthreats targeting the SMEs are currently on the rise, with increasing attack numbers especially during the COVID-19 pandemic and attack techniques growing more sophisticated. There is a team of threat actors that are actively exploiting the SME's IT infrastructure with ransomwares targeting businesses in the education, finance, event planning, and manufacturing sectors [1]. To avoid becoming a victim of cybercriminals, these organizations must understand about emerging threats and deploy adequate cybersecurity measures for their IT systems.

The SMEs plays a major role to the Singapore economic; however, many are not well prepared to deal with cyberattacks and their impact. A study has shown that within the 30% of the SMEs that were forced to adopt digitalization during the COVID-19 pandemic [2], the majority of them had to postpone the implementation of digitalization mainly due to high cost and lack of digitally skilled workforce. Thus, SMEs need to proactively focus on the aspect of cybersecurity as a security breach can result in financial losses, damages to reputation, and loss of client's trust.

Honeypot is a security mechanism widely used to detect potential system vulnerabilities, so security engineers and/or system administrators can study to improve the security policies. It is known that majority of the existing honeypot projects do not focus on the SMEs. Similarly, there is a lack of studies measuring the impact of threat actors targeting the SMEs. At present, there is no honeypot projects that focus on imitating a genuine SME website, setup to attract attacks from threat actors and study the outcomes. There is no existing dashboard to assist the system administrators of the SMEs to identify high-risk vulnerabilities by determining the Common Vulnerabilities and Exposures (CVE) score. This results in the system administrators not being able to patch the critical vulnerabilities in time for their live web portals, exposing these portals to additional unnecessary cyberattacks which could have been avoided.

In this paper, we propose a honeypot to gather intelligence about current cybersecurity attacks targeting the web portals of the SMEs. The proposed honeypot aims to provide the SMEs more targeted help in the aspect of cybersecurity while reducing the resources required to identify and prevent cyberthreats with application of threat intelligence. This paper is organized as follows. Section 1 is the introduction, and Sect. 2 as followed is the literature review. Section 3 will present the design and approach of the honeypot. Section 4 illustrates the implementation for the components of the honeypot. Section 5 introduces the results of the individual deployment phases of the honeypot. The last Sect. 6 will conclude the paper.

2 Literature Review

A honeypot is a computer system which mimics a vulnerable system to lure attackers to believe the targeted system is genuine [3]. In businesses, implementation of the honeypot technologies aims to divert attacker's attention away from production networks that facilitate business operations. Valuable information including attack methodology and methods can also be collected from attackers when they target the honeypot networks.

The primary design of the honeypot is based on the research type containing trackable data which allows tracing to compromised areas for analyzing the attacks. Both high- and low-interaction honeypots provide retrieval of forensic data during deployment [4].

Within businesses, honeypot is used for years to protect business operations. Honeypots cover many areas which traditional intrusion-detection systems (IDS) are lacking according to security experts [5]. However, honeypots are currently used by firms in the technology and defense sectors but are not adopted among the SMEs. Intelligence derived from the honeypots designs for SMEs will provide an attractive low-cost method to improve their cybersecurity protection.

After some trials, a honeypot was set up using an online web hosting service call DreamHost. The honeypot mimics an academy portal to attract hackers to strike. The web portal's activity log files are retrieved regularly for analysis with distinctive vulnerabilities selected to be implemented onto the web portal with the aim to attract more attackers.

Instead of the commonly used Indicator of Compromise (IoC) methodology to represent cybersecurity threat intelligence, the Indicator of Attack (IoA) methodology is used. IoC relates forensic data collected after a cyberattack has taken place to understand how the attack took place and to identify the attackers. As the focus of our study is on the detection of attacks and to alert the administrator to the presence of attempts to compromise the SME networks [6], we adopt the IoA methodology in our study.

3 Design and Approach of the Honeypot

This paper proposes a honeypot using a web portal with vulnerabilities derived from the detected attack results analysis to maximize the effectiveness and exposure to the threat actors. The system targets to study the behavior of threat actors and the latest types of cyberthreats which generates the threat intelligence that will help to improve the SME's cybersecurity protection.

3.1 Architecture of the Honeypot

System architecture of the proposed honeypot consists of the web hosting service, acquiring of the Domain Name Service (DNS), a website imitating a SME organization, and the Threat Intelligence Dashboard (Fig. 1).

Similar to typical SMEs, the honeypot leverages hosting service provider to host web contents to save costs. The honeypot implemented a web portal offering primary level education, kindergarten, and child-care services. For a start, the web portal is intentionally designed to be accessed via non-secure Hypertext Transfer Protocol (HTTP) to attract attackers to attack the system containing fictional student particulars and school fees payment details.

During the deployment of the honeypot, the log files are regularly retrieved using the web logging feature for carrying out analysis to gain threat intelligence. The log entries are viewable on real time using Secure Shell (SSH) method, and the log files are downloadable by using Secure File Transfer Protocol (SFTP).

A web domain name is acquired to make the honeypot look more legitimate. The domain name for the initial website setup of the honeypot is "www.wonderfulraylearning.com" featuring a fictional local educational academy known as Wonderful Ray Learning Centre.

As limited attacks are observed for the initial design, the honeypot is revamped into a mock online Bookstore website, using the non-secure HTTP Internet Protocol with the domain name changed to "www.bigraybookstore.com". It features typical bookstore website theme, includes login and payment function, and a section with a list of popular authentic book titles published by well-known authors which aims to attract threat actors more effectively.

The Threat Intelligence Dashboard illustrates the cyberthreats gathered from all deployment phases and is divided into three sections with each representing an individual deployment phase of the web portal honeypot. The gathered findings from

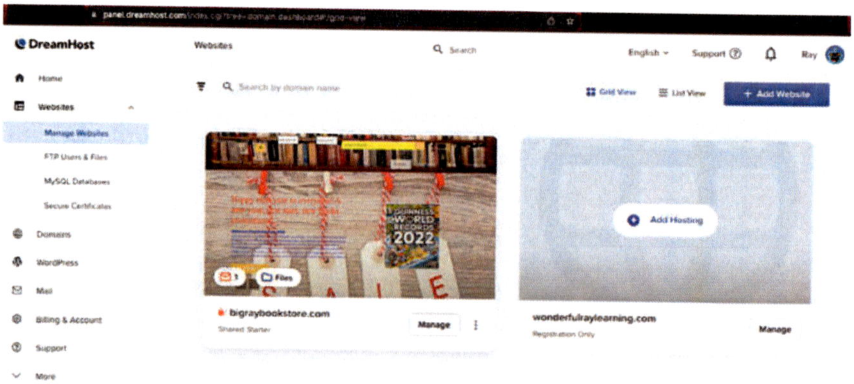

Fig. 1 Honeypot using a web portal

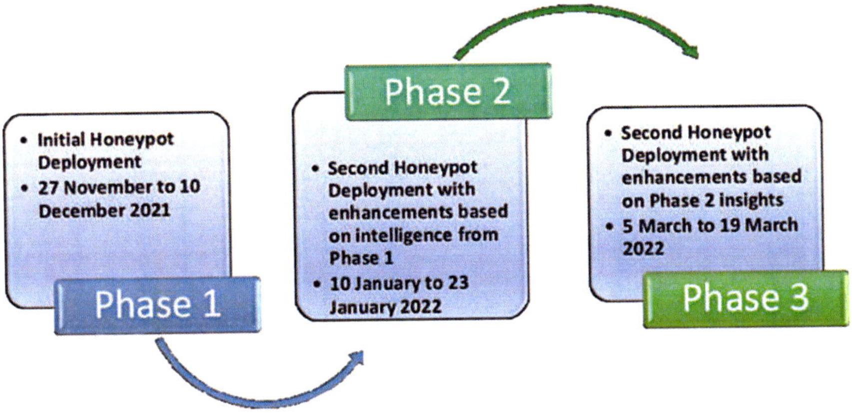

Fig. 2 Web portal honeypot deployment phases

each deployment phases will be generalized in point forms on the dashboard main page with accessing further details of individual vulnerabilities available through clickable links. It contains the analysis of the threat actors IP addresses for further understanding of their behavior.

3.2 *Approach*

As there is little prior literature on honeypot for SMEs, an adaptive 3-phase approach is used—by deploying the honeypot for approximately 2 weeks, enhancements will be made to the setup to lure more attackers in order to gain richer threat intelligence. Figure 2 illustrates the overall approach taken.

4 Implementation

Implementation of the web portal honeypot includes various hardware setups and software development for web portal, data store, process, and presentation.

4.1 *Web Hosting*

The web hosting service, web logging feature, and a simulated website are used to achieve an operatable honeypot. The DreamHost online web hosting service is selected to host the honeypot website due to its cost-efficiency and reliability. For the collection of cyberthreat feeds, the web logs are frequently reviewed to track for any data feeds detected from threat actors.

During the trial phase, the initial web honeypot is deployed to stabilize for a period of three days for web crawlers to pick up and DNS services to register the domain name. Proceeding to the next step can be carried out only when the hunt for attacks from threat actors is confirmed. Upon retrieving the web logs after a few days of deployment, the data feeds from the web logs detected traces of potential attacks. After researching on these data feeds, it is evident that at least three attacks are detected.

These detected attacks enable enhancements to the honeypot through identifying the vulnerabilities targeted by these attacks as feedback and then implementing the same vulnerabilities into the honeypot with the aim to increase the attack numbers. There are three deployment phases.

4.2 *Dashboard*

After the results for each deployment phase of the web portal honeypot are concluded, the grouping of attacks, assignment to appropriate category is performed before compiling the data into the Threats Intelligence Dashboard. The dashboard is built using HTML5. The results from each phase are divided into two groups, association with CVE or without CVEs. The dashboard is tailored for SME system administrators to view the details of each detected type of attack and the analysis of the threat actors' IP addresses.

5 Results

After implementing the web honeypot, the daily web logs from the each of the honeypot deployment phase are thoroughly analyzed to capture all the data feeds which present potential threats. These data are then categorized according to threat types before finalizing the findings as conclusion for the honeypot phase. Below represents the findings for each deployment phases.

5.1 *Phase 1*

Before the first phase, the web honeypot is deployed on November 20, 2021, featuring a website of a fictitious local educational academy. Upon detecting suspicious requests during regular tracking on the web logs, we derived three vulnerabilities, which are then implemented to enhance the honeypot to attract attackers during the first deployment phase.

Using the DreamHost web portal, the first phase of the web honeypot is deployed for two weeks in December 2021 with the implementation of the above-mentioned vulnerabilities. Analysis of the logs reveals that the majority of cyberattacks consists of WordPress application attacks while a handful of attacks target a vulnerability by requesting a list of directory names (see Fig. 4). Nearly all of the captured cyberattacks are carried out by botnets based on the identified attacks characteristics.

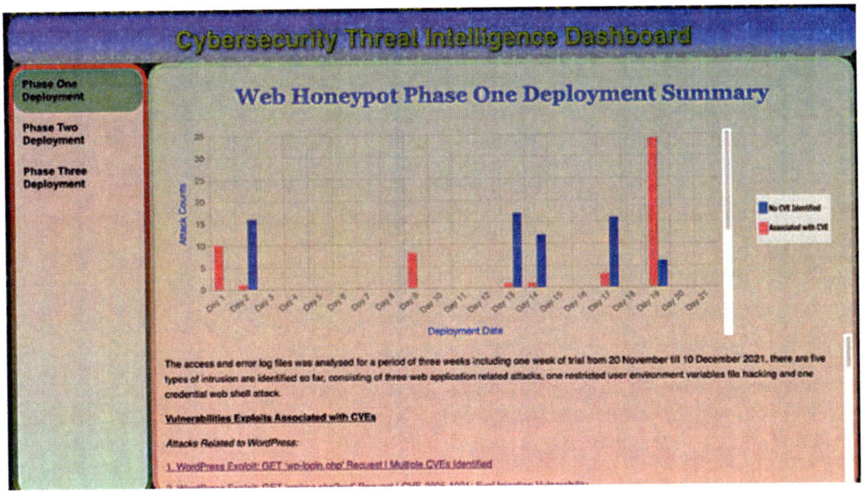

Fig. 3 Cybersecurity threat intelligence dashboard

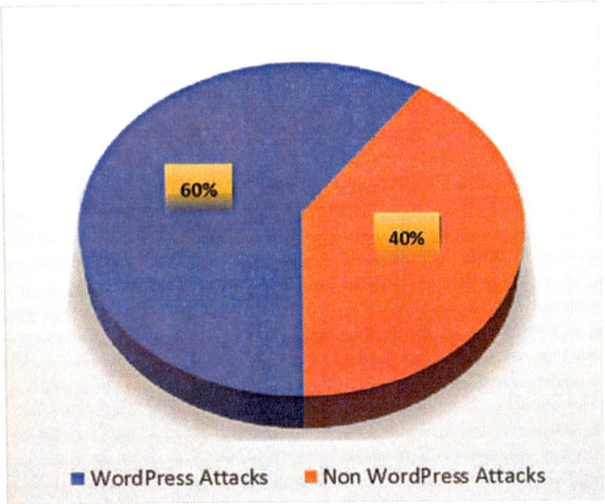

Fig. 4 Number of types of attacks in Phase 1

5.2 Phase 2

After analysis of the first honeypot deployment phase is concluded, four enhancements were derived with the objective to discover new attack types and to increase attack frequency and attackers to targets the web portal.

Enhancement 1—As the learning academy web portal is not very attractive, the portal is revamped into an online Bookstore portal.

Enhancement 2—From Fig. 3, WordPress is the most targeted application by threat actors. Fake WordPress application source files are uploaded onto the web portal as decoys. These files attract threat actors to realize the potential of the web portal for exploitation.

Enhancement 3—In Phase 1, attempts to exploit vulnerabilities relating the ".env" files are observed. To take advantage of this finding, a mock ".env" file is added to the portal's root directory to gain more insight on such attacks [7]. The presence of ".env" file, which is used by web application with database credentials, gives the honeypot an appearance of a misconfigured SME target, increasing its attractiveness to attackers.

Enhancement 4—In Phase 1, a few instances of attack are observed targeting WordPress PHP files. Some fake PHP files are created to attract threat actor's attentions for the account login and payment feature [8].

After enhancements to the honeypot, the second phase of the web honeypot is deployed for two weeks in January 2022. A notable increase on attack types and frequency is observed. Two-third of the detected attack types still targets WordPress, which further confirms that WordPress is a highly vulnerable application to cyberattacks. Vulnerabilities link to ".env" file continue to be targeted, including different attack methods and with increased attack request counts. A new web application attack on a web-based file manager was discovered, supporting the finding that attacks that are targeting web applications are dominant. A new rare non-botnet, likely a human actor was observed targeting the decoy "sitemap.xml" file. Although "sitemap.xml" is utilized by search engines for indexing, the decoy contains content relating to a fictitious sensitive payment related entry [9]. Analyzing the time interval from attacks originating from same IP address reveals that typically micro-seconds separate individual attacks. This indicates that most of the attacks are carried out by botnets.

5.3 Phase 3

Four enhancements are implemented to the honeypot with the objectives revised to further narrow down on attack types, understanding threat actor's behavior on which kind of vulnerabilities are specifically targets at present.

Enhancement 1—The first enhancement relates to four vulnerabilities, including three WordPress related vulnerabilities, namely WordPress Users Enumeration Scanning vulnerability, WordPress User Enumeration via REST API Vulnerability, and WordPress Information Disclosure Vulnerability, as well as the web application Responsive File Manager XSS vulnerability.

Enhancement 2—With the observation of attacks targeting "sitemap.xml", all endpoint entries contained in the common files "robots.txt" and "sitemap.xml" are implemented with dummy files. This removes HTTP 404 "File Not Found" error when attackers try to attack the decoys in "sitemap.xml".

Enhancement 3—To study injection and other type of attacks targeting HTML forms, a fake form is created in the newsletter section of the bookstore.

Enhancement 4—A new login function of the honeypot portal is implemented. The data is stored separately in a database so that a study can be done on common password-driven techniques initiated from botnets.

After these further enhancements to the honeypot, the third phase of web honeypot is deployed for two weeks in March 2022. The results include an assortment of attacks, threats, and suspicious activities. The usual WordPress attacks remained, with a particular XML-RPC brute-force attack shown increased attack requests after implementation of the above four vulnerabilities. Two new web applications are being attacked, affirming web applications remained popular target by threat actors. A new attack on a web-shell with exploitable vulnerabilities, called Alfa Team Shell, is identified. This malware web-shell is designed to be only used by hackers to attack other web portals, evident that threat actors are exploring new methods of exploitation [9]. Another new potential attack on the Flask Web Framework was detected, and exhibiting web application framework is also under threat actor's radar [10]. There is a newly discovered Chinese web applications attacks with threat actors targeting at least 40 vulnerabilities all at once. Research has shown that Chinese web applications are highly susceptible to security risks, illustrating that these applications are popular targets of threat actors [11]. Also, a newly re-infector attack is discovered where attackers look for a particular Indication of Compromise (IoC) backdoor entry into web portals, which signals that the web portal was compromised with persistent attack that allows long-term backdoor access for threat actors to hack the compromised web portal [12]. There also are indications that both the decoy form and login function have been attacked.

5.4 Summary

As shown in Figs. 5 and 6, the honeypot can lure more attackers as enhancements are introduced. More severe exploits (based on commonly used CVSS score [13]) are used by the attackers as the honeypot is enhanced.

Fig. 5 Number of types of attacks detected

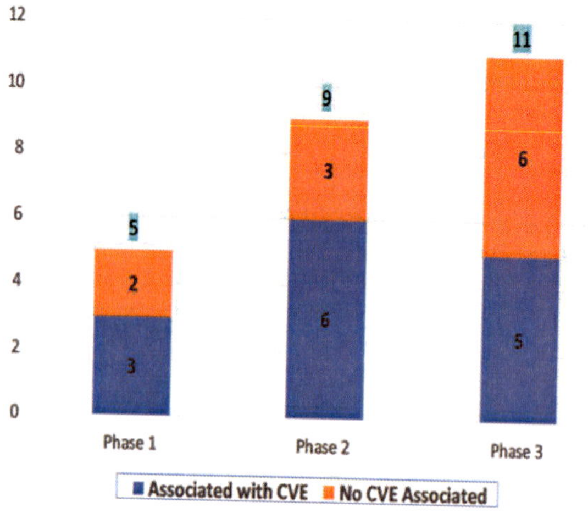

Fig. 6 Number of attack types detected categorized by severity

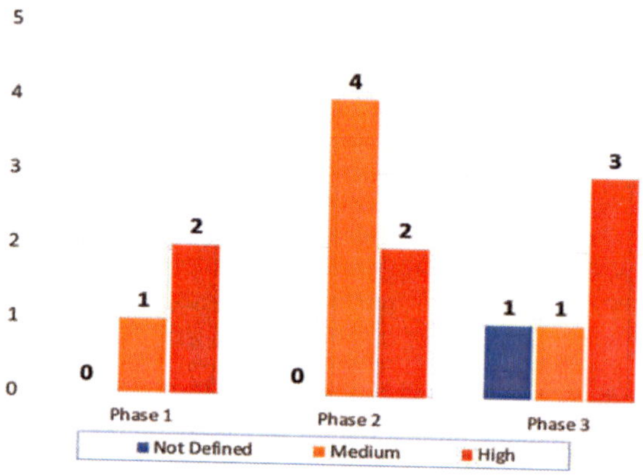

Figure 7 shows the distribution of the attackers' IP addresses targeting vulnerabilities with known CVEs. It is not surprising that most attacks originate from United States and Singapore with the high Internet adoption rate in both countries.

Figure 8 shows the attackers' IP addresses that appear on DNS Blacklists [14]. The huge increase in blacklisted adversaries attacking the honeypot in Phase 2 can be attributed to the enhancements made, especially in revamping it from an academy to bookstore. As the honeypot stays online for longer period of time, adversaries may lose interest in the honeypot after unsuccessful attempts. This is the likely reason behind the decrease in the attacks detected from Phase 2 to Phase 3.

The results of the three phases are compiled into the component of the Threat Intelligence Dashboard.

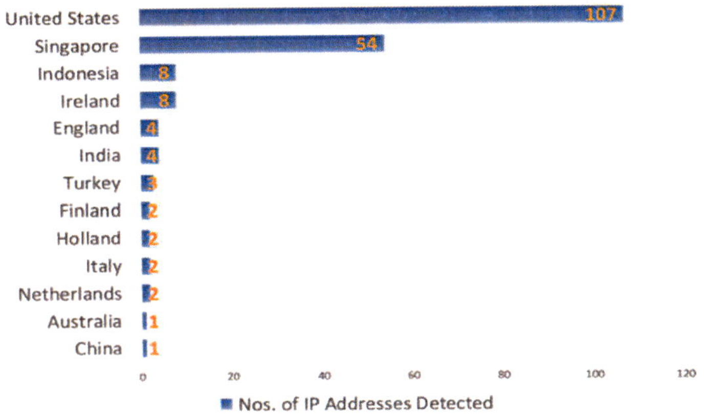

Fig. 7 Countries where attacks originate

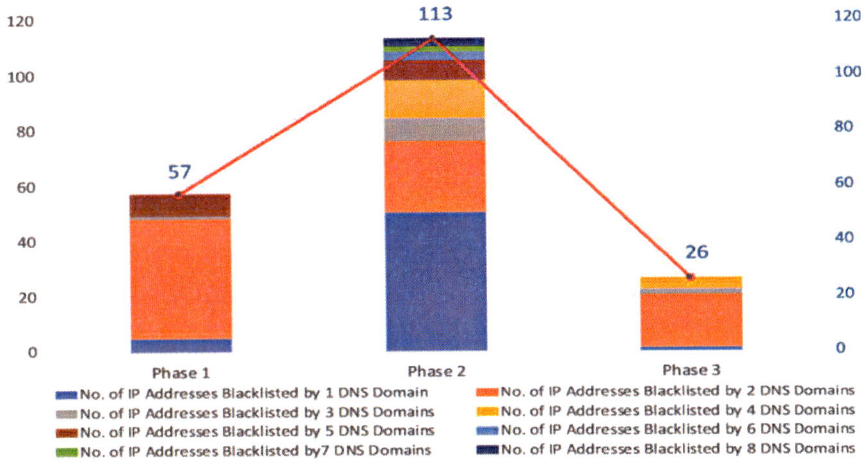

Fig. 8 No. of IP addresses blacklisted by DNS domains

6 Conclusions

This paper presents the deployment of web portal honeypot of the SMEs to gather cybersecurity threat intelligence. Analysis of the results from deploying the honeypot allows SMEs to understand about latest kind of emerging threats, which inculcates the importance of cybersecurity threat protections to the SMEs. This knowledge is useful to assist system administrators for making decisions for implementing cyberthreat security solutions or understanding their organizational needs in terms of cybersecurity.

After the proper design and implementation, the honeypot has successfully deployed for three phases, each harvested different results that allows SMEs to achieve a thorough understanding of the most recent behavior of threat actors toward the Internet. With greater insights gained from each phase, the adaptive approach used is shown to be useful to make the honeypot more effective.

Further research to identifying new undiscovered techniques and methods of web exploitations is explorable to stay current on the constantly changing threat actors' behaviors and attack pattern similarities.

References

1. Microsoft Security Blog. (2022). *North Korean threat actor targets small and midsize businesses with h0lygh0st ransomware* (Online). Available at: https://www.microsoft.com/security/blog/2022/07/14/north-korean-threat-actor-targets-small-and-midsize-businesses-with-h0lygh0st-ransomware/
2. Singapore News Center. (2020). *Over 80% of Singapore SMEs embrace digital transformation; more than half report slowdowns due to covid-19: ASME-Microsoft Study 2020* (Online) Available at: https://news.microsoft.com/en-sg/2020/10/22/over-80-of-singapore-smes-embrace-digital-transformation-more-than-half-report-slowdowns-due-to-covid-19-asme-microsoft-study-2020/
3. Fruhlinger, J. (2022). *What is a honeypot? A trap for catching hackers in the act* (Online). Available at: https://www.csoonline.com/article/3384702/what-is-a-honeypot-a-trapfor-catching-hackers-in-the-act.html
4. Imperva Blog. (2021). *Honeypot* (Online). Available at: https://www.imperva.com/learn/application-security/honeypot-honeynet/
5. Harrison, J. (2003). *Honeypots: The sweet spot in network security* (Online). Available at: https://www.computerworld.com/article/2573345/honeypots--the-sweet-spot-in-network-security.html
6. IOA VS IOC. (2021). (Online) Available at: https://www.crowdstrike.com/cybersecurity-101/indicators-of-compromise/ioa-vs-ioc/
7. Waeosri, W. (2021). *How to separate your credentials, secrets, and configurations from your source code with environment variables* (Online). Available at: https://developers.refinitiv.com/en/article-catalog/article/how-to-separate-your-credentials--secrets--and-configurations-fr
8. *Essential security tips to protect your PHP website from hackers*, 2022 (Online). Available at: https://geekflare.com/php-site-security-tips/
9. *Alfa Team Shell Tesla 4.1 remote code execution* (Online). Available at: https://packetstormsecurity.com/files/165366
10. Dyouri, A. (2020). *How to make a web application using flask in Python 3* (Online). Available at: https://www.digitalocean.com/community/tutorials/how-to-make-a-web-application-using-flask-in-python-3
11. Paganini, P. (2014). *Percent of government Chinese websites are vulnerable* (Online). Available at: https://securityaffairs.co/wordpress/30907/hacking/93-percent-chinese-websites-vulnerable.html
12. Morrow, M. (2021). *PHP re-infectors—The Malware that keeps on giving* (Online). Available at: https://securityboulevard.com/2021/12/php-re-infectors-the-malware-that-keeps-on-giving/
13. *US NIST National Vulnerability Database* (Online). Available at: https://nvd.nist.gov/vuln-metrics/cvss
14. *WhatIsMyIP.com IP Address blacklist check* (Online). Available at: https://www.whatismyip.com/blacklist-check

Automatic Detection of Age-Related Macular Degeneration from Optical Coherence Tomography Images

Camilla Wan Qi Zheng, Hoy Ee Genevieve Mah, and Ruchir Srivastava

Abstract Age-related Macular Degeneration (AMD) is the leading medical condition causing blindness for the older population due to damaged macular. This paper compares the performance of two models with differing numbers of Convolutional Neural Network (CNN) layers in Deep Learning in detecting AMD through Optical Coherence Tomography (OCT) images of the human retina. This will potentially be able to provide assistance to clinical specialists for quicker diagnosis and earlier intervention. ResNet50 with 50 CNN layers and a custom model (OwnModel) with five CNN layers were tested using fourfold cross-validation and on a separate data set. Results were then collected in two stages from the same pre-labelled public data set: 1000 images for Stage 1 on a personal machine and another 5000 for Stage 2 on the mentor's machine. ResNet50 produced more satisfactory results than OwnModel in Stage 2 with accuracies of 98.09% $\pm$ 1.00% and 96.62% $\pm$ 2.00%, respectively, after cross-validation. Similarly, ResNet50 yielded a higher accuracy at 96.90 $\pm$ 2.00% than OwnModel's at 90.00 $\pm$ 2.00% through testing on a separate data set. A higher number of CNN layers improved the machine's performance in detecting AMD. Analysis of results has proven that AI has great potential to act as future diagnostic aids based on high accuracy rates comparable to their human counterparts. This reduces the workload of clinicians and allows more time for patient care, paving the way for improved patient outcomes.

Keywords Age-related macular degeneration · Machine learning · Artificial intelligence

C. W. Q. Zheng (✉) · H. E. G. Mah
Nanyang Girls' High School, Bukit Timah, Singapore

R. Srivastava
Institute for Infocomm Research, A*STAR, Connexis, Singapore
e-mail: srivastavar@i2r.a-star.edu.sg

H. Guo et al. (eds.), *IRC-SET 2022*,
https://doi.org/10.1007/978-981-19-7222-5_45

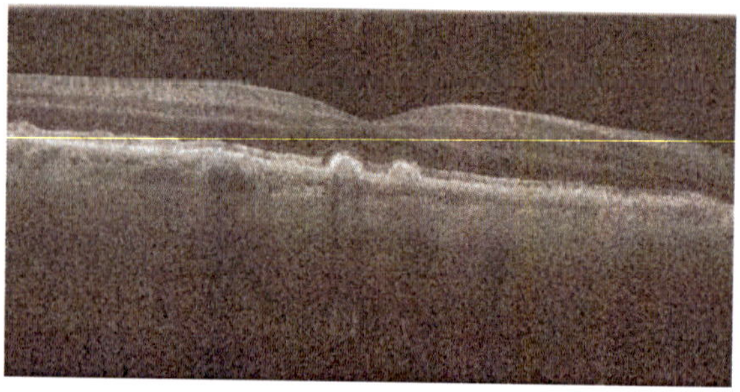

Fig. 1 Sample of AMD OCT

1 Introduction

1.1 Background and Purpose of Research Area

Age-related Macular Degeneration (AMD) is the leading cause of permanent vision loss for the older population due to damaged macula, blurring the central vision of patients. An impaired visual function has a tangible influence on affected individuals, with its association to impact the quality of life, worsen general and mental health, curtail life chances, and increase all-cause mortality [1]. Early intervention is therefore of great priority to ensure that individuals with AMD are correctly diagnosed and treated to slow the deterioration of their eyesight. This paper utilizes Deep Learning to assist clinical specialists in the detection of AMD from Optical Coherence Tomography (OCT) images of the human retina. This allows for a quicker and more efficient diagnosis of patients.

Two models with differing numbers of Convolutional Neural Network (CNN) layers were used: ResNet50 with 50 and a custom model written by a mentor with 5, named OwnModel for simplicity. Both models were trained and tested on a preprocessed data set of AMD patients and normal subjects. All OCT images adopted human graders as the reference standards for the labels of the images. This paper analyses the performance between the two models and identifies areas of improvement to ensure a satisfactory AMD diagnosis. Figures 1 and 2 are samples of OCT images of AMD patients and normal subjects in the data used.

1.2 Hypothesis of Research

The CNN with 50 layers (ResNet50) is predicted to perform better and lead to higher accuracy in AMD diagnosis than the CNN with five layers (OwnModel). A model with more CNN layers can recognize minute details and is expected to identify the varying thicknesses of the retina in OCT images of AMD patients and normal subjects. As a whole, both models have been envisioned to be of use to medical staff in increasing the efficiency of diagnostics in ocular clinics.

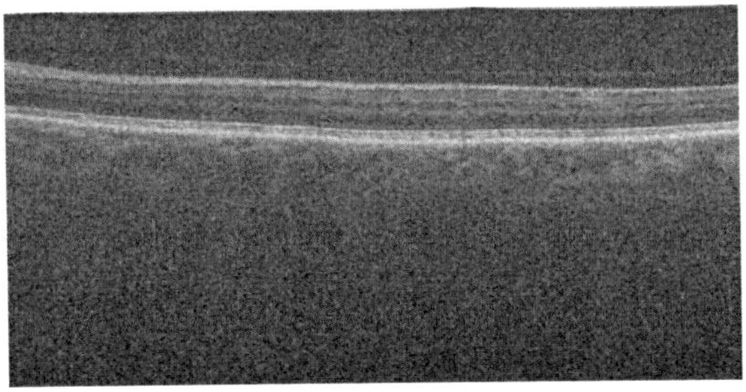

Fig. 2 Sample of normal OCT

This paper does take into consideration that there are several other key factors influencing diagnostic accuracies, such as the problem complexity and the parameter differences between the two models.

2 Methodology

ResNet50 without pre-trained weights was used and the collection of results was executed in two stages. Stage 1 was conducted on an institution's machine, while Stage 2 was conducted on the mentor's machine which had more storage and a faster central processing unit (CPU). ResNet50 required large amounts of computing power which Stage 1 was limited in, so Stage 2 had to be conducted as well.

2.1 Image Requisition and Data Processing

From a pre-labelled public OCT data set by Duke University with 269 AMD patients and 115 normal subjects, 5000 images were selected for Stage 2 and of which, another 1000 images were randomly selected for Stage 1. Both stages had an equal number of AMD and normal OCT images.

2.2 Data Collection

Both models were trained and tested with their respective-sized data sets using 4-fold cross-validation and subsequently, tested on a separate data set in Stages 1 and 2.

4-Fold Cross-Validation. ResNet50 and OwnModel were trained on three-fourths of data set in Stages 1 and 2. Attempts were made to transform and augment data for Stage 1 to improve its performance by increasing its training size as it has a comparatively smaller data set. Parameters tuned were rotation, horizontal and

```
Epoch 00010: val_acc did not improve from 0.91350
Evaluating network ...
              precision    recall  f1-score   support

           0       0.95      0.88      0.91       689
           1       0.84      0.93      0.88       476

    accuracy                           0.90      1165
   macro avg       0.90      0.91      0.90      1165
weighted avg       0.90      0.90      0.90      1165

Using real-time data augmentation.
training head ...
Epoch 1/20
64/64 [==============================] - 18s 277ms/step - loss: 0.0270 - acc: 0.9927 - val_l
- val_acc: 0.9479
```

Fig. 3 Stage 1 results in Run 1 after cross-validation

vertical flip. This produced negligible improvement in results which had limited feasibility moving forward, and hence, more focus was placed on the model instead. Both models were then tested on the remaining one-fourth of their data set.

Test on Separate Data set. Previously trained models were tested on a separate data set with different AMD patients and normal subjects.

3 Results

3.1 *Stage 1*

ResNet50 was not conducted in this stage due to the high computing power needed and incompatibility of TensorFlow and Keras. Cross-validation on OwnModel yielded an average accuracy of 90.00% ± 3.00, average precision of 90.00 ± 5.00% and average recall of 90.00 ± 3.00%. Testing on a separate data set on OwnModel produced an accuracy of around 40.74 ± 2.00% and precision and recall of 41.00 ± 2.00%. Figures 3 and 4 are examples of results taken from Stage 1 after cross-validation and testing on separate data set, respectively.

3.2 *Stage 2*

During cross-validation, OwnModel's test had accuracy average of 0.96 ± 0.02 or 96.62 ± 2.00% in variation, only slightly different from ResNet50. The precision and recall figures were both generally 0.97 ± 0.02 for precision and 97.0 ± 3.00% for recall. Similarly, ResNet50 performed satisfactorily overall with test run accuracy averages of 98.09 ± 1.00%, precision of 98.00% ± 1.00 and recall of 98.50 ± 0.50%.

When tested on a separate data set, OwnModel had an average accuracy of around 90.00 ± 2.00%, average precision of 90.00% ± 6.00 and average recall of 90.00 ±

```
Epoch 00020: val_acc did not improve from 0.98662
Evaluating network ...
              precision    recall  f1-score   support

           0       0.99      0.98      0.98       652
           1       0.97      0.99      0.98       501

    accuracy                           0.98      1153
   macro avg       0.98      0.98      0.98      1153
weighted avg       0.98      0.98      0.98      1153

1153/1153 [==============================] - 3s 3ms/step
Test loss: 0.06509187956212015
Test accuracy: 0.9809193408499567
resnet_code_trf_learning_AMDDetection.py:50: H5pyDeprecationWarning: The defa
h5py.File(), or set the global default h5.get_config().default_file_mode, or
the docs for details.
  x_train=h5py.File('train_x' + str(runcnt) + '.mat')
Version 1, run 3
resnet_code_trf_learning_AMDDetection.py:67: H5pyDeprecationWarning: The defa
```

Fig. 4 Stage 1 results in Run 1 after test on separate data set

3.00%. ResNet50 performed slightly better, with an average accuracy of 96.90 ± 2.00%, precision of 97.00 ± 1.00%, and recall of 97.00 ± 1.00%

A precision of 0.97 implies that in 100 instances of images labelled as an AMD being present in the OCT image, 97 were correctly identified as AMD images. Precision is the ratio between the True Positives and all the Positives. A recall of 0.97 means that the model correctly identified 97 True Positives for AMD out of 100 instances of the machine classifying the image as AMD positive.

Test loss was also generally less than 0.1 or 10% throughout the results. Test loss is not a percentage. It is a summation of the penalties assigned to the errors made for each incorrect diagnosis in training and validation sets. Generally, a decreasing test loss across multiple runs is the most ideal as this shows the machine was able to learn from previous mistakes and improve upon its judgement.

Calculation of the accuracies, precisions and recall is done using the following formulae in Fig. 5:

The precision is the ratio between True Positives and all Positives, recall is the percentage of correct classifications of True Positives of all instances of classification

$$\textbf{Precision} = \frac{\text{True Positive}}{\text{Actual Results}} \quad \text{or} \quad \frac{\text{True Positive}}{\text{True Positive + False Positive}}$$

$$\textbf{Recall} = \frac{\text{True Positive}}{\text{Predicted Results}} \quad \text{or} \quad \frac{\text{True Positive}}{\text{True Positive + False Negative}}$$

$$\textbf{Accuracy} = \frac{\text{True Positive + True Negative}}{\text{Total}}$$

Fig. 5 Formulae used to calculate accuracy, precision and recall

```
              precision    recall  f1-score   support

           0       0.96      1.00      0.98       689
           1       0.99      0.93      0.96       476

    accuracy                           0.97      1165
   macro avg       0.97      0.96      0.97      1165
weighted avg       0.97      0.97      0.97      1165

1165/1165 [==============================] - 4s 3ms/step
Test loss: 0.09079224900632313
Test accuracy: 0.9699570815450643
```

Fig. 6 Stage 2 results in Run 1 of OwnModel after cross-validation

```
              precision    recall  f1-score   support

           0       0.99      0.95      0.97       689
           1       0.93      0.99      0.96       476

    accuracy                           0.97      1165
   macro avg       0.96      0.97      0.97      1165
weighted avg       0.97      0.97      0.97      1165

1165/1165 [==============================] - 3s 3ms/step
Test loss: 0.14362209956508926
Test accuracy: 0.9682403433476395
```

Fig. 7 Stage 2 results in Run 1 of ResNet50 after cross-validation

as positive and test loss is the summation of penalties assigned to the errors made for each incorrect diagnosis in training and validation sets. Figures 6 and 7 are examples of results taken from Stage 2 using OwnModel and ResNet50 after cross-validation, respectively.

4 Conclusion

4.1 Discussion of Results

From the results collected, the number of convolutional layers does seem to marginally improve the accuracy of the machine's performance in diagnosing AMD in subjects. In Stage 2, OwnModel which has one-tenth of ResNet50's CNN layers, achieved an average of 96.62% accuracy as to ResNet50's 98.09% during cross-validation. The value loss for ResNet50 reflected the steady reduction from 1.5 to approximately 0.3 after 5 runs. Meanwhile, the value loss for OwnModel has remained fluctuating between the highest value of around 7.6 and the lowest value of 0.11, remaining unstable throughout 5 runs. This is most likely due to a smaller data set to train OwnModel and had previously changed the batch size to 2, which was too small and, hence, resulted in fluctuations of the value losses. With enough runs, the model is estimated to reach an accuracy of close to 100%, but this was not done in this paper due to the lack of time.

After a few runs on the two models in Stage 1, the accuracy achieved on the personal machine had been limited due to overfitting which may have been caused by the small data set. To improve on that, attempts to tune the parameters of the data set were made to increase training data set for the models to have a larger data set to run on. Tuning these parameters brought about negligible improvements to the accuracy of the machine which had limited feasibility moving forward. Thereafter, the value of epochs and the learning rate of the machine were reduced which made improvements to the accuracy later. The smaller training data set was due to the fact that the institution computer prescribed to us did not render enough storage, computing power or an up-to-date system for a large data set. In such challenging conditions, the machine only managed to achieve an accuracy of approximately 40.74%. There is great potential in these models as their performance exceeded expectations under better conditions in Stage 2.

Mini batch learning could also be a solution to this problem. However, this was not implemented in this paper due to its computation intensive nature. Implementation of batch learning instead could allow for a larger data set to be computed and, hence, could increase the accuracy results. Varying the test-train split could have also served as an improvement in this paper.

The results that were discussed in Stage 2 ResNet50's detection had a higher average accuracy of 96.9% when tested on a separate data set, which surpasses medical students at 69.4% and perform as well as retinal specialists at 92.73% [2]. The AI model that used in this paper would indeed be of significant help to the medical staff in increasing the efficiency of diagnostics.

4.2 Implications

AI has been proven to be a major aide to many professions. Deep learning algorithms with CNNs have been increasing in usage throughout diagnostics in various branches of medicine in recent years and have the potential to significantly reduce the workload of clinicians, hence improving efficiency of disease diagnostics.

Although numerous studies have shown that the use of AI would be extremely beneficial to both medical staff efficiency and patients' outcomes, the topic of AI in the medical field still remains a topic of controversy.

AI such as the ones used in this study can act as accurate diagnostic tools, allowing doctors to have more time and capacity to improve the quality of patient care, and hence, patient outcomes. This paper has shown that AI machines were able to perform just as well as, if not better than human medical specialists at diagnosing AMD in OCT images when its environment and training is under optimal conditions, regardless of what model is in usage. AI diagnostics in future of medicine in support of human diagnosis of ocular diseases is fundamental in improving the efficiency of hospitals and could be a game changer in increasing the number of ocular disease diagnosis and early intervention measures to prevent the advancement of these diseases into their late stages which could be detrimental to patients' vision and overall well-being.

Some may be concerned about AI diagnosing patients incorrectly which might lead to a snowball effect of delayed treatment and accurate diagnosis for the patient, severely impacting their well-being. Readers should note that no individual or machine could realistically have a guaranteed accurate diagnostic ability. Instead of focusing on one or two things that go poorly, more weight should be put on the benefits provided by AI usage in diagnostic medicine, as they generally outweigh the cost and is seen in numerous studies to be overall beneficial for both patients and medical staff.

If the models used in this paper were implemented, it would be best if some improvements be made upon increasing training data size and implementing more regularisation measures to increase the accuracy of the diagnosis by the machine. This directly impacts a person's well-being and determines the patient's access to early intervention to prevent further eyesight degradation.

4.3 Limitations

In all, the research done in this paper had some limitations, the comparison of accuracy in subject diagnosis between medical students, specialists [2] and machine was not made on the same data set. The training data sets may have also contained noise.

Specific Stage 1 Limitations. The machine used in Stage 1 had limited storage and computational abilities. This induced a reduced training data set of 1000 to be used instead of ideal 5000, which may have resulted in overfitting. A reduced number of runs on training data sets due to limited duration of access to machine. Due to the incompatibility of Tensorflow and Keras on the prescribed machine, the ResNet50 model was not run on the institution's machine, which could have provided a more comprehensive picture for comparison.

References

1. Cumberland, P. M., & Rahi, J. S. (2016, September). Visual function, social position, and health and social outcomes. *JAMA Ophthalmology*. https://jamanetwork.com/journals/jamaophthalmology/fullarticle/2535224. Last accessed on January 19, 2022.
2. Hwang, D. K., Hsu, C. C., Chang, K. J., Chao, D., Sun, C. H., Jheng, Y. C., Yarmishyn, A. A., Wu, J. C., Tsai, C. Y., Wang, M. L., Peng, C. H., Chien, K. H., Kao, C. L., Lin, T. C., Woung, L. C., Chen, S. J., & Chiou, S. H. (2019). Artificial intelligence-based decision-making for age-related macular degeneration. *Theranostics, 9*(1), 232–245. https://www.thno.org/v09p0232.htm. Last accessed on January 19, 2022.